P9-AQP-896

Methods in Enzymology

Volume 377
CHROMATIN AND CHROMATIN
REMODELING ENZYMES
Part C

METHODS IN ENZYMOLOGY

EDITORS-IN-CHIEF

John N. Abelson   Melvin I. Simon

DIVISION OF BIOLOGY
CALIFORNIA INSTITUTE OF TECHNOLOGY
PASADENA, CALIFORNIA

FOUNDING EDITORS

Sidney P. Colowick and Nathan O. Kaplan

Methods in Enzymology

Volume 377

Chromatin and Chromatin Remodeling Enzymes

Part C

EDITED BY

C. David Allis

THE ROCKEFELLER UNIVERSITY
NEW YORK, NEW YORK

Carl Wu

NATIONAL CANCER INSTITUTE
BETHESDA, MARYLAND

ELSEVIER
ACADEMIC
PRESS

AMSTERDAM • BOSTON • HEIDELBERG • LONDON
NEW YORK • OXFORD • PARIS • SAN DIEGO
SAN FRANCISCO • SINGAPORE • SYDNEY • TOKYO
Academic Press is an imprint of Elsevier

Elsevier Academic Press
525 B Street, Suite 1900, San Diego, California 92101-4495, USA
84 Theobald's Road, London WC1X 8RR, UK

This book is printed on acid-free paper. ∞

ISBN: 0-12-182781-X

PRINTED IN THE UNITED STATES OF AMERICA
04 05 06 07 08   9 8 7 6 5 4 3 2 1

Table of Contents

Section I. Genetic Assays of Chromatin Modification and Remodeling

Section II: Histone Modifying Enzymes

Section III. ATP-Dependent Chromatin Remodeling Enzymes

Section IV. Transcription and Other Transactions on Chromatin Templates

Contributors to Volume 377

Article numbers are in parentheses and following the names of contributors.
Affiliations listed are current.

WOOJIN AN (30), *Laboratory of Biochemistry and Molecular Biology, The Rockefeller University, New York, New York 10021*

JENNIFER A. ARMSTRONG (4), *Department of Molecular, Cell and Developmental Biology, University of California, Santa Cruz, Santa Cruz, California 95064*

ORR G. BARAK (25), *The Wistar Institute, Philadelphia, Pennsylvania 19104*

BRIAN C. BEARD (32), *Department of Biochemistry and Biophysics, School of Molecular Biosciences, Washington State University, Pullman, Washington 99164 –4660*

PETER B. BECKER (21), *Adolf-Butenandt-Institut, Molekularbiologie, München D-80336, Germany*

SHELLEY L. BERGER (7), *The Wistar Institute, Philadelphia, Pennsylvania 19104*

TIZIANA BONALDI (6), *Protein Analysis Unit, Adolf-Butenandt Institut, Ludwig Maximillians Universität, München, 80336 München, Germany*

LUDMILA BOZHENOK (24), *Chromatin Lab, Marie Curie Research Institute, Surrey RH8 0TL, United Kingdom*

ELI CANAANI (15), *Department of Molecular Cell Biology, Weizmann Institute of Science, Rehovot 76100, Israel*

BRAD CAIRNS (20), *University of Utah School of Medicine, Department of Oncological Sciences, Howard Hughes Medical Institute and Huntsman Cancer Institute, Salt Lake City, Utah 84112*

YUH-LONG CHANG (16), *Institute of Molecular Biology, Academia Sinica, Taiwan 115, Republic of China*

GILLIAN E. CHALKLEY (28), *Gene Regulation Laboratory, Center for Biomedical Genetics, Department of Molecular and Cell Biology, Leiden University Medical Center, 2300 RA Leiden, The Netherlands*

NADINE COLLINS (24), *Chromatin Lab, Marie Curie Research Institute, Surrey RH8 0TL, United Kingdom[†]*

DAVIDE F. V. CORONA (4), *Department of Molecular, Cell and Developmental Biology, University of California, Santa Cruz, Santa Cruz, California 95064*

JACQUES CÔTÉ (8), *Laval University Cancer Research Center, Quebec, GIR 2J6 Canada*

TIANHUAI CHI (18), *Howard Hughes Medical Institute, Stanford University, Stanford, California 94305[†]*

CARLO M. CROCE (15), *Kimmel Cancer Center, Thomas Jefferson University, Philadelphia, Pennsylvania 19107*

Current Affiliation: Joint Science Department, W. M. Keck Sceince Center, The Claremont Colleges, Claremont, California 91711
[†]*Current Affiliation: Cellular Pathology, Royal Surrey County Hospital, Guildford, United Kingdom*
[†]*Current Affiliation: Section of Immunology, Yale University School of Medicine, New Haven, Connecticut 06520*

FRANCK DEQUIEDT (10), *Molecular and Cellular Biology Unit, Faculty of Agronomy, Gembloux B-5030, Belgium*

JIM DOVER (13), *Department of Genetics, Washington University School of Medicine, St. Louis, Missouri 63110*

YANNICK DOYON (8), *Laval University Cancer Research Center, Quebec, GIR 2J6 Canada*

ANTON EBERHARTER (21), *Adolf-Butenandt-Institut, Molekularbiologie, München D-80336, Germany*

STUART ELGAR (23), *Emory University School of Medicine, Department of Pathology and Laboratory Medicine, Atlanta, Georgia 30322*

YUHONG FAN (5), *Department of Cell Biology, Albert Einstein College of Medicine, Bronx, New York 10461*

JIA FANG (12), *Lineberger Comprehensive Cancer Center, Department of Biochemistry and Biophysics, University of North Carolina at Chapel Hill, Chapel Hill, North Carolina 27599–7295*

WOLFGANG FISCHLE (10), *Laboratory of Chromatin Biology, The Rockefeller University, New York, New York 10021*

ROY FRYE (10), *VA Medical Center, Pittsburgh, Pennsylvania 15240*

SUNIL GANGADHARAN (14), *National Institute of Child Health and Human Development, Unit on Chromatin and Transcription, Bethesda, Maryland 20892*

SONJA GHIDELLI (14), *National Institute of Child Health and Human Development, Unit on Chromatin and Transcription, Bethesda, Maryland 20892[§]*

PATRICK A. GRANT (8), *University of Virgina School of Medicine, Charlottesville, Virginia 22908*

KARIEN HAMER (17), *Swammerdam Institute for Life Sciences, University of Amsterdam, 1018 TV Amsterdam, The Netherlands*

ALI HAMICHE (22), *Institut Andre Lwoff, 94800 Villejuif, France*

SHU HE (31), *Johnson Research Foundation, Department of Biochemistry and Biophysics, University of Pennsylvania School of Medicine, Philadelphia, Pennsylvania 19104–6059*

KARL W. HENRY (7), *The Wistar Institute, Philadelphia, Pennsylvania 19104*

DER HWA-HUANG (16), *Institute of Molecular Biology, Academia Sinica, Taiwan 115, Republic of China*

AXEL IMHOF (6), *Histone Modifications Group, Adolf-Butenandt Institut, Ludwig Maximillians Universität, München, 80336 München, Germany*

SANDRA J. JACOBSON (1), *Department of Biology, University of California, San Diego, La Jolla, California 92093–0347*

MARK JOHNSTON (13), *Department of Genetics, Washington University School of Medicine, St. Louis, Missouri 63110*

ROHINTON T. KAMAKAKA (14), *National Institute of Child Health and Human Development, Unit on Chromatin and Transcription, Bethesda, Maryland 20892*

MIKHAIL KASHLEV (29), *National Cancer Institute Center for Cancer Research, National Cancer Institute-Frederick Cancer Research and Development Center, Frederick, Maryland 21702*

JAMES A. KENNISON (3), *Laboratory of Molecular Genetics, National Institute of Child Health and Human Development, National Institutes of Health, Bethesda, Marlyland, 20892–2785*

[§]*Current Affiliation: Cellzome AG, 69117 Heidelberg, Germany*

ROGER D. KORNBERG (19), *Department of Structural Biology, Stanford University School of Medicine, Stanford, California 94305*

WLADYSLAW KRAJEWSKI (15), *Kimmel Cancer Center, Thomas Jefferson University, Philadelphia, Pennsylvania 19107*¶

TED H. J. KWAKS (17), *Swammerdam Institute for Life Sciences, University of Amsterdam, 1018 TV Amsterdam, The Netherlands*

GERNOT LÄNGST (21), *Adolf-Butenandt-Institut, Molekularbiologie, München D-80336, Germany*

PATRICIA M. LAURENSON (1), *Department of Biology, University of California, San Diego, La Jolla, California 92093-0347*

HONG LIU (27), *Laboratory of Molecular Immunology, National Institutes of Health, Bethesda, Maryland 20892-1674*

WAN-SHENG LO (7), *The Wistar Institute, Philadelphia, Pennsylvania 19104*

LORRAINE PILLUS (1), *Department of Biology, University of California, San Diego, La Jolla, California 92093-0347*

YAHLI LORCH (19), *Department of Structural Biology, Stanford University School of Medicine, Stanford, California 94305*

ROMAIN LOURY (11), *Institut de Génétique et de Biologie Moleculaire et Cellulaire, 67404 Illkirch, Strasbourg, France*

ALEJANDRA LOYOLA (31), *Howard Hughes Medical Institute, Division of Nucleic Acids Enzymology, Department of Biochemistry, University of Medicine and Dentistry of New Jersey, Robert Wood Johnson Medical School, Piscataway, New Jersey 08854-5635*

BRETT MARSHALL (10), *Gladstone Institute of Virology and Immunology, University of California, San Francisco, San Francisco, California 94103*

ALXANDER MAZO (15), *Kimmel Cancer Center, Department of Microbiology and Immunology, Thomas Jefferson University, Philadelphia, Pennsylvania 19107*

STACEY MCMAHON (8), *University of Virgina School of Medicine, Charlottesville, Virginia 22908*

DEWEY G. MCCAFFERTY (31), *Johnson Research Foundation, Department of Biochemistry and Biophysics, University of Pennsylvania School of Medicine, Philadelphia, Pennsylvania 19104-6059*

TATSUYA NAKAMURA (15), *Kimmel Cancer Center, Department of Microbiology and Immunology, Thomas Jefferson University, Philadelphia, Pennsylvania 19107*

BRIAN NORTH (10), *Gladstone Institute of Virology and Immunology, University of California, San Francisco, San Francisco, California 94103*

SANTAEK OH (31), *Howard Hughes Medical Institute, Division of Nucleic Acids Enzymology, Department of Biochemistry, University of Medicine and Dentistry of New Jersey, Robert Wood Johnson Medical School, Piscataway, New Jersey 08854-5635*

ERIN K. O'SHEA (2), *Howard Hughes Medical Institute, University of California, San Francisco, Department. of Biochemistry and Biophysics, San Francisco, California 94143-2240*

ARIE P. OTTE (17), *Swammerdam Institute for Life Sciences, University of Amsterdam, 1018 TV Amsterdam, The Netherlands*

MATTHEW B. PALMER (23), *Emory University School of Medicine, Department of Pathology and Laboratory Medicine, Atlanta, Georgia 30322*

SVETLANA PETRUK (15), *Kimmel Cancer Center, Thomas Jefferson University, Philadelphia, Pennsylvania 19107*

¶*Current Affiliation: Institute of Developmental Biology, Moscow 117808, Russia*

RAYMOND POOT (24), *Chromatin Lab, Marie Curie Research Institute, Surrey RH8 0TL, United Kingdom*

DANNY REINBERG (31), *Howard Hughes Medical Institute, Division of Nucleic Acids Enzymology, Department of Biochemistry, University of Medicine and Dentistry of New Jersey, Robert Wood Johnson Medical School, Piscataway, New Jersey 08854–5635*

JÖRG T. REGULA (6), *Protein Analysis Unit, Adolf-Butenandt Institut, Ludwig Maximillians Universität, München, 80336 München, Germany*

NATALIE REZAI-ZADEH (9), *H. Lee Moffitt Cancer Center and Research Institute, University of South Florida, Tampa, Florida 33612*

ROBERT ROEDER (30), *Head, Laboratory of Biochemistry and Molecular Biology, The Rockefeller University, New York, New York 10021*

ANJANABHA SAHA (20), *University of Utah School of Medicine, Department of Oncological Sciences, Howard Hughes Medical Institute and Huntsman Cancer Institute, Salt Lake City, Utah 84112*

PAOLO SASSONE-CORSI (11), *Institut de Génétique et de Biologie Moleculaire et Cellulaire, 67404 Illkirch, Strasbourg, France*

JESSICA SCHNEIDER (13), *Saint Louis University School of Medicine, Department of Biochemistry, St. Louis, Missouri 63104*

MARC F. SCHWARTZ (7), *The Wistar Institute, Philadelphia, Pennsylvania 19104*

YURII SEDKOV (15), *Kimmel Cancer Center, Thomas Jefferson University, Philadelphia, Pennsylvania 19107*

EDWARD SETO (9), *H. Lee Moffitt Cancer Center and Research Institute, University of South Florida, Tampa, Florida 33612*

RICHARD G. A. B. SEWALT (17), *Swammerdam Institute for Life Sciences, University of Amsterdam, 1018 TV Amsterdam, The Netherlands*

XUETONG SHEN (26), *Department of Carcinogenesis, University of Texas, M.D. Anderson Cancer Center, Science Park Research Division, Smithville, Texas 78957*

RAMIN SHIEKHATTAR (25), *Gene Expression and Regulation Program, The Wistar Institute, Philadelphia, Pennsylvania 19104*

ALI SHILATIFARD (13), *Saint Louis University School of Medicine, Department of Biochemistry, St. Louis, Missouri 63104*

ARTHUR I. SKOULTCHI (5), *Department of Cell Biology, Albert Einstein College of Medicine, Bronx, New York 10461*

MICK SMERDON (32), *Department of Biochemistry and Biophysics, School of Molecular Biosciences, Washington State University, Pullman, Washington 99164–4660*

SHERYL T. SMITH (15), *Kimmel Cancer Center, Thomas Jefferson University, Philadelphia, Pennsylvania 19107*

DAVID J. STEGER (2), *Howard Hughes Medical Institute, University of California, San Francisco, Department of Biochemistry and Biophysics, San Francisco, California 94143–2240*

VASSILY M. STUDITSKY (29), *Department of Biochemistry and Molecular Biology Wayne State University School of Medicine, Detroit, Michigan 4820***

JOHN W. TAMKUN (4), *Department of Molecular, Cell and Developmental Biology, University of California, Santa Cruz, Santa Cruz, California 95064*

***Current Affiliation: Department of Pharmacology, University of Medicine and Dentistry of New Jersey, Robert Wood Johnson Medical School, Piscataway, New Jersey 08854*

SHIH-CHANG TSAI (9), *H. Lee Moffitt Cancer Center and Research Institute, University of South Florida, Tampa, Florida 33612*

PATRICK VARGA-WEISZ (24), *Chromatin Lab, Marie Curie Research Institute, Surrey RH8 0TL, United Kingdom*

ERIC VERDIN (10), *Gladstone Institute of Virology and Immunology, University of California, San Francisco, San Francisco, California 94103*

C. PETER VERRIJZER (28), *Gene Regulation Laboratory, Center for Biomedical Genetics, Department of Molecular and Cell Biology, Leiden University Medical Center, 2300 RA Leiden, The Netherlands*

PAUL A. WADE (23), *Emory University School of Medicine, Department of Pathology and Laboratory Medicine, Atlanta, Georgia 30322*

WENDY WALTER (29), *Center for Molecular Medicine and Genetics, Wayne State University School of Medicine, Detroit, Michigan 48201*

HENGBIN WANG (12), *Lineberger Comprehensive Cancer Center, Department of Biochemistry and Biophysics, University of North Carolina at Chapel Hill, Chapel Hill, North Carolina 27599-7295*

WEI-DONG WANG (18), *Laboratory of Genetics, National Institute on Aging, National Institute of Health, Baltimore, Maryland 21224*

YU-DER WEN (9), *H. Lee Moffitt Cancer Center and Research Institute, University of South Florida, Tampa, Florida 33612*

JACQUELINE WITTMEYER (20), *University of Utah School of Medicine, Department of Oncological Sciences, Howard Hughes Medical Institute and Huntsman Cancer Institute, Salt Lake City, Utah 84112*

HUA XIAO (22), *Laboratory of Molecular Cell Biology, National Institute of Health, Bethesda, Maryland 20892-4255*

YUTONG XUE (18), *Laboratory of Genetics, National Institute on Aging, National Institute of Health, Baltimore, Maryland 21224*

ZHIJIANG YAN (18), *Laboratory of Genetics, National Institute on Aging, National Institute of Health, Baltimore, Maryland 21224*

WEN-MING YANG (9), *H. Lee Moffitt Cancer Center and Research Institute, University of South Florida, Tampa, Florida 33612*

YA-LI YAO (9), *H. Lee Moffitt Cancer Center and Research Institute, University of South Florida, Tampa, Florida 33612*

YI ZHANG (12), *Lineberger Comprehensive Cancer Center, Department of Biochemistry and Biophysics, University of North Carolina at Chapel Hill, Chapel Hill, North Carolina 27599-7295*

KEJI ZHAO (27), *Laboratory of Molecular Immunology, National Institutes of Health, Bethesda, Maryland 20892-1674*

Preface

A central challenge of the post-genomic era is to understand how the 30,000 to 40,000 unique genes in the human genome are selectively expressed or silenced to coordinate cellular growth and differentiation. The packaging of eukaryotic genomes in a complex of DNA, histones, and nonhistone proteins called chromatin provides a surprisingly sophisticated system that plays a critical role in controlling the flow of genetic information. This packaging system has evolved to index our genomes such that certain genes become readily accessible to the transcription machinery, while other genes are reversibly silenced. Moreover, chromatin-based mechanisms of gene regulation, often involving domains of covalent modifications of DNA and histones, can be inherited from one generation to the next. The heritability of chromatin states in the absence of DNA mutation has contributed greatly to the current excitement in the field of epigenetics.

The past 5 years have witnessed an explosion of new research on chromatin biology and biochemistry. Chromatin structure and function are now widely recognized as being critical to regulating gene expression, maintaining genomic stability, and ensuring faithful chromosome transmission. Moreover, links between chromatin metabolism and disease are beginning to emerge. The identification of altered DNA methylation and histone acetylase activity in human cancers, the use of histone deacetylase inhibitors in the treatment of leukemia, and the tumor suppressor activities of ATP-dependent chromatin remodeling enzymes are examples that likely represent just the tip of the iceberg.

As such, the field is attracting new investigators who enter with little first hand experience with the standard assays used to dissect chromatin structure and function. In addition, even seasoned veterans are overwhelmed by the rapid introduction of new chromatin technologies. Accordingly, we sought to bring together a useful "go-to" set of chromatin-based methods that would update and complement two previous publications in this series, Volume 170 (Nucleosomes) and Volume 304 (Chromatin). While many of the classic protocols in those volumes remain as timely now as when they were written, it is our hope the present series will fill in the gaps for the next several years.

This 3-volume set of *Methods in Enzymology* provides nearly one hundred procedures covering the full range of tools—bioinformatics, structural biology, biophysics, biochemistry, genetics, and cell biology—employed in chromatin research. Volume 375 includes a histone database, methods for preparation of histones, histone variants, modified histones and defined chromatin segments, protocols for nucleosome reconstitution and analysis, and cytological methods for imaging chromatin functions *in vivo*. Volume 376 includes electron microscopy and biophysical protocols for visualizing chromatin and detecting chromatin interactions, enzymological assays for histone modifying enzymes, and immunochemical protocols for the *in situ* detection of histone modifications and chromatin proteins. Volume 377 includes genetic assays of histones and chromatin regulators, methods for the preparation and analysis of histone modifying and ATP-dependent chromatin remodeling enzymes, and assays for transcription and DNA repair on chromatin templates. We are exceedingly grateful to the very large number of colleagues representing the field's leading laboratories, who have taken the time and effort to make their technical expertise available in this series.

Finally, we wish to take the opportunity to remember Vincent Allfrey, Andrei Mirzabekov, Harold Weintraub, Abraham Worcel, and especially Alan Wolffe, co-editor of Volume 304 (Chromatin). All of these individuals had key roles in shaping the chromatin field into what it is today.

C. DAVID ALLIS
CARL WU

Editors' Note: Additional methods can be found in Methods in Enzymology, Vol. 371 (RNA Polymerases and Associated Factors, Part D) Section III Chromatin, *Sankar L. Adhya and Susan Garges, Editors.*

METHODS IN ENZYMOLOGY

VOLUME I. Preparation and Assay of Enzymes
Edited by SIDNEY P. COLOWICK AND NATHAN O. KAPLAN

VOLUME II. Preparation and Assay of Enzymes
Edited by SIDNEY P. COLOWICK AND NATHAN O. KAPLAN

VOLUME III. Preparation and Assay of Substrates
Edited by SIDNEY P. COLOWICK AND NATHAN O. KAPLAN

VOLUME IV. Special Techniques for the Enzymologist
Edited by SIDNEY P. COLOWICK AND NATHAN O. KAPLAN

VOLUME V. Preparation and Assay of Enzymes
Edited by SIDNEY P. COLOWICK AND NATHAN O. KAPLAN

VOLUME VI. Preparation and Assay of Enzymes (*Continued*)
Preparation and Assay of Substrates
Special Techniques
Edited by SIDNEY P. COLOWICK AND NATHAN O. KAPLAN

VOLUME VII. Cumulative Subject Index
Edited by SIDNEY P. COLOWICK AND NATHAN O. KAPLAN

VOLUME VIII. Complex Carbohydrates
Edited by ELIZABETH F. NEUFELD AND VICTOR GINSBURG

VOLUME IX. Carbohydrate Metabolism
Edited by WILLIS A. WOOD

VOLUME X. Oxidation and Phosphorylation
Edited by RONALD W. ESTABROOK AND MAYNARD E. PULLMAN

VOLUME XI. Enzyme Structure
Edited by C. H. W. HIRS

VOLUME XII. Nucleic Acids (Parts A and B)
Edited by LAWRENCE GROSSMAN AND KIVIE MOLDAVE

VOLUME XIII. Citric Acid Cycle
Edited by J. M. LOWENSTEIN

VOLUME XIV. Lipids
Edited by J. M. LOWENSTEIN

VOLUME XV. Steroids and Terpenoids
Edited by RAYMOND B. CLAYTON

VOLUME XVI. Fast Reactions
Edited by KENNETH KUSTIN

VOLUME XVII. Metabolism of Amino Acids and Amines (Parts A and B)
Edited by HERBERT TABOR AND CELIA WHITE TABOR

VOLUME XVIII. Vitamins and Coenzymes (Parts A, B, and C)
Edited by DONALD B. MCCORMICK AND LEMUEL D. WRIGHT

VOLUME XIX. Proteolytic Enzymes
Edited by GERTRUDE E. PERLMANN AND LASZLO LORAND

VOLUME XX. Nucleic Acids and Protein Synthesis (Part C)
Edited by KIVIE MOLDAVE AND LAWRENCE GROSSMAN

VOLUME XXI. Nucleic Acids (Part D)
Edited by LAWRENCE GROSSMAN AND KIVIE MOLDAVE

VOLUME XXII. Enzyme Purification and Related Techniques
Edited by WILLIAM B. JAKOBY

VOLUME XXIII. Photosynthesis (Part A)
Edited by ANTHONY SAN PIETRO

VOLUME XXIV. Photosynthesis and Nitrogen Fixation (Part B)
Edited by ANTHONY SAN PIETRO

VOLUME XXV. Enzyme Structure (Part B)
Edited by C. H. W. HIRS AND SERGE N. TIMASHEFF

VOLUME XXVI. Enzyme Structure (Part C)
Edited by C. H. W. HIRS AND SERGE N. TIMASHEFF

VOLUME XXVII. Enzyme Structure (Part D)
Edited by C. H. W. HIRS AND SERGE N. TIMASHEFF

VOLUME XXVIII. Complex Carbohydrates (Part B)
Edited by VICTOR GINSBURG

VOLUME XXIX. Nucleic Acids and Protein Synthesis (Part E)
Edited by LAWRENCE GROSSMAN AND KIVIE MOLDAVE

VOLUME XXX. Nucleic Acids and Protein Synthesis (Part F)
Edited by KIVIE MOLDAVE AND LAWRENCE GROSSMAN

VOLUME XXXI. Biomembranes (Part A)
Edited by SIDNEY FLEISCHER AND LESTER PACKER

VOLUME XXXII. Biomembranes (Part B)
Edited by SIDNEY FLEISCHER AND LESTER PACKER

VOLUME XXXIII. Cumulative Subject Index Volumes I-XXX
Edited by MARTHA G. DENNIS AND EDWARD A. DENNIS

VOLUME XXXIV. Affinity Techniques (Enzyme Purification: Part B)
Edited by WILLIAM B. JAKOBY AND MEIR WILCHEK

VOLUME 90. Carbohydrate Metabolism (Part E)
Edited by WILLIS A. WOOD

VOLUME 91. Enzyme Structure (Part I)
Edited by C. H. W. HIRS AND SERGE N. TIMASHEFF

VOLUME 92. Immunochemical Techniques (Part E: Monoclonal Antibodies and General Immunoassay Methods)
Edited by JOHN J. LANGONE AND HELEN VAN VUNAKIS

VOLUME 93. Immunochemical Techniques (Part F: Conventional Antibodies, Fc Receptors, and Cytotoxicity)
Edited by JOHN J. LANGONE AND HELEN VAN VUNAKIS

VOLUME 94. Polyamines
Edited by HERBERT TABOR AND CELIA WHITE TABOR

VOLUME 95. Cumulative Subject Index Volumes 61–74, 76–80
Edited by EDWARD A. DENNIS AND MARTHA G. DENNIS

VOLUME 96. Biomembranes [Part J: Membrane Biogenesis: Assembly and Targeting (General Methods; Eukaryotes)]
Edited by SIDNEY FLEISCHER AND BECCA FLEISCHER

VOLUME 97. Biomembranes [Part K: Membrane Biogenesis: Assembly and Targeting (Prokaryotes, Mitochondria, and Chloroplasts)]
Edited by SIDNEY FLEISCHER AND BECCA FLEISCHER

VOLUME 98. Biomembranes (Part L: Membrane Biogenesis: Processing and Recycling)
Edited by SIDNEY FLEISCHER AND BECCA FLEISCHER

VOLUME 99. Hormone Action (Part F: Protein Kinases)
Edited by JACKIE D. CORBIN AND JOEL G. HARDMAN

VOLUME 100. Recombinant DNA (Part B)
Edited by RAY WU, LAWRENCE GROSSMAN, AND KIVIE MOLDAVE

VOLUME 101. Recombinant DNA (Part C)
Edited by RAY WU, LAWRENCE GROSSMAN, AND KIVIE MOLDAVE

VOLUME 102. Hormone Action (Part G: Calmodulin and Calcium-Binding Proteins)
Edited by ANTHONY R. MEANS AND BERT W. O'MALLEY

VOLUME 103. Hormone Action (Part H: Neuroendocrine Peptides)
Edited by P. MICHAEL CONN

VOLUME 104. Enzyme Purification and Related Techniques (Part C)
Edited by WILLIAM B. JAKOBY

VOLUME 105. Oxygen Radicals in Biological Systems
Edited by LESTER PACKER

VOLUME 106. Posttranslational Modifications (Part A)
Edited by FINN WOLD AND KIVIE MOLDAVE

VOLUME 107. Posttranslational Modifications (Part B)
Edited by FINN WOLD AND KIVIE MOLDAVE

VOLUME 108. Immunochemical Techniques (Part G: Separation and Characterization of Lymphoid Cells)
Edited by GIOVANNI DI SABATO, JOHN J. LANGONE, AND HELEN VAN VUNAKIS

VOLUME 109. Hormone Action (Part I: Peptide Hormones)
Edited by LUTZ BIRNBAUMER AND BERT W. O'MALLEY

VOLUME 110. Steroids and Isoprenoids (Part A)
Edited by JOHN H. LAW AND HANS C. RILLING

VOLUME 111. Steroids and Isoprenoids (Part B)
Edited by JOHN H. LAW AND HANS C. RILLING

VOLUME 112. Drug and Enzyme Targeting (Part A)
Edited by KENNETH J. WIDDER AND RALPH GREEN

VOLUME 113. Glutamate, Glutamine, Glutathione, and Related Compounds
Edited by ALTON MEISTER

VOLUME 114. Diffraction Methods for Biological Macromolecules (Part A)
Edited by HAROLD W. WYCKOFF, C. H. W. HIRS, AND SERGE N. TIMASHEFF

VOLUME 115. Diffraction Methods for Biological Macromolecules (Part B)
Edited by HAROLD W. WYCKOFF, C. H. W. HIRS, AND SERGE N. TIMASHEFF

VOLUME 116. Immunochemical Techniques (Part H: Effectors and Mediators of Lymphoid Cell Functions)
Edited by GIOVANNI DI SABATO, JOHN J. LANGONE, AND HELEN VAN VUNAKIS

VOLUME 117. Enzyme Structure (Part J)
Edited by C. H. W. HIRS AND SERGE N. TIMASHEFF

VOLUME 118. Plant Molecular Biology
Edited by ARTHUR WEISSBACH AND HERBERT WEISSBACH

VOLUME 119. Interferons (Part C)
Edited by SIDNEY PESTKA

VOLUME 120. Cumulative Subject Index Volumes 81–94, 96–101

VOLUME 121. Immunochemical Techniques (Part I: Hybridoma Technology and Monoclonal Antibodies)
Edited by JOHN J. LANGONE AND HELEN VAN VUNAKIS

VOLUME 122. Vitamins and Coenzymes (Part G)
Edited by FRANK CHYTIL AND DONALD B. McCORMICK

VOLUME 123. Vitamins and Coenzymes (Part H)
Edited by FRANK CHYTIL AND DONALD B. McCORMICK

VOLUME 124. Hormone Action (Part J: Neuroendocrine Peptides)
Edited by P. MICHAEL CONN

VOLUME 125. Biomembranes (Part M: Transport in Bacteria, Mitochondria, and Chloroplasts: General Approaches and Transport Systems)
Edited by SIDNEY FLEISCHER AND BECCA FLEISCHER

VOLUME 126. Biomembranes (Part N: Transport in Bacteria, Mitochondria, and Chloroplasts: Protonmotive Force)
Edited by SIDNEY FLEISCHER AND BECCA FLEISCHER

VOLUME 127. Biomembranes (Part O: Protons and Water: Structure and Translocation)
Edited by LESTER PACKER

VOLUME 128. Plasma Lipoproteins (Part A: Preparation, Structure, and Molecular Biology)
Edited by JERE P. SEGREST AND JOHN J. ALBERS

VOLUME 129. Plasma Lipoproteins (Part B: Characterization, Cell Biology, and Metabolism)
Edited by JOHN J. ALBERS AND JERE P. SEGREST

VOLUME 130. Enzyme Structure (Part K)
Edited by C. H. W. HIRS AND SERGE N. TIMASHEFF

VOLUME 131. Enzyme Structure (Part L)
Edited by C. H. W. HIRS AND SERGE N. TIMASHEFF

VOLUME 132. Immunochemical Techniques (Part J: Phagocytosis and Cell-Mediated Cytotoxicity)
Edited by GIOVANNI DI SABATO AND JOHANNES EVERSE

VOLUME 133. Bioluminescence and Chemiluminescence (Part B)
Edited by MARLENE DELUCA AND WILLIAM D. MCELROY

VOLUME 134. Structural and Contractile Proteins (Part C: The Contractile Apparatus and the Cytoskeleton)
Edited by RICHARD B. VALLEE

VOLUME 135. Immobilized Enzymes and Cells (Part B)
Edited by KLAUS MOSBACH

VOLUME 136. Immobilized Enzymes and Cells (Part C)
Edited by KLAUS MOSBACH

VOLUME 137. Immobilized Enzymes and Cells (Part D)
Edited by KLAUS MOSBACH

VOLUME 138. Complex Carbohydrates (Part E)
Edited by VICTOR GINSBURG

VOLUME 139. Cellular Regulators (Part A: Calcium- and Calmodulin-Binding Proteins)
Edited by ANTHONY R. MEANS AND P. MICHAEL CONN

VOLUME 140. Cumulative Subject Index Volumes 102–119, 121–134

VOLUME 141. Cellular Regulators (Part B: Calcium and Lipids)
Edited by P. MICHAEL CONN AND ANTHONY R. MEANS

VOLUME 142. Metabolism of Aromatic Amino Acids and Amines
Edited by SEYMOUR KAUFMAN

VOLUME 143. Sulfur and Sulfur Amino Acids
Edited by WILLIAM B. JAKOBY AND OWEN GRIFFITH

VOLUME 144. Structural and Contractile Proteins (Part D: Extracellular Matrix)
Edited by LEON W. CUNNINGHAM

VOLUME 145. Structural and Contractile Proteins (Part E: Extracellular Matrix)
Edited by LEON W. CUNNINGHAM

VOLUME 146. Peptide Growth Factors (Part A)
Edited by DAVID BARNES AND DAVID A. SIRBASKU

VOLUME 147. Peptide Growth Factors (Part B)
Edited by DAVID BARNES AND DAVID A. SIRBASKU

VOLUME 148. Plant Cell Membranes
Edited by LESTER PACKER AND ROLAND DOUCE

VOLUME 149. Drug and Enzyme Targeting (Part B)
Edited by RALPH GREEN AND KENNETH J. WIDDER

VOLUME 150. Immunochemical Techniques (Part K: *In Vitro* Models of B and T Cell Functions and Lymphoid Cell Receptors)
Edited by GIOVANNI DI SABATO

VOLUME 151. Molecular Genetics of Mammalian Cells
Edited by MICHAEL M. GOTTESMAN

VOLUME 152. Guide to Molecular Cloning Techniques
Edited by SHELBY L. BERGER AND ALAN R. KIMMEL

VOLUME 153. Recombinant DNA (Part D)
Edited by RAY WU AND LAWRENCE GROSSMAN

VOLUME 154. Recombinant DNA (Part E)
Edited by RAY WU AND LAWRENCE GROSSMAN

VOLUME 155. Recombinant DNA (Part F)
Edited by RAY WU

VOLUME 156. Biomembranes (Part P: ATP-Driven Pumps and Related Transport: The Na, K-Pump)
Edited by SIDNEY FLEISCHER AND BECCA FLEISCHER

VOLUME 157. Biomembranes (Part Q: ATP-Driven Pumps and Related Transport: Calcium, Proton, and Potassium Pumps)
Edited by SIDNEY FLEISCHER AND BECCA FLEISCHER

VOLUME 175. Cumulative Subject Index Volumes 135–139, 141–167

VOLUME 176. Nuclear Magnetic Resonance (Part A: Spectral Techniques and Dynamics)
Edited by NORMAN J. OPPENHEIMER AND THOMAS L. JAMES

VOLUME 177. Nuclear Magnetic Resonance (Part B: Structure and Mechanism)
Edited by NORMAN J. OPPENHEIMER AND THOMAS L. JAMES

VOLUME 178. Antibodies, Antigens, and Molecular Mimicry
Edited by JOHN J. LANGONE

VOLUME 179. Complex Carbohydrates (Part F)
Edited by VICTOR GINSBURG

VOLUME 180. RNA Processing (Part A: General Methods)
Edited by JAMES E. DAHLBERG AND JOHN N. ABELSON

VOLUME 181. RNA Processing (Part B: Specific Methods)
Edited by JAMES E. DAHLBERG AND JOHN N. ABELSON

VOLUME 182. Guide to Protein Purification
Edited by MURRAY P. DEUTSCHER

VOLUME 183. Molecular Evolution: Computer Analysis of Protein and Nucleic Acid Sequences
Edited by RUSSELL F. DOOLITTLE

VOLUME 184. Avidin-Biotin Technology
Edited by MEIR WILCHEK AND EDWARD A. BAYER

VOLUME 185. Gene Expression Technology
Edited by DAVID V. GOEDDEL

VOLUME 186. Oxygen Radicals in Biological Systems (Part B: Oxygen Radicals and Antioxidants)
Edited by LESTER PACKER AND ALEXANDER N. GLAZER

VOLUME 187. Arachidonate Related Lipid Mediators
Edited by ROBERT C. MURPHY AND FRANK A. FITZPATRICK

VOLUME 188. Hydrocarbons and Methylotrophy
Edited by MARY E. LIDSTROM

VOLUME 189. Retinoids (Part A: Molecular and Metabolic Aspects)
Edited by LESTER PACKER

VOLUME 190. Retinoids (Part B: Cell Differentiation and Clinical Applications)
Edited by LESTER PACKER

VOLUME 191. Biomembranes (Part V: Cellular and Subcellular Transport: Epithelial Cells)
Edited by SIDNEY FLEISCHER AND BECCA FLEISCHER

VOLUME 192. Biomembranes (Part W: Cellular and Subcellular Transport: Epithelial Cells)
Edited by SIDNEY FLEISCHER AND BECCA FLEISCHER

VOLUME 316. Vertebrate Phototransduction and the Visual Cycle (Part B)
Edited by KRZYSZTOF PALCZEWSKI

VOLUME 317. RNA–Ligand Interactions (Part A: Structural Biology Methods)
Edited by DANIEL W. CELANDER AND JOHN N. ABELSON

VOLUME 318. RNA–Ligand Interactions (Part B: Molecular Biology Methods)
Edited by DANIEL W. CELANDER AND JOHN N. ABELSON

VOLUME 319. Singlet Oxygen, UV-A, and Ozone
Edited by LESTER PACKER AND HELMUT SIES

VOLUME 320. Cumulative Subject Index Volumes 290–319

VOLUME 321. Numerical Computer Methods (Part C)
Edited by MICHAEL L. JOHNSON AND LUDWIG BRAND

VOLUME 322. Apoptosis
Edited by JOHN C. REED

VOLUME 323. Energetics of Biological Macromolecules (Part C)
Edited by MICHAEL L. JOHNSON AND GARY K. ACKERS

VOLUME 324. Branched-Chain Amino Acids (Part B)
Edited by ROBERT A. HARRIS AND JOHN R. SOKATCH

VOLUME 325. Regulators and Effectors of Small GTPases (Part D: Rho Family)
Edited by W. E. BALCH, CHANNING J. DER, AND ALAN HALL

VOLUME 326. Applications of Chimeric Genes and Hybrid Proteins (Part A: Gene Expression and Protein Purification)
Edited by JEREMY THORNER, SCOTT D. EMR, AND JOHN N. ABELSON

VOLUME 327. Applications of Chimeric Genes and Hybrid Proteins (Part B: Cell Biology and Physiology)
Edited by JEREMY THORNER, SCOTT D. EMR, AND JOHN N. ABELSON

VOLUME 328. Applications of Chimeric Genes and Hybrid Proteins (Part C: Protein–Protein Interactions and Genomics)
Edited by JEREMY THORNER, SCOTT D. EMR, AND JOHN N. ABELSON

VOLUME 329. Regulators and Effectors of Small GTPases (Part E: GTPases Involved in Vesicular Traffic)
Edited by W. E. BALCH, CHANNING J. DER, AND ALAN HALL

VOLUME 330. Hyperthermophilic Enzymes (Part A)
Edited by MICHAEL W. W. ADAMS AND ROBERT M. KELLY

VOLUME 331. Hyperthermophilic Enzymes (Part B)
Edited by MICHAEL W. W. ADAMS AND ROBERT M. KELLY

VOLUME 332. Regulators and Effectors of Small GTPases (Part F: Ras Family I)
Edited by W. E. BALCH, CHANNING J. DER, AND ALAN HALL

VOLUME 333. Regulators and Effectors of Small GTPases (Part G: Ras Family II)
Edited by W. E. BALCH, CHANNING J. DER, AND ALAN HALL

VOLUME 334. Hyperthermophilic Enzymes (Part C)
Edited by MICHAEL W. W. ADAMS AND ROBERT M. KELLY

VOLUME 335. Flavonoids and Other Polyphenols
Edited by LESTER PACKER

VOLUME 336. Microbial Growth in Biofilms (Part A: Developmental and Molecular Biological Aspects)
Edited by RON J. DOYLE

VOLUME 337. Microbial Growth in Biofilms (Part B: Special Environments and Physicochemical Aspects)
Edited by RON J. DOYLE

VOLUME 338. Nuclear Magnetic Resonance of Biological Macromolecules (Part A)
Edited by THOMAS L. JAMES, VOLKER DÖTSCH, AND ULI SCHMITZ

VOLUME 339. Nuclear Magnetic Resonance of Biological Macromolecules (Part B)
Edited by THOMAS L. JAMES, VOLKER DÖTSCH, AND ULI SCHMITZ

VOLUME 340. Drug–Nucleic Acid Interactions
Edited by JONATHAN B. CHAIRES AND MICHAEL J. WARING

VOLUME 341. Ribonucleases (Part A)
Edited by ALLEN W. NICHOLSON

VOLUME 342. Ribonucleases (Part B)
Edited by ALLEN W. NICHOLSON

VOLUME 343. G Protein Pathways (Part A: Receptors)
Edited by RAVI IYENGAR AND JOHN D. HILDEBRANDT

VOLUME 344. G Protein Pathways (Part B: G Proteins and Their Regulators)
Edited by RAVI IYENGAR AND JOHN D. HILDEBRANDT

VOLUME 345. G Protein Pathways (Part C: Effector Mechanisms)
Edited by RAVI IYENGAR AND JOHN D. HILDEBRANDT

VOLUME 346. Gene Therapy Methods
Edited by M. IAN PHILLIPS

VOLUME 347. Protein Sensors and Reactive Oxygen Species (Part A: Selenoproteins and Thioredoxin)
Edited by HELMUT SIES AND LESTER PACKER

VOLUME 348. Protein Sensors and Reactive Oxygen Species (Part B: Thiol Enzymes and Proteins)
Edited by HELMUT SIES AND LESTER PACKER

VOLUME 349. Superoxide Dismutase
Edited by LESTER PACKER

VOLUME 350. Guide to Yeast Genetics and Molecular and Cell Biology (Part B)
Edited by CHRISTINE GUTHRIE AND GERALD R. FINK

VOLUME 351. Guide to Yeast Genetics and Molecular and Cell Biology (Part C)
Edited by CHRISTINE GUTHRIE AND GERALD R. FINK

VOLUME 352. Redox Cell Biology and Genetics (Part A)
Edited by CHANDAN K. SEN AND LESTER PACKER

VOLUME 353. Redox Cell Biology and Genetics (Part B)
Edited by CHANDAN K. SEN AND LESTER PACKER

VOLUME 354. Enzyme Kinetics and Mechanisms (Part F: Detection and Characterization of Enzyme Reaction Intermediates)
Edited by DANIEL L. PURICH

VOLUME 355. Cumulative Subject Index Volumes 321–354

VOLUME 356. Laser Capture Microscopy and Microdissection
Edited by P. MICHAEL CONN

VOLUME 357. Cytochrome P450, Part C
Edited by ERIC F. JOHNSON AND MICHAEL R. WATERMAN

VOLUME 358. Bacterial Pathogenesis (Part C: Identification, Regulation, and Function of Virulence Factors)
Edited by VIRGINIA L. CLARK AND PATRIK M. BAVOIL

VOLUME 359. Nitric Oxide (Part D)
Edited by ENRIQUE CADENAS AND LESTER PACKER

VOLUME 360. Biophotonics (Part A)
Edited by GERARD MARRIOTT AND IAN PARKER

VOLUME 361. Biophotonics (Part B)
Edited by GERARD MARRIOTT AND IAN PARKER

VOLUME 362. Recognition of Carbohydrates in Biological Systems (Part A)
Edited by YUAN C. LEE AND REIKO T. LEE

VOLUME 363. Recognition of Carbohydrates in Biological Systems (Part B)
Edited by YUAN C. LEE AND REIKO T. LEE

VOLUME 364. Nuclear Receptors
Edited by DAVID W. RUSSELL AND DAVID J. MANGELSDORF

VOLUME 365. Differentiation of Embryonic Stem Cells
Edited by PAUL M. WASSAUMAN AND GORDON M. KELLER

VOLUME 366. Protein Phosphatases
Edited by SUSANNE KLUMPP AND JOSEF KRIEGLSTEIN

VOLUME 367. Liposomes (Part A)
Edited by NEJAT DÜZGÜNEŞ

VOLUME 368. Macromolecular Crystallography (Part C)
Edited by CHARLES W. CARTER, JR., AND ROBERT M. SWEET

VOLUME 369. Combinational Chemistry (Part B)
Edited by GUILLERMO A. MORALES AND BARRY A. BUNIN

VOLUME 370. RNA Polymerases and Associated Factors (Part C)
Edited by SANKAR L. ADHYA AND SUSAN GARGES

VOLUME 371. RNA Polymerases and Associated Factors (Part D)
Edited by SANKAR L. ADHYA AND SUSAN GARGES

VOLUME 372. Liposomes (Part B)
Edited by NEGAT DÜZGÜNEŞ

VOLUME 373. Liposomes (Part C)
Edited by NEGAT DÜZGÜNEŞ

VOLUME 374. Macromolecular Crystallography (Part D)
Edited by CHARLES W. CARTER, JR., AND ROBERT W. SWEET

VOLUME 375. Chromatin and Chromatin Remodeling Enzymes (Part A)
Edited by C. DAVID ALLIS AND CARL WU

VOLUME 376. Chromatin and Chromatin Remodeling Enzymes (Part B)
Edited by C. DAVID ALLIS AND CARL WU

VOLUME 377. Chromatin and Chromatin Remodeling Enzymes (Part C)
Edited by C. DAVID ALLIS AND CARL WU

VOLUME 378. Quinones and Quinone Enzymes (Part A) (in preparation)
Edited by HELMUT SIES AND LESTER PACKER

VOLUME 379. Energetics of Biological Macromolecules (Part D)
(in preparation)
Edited by JO M. HOLT, MICHAEL L. JOHNSON, AND GARY K. ACKERS

VOLUME 380. Energetics of Biological Macromolecules (Part E)
(in preparation)
Edited by JO M. HOLT, MICHAEL L. JOHNSON, AND GARY K. ACKERS

VOLUME 381. Oxygen Sensing (in preparation)
Edited by CHANDAN K. SEN AND GREGG L. SEMENZA

VOLUME 382. Quinones and Quinone Enzymes (Part B) (in preparation)
Edited by HELMUT SIES AND LESTER PACKER

VOLUME 383. Numerical Computer Methods, (Part D) (in preparation)
Edited by LUDWIG BRAND AND MICHAEL L. JOHNSON

VOLUME 384. Numerical Computer Methods, (Part E) (in preparation)
Edited by LUDWIG BRAND AND MICHAEL L. JOHNSON

VOLUME 385. Imaging in Biological Research, (Part A) (in preparation)
Edited by P. MICHAEL CONN

VOLUME 386. Imaging in Biological Research, (Part B) (in preparation)
Edited by P. MICHAEL CONN

Section I

Genetic Assays of Chromatin Modification and Remodeling

[1] Functional Analyses of Chromatin Modifications in Yeast

By Sandra J. Jacobson, Patricia M. Laurenson, and Lorraine Pillus

Site-specific modification of histones is fundamental to chromatin function. The enzymes that perform these modifications include protein acetyltransferases and deacetylases, methyltransferases, kinases and phosphatases, and ubiquitin-conjugating enzymes that are highly conserved from yeast to humans. Histone modifying enzymes often reside in multi-protein complexes whose subunits target and/or regulate the respective enzymatic activity. A number of these complexes have now been biochemically purified and analyzed. In the budding yeast, *Saccharomyces cerevisiae*, biochemical approaches are readily combined with genetic analyses to coordinate understanding of histone modifying complexes, their *in vivo* substrate specificity, their target genetic loci, and the functional consequences of their activity.

Here we present principles and mechanics of using *S. cerevisiae* to analyze the function of histones and histone modifiers. We depict the well-studied, posttranslational modifications of yeast histone residues (see Fig. 1) and the histone genes (see Fig. 2), and outline the enzymes responsible for histone modifications and their known cellular functions (see Table II). We present experimental strategies for studying chromatin modifiers and histone mutants (see Fig. 3; Table III) with a case study (see Table IV) and examples from the literature. This is accompanied by methods for studying chromatin-related functions, including chromatin-related assays (see Table V) and silencing assays (see Table VI, Fig. 4). Beyond these studies, *S. cerevisiae* is valuable for examining chromatin-related functions of a favorite protein from multicellular eukaryotes. We consider briefly human chromatin modifier genes associated with disease (see Table VII) and methods for analyzing their functions in yeast (see Figs. 5 and 6). Finally, we include a discussion of genomics tools and resources currently available in yeast (see Table VIII; Table I) and how these may be used to complement more traditional genetic approaches.

METHODS IN ENZYMOLOGY, VOL. 377

Histone Genetics in *S. cerevisiae*

Nucleosome Structure

An underlying theme in considering chromatin modifications is that they provide mechanisms for dynamic regulation of gene expression. Such dynamism, which correlates with epigenetic aspects of regulation, is critical because it constitutes a framework for developmental switches and environmental responses without changes in primary DNA sequence. Understanding how histone modifications contribute to biological regulation ultimately relies on coordinated biochemical and genetic approaches that are readily accessible in yeast. Experimental dissection of chromatin function has gained momentum with the availability of high-resolution structural data of chromatin proteins and their modifiers, which help guide the construction and interpretation of mutants. The X-ray crystallographic structure of the nucleosome core particle at 2.8A resolution provided key details of the precise spatial orientation of histones with each other and with DNA.[1,2] This image of the nucleosome showed amino acids that were poised for post-translational modification as well as those that were likely to support the structural integrity of the nucleosome. It has become the bench-side companion of investigators designing and interpreting chromatin-related experiments.

Many studies have focused on understanding the significance of post-translational modifications of N-terminal histone tails. These solvent exposed tails are modified at discrete sites through the covalent addition of acetyl, methyl, phosphate or ubiquitin groups (see Fig. 1). The marks have significant effects on chromatin structure and function where they may alter nucleosome structure or inter-nucleosomal interactions and regulate binding of chromatin-associated proteins.

The role of chromatin modifications in the process of DNA transcription has been studied in detail, particularly that of acetylation which impacts basal transcription levels and reversible activation of genes (reviewed in Kurdistani and Grunstein[3]). Genome-wide screening of histone acetylation and RNA transcript profiles in acetylase and deacetylase mutants has revealed that histone acetylation can exert long range effects to create chromosomal domains.[4-9] In other cases, acetylation may affect only several neighboring nucleosomes to facilitate binding of regulatory

[1] K. Luger, A. W. Mader, R. K. Richmond, D. F. Sargent, and T. J. Richmond, *Nature* **389,** 251 (1997).
[2] C. L. White, R. K. Suto, and K. Luger, *EMBO J.* **20,** 5207 (2001).
[3] S. K. Kurdistani and M. Grunstein, *Nat. Rev. Mol. Cell. Biol.* **4,** 276 (2003).
[4] M. Vogelauer, J. Wu, N. Suka, and M. Grunstein, *Nature* **408,** 495 (2000).

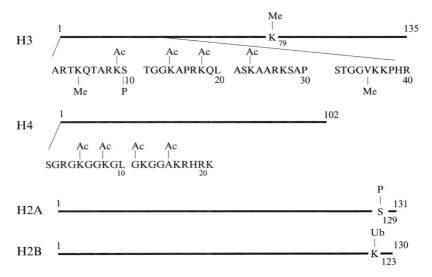

FIG. 1. Well-characterized sites of modifications in yeast core histones. Histones H3, H4, H2A, and H2B are represented as lines, with amino acid sequence of N-terminal tails of H3 and H4 included as detailed insets. Numbers refer to amino acid position. K, lysine, S, serine. Post-translational modifications are designated as follows: Me, methylated, Ac, acetylated, P, phosphorylated, Ub, ubiquitinated. Sites illustrated do not include all known sites of modification on yeast histones, but rather those that have been closely tied to a cellular function (Table II).

proteins at particular DNA sequences.[10] Thus, gene-specific transcriptional regulation may in some cases be closely tied to the chromosomal context of a gene. Adding to the complexity is that multiple modifications can exist simultaneously on histone tails. Such combinatorial modification raises the possibility of a histone code[11] or particular histone surfaces[3] that program precise functional outputs. As strains harboring mutations in histones

[5] B. E. Bernstein, J. K. Tong, and S. L. Schreiber, *Proc. Natl. Acad. Sci. USA* **97,** 13708 (2000).

[6] N. Suka, Y. Suka, A. A. Carmen, J. Wu, and M. Grunstein, *Mol. Cell* **8,** 473 (2001).

[7] A. Kimura, T. Umehara, and M. Horikoshi, *Nat. Genet.* **32,** 370 (2002).

[8] D. Robyr, Y. Suka, I. Xenarios, S. K. Kurdistani, A. Wang, N. Suka, and M. Grunstein, *Cell* **109,** 437 (2002).

[9] J. J. Wyrick, F. C. Holstege, E. G. Jennings, H. C. Causton, D. Shore, M. Grunstein, E. S. Lander, and R. A. Young, *Nature* **402,** 418 (1999).

[10] M. H. Kuo, J. Zhou, P. Jambeck, M. E. Churchill, and C. D. Allis, *Genes Dev.* **12,** 627 (1998).

[11] T. Jenuwein and C. D. Allis, *Science* **293,** 1074 (2001).

and histone modifiers are examined further, we will gain an even better understanding of the relationships among modifications in addition to understanding less well-studied chromatin-dependent processes like replication, repair, recombination, and chromosomal segregation.

Using S. cerevisiae to Study Histone Function

For researchers interested in chromatin-related processes, *S. cerevisiae* has many attributes that promote insightful genetic studies of histone gene function (see Smith and Santisteban[12] for a more extensive review). Most importantly, there are only two copies of each major core histone gene, so phenotypes of recessive as well as dominant mutations in the histone genes can be examined. The large number of copies of histone genes in many eukaryotes makes a similar analysis difficult if not impossible. In

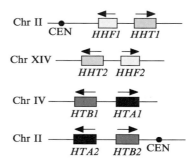

Fig. 2. *S. cerevisiae* major core histone genes. The histone genes are duplicated, and are present in divergently-transcribed, nonallelic pairs. The genes are depicted as boxes on a linear chromosome, with the direction of transcription indicated by the arrows. The histone gene names are *HHT1* and *HHT2* (for histone *H* three), *HHF1* and *HHF2* (for histone *H* four), *HTA1* and *HTA2* (for histone *H* two A) and *HTB1* and *HTB2* (for histone *H* two B). *HHT1* and *HHT2* encode identical H3 proteins, and *HHF1* and *HHF2* encode identical H4 proteins. *HTA1* and *HTA2*, however, encode proteins that differ by two amino acids. Likewise, *HTB1* and *HTB2* encode proteins that differ by four amino acids. The figure is not to scale and does not show the genes in the sequences between the histone genes on chromosome II and its centromere. Strains are available [see M. M. Smith and M. S. Santisteban, *Methods* **15**, 269 (1998)] in which both sets of gene pairs are deleted (e.g., *hht1-hhf1*Δ; *hht2-hhf2*Δ) and the strain is kept alive by a plasmid containing one gene pair (e.g., *HHT1-HHF1*). Alternatively, strains are available [see M. M. Smith and M. S. Santisteban, *Methods* **15**, 269 (1998)] in which both gene copies are deleted (e.g. *hhf1*Δ; *hhf2*Δ) and the strain is kept alive by a plasmid containing a copy of one of the genes (e.g., *HHF1*). For excellent basic reviews on getting started with yeast, see F. Sherman, *Methods Enzymol.* **350**, 3 (2002) and C. Styles, *Methods Enzymol.* **350**, 42 (2002).

---

[12] M. M. Smith and M. S. Santisteban, *Methods* **15**, 269 (1998).

yeast, the major histone genes occur chromosomally in pairs, and are divergently transcribed (see Fig. 2). Histone mutations are often studied by creating strains that lack both chromosomal sets of the wild-type histone gene pairs (e.g., deletion of *HHT1-HHF1* and *HHT2-HHF2*) but survive by carrying a mutated copy of one of the histone gene pairs (e.g., *hht1-HHF1*) on a plasmid or replaced into the chromosome. Mutant versions of histone genes are typically generated by site-directed or random mutagenesis to construct strains that can be analyzed in a variety of assays (see later). A further advantage of yeast is that phenotypes caused by histone mutants can be examined coordinately with mutations in genes encoding the cognate histone modifier or chromatin-associated protein.

There are several strategies to create or isolate strains containing histone mutations.[12] One approach combines traditional genetic techniques with a more modern twist called the plasmid shuffle[13] (outlined in Fig. 3). This approach relies on the observation that just one copy of each histone gene is sufficient for cell viability. As a starting point, one chromosomal copy of a histone gene pair is deleted in one haploid strain and the other chromosomal copy is deleted in a second haploid strain. The two haploid strains are crossed and the resulting diploid is transformed with a plasmid containing a wild-type copy of one of the histone gene pairs and a counter-selectable marker such as *URA3* (see Table I for counterselectable markers). The diploid is sporulated and dissected to yield four haploid segregants. Approximately one-quarter of the segregants will have knockouts of both chromosomal copies of the histone gene pairs and will be Ura$^+$ due to the requirement for the plasmid. These segregants then can be used to do the plasmid shuffle (see Fig. 3, lower half). The strain is transformed with a second plasmid that has a different selectable marker and contains a mutated version of the histone gene pair of interest. Thus, the transformants have chromosomal deletions of the histone gene pairs and bear two plasmids, one with a wild-type copy of the histone gene pair and one with a mutagenized copy of the histone gene pair. Cells that have lost the *URA3*-marked wild-type plasmid are recovered by plating on 5-FOA, so that the only copy of the histone gene pair present in the 5-FOA-resistant isolates comes from the mutagenized histone gene pair on the second plasmid.

When working with strains containing mutations in the histone genes, it is important to take precautions to ensure that the genotype is stable. Strains with histone mutations show varying degrees of chromosomal instability: they may spontaneously diploidize or accumulate suppressors and chromosomal rearrangements. Also, diploids carrying mutations in both copies of the *HTA1* and *HTB1* genes and lacking a covering plasmid

---

[13] J. D. Boeke, J. Trueheart, G. Natsoulis, and G. R. Fink, *Methods Enzymol.* **154,** 164 (1987).

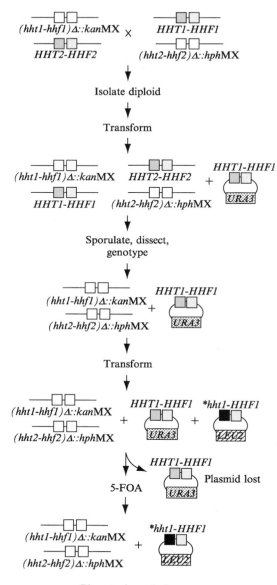

Phenotypic analysis

Fig. 3. Histone functional analysis flowchart. This figure outlines one strategy to construct or isolate strains containing mutations in a histone gene. Strains with similar genotypes are described [J. H. Park, M. S. Cosgrove, E. Youngman, C. Wolberger, and J. Boeke, *Nat. Genet.*

TABLE I
DRUGS FOR MARKING DELETIONS AND FOR NEGATIVE SELECTION IN SILENCING

Drug^a	Resistance or Target gene^b	Concentration	Recipe^c
G418 (geneticin)	kanMX^d (Tn 903)	200 mg/L^e	(1)
ClonNAT (nourseothricin)	$nat1^d$ (S. noursei)	100 mg/L	(2)^f
5-FOA	URA3^g	1 g/L	(3)
6-AU	URA3^g	3–20 mg/L^e	(3)
3-AT	HIS3^g	5–100 mM^e	(3)
Canavanine	CAN1^g	8–80 mg/L^e	(3)
5-FAA	TRP1^g	0.5–1 g/L^e	(4)

^{a} Drugs are available from various sources including Sigma; Life Technologies, Rockville, MD (Geneticin); Werner BioAgents, Jena-Cospeda, Germany (ClonNAT); USB, Cleveland OH (5-FOA and 3-AT; substantially discounted pricing available on 5-FOA to members of the Genetics Society of America); Aldrich Chemical.

^{b} Genes are from S. cerevisiae, except where (indicated).

^{c} Details for media preparation vary depending on the drug. Detailed recipes are given in the indicated references.

Key to references:
(1) M. Johnston, L. Riles, and J. H. Hegemann, Methods Enzymol. 350, 290 (2002);
(2) A. L. Goldstein and J. H. McCusker, Yeast 15, 1541 (1999);
(3) F. van Leeuwen and D. E. Gottschling, Methods Enzymol. 350, 165 (2002);
(4) J. H. Toyn, P. L. Gunyuzlu, W. H. White, L. A. Thompson, and G. F. Hollis, Yeast 16, 553 (2000).

^{d} Expression of the gene provides resistance to the drug indicated.

^{e} Note that the bio-activity of G418 varies among lots. When selecting transformants, it is sometimes useful to initially plate on 100 μg/ml drug, then do a secondary selection on 200 μg/ml. Concentration ranges are given for these compounds because some strains or sites of reporter gene insertion have differing sensitivities. The optimal dynamic range should be established in pilot experiments.

^{f} Note that in addition to nourseothricin, alternative drug resistance cassettes to hygromycin B and bialaphos are presented. Although less widely used to date, they provide additional possibilities for selection.

^{g} Expression of the gene results in sensitivity to the drug indicated.

32, 273 (1998)]. A genetic cross between two strains is indicated by an X. The notation + indicates that the strain bears a plasmid. Genes are depicted as boxes on linear chromosomes or circular plasmids. Open boxes indicate that the histone genes are deleted and replaced with marker genes, whereas shaded boxes indicate that the gene or gene pair is present. Each wild-type HHT1-HHF1 or HHT2-HHF2 gene pair uses the same shading scheme as in Fig. 2; for simplicity the orientation of the HHT1-HHF1 gene pair is switched. The * denotes a mutagenized version of HHT1, which is depicted as a black box. Different selectable markers (URA3 or LEU2) are present in plasmids and are maintained by growth in a medium lacking these supplements. 5-FOA refers to the counterselectable medium used to identify isolates that have lost the URA3-containing plasmid.

sporulate poorly.[14] Accordingly, frozen stocks should be made at sequential points during strain construction: the diploid prior to sporulation, the haploid carrying both wild-type and mutant plasmids prior to plasmid shuffle, and the haploids obtained after the plasmid shuffle. Frozen stocks are prepared from cells grown in medium selecting for one or both plasmids, as appropriate. DMSO (methyl sulfoxide, Sigma) is added to a final concentration of 7%, and the cell suspension is frozen in a cryovial at $-70°$. Yeast frozen in this manner are readily recovered for future experiments by simply scraping a toothpick over the frozen stock and depositing the ice chips on a fresh agar plate.

Despite the fact that histones are essential proteins, a large number of mutations in histone genes yield informative, viable phenotypes. Thus, it has been possible to study nonessential chromatin-related processes such as transcriptional silencing (see below) using histone mutants. However, strains carrying histone mutations affecting essential chromatin functions are not likely to survive. When studying an essential process such as DNA replication, it may be necessary to isolate conditional alleles or utilize conditional expression of the mutant histone.

Validating a Correlation Between Histone Modification and Cellular Function

Yeast offers the opportunity to combine biochemical and genetic approaches to evaluate the functional consequences of histone modifications *in vivo*. In the simplest cases, alteration of one or two histone amino acids, or deletion of the corresponding histone-modifying enzyme, disrupts a particular cellular function. In other cases, genetic redundancy, functional overlap of chromatin modifiers, or incomplete experimental analysis does not yield a clear correlation between histone modification and cellular function. Figure 1 and Table II list histone modifications and their corresponding chromatin modifiers that have been studied in sufficient detail (as outlined in Table III) to warrant a high degree of confidence in their assignment to a particular function.

Yeast Histone Mutants: What We Have Learned about Histone Function From Mutational Analysis

This section describes posttranslational modifications of histones H3, H4, H2A, H2B, and histone variants that have been investigated through mutational analysis. The emphasis on acetylation of the N-terminal tails of histones H3 and H4 reflects the prominence of this modification in

---

[14] K. Tsui, L. Simon, and D. Norris, *Genetics* **145,** 647 (1997).

TABLE II
HISTONE MODIFYING ENZYMES^a

Histone^b	Amino acid^b	Modification^b	Enzyme	Phenotype of mutant	Ref^c
H3	K4	Me	Set 1	slow growth, rDNA silencing defect, telomeric silencing and/or telomeric length defect	$(1)^d$ (2–6)
				transcriptional activation and/or elongation defect	(7–10)
	K9, K14	Ac	Gcn5^e	transcriptional activation defect	(11–16)
	K9/14	deAc	Rpd3	transcriptional repression defect^f	(17–18)
	K9, 14, 18, 23, 27	deAc	Hda1	transcriptional repression defect^f	19
	K14	deAc	Sir2	transcriptional repression defect^f	(20–23)
			Sas3	6-AU sensitivity of $sas3\ \Delta/spt1\text{-}\Delta922$	24
				synthetic lethality with $gcn5\ \Delta$	25
	S10	P	Snf1	transcriptional activation defect	16
	K36	Me	Set2	transcriptional activation and/or elongation defect	(9,26–28)
				transcriptional repression defect^g	29
	K79	Me	Dot1	telomeric silencing defect	(30,31,9,10)
H4	K5,8,12,16	Ac	Esa1^h	G2/M cell cycle block, nucleolar disruption, transcriptional activation defect	32
					(33–36)
				DNA double-strand break repair	37
	K5, 12^f	deAc	Rpd3	transcriptional repression defect^f	17
	K12	Ac	Hat1	telomeric silencing defect^i	38
				DNA repair defect	39
	K16	Ac	Sas2	telomeric and HM silencing defect	(40–43)
	K16	deAc	Sir2	transcriptional silencing defect^f	(20–23)

(continued)

TABLE II *(continued)*

Histone^b	Amino acid^b	Modification^b	Enzyme	Phenotype of mutant	Ref^c
H2A	S129	P	Mec1	ds DNA damage repair defect	44
H2B	K123	Ub	Rad6	UV-induced DNA repair defect	45
	K11, 16	deAc	Hda1	transcriptional repression defect^f	19

^{a} The table is constructed to highlight yeast core histones and their modifying enzymes where functional correlations have been validated *in vivo*. Additional enzymes and modifications that have been less well-studied to date are discussed in the text.

^{b} Key to the core histone, amino acid, and histone modification designations are as in Fig. 1.

^{c} Note that references highlight representative studies combining biochemical and functional experiments. They are not complete citations for the enzyme/modification site. Key to references:

(1) A. Roguev, D. Schaft, A. Shevchenko, W. W. Pijnappel, M. Wilm, R. Aasland, and A. F. Stewart, *EMBO J.* **20**, 7137 (2001).

(2) S. D. Briggs, M. Bryk, B. D. Strahl, W. L. Cheung, J. K. Davie, S. Y. Dent, F. Winston, and C. D. Allis, *Genes Dev.* **15**, 3286 (2001).

(3) T. Miller, N. J. Krogan, J. Dover, H. Erdjument-Bromage, P. Tempst, M. Johnston, J. F. Greenblatt, and A. Shilatifard, *Proc. Natl. Acad. Sci. USA* **98**, 12902 (2001).

(4) M. Bryk, S. D. Briggs, B. D. Strahl, M. J. Curcio, C. D. Allis, and F. Winston, *Curr. Biol.* **12**, 165 (2002).

(5) P. L. Nagy, J. Griesenbeck, R. D. Kornberg, and M. L. Cleary, *Proc. Natl. Acad. Sci. USA* **99**, 90 (2002).

(6) N. J. Krogan, J. Dover, S. Khorrami, J. F. Greenblatt, J. Schneider, M. Johnston, and A. Shilatifard, *J. Biol. Chem.* **277**, 10753 (2002).

(7) B. E. Bernstein, E. L. Humphrey, R. L. Erlich, R. Schneider, P. Bouman, J. S. Liu, T. Kouzarides, and S. L. Schreiber, *Proc. Natl. Acad. Sci. USA* **99**, 8695 (2002).

(8) H. Santos-Rosa, R. Schneider, A. J. Bannister, J. Sherriff, B. E. Bernstein, N. C. Emre, S. L. Schreiber, J. Mellor, and T. Kouzarides, *Nature* **419**, 407 (2002).

(9) N. J. Krogan, J. Dover, A. Wood, J. Schneider, J. Heidt, M. A. Boateng, K. Dean, O. W. Ryan, A. Golshani, M. Johnston, J. F. Greenblatt, and A. Shilatifard, *Mol. Cell* **11**, 721 (2003).

(10) H. H. Ng, F. Robert, R. A. Young, and K. Struhl, *Mol. Cell* **11**, 709 (2003).

(11) M. H. Kuo, J. E. Brownell, R. E. Sobel, T. A. Ranalli, R. G. Cook, D. G. Edmondson, S. Y. Roth, and C. D. Allis, *Nature* **383**, 269 (1996)

(12) P. A. Grant, A. Eberharter, S. John, R. G. Cook, B. M. Turner, and J. L. Workman, *J. Biol. Chem.* **274**, 5895 (1999).

(13) M. H. Kuo, J. Zhou, P. Jambeck, M. E. Churchill, and C. D. Allis, *Genes Dev.* **12**, 627 (1998).

(14) W. Zhang, J. R. Bone, D. G. Edmondson, B. M. Turner, and S. Y. Roth, *EMBO J.* **17**, 3155 (1998).

(15) K. Ikeda, D. J. Steger, A. Eberharter, and J. L. Workman, *Mol. Cell. Biol.* **19**, 855 (1999).

(continued)

TABLE II   (continued)

(16) W. S. Lo, R. C. Trievel, J. R. Rojas, L. Duggan, J. Y. Hsu, C. D. Allis, R. Marmorstein, and S. L. Berger, Mol. Cell 5, 917 (2000).

(17) D. Kadosh and K. Struhl, Mol. Cell. Biol. 18, 5121 (1998).

(18) M. Vogelauer, J. Wu, N. Suka, and M. Grunstein, Nature 408, 495 (2000).

(19) J. Wu, N. Suka, M. Carlson, and M. Grunstein, Mol. Cell 7, 117 (2001).

(20) J. C. Tanny, G. J. Dowd, J. Huang, H. Hilz, and D. Moazed, Cell 99, 735 (1999).

(21) S. Imai, C. M. Armstrong, M. Kaeberlein, and L. Guarente, Nature 403, 795 (2000).

(22) C. M. Armstrong, M. Kaeberlein, S. I. Imai, and L. Guarente, Mol. Biol. Cell 13, 1427 (2002).

(23) S. N. Garcia and L. Pillus, Genetics 162, 721 (2002)

(24) S. John, L. Howe, S. T. Tafrov, P. A. Grant, R. Sternglanz, and J. L. Workman, Genes Dev. 14, 1196 (2000).

(25) L. Howe, D. Auston, P. Grant, S. John, R. G. Cook, J. L. Workman, and L. Pillus, Genes Dev. 15, 3144 (2001).

(26) J. Li, D. Moazed, and S. P. Gygi, J. Biol. Chem. 277, 49383 (2002).

(27) B. Li, L. Howe, S. Anderson, J. R. Yates, III, and J. L. Workman, J. Biol. Chem. 278, 8897 (2003).

(28) T. Xiao, H. Hall, K. O. Kizer, Y. Shibata, M. C. Hall, C. H. Borchers, and B. D. Strahl, Genes Dev. 17, 654 (2003).

(29) B. D. Strahl, P. A. Grant, S. D. Briggs, Z. W. Sun, J. R. Bone, J. A. Caldwell, S. Mollah, R. G. Cook, J. Shabanowitz, D. F. Hunt, and C. D. Allis, Mol. Cell. Biol. 22, 1298 (2002).

(30) F. van Leeuwen, P. R. Gafken, and D. E. Gottschling, Cell 109, 745 (2002).

(31) H. H. Ng, Q. Feng, H. Wang, H. Erdjument-Bromage, P. Tempst, Y. Zhang, and K. Struhl, Genes Dev. 16, 1518 (2002).

(32) A. S. Clarke, J. E. Lowell, S. J. Jacobson, and L. Pillus, Mol. Cell. Biol. 19, 2515 (1999).

(33) S. Allard, R. T. Utley, J. Savard, A. Clarke, P. Grant, C. J. Brandl, L. Pillus, J. L. Workman, and J. Cote, EMBO J. 18, 5108 (1999).

(34) L. Galarneau, A. Nourani, A. A. Boudreault, Y. Zhang, L. Heliot, S. Allard, J. Savard, W. S. Lane, D. J. Stillman, and J. Cote, Mol. Cell 5, 927 (2000).

(35) A. Eisen, R. T. Utley, A. Nourani, S. Allard, P. Schmidt, W. S. Lane, J. C. Lucchesi, and J. Côté J. Biol. Chem. 276, 3484 (2001).

(36) J. L. Reid, V. R. Iyer, P. O. Brown, and K. Struhl, Mol. Cell 6, 1297 (2000).

(37) A. W. Bird, D. Y. Yu, M. G. Pray-Grant, Q. Qiu, K. E. Harmon, P. C. Megee, P. A. Grant, M. M. Smith, and M. F. Christman, Nature 419, 411 (2002).

(38) T. Kelly, S. Qin, D. E. Gottschling, and M. R. Parthun, Mol. Cell. Biol. 20, 7051 (2000).

(39) S. Qin and M. R. Parthun, Mol. Cell. Biol. 22, 8353 (2002).

(40) S. H. Meijsing and A. E. Ehrenhofer-Murray, Genes Dev. 15, 3169 (2001).

(41) S. Osada, A. Sutton, N. Muster, C. E. Brown, J. R. Yates, III, R. Sternglanz, and J. L. Workman, Genes Dev. 15, 3155 (2001).

(42) A. Kimura, T. Umehara, and M. Horikoshi, Nat. Genet. 32, 370 (2002).

(43) N. Suka, K. Luo, and M. Grunstein, Nat. Genet. 32, 378 (2002).

(44) J. A. Downs, N. F. Lowndes, and S. P. Jackson, Nature 408, 1001 (2000).

(continued)

TABLE II　*(continued)*

(45) K. Robzyk, J. Recht, and M. A. Osley, *Science* **287**, 501 (2000).

[d] Defective methyltransferase activity of the TAP-Set1 protein was inferred from its lack of methyltransferase activity *in vitro*.

[e] Gcn5p *in vitro* substrate specificity. Recombinant Gcn5 acetylates H3 primarily on K14 with free histones, not nucleosomes.[11] Gcn5p in context of SAGA HAT complex acetylates H3 K14 > K18 > K9 = K23 and H4 K8 > K16 on free histones, nucleosomes and/or N-terminal histone peptides.[12] Accompanying references highlight studies that defined key aspects of Gcn5p function. Gcn5p HAT activity required for transcriptional activation of HIS3 *in vivo*.[13] Coordinate phenotypic analysis of *gcn5* mutants and histone H3 and H4 N-terminal lysine mutants.[14] Gcn5p HAT activity in context of SAGA and Ada HAT complexes *in vitro*.[12] Gcn5p HAT complex required for transcriptional activation *in vitro*.[15] Gcn5p HAT mutant and cognate histone H3 mutant defective in SAGA-dependent gene activation *in vivo*.[16]

[f] The phenotypes of several deacetylase mutants are presented in cases where individual target genes of the modifying enzymes have been studied in detail in terms of histone acetylation changes and transcriptional regulation. Studies in which histone acetylation states and/or steady-state RNA levels have been surveyed in deacetylase mutants on a genome-wide scale are covered in the text in the Deacetylase section and are not included in this table. Although these studies offer a wealth of information, it is difficult to assess the contribution of secondary effects which obscures a clear functional assignment. For example, genomic RNA profiling data in an *rpd*Δ mutant[7] suggested both activating and repressing roles for Rpd3p. Comparison of this data set with that derived from genomic Ac ChIP in an *rpd3*Δ mutant supports a role for Rpd3p primarily in repression [D. Robyr, Y. Suka, I. Xenarios, S. K. Kurdistani, A. Wang, N. Suka, and M. Grunstein, *Cell* **109**, 437 (2002)].

[g] A Set2 fusion protein was targeted to the promoter of a heterologous gene and the transcriptional output was measured.

[h] Esa1 is the only HAT encoded by an essential gene in yeast.
Esa1p in vitro substrate specificity. Recombinant Esa1p acetylates primariy H4 K5>K8>K12 and to a lesser extent H3 K4 and H2A K4 and K7 on free histones[18] and E. R. Smith, A. Eisen, W. Gu, M. Sattah, A. Pannuti, J. Zhou, R. G. Cook, J. C. Lucchesi and C. D. Allis, *Proc. Natl. Acad. Sci. USA* **95**, 3561 (1998). Esa1p in context of NuA4HAT complex: similar substrate specificity, except H4 K5,K8,K12,K16 on free histones and nucleosomes.[19]

[i] A *hat1* Δ strain has a telomeric silencing defect only in combination with N-terminal mutations in histone H3.

chromatin function, and the experimental focus of many labs in recent years. Although acetylation has been analyzed primarily as it affects transcriptional regulation, other processes such as DNA repair are now under scrutiny. The deacetylases that are an integral part of gene-specific and genome-wide acetylation states are discussed in the last part of this section. We also discuss recent observations regarding methylation of histone H3,

TABLE III
CORRELATING HISTONE MODIFICATION AND CELLULAR FUNCTION

Aim	Method
Demonstrate *in vitro* enzymatic activity using histone substrates	*In vitro* chemical transfer reaction
Correlate enzymatic activity with amino acid modification *in vivo*	Isolate modified substrates for mass spectrometry[a]
	To identify histone substrate *in vivo:* mutate candidate histone modified amino acid and look for change in histone modification in the cell by Western, ChIP or TAU gel[b]
	To identify histone-modifying enzyme *in vivo:* mutate ORF or catalytic domain of candidate enzyme and look for change in histone modification as above[c]
Correlate enzymatic activity with cellular function	Mutate candidate enzyme and assay phenotype[d] (see Table V for list of assays and references)
	Mutate histone amino acid(s) and assay phenotype[e]

[a] *In vitro* substrate specificity data may not strictly correlate with those *in vivo*, but can guide construction of histone mutants whose phenotypes can be examined.

[b] In the case of acetylation, lysines are usually mutated to arginine (R) to block acetylation or to glutamine (Q) to mimic the acetylated state. In the case of phosphorylation, serines are usually mutated to threonine (T) as a potential phosphorylation site or to glutamic acid (E) to mimic the phosphorylated state.

[c] In cases where two related proteins have overlapping substrate specificity, multiple genetic mutations may be required to eliminate a histone modification.

[d] If the enzyme resides in one or more complexes, other subunits may need to be deleted to distinguish which complex the enzyme is working through. An additional control is to express domains of the candidate enzyme and correlate recovery of histone modification with suppression of the mutant phenotype.

[e] Mutation of multiple histone amino acids may be required to produce a phenotype.

phosphorylation of histone H2A and ubiquitination of histone H2B in the relevant subsection, as these studies are excellent examples of the methodological approaches and interpretations applicable to studying other histone modifications in yeast.

Histone H3

Early genetic studies on histone H3 demonstrated that deletion of its N-terminus was not lethal,[15] but caused aberrant transcription of several genes involved in carbon source utilization[16] and caused transcriptional

[15] B. A. Morgan, B. A. Mittman, and M. M. Smith, *Mol. Cell. Biol.* **11,** 4111 (1991).
[16] R. K. Mann and M. Grunstein, *EMBO J.* **11,** 3297 (1992).

activation of normally silenced regions of the genome.[17] These phenotypes can now be considered in light of known sites of posttranslational modification within the first 40 amino acids of histone H3 (see Fig. 1, Table II).

The histone H3 N-terminal tails are reversibly acetylated on lysines (K) 9,14,18, and 23. Histone H3 K14 is the preferred target for acetylation by the well-studied histone acetyltransferase (HAT) Gcn5p *in vitro* and *in vivo*. Additionally, Gcn5p also can contribute to H3 K9 and K18 acetylation.[18,19] Gcn5p, a member of the GNAT family of HATs, is required for activation of several classes of genes involved in carbon source utilization, phosphate metabolism, phospholipid synthesis, amino acid synthesis, and cell-type identity (reviewed in Sterner and Berger[20]). A plasmid shuffle assay (see Fig. 3) was used by Zhang *et al.*,[18] in which specifically mutated lysines in the N-termini of histones H3 and H4 were constructed and expressed in a yeast strain deleted for chromosomal H3 and H4 genes. The phenotypes of such mutants were examined in the presence or absence of functional *GCN5*. This study confirmed that Gcn5p preferentially acetylates H3 K9 and K14 and identified histone amino acid modifications that were important for cell growth and transcriptional activation as measured by a targeted Gal4-VP16 assay.

The roles for Gcn5p in chromatin function are apparently complex. From the extensive studies of Gcn5p, several principles about histone modifier function have emerged. These principles are outlined in Table IV and should prove useful in guiding experiments with other modifiers.

Another site of modification in the H3 N-terminal tail is at serine 10. Phosphorylation of S10 is required at some, but not all, SAGA-dependent genes for maximal gene induction.[21,22] Of those it does affect, S10 phosphorylation can increase the acetylation of H3 K14 both *in vitro* and *in vivo*, which correlates with increased transcription. S10 is also phosphorylated during mitosis by Ipl1p.[23] The functional significance of this is

[17] J. S. Thompson, X. Ling, and M. Grunstein, *Nature* **369**, 245 (1994).

[18] W. Zhang, J. R. Bone, D. G. Edmondson, B. M. Turner, and S. Y. Roth, *EMBO J.* **17**, 3155 (1998).

[19] P. A. Grant, A. Eberharter, S. John, R. G. Cook, B. M. Turner, and J. L. Workman, *J. Biol. Chem.* **274**, 5895 (1999).

[20] D. E. Sterner and S. L. Berger, *Microbiol. Mol. Biol. Rev.* **64**, 435 (2000).

[21] W. S. Lo, R. C. Trievel, J. R. Rojas, L. Duggan, J. Y. Hsu, C. D. Allis, R. Marmorstein, and S. L. Berger, *Mol. Cell* **5**, 917 (2000).

[22] W. S. Lo, L. Duggan, N. C. Tolga, N. C. Emre, R. Belotserkovskya, W. S. Lane, R. Shiekhattar, and S. L. Berger, *Science* **293**, 1142 (2001).

[23] J. Y. Hsu, Z. W. Sun, X. Li, M. Reuben, K. Tatchell, D. K. Bishop, J. M. Grushcow, C. J. Brame, J. A. Caldwell, D. F. Hunt, R. Lin, M. M. Smith, and C. D. Allis, *Cell* **102**, 279 (2000).

TABLE IV

$GCN5$: A Case Study in Principles of Histone Modification and Function

Chromatin-modifier characteristic	Gcn5p example
1. Substrate specificity of chromatin-modifying enzyme may vary	Gcn5p *in vitro* substrate specificity depends on the source of enzyme used (recombinant or purified as a complex from cell lysates) and whether free histones, nucleosomes or synthetic peptides are used as substrates[a]
2. Chromatin modifier may exert short-range gene-specific effects and long-range effects on genome-wide chromatin structure	Gcn5p can be selectively recruited to target genes by transcriptional activators in a gene-specific manner[b]. In the case of $HIS3$ induction, acetylation of H3 K14 is limited to several promoter proximal nucleosomes[c]. In contrast, deletion of $GCN5$ causes significant genome-wide loss of histone H3 acetylation[d,e]
3. Histone modification may be part of a temporal process of nucleosome altering events	Transcriptional activation of the HO gene requires the sequential activity of the sequence-specific DNA-binding protein Swi5p, followed by the chromatin-remodeling complex SWI/SNF, then the Gcn5p-containing SAGA complex[f]
4. Chromatin-modifying enzyme may reside in multiple complexes with different substrates, target genes and/or enzymatic activity	Gcn5p resides in at least three distinct yeast HAT complexes: SAGA[g], Ada[h], and SLIK/SALSA[i,j]
5. Chromatin-modifying enzymes may have overlapping functions with other modifiers	Deletion of $GCN5$ is synthetically lethal with deletion of $SAS3$, which encodes a MYST family HAT whose cellular function is unknown but has similar *in vitro* HAT substrate specificity[k]
6. Post-translational modification of one histone amino acid can affect modification of a neighboring residue	Phosphorylation of histone H3 Ser10 is required for maximal gene induction at a subset of SAGA-dependent genes. S10 phosphorylation increases acetylation of H3 K14, which correlates with increased DNA transcription[l,m]
7. Variation in promoter architecture may elicit different subunit requirements from the same chromatin-modifying complex	Transcriptional activation of the $HIS3$ gene requires Gcn5p enzymatic HAT activity[c]. However, transcriptional activation of the $GAL1$ gene is SAGA-dependent but Gcn5p-independent[n,o]
8. Chromatin-modifying enzyme may have closely related counter-parts in multicellular organisms	Proteins related to yeast Gcn5p and Gcn5p-associated HAT complex subunits have been identified in organisms ranging from yeast to humans[a] (see Table V). This evolutionary conservation encourages cross-species complementation and pharmacological studies in yeast (Fig. 5)

[a] Reviewed in D. E. Sterner and S. L. Berger, *Microbiol. Mol. Biol. Rev.* **64**, 435 (2000) and see Table II.

(continued)

unclear in *S. cerevisiae*, although this modification is required for proper chromosomal segregation in mammals.[24]

In addition to the role played by histone H3 acetylation, recent evidence also points to a critical contribution of histone H3 methylation on residues 4, 36, and 79 to chromatin function. Set1p methylates histone H3 at lysine 4,[25–30] which may in part explain the silencing defects observed in the early H3 N-terminal deletion studies. Deletion of *SET1* causes

[24] K. B. Shannon and E. D. Salmon, *Curr. Biol.* **12,** R458 (2002).

[25] A. Roguev, D. Schaft, A. Shevchenko, W. W. Pijnappel, M. Wilm, R. Aasland, and A. F. Stewart, *EMBO J.* **20,** 7137 (2001).

[26] S. D. Briggs, M. Bryk, B. D. Strahl, W. L. Cheung, J. K. Davie, S. Y. Dent, F. Winston, and C. D. Allis, *Genes Dev.* **15,** 3286 (2001).

[27] N. J. Krogan, J. Dover, S. Khorrami, J. F. Greenblatt, J. Schneider, M. Johnston, and A. Shilatifard, *J. Biol. Chem.* **277,** 10753 (2002).

[28] P. L. Nagy, J. Griesenbeck, R. D. Kornberg, and M. L. Cleary, *Proc. Natl. Acad. Sci. USA* **99,** 90 (2002).

[29] J. Dover, J. Schneider, M. A. Tawiah-Boateng, A. Wood, K. Dean, M. Johnston, and A. Shilatifard, *J. Biol. Chem.* **277,** 28368 (2002).

[30] H. Santos-Rosa, R. Schneider, A. J. Bannister, J. Sherriff, B. E. Bernstein, N. C. Emre, S. L. Schreiber, J. Mellor, and T. Kouzarides, *Nature* **419,** 407 (2002).

[b] Reviewed in O. E. Brown, T. Lechnen, E. Rowe, and J. L. Workman, *Trends Biochem. Sci.* **25,** 15 (2000).

[c] M. H. Kuo, J. Zhou, P. Jambeck, M. E. Churchill, and C. D. Allis, *Genes Dev.* **12,** 627 (1998).

[d] M. Vogelauer, J. Wu, N. Suka, and M. Grunstein, *Nature* **408,** 495 (2000).

[e] L. Howe, D. Auston, P. Grant, S. John, R. G. Cook, J. L. Workman, and L. Pillus, *Genes Dev.* **15,** 3144 (2001).

[f] M. P. Cosma, T. Tanaka, and K. Nasmyth, *Cell* **97,** 299 (1999).

[g] P. A. Grant, L. Duggan, J. Cote, S. M. Roberts, J. E. Brownell, R. Candau, R. Ohba, T. Owen-Hughes, C. D. Allis, F. Winston, S. L. Berger, and J. L. Workman, *Genes Dev.* **11,** 1640 (1997).

[h] A. Eberharter, D. E. Sterner, D. Schieltz, A. Hassan, J. R. Yates, III, S. L. Berger, and J. L. Workman, *Mol. Cell. Biol.* **19,** 6621 (1999).

[i] M. G. Pray-Grant, D. Schieltz, S. J. Mcmahon, J. M. Wood, E. L. Kennedy, R. G. Cook, J. L. Workman, J. R. Yates, III, and P. A. Grant, *Mol. Cell. Biol.* **22,** 8774 (2002).

[j] D. E. Sterner, R. Belotserkovskaya, and S. L. Berger, *Proc. Natl. Acad. Sci. USA* **99,** 11622 (2002).

[k] L. Howe, D. Auston, P. Grant, S. John, R. G. Cook, J. L. Workman, and L. Pillus, *Genes Dev.* **15,** 3144 (2001).

[l] W. S. Lo, R. C. Trievel, J. R. Rojas, L. Duggan, J. Y. Hsu, C. D. Allis, R. Marmorstein, and S. L. Berger, *Mol. Cell* **5,** 917 (2000).

[m] W. S. Lo, L. Duggan, N. C. Tolga, N. C. Emre, R. Belotserkovskya, W. S. Lane, R. Shiekhattar, and S. L. Berger, *Science* **293,** 1142 (2001).

[n] E. Larschan and F. Winston, *Genes Dev.* **15,** 1946 (2001).

[o] S. R. Bhaumik and M. R. Green, *Genes Dev.* **15,** 1935 (2001).

pleiotropic effects in yeast,[31] including slow growth, transcriptional silencing defects, and transcriptional activation defects (see Table II and references therein).

A mechanistic explanation for these phenotypes is now emerging from converging biochemical and genetic approaches. Biochemical tools, including epitope-tagged proteins, highly specific antisera, chromatin immunoprecipitation (ChIP) experiments, and *in vitro* methyltransferase assays, have been combined with genetic tools, including strain construction and mutant analysis, to solidify the correlation between Set1p-dependent histone H3 K4 methylation and transcriptional regulation (see Tables II and III). For example, *set1* mutants lost detectable histone H3 K4 methylation as determined by immunoblotting of whole cell protein extracts using an antiserum specific for methylated H3 K4.[26]

Consistent with this, K4 methylation was blocked in cells expressing mutant histone H3 alleles (*hht1*-K4R or *hht1*-K4A). Importantly, these histone mutant strains exhibited growth defects reminiscent of the growth defects seen in *set1* null strains.[31] In an independent study, a *set1* mutant was recovered from a genetic screen for factors involved in rDNA silencing.[32] Transcriptional silencing of rDNA-embedded reporter genes (see Table VI and Fig. 4A) was abolished in a *set1* null or H3 K4R strains and correlated with loss of H3 K4 methylation as determined by ChIP. This silencing defect was rescued by expressing *SET1* gene fragments containing the methyltransferase domain that restored histone H3 K4 methylation.[24] Furthermore, deletion of genes encoding subunits of Set1-containing complexes[33,25,28] caused telomeric silencing defects also characteristic of a *set1* null strain (see Table II).

One explanation for the silencing defects of a *set1* mutant strain is that histone H3 K4 methylation enhances the binding or affinity of silencing factors at the silenced locus. However, recent observations suggest that the silencing defects of *set1* strains may arise indirectly. The Paf1 protein complex, which associates with RNA Pol II,[34] is important in mediating the methylation of histone H3 K4 (and K79), apparently by recruiting the Set1p containing methyltransferase complex to Pol I.[35,36] This association

[31] C. Nislow, E. Ray, and L. Pillus, *Mol. Biol. Cell* **8,** 2421 (1997).

[32] M. Bryk, S. D. Briggs, B. D. Strahl, M. J. Curcio, C. D. Allis, and F. Winston, *Curr. Biol.* **12,** 165 (2002).

[33] T. Miller, N. J. Krogan, J. Dover, H. Erdjument-Bromage, P. Tempst, M. Johnston, J. F. Greenblatt, and A. Shilatifard, *Proc. Natl. Acad. Sci. USA* **98,** 12902 (2001).

[34] P. A. Wade, W. Werel, R. C. Fentzke, N. E. Thompson, J. F. Leykam, R. R. Burgess, J. A. Jaehning, and Z. F. Burton, *Protein Expr. Purif.* **8,** 85 (1996).

[35] N. J. Krogan, J. Dover, A. Wood, J. Schneider, J. Heidt, M. A. Boateng, K. Dean, O. W. Ryan, A. Golshani, M. Johnston, J. F. Greenblatt, and A. Shilatifard, *Mol. Cell* **11,** 721 (2003).

[36] H. H. Ng, F. Robert, R. A. Young, and K. Struhl, *Mol. Cell* **11,** 709 (2003).

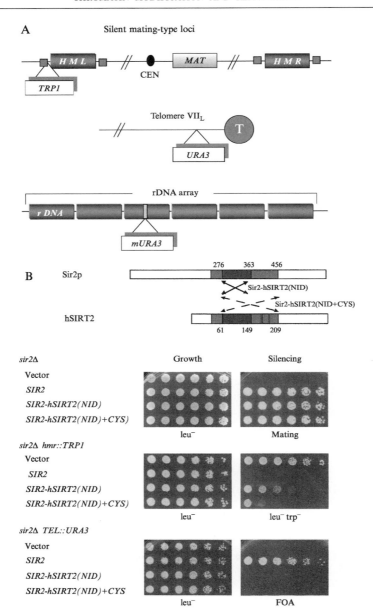

FIG. 4. (A) Three well-characterized chromosomal domains in yeast that confer transcriptional silencing on embedded reporter genes. Top: the silent mating-type loci *HML* and *HMR* (dark boxes) on chromosome III are flanked by *cis*-acting silencer sequences

of methylation with transcriptional activity is consistent with recent micro-array and ChIP data revealing a Set1p-dependent gradient of di- and tri-methylated histone H3 K4 in transcriptionally active coding regions.[30,36,37] By analogy to the model proposed for the role of histone H3 K79 methylation (see later), methylation of histone H3 K4 may be a mark of active chromatin that restricts binding of proteins that repress transcription.[38] According to this model, when appropriate methylation patterns are

[37] B. E. Bernstein, E. L. Humphrey, R. L. Erlich, R. Schneider, P. Bouman, J. S. Liu, T. Kouzarides, and S. L. Schreiber, *Proc. Natl. Acad. Sci. USA* **99**, 8695 (2002).

(small boxes) and silence nearby reporter genes (e.g., *TRP1*). *HML* and *HMR* are normally silenced but contain divergent transcription units encoding $\alpha 1/\alpha 2$ and $\mathbf{a}1/\mathbf{a}2$ proteins. In contrast, the *MAT* locus is constitutively expressed and contains either $\mathbf{a}$ or α information, as at *HMR* and *HML*, that specifies mating-type identity. CEN, centromere. Middle: telomeres (T) confer silencing upon reporter genes such as *URA3* (shown). Bottom: the rDNA array on chromosome XII consists of 100–200 repeat units of [5′ NTS1–5S–NTS2–18S–5.8S–25S 3′] which are diagrammed as individual units (dark boxes). Reporter genes, such as *mURA3*, are inserted at various sites within a single repeat and exhibit variability both in sensitivity to deletion of different chromatin modifiers and in stability of the integrated gene. For each region, boxes represent relative positions of representative reporter genes although diagrams are not to scale. See Table VI for a list of available markers and references. (B) A representative dilution assay demonstrating locus-specific rescue of *sir2*Δ silencing defects by two y*SIR2*-hSIRT2 chimeric proteins. Yeast Sir2p core domain amino acids (dark boxes) were substituted with conserved human sequences (boundaries denoted by arrows) at two genetically defined motifs (dark boxes). The NID motif spans Sir2p amino acids 276–363, solid arrows. The NID + CYS motif includes the NID motif plus conserved cysteines (red narrow boxes), dashed arrows. Numbers above boxes are relevant amino acid numbers of the yeast and human proteins. Chimeric proteins were expressed under the control of the endogenous *SIR2* promoter on *LEU2* marked high copy-number plasmids. Lower panels: five-fold serial dilutions of *sir2*Δ cells expressing wild-type *SIR2* (*SIR2*) as a positive control, *SIR2-hSIRT2(NID)* or *SIR2-hSIRT2(NID+CYS)*. Cells were plated on selective medium lacking leucine (leu⁻) as a growth control (left panels) or medium lacking leucine and containing another selectable condition to quantitatively measure silencing (right panels). Silencing at *HMR* (top right panel) was measured by mating efficiency, which was undetectable in a *sir2*Δ strain carrying a vector negative control, but was rescued comparably by wild-type Sir2p and both Sir2 chimeric proteins as visualized by equal number of colonies in these rows. Silencing at *HMR* (middle right panel) was also measured by expression of a *TRP1* reporter gene near the *HMR* silenced locus. *TRP1* silencing was disrupted in a *sir2*Δ strain carrying vector alone, allowing it to grow on medium lacking tryptophan. Silencing was fully restored in cells expressing ySir2p, partially restored with Sir2-hSIRT2(NID), and poorly rescued with Sir2-hSIRT2(NID+CYS) proteins. Telomeric silencing in a *sir2*Δ strain (bottom right panel) was restored in cells expressing Sir2p, but not the chimeric proteins. Panel B is modified from J. M. Sherman, E. M, Stone, L. L. Freeman-Cook, C. B. Brachmann, J. D. Boeke, and L. Pillus, *Mol. Biol. Cell* **10**, 3045 (1999) and is reprinted from *Molecular Biology of the Cell* (1999, vol. 10, p. 3045–3059) with permission by the American Society for Cell Biology.

abolished, silencing proteins such as the *S*ilent *I*nformation *R*egulator (Sir) proteins can bind to broader regions of the genome, thereby delocalizing silencing proteins from normally silenced regions and disrupting efficient silencing.

In addition to N-terminal methylation, core domain residues K36 and K79 of histone H3 are also methylated by distinct enzymes (see Fig. 1, Table II). Histone H3 K36 is methylated by Set2p.[39] Like Set1p, Set2p can also associate with RNA Pol II and preferentially binds the phosphorylated form of the carboxy-terminal domain (CTD) of elongating Pol II.[35,40–43]

Histone H3 K79 is methylated by Dot1p,[44–48] a protein originally identified as having dosage-dependent effects on telomeric silencing.[49] Mutations that altered H3 K79 and those that abolished Dot1p catalytic activity disrupted telomeric silencing.[44,46] Interestingly, *dot1* null strains showed decreased association of silencing proteins Sir2p and Sir3p with telomeres *in vivo*,[44,46,50] and it was proposed that methylation of K79 may regulate the binding of silencing factors to histones.[44,46]

Additional support for this idea came from three independent screens in which randomly generated mutations in histones H3 and/or H4 were evaluated for silencing defects[51,52] or enhanced silencing.[53] Mutations

[38] F. van Leeuwen and D. E. Gottschling, *Curr. Opin. Cell Biol.* **14,** 756 (2002).

[39] B. D. Strahl, P. A. Grant, S. D. Briggs, Z. W. Sun, J. R. Bone, J. A. Caldwell, S. Mollah, R. G. Cook, J. Shabanowitz, D. F. Hunt, and C. D. Allis, *Mol. Cell. Biol.* **22,** 1298 (2002).

[40] J. Li, D. Moazed, and S. P. Gygi, *J. Biol. Chem.* **277,** 49383 (2002).

[41] T. Xiao, H. Hall, K. O. Kizer, Y. Shibata, M. C. Hall, C. H. Borchers, and B. D. Strahl, *Genes Dev.* **17,** 654 (2003).

[42] B. Li, L. Howe, S. Anderson, J. R. R. Yates, and J. L. Workman, *J. Biol. Chem.* **278,** 8897 (2003).

[43] D. Schaft, A. Roguev, K. M. Kotovic, A. Shevchenko, M. Sarov, K. M. Neugebauer, and A. F. Stewart, *Nucleic Acids Res.* **31,** 2475 (2003).

[44] F. van Leeuwen, P. R. Gafken, and D. E. Gottschling, *Cell* **109,** 745 (2002).

[45] Q. Feng, H. Wang, H. H. Ng, H. Erdjument-Bromage, P. Tempst, K. Struhl, and Y. Zhang, *Curr. Biol.* **12,** 1052 (2002).

[46] H. H. Ng, Q. Feng, H. Wang, H. Erdjument-Bromage, P. Tempst, Y. Zhang, and K. Struhl, *Genes Dev.* **16,** 1518 (2002).

[47] N. Lacoste, R. T. Utley, J. M. Hunter, G. G. Poirier, and J. Cote, *J. Biol. Chem.* **277,** 30421 (2002).

[48] S. D. Briggs, T. Xiao, Z. W. Sun, I. J. A. Caldwel, J. Shabanowitz, D. F. Hunt, S. C. D. Alli, and B. D. Strahl, *Nature* **418,** 498 (2002).

[49] M. S. Singer, A. Kahana, A. J. Wolf, L. L. Meisinger, S. E. Peterson, C. Goggin, M. Mahowald, and D. E. Gottschling, *Genetics* **150,** 613 (1998).

[50] P. A. San-Segundo and G. S. Roeder, *Mol. Biol. Cell* **11,** 3601 (2000).

[51] J. S. Thompson, M. L. Snow, S. Giles, L. E. McPherson, and M. Grunstein, *Genetics* **163,** 447 (2003).

were recovered in the N-terminal domains, as expected, but were also found to cluster in the H3 globular core domain. These fall within the H3 alpha 1 (residues 64–78)/L1 loop (residues 79–85),[51-53] and the H4 L2 region[52] at the H3/H4 dimer interface. These mutations alter histone H3 K79 and surrounding residues, and cause differential effects on *HML*, telomeric, and rDNA silencing. The histone H3 and H4 N-termini, which interact with Sir3p and Sir4p,[54] lie in close proximity to the H3 alpha 1/L1 domain.

Several models have been proposed to explain the silencing defects caused by these H3 mutants. This histone H3/H4 interface may be a surface that enables binding of silencing proteins, or it may be a mark that excludes binding of silencing proteins. In this latter model, loss of K79 methylation, which may be influenced by mutation of adjoining residues of H3 and H4, causes promiscuous binding of silencing proteins to effectively delocalize silencing proteins from their normal location.[38] Consistent with this, about 90% of histone H3 K79 is methylated by Dot1p in wild-type cells,[44] and *SIR*-dependent silenced regions (telomeres and the *HM* loci) preferentially contain hypomethylated H3 K79.[55] H3 K79 methylation state at the telomeres and adjoining regions is influenced by histone H4 N-terminal acetylation state.[55] It will be important to determine the role of the highly conserved H3/H4 junction and how K79 and surrounding residues contribute to locus-specific silencing functions.

Histone H4

N-terminal deletion of histone H4 does not affect viability,[15] but simultaneous deletion of H3 and H4 N-termini is lethal, suggesting a functional overlap of these regions. N-terminal H4 deletions or those that change the four conserved N-terminal lysines at positions 5, 8, 12, and 16 to glutamine have pleiotropic effects including mating defects, an extended G2/M stage of the cell cycle, a temperature sensitive phenotype, and prolonged DNA replication.[15,56-58] These and other early studies (reviewed in Smith[59])

[52] J. H. Park, M. S. Y. Cosgrove, E. Youngman, C. Wolberger, and J. D. Boeke, *Nat. Genet.* **32,** 273 (2002).

[53] C. M. Smith, Z. W. Haimberger, C. O. Johnson, A. J. Wolf, P. R. Gafken, Z. Zhang, M. R. Parthun, and D. E. Gottschling, *Proc. Natl. Acad. Sci. USA* **99**(Suppl 4), 16454 (2002).

[54] M. Grunstein, *Curr. Opin. Cell Biol.* **9,** 383 (1997).

[55] H. H. Ng, D. N. Ciccone, K. B. Morshead, M. A. Oettinger, and K. Struhl, *Proc. Natl. Acad. Sci. USA* **100,** 1820 (2003).

[56] P. S. Kayne, U. J. Kim, M. Han, J. R. Mullen, F. Yoshizaki, and M. Grunstein, *Cell* **55,** 27 (1988).

[57] P. C. Megee, B. A. Morgan, B. A. Mittman, and M. M. Smith, *Science* **247,** 841 (1990).

[58] P. C. Megee, B. A. Morgan, and M. M. Smith, *Genes Dev.* **9,** 1716 (1995).

[59] M. M. Smith, *Curr. Opin. Cell Biol.* **3,** 429 (1991).

established a recurring theme that there is some redundancy in the functional contributions of individual N terminal H4 lysines.

The MYST family HAT Esa1p acetylates the four N-terminal lysines at positions 5, 8, 12, and 16 $in\ vitro^{60,61}$ and in the context of the NuA4 HAT complex.[62] Temperature-sensitive alleles of $ESA1$ lack HAT activity $in\ vitro$ and lose most detectable H4 acetylation at the non-permissive temperature.[60,62] This loss of acetylation by Esa1p correlates with a G2/M block and nucleolar disruption.[60]

Although the precise defects underlying these phenotypes remain to be elucidated, Esa1 activity and N-terminal histone H4 acetylation are apparently important in both a locus-specific and global manner in a variety of cellular functions. Chromatin immunoprecipitation analysis across large regions of the genome demonstrated that Esa1p contributed to H4 acetylation state.[4] Furthermore, acetylation of histone H4 by Esa1p was recently found to be required for DNA double-strand break repair $in\ vitro$ and $in\ vivo$.[63] This aspect of Esa1p function is particularly intriguing because a closely related human protein, Tip60, is present in a complex with several repair-related subunits,[64] homologous to those in yeast.

A number of studies have now demonstrated that Esa1p functions in transcriptional activation.[62,65–67] Notably, Esa1p regulates transcription of several ribosomal protein genes.[67] This raises the possibility that the N-terminal region of H4 may interpret environmental cues and is consistent with the observations that strains with N-terminal histone H4 deletions have decreased $GAL1$ and $PHO5$ gene transcription.[66,68]

Multiple chromatin modifiers and multiple regions of histone H4 are likely to contribute to histone H4 function. For example, another histone acetyltransferase, Elp3p, is implicated in transcriptional elongation and targets H4 K8 and H3 K14 $in\ vitro^{69}$ or all four histones in an in-gel

[60] A. S. Clarke, J. E. Lowell, S. J. Jacobson, and L. Pillus, *Mol. Cell. Biol.* **19**, 2515 (1999).

[61] E. R. Smith, A. Eisen, W. Gu, M. Sattah, A. Pannuti, J. Zhou, R. G. Cook, J. C. Lucchesi, and C. D. Allis, *Proc. Natl. Acad. Sci. USA* **95**, 3561 (1998).

[62] S. Allard, R. T. Utley, J. Savard, A. Clarke, P. Grant, C. J. Brandl, L. Pillus, J. L. Workman, and J. Côté, *EMBO J.* **18**, 5108 (1999).

[63] A. W. Bird, D. Y. Yu, M. G. Pray-Grant, Q. Qiu, K. E. Harmon, P. C. Megee, P. A. Grant, M. M. Smith, and M. F. Christman, *Nature* **419**, 411 (2002).

[64] T. Ikura, V. V. Ogryzko, M. Grigoriev, R. Groisman, J. Wang, M. Horikoshi, R. Scully, J. Qin, and Y. Nakatani, *Cell* **102**, 463 (2000).

[65] L. Galarneau, A. Nourani, A. A. Boudreault, Y. Zhang, L. Heliot, S. Allard, J. Savard, W. S. Lane, D. J. Stillman, and J. Côté, *Mol. Cell* **5**, 927 (2000).

[66] A. Eisen, R. T. Utley, A. Nourani, S. Allard, P. Schmidt, W. S. Lane, J. C. Lucchesi, and J. Côté, *J. Biol. Chem.* **276**, 3484 (2001).

[67] J. L. Reid, V. R. Iyer, P. O. Brown, and K. Struhl, *Mol. Cell* **6**, 1297 (2000).

[68] L. K. Durrin, R. K. Mann, P. S. Kayne, and M. Grunstein, *Cell* **65**, 1023 (1991).

HAT assay using purified histones or synthetic N-terminal histone tails.[70] Additionally, mutation of H4 tyrosines 72, 88, or 98 at the H4-H2A/H2B interface[71] causes transcriptional and chromosomal segregation defects, revealing core domain structural requirements for H4 function.

Normal chromatin structure and DNA repair[72] may also be mediated by Hat1p, which acetylates free histones primarily on H4 K12 *in vitro*.[73,74] Based on its substrate specificity, and the fact that all eukaryotes examined to date have newly synthesized H4 di-acetylated at K5 and K12, it was suspected that Hat1p may be important for chromatin assembly. However, mutation of H4 K5 and K12 in an H3 N-terminal deletion strain yielded no obvious chromatin assembly defects.[75] A role for Hat1p in telomeric silencing, primarily mediated through acetylation of lysine H4 K12, has been reported.[76]

Although the importance of H4 K12 acetylation in silencing remains under study, H4 K16 acetylation is clearly important for silencing. The MYST family HAT, Sas2p, acetylates histone H4 K16 *in vitro*[77] in the context of the SAS complex[78,79] containing Sas4p and Sas5p. Deletion of *SAS2* causes pronounced telomeric silencing defects[80,81] that are phenocopied by changing H4 K16 to arginine.[78] Recent genome-wide analysis of acetylation states in various HAT and deacetylase mutants suggests that the balance of acetylation in silenced regions may be a critical determinant for chromatin structure at transcriptionally silent regions.[7,82]

[69] G. S. Winkler, A. Kristjuhan, H. Erdjument-Bromage, P. Tempst, and J. Q. Svejstrup, *Proc. Natl. Acad. Sci. USA* **99**, 3517 (2002).

[70] B. O. Wittschieben, G. Otero, T. de Bizemont, J. Fellows, H. Erdjument-Bromage, R. Ohba, Y. Li, C. D. Allis, P. Tempst, and J. Q. Svejstrup, *Mol. Cell* **4**, 124 (1999).

[71] M. S. Santisteban, G. Arents, E. N. Moudrianakis, and M. M. Smith, *EMBO J.* **16**, 2493 (1997).

[72] S. Qin and M. R. Parthun, *Mol. Cell. Biol.* **22**, 8353 (2002).

[73] S. Kleff, E. D. Andrulis, C. W. Anderson, and R. Sternglanz, *J. Biol. Chem.* **270**, 24674 (1995).

[74] M. R. Parthun, J. Widom, and D. E. Gottschling, *Cell* **87**, 85 (1996).

[75] X. J. Ma, J. Wu, B. A. Altheim, M. C. Schultz, and M. Grunstein, *Proc. Natl. Acad. Sci. USA* **95**, 6693 (1998).

[76] T. Kelly, S. Qin, D. E. Gottschling, and M. R. Parthun, *Mol. Cell. Biol.* **20**, 7051 (2000).

[77] A. Sutton, W. J. Shia, D. Band, P. D. Kaufman, S. Osada, J. L. Workman, and R. Sternglanz, *J. Biol. Chem.* **278**, 16887 (2003).

[78] S. H. Meijsing and A. E. Ehrenhofer-Murray, *Genes Dev.* **15**, 3169 (2001).

[79] S. Osada, A. Sutton, N. Muster, C. E. Brown, J. R. Yates, 3rd, R. Sternglanz, and J. L. Workman, *Genes Dev.* **15**, 3155 (2001).

[80] C. Reifsnyder, J. Lowell, A. Clarke, and L. Pillus, *Nat. Genet.* **14**, 42 (1996).

[81] A. E. Ehrenhofer-Murray, D. H. Rivier, and J. Rine, *Genetics* **145**, 923 (1997).

[82] N. Suka, K. Luo, and M. Grunstein, *Nat. Genet.* **32**, 378 (2002).

Histone H2A

Phosphorylation of serine 129 in the C-terminal tail of H2A is required for viability in response to DNA damage.[83] Based on the presence of a putative SQE phosphatidylinositol-3-OH kinase (PIKK) motif in the C-terminus of histone H2A, a plasmid shuffle (e.g., Fig. 3) was performed to test the ability of H2A C-terminal mutants to survive treatment with DNA damaging agents. This thorough study illustrates the power of combining biochemistry and genetics to establish a tight correlation between a histone modification and a cellular function (as outlined in Table III). A strain with both sets of H2A- and H2B-encoding genes deleted that was kept alive by a low-copy $URA3$-marked plasmid containing wild-type $HTA1/HTB1$ genes[84] was transformed with a $HIS3$-marked plasmid containing an $hta1$-S129 mutant allele. The double transformants were then plated on 5-FOA to select for cells that had lost the $URA3$-marked plasmid containing the wild-type $HTA1/HTB1$ genes. Cells containing exclusively the H2A mutant were then analyzed phenotypically.

Substitution of histone H2A S129 with alanine ($hta1$-S129A) caused decreased growth in response to the DNA damaging agents methyl methane-sulfonate (MMS) and phleomycin. The importance of S129 phosphorylation in this process was further established by substitution of S129 with threonine (S129T) or glutamic acid (S129E), amino acids that can be phosphorylated or mimic constitutive phosphorylation, respectively. These mutants did not exhibit sensitivity to MMS. The inferred DNA repair defect was then further analyzed (see Table V for assays typically used). The $hta1$-S129A strain was not sensitive to ultraviolet irradiation or EMS treatment, suggesting a selective defect in repair of double-strand (ds) DNA breaks. Double mutant combinations of $hta1$-S129A with mutations in genes involved in other ds break DNA repair pathway genes, $RAD52$ and $yKU80$, indicated that the nonhomologous end-joining (NHEJ) pathway was selectively impaired.

Immunoprecipitation studies using an antiserum specific for the phosphorylated form of H2A demonstrated increased H2A phosphorylation upon DNA damage. This response was blocked in strains carrying mutations in $MEC1$, the ATM-dependent damage response kinase. A role for Mec1p in H2A-dependent DNA repair was further strengthened by immunoprecipitation studies using antiserum specific for Mec1p in which the immunoprecipitated material phosphorylated an H2A C-terminal peptide

[83] J. A. Downs, N. F. Lowndes, and S. P. Jackson, *Nature* **408,** 1001 (2000).
[84] J. N. Hirschhorn, A. L. Bortvin, S. L. Ricupero-Hovasse, and F. Winston, *Mol. Cell. Biol.* **15,** 1999 (1995).

TABLE V
ASSAYS FOR CHROMATIN FUNCTIONS

Category	Assay description	Ref.^{a}
Transcription	RNase protection assay, RNA blot hybridization or RT-PCR analysis to determine steady-state RNA levels	(1)
	Reporter gene expression (e.g., $lacZ$ expression, auxotrophic to prototrophic conversion)	(2–4)
	Nuclear run-on assay	(1)
	Sensitivity to 6-azauracil or mycophenolic acid (transcriptional elongation)	(5–7)
	Microarrays or SAGE for genome-wide RNA profiling	(8,9)
	Mating ability to evaluate silencing of HML or HMR	(10)
	Growth of strains harboring reporter genes at the telomeres, rDNA or silent mating-type loci	(10) Table VI and Fig. 4
	Transcriptional insulator assays/boundary assays	(11–14)
Chromatin assembly/structure	TAU gels to separate histones and assess global modifications	(1)
	Micrococcal nuclease digestion	(15,16)
	Methylation by ectopically expressed DNA methyl transferases	(17,18)
	Two micron plasmid topoisomer distributions by chloroquine gels	(16)
	Bulk histone acetylation at specific residues by chemical isotopic acetylation of histones followed by trypsin cleavage and mass spectroscopy	(19)
	Isolation of specific chromatin fragments	(20)
	$In\ vivo$ footprinting	(21)
DNA replication/ segregation	2D gels to assess origin firing/direction of fork movement	(22)
	Chromosome/plasmid/artificial chromosome loss rates	(23–29)
	Flow cytometry	(30,31)
	Replication timing	(22)
	Sensitivity to microtubule destabilizing agents	(32,33)
	Sensitivity to hydroxyurea	(33)
	Centromere localization/function by integrated lac operator tandem array phenotypic analysis	(34–36)
Recombination/ repair	Detection of different classes of recombinants at the $SUP4$ locus	(37,38)
	Mitotic recombination in diploids cis-heterozygous for linked markers on either side of a centromere	(29)
	Characterization of mutation types in spontaneous can^R strains	(29)
	Decreased viability/growth after induced DNA damage	(16,33)

(continued)

TABLE V *(continued)*

Category	Assay description	Ref.^{a}
	Repair, homologous recombination, gene conversion, non-homologous end joining or cell cycle arrest after HO endonuclease cleavage	(39,40)
	Homologous recombination and double-strand break repair	(40)
	Oxidative DNA damage/repair	(41)
	Mutation spectra assays	(42)
Condensation	Integrated *lac* operator tandem array phenotypic analysis	(34)
General	Microscopic examination	(43–46)
	Cell cycle phenotypes	(33,47)
	Mass spectroscopy to characterize protein complexes and protein modifications	(48,49)
	Chromatin immunoprecipitations for presence of *trans*-acting factors and for post-translational modifications of histones at a specific locus	(50)
	Chromatin immunoprecipitations on a microchip	(51)
	Flow cytometry	(30,31)
Telomere function	*De novo* telomere addition	(52–54)
	Telomere length	(55)
	Telomere protection	(56)
	Telomeric tract rapid deletion and movement	(57)

^{a} References are not comprehensive; they include representative articles in which an assay was used, the first use of an assay, and reviews of a particular assay. Key to references:
(1) F. M. Ausubel, "Current Protocols in Molecular Biology." Wiley, New York, 2001.
(2) R. C. Mount, B. E. Jordan, and C. Hadfield, *Methods Mol. Biol.* **53,** 239 (1996).
(3) C. Mateus and S. V. Avery, *Yeast* **16,** 1313 (2000).
(4) S. Rupp, *Methods Enzymol.* **350,** 112 (2002).
(5) J. Archambault, F. Lacroute, A. Ruet, and J. D. Friesen, *Mol. Cell. Biol.* **12,** 4142 (1992).
(6) F. Exinger and F. Lacroute, *Curr. Genet.* **22,** 9 (1992).
(7) S. M. Uptain, C. M. Kane, and M. J. Chamberlin, *Annu. Rev. Biochem.* **66,** 117 (1997).
(8) A. P. Gasch, *Methods Enzymol.* **350,** 393 (2002).
(9) M. A. Basrai and P. Hieter, *Methods Enzymol.* **350,** 414 (2002).
(10) F. Van Leeuwen and D. E. Gottschling, *Methods Enzymol.* **350,** 165 (2002).
(11) M. Oki and R. T. Kamakaka, *Curr. Opin. Cell Biol.* **14,** 299 (2002).
(12) K. Ishii and U. K. Laemmli, *Mol. Cell* **11,** 237 (2003).
(13) P. A. Defossez and E. Gilson, *Nucleic Acids Res.* **30,** 5136 (2002).
(14) Q. Yu, R. Qiu, T. B. Foland, D. Griesen, C. S. Galloway, Y. H. Chiu, J. Sandmeier, J. R. Broach, and X. Bi, *Nucleic Acids Res.* **31,** 1224 (2003).
(15) A. Ravindra, K. Weiss, and R. T. Simpson, *Mol. Cell. Biol.* **19,** 7944 (1999).
(16) E. M. Martini, S. Keeney, and M. A. Osley, *Genetics* **160,** 1375 (2002).
(17) M. P. Kladde, M. Xu, and R. T. Simpson, *Methods Enzymol.* **304,** 431 (1999).
(18) M. P. Kladde, M. Xu, and R. T. Simpson, *Methods Mol. Biol.* **119,** 395 (1999).
(19) C. M. Smith, P. R. Gafken, Z. Zhang, D. E. Gottschling, J. B. Smith, and D. L. Smith, *Anal. Biochem.* **316,** 23 (2003).

(continued)

TABLE V *(continued)*

(20) A. Ansari, T. H. Cheng, and M. R. Gartenberg, *Methods* **17**, 104 (1999).

(21) P. D. Gregory, S. Barbaric, and W. Horz, *Methods* **15**, 295 (1998).

(22) A. J. Van Brabant and M. K. Raghuraman, *Methods Enzymol.* **351**, 539 (2002).

(23) J. H. Hegemann, S. Klein, S. Heck, U. Guldener, R. K. Niedenthal, and U. Fleig, *Yeast* **15**, 1009 (1999).

(24) F. Spencer, S. L. Gerring, C. Connelly, and P. Hieter, *Genetics* **124**, 237 (1990).

(25) J. H. Shero, M. Koval, F. Spencer, R. E. Palmer, P. Hieter, and D. Koshland, *Methods Enzymol.* **194**, 749 (1991).

(26) J. A. Huberman, *Methods* **18**, 356 (1999).

(27) P. Hieter, C. Mann, M. Snyder, and R. W. Davis, *Cell* **40**, 381 (1985).

(28) D. Koshland, J. C. Kent, and L. H. Hartwell, *Cell* **40**, 393 (1985).

(29) R. J. Craven, P. W. Greenwell, M. Dominska, and T. D. Petes, *Genetics* **161**, 493 (2002).

(30) K. J. Hutter and H. E. Eipel, *J. Gen. Microbiol.* **113**, 369 (1979).

(31) S. B. Haase and S. I. Reed, *Cell Cycle* **1**, 132 (2002).

(32) T. Stearns, M. A. Hoyt, and D. Botstein, *Genetics* **124**, 251 (1990).

(33) M. Hampsey, *Yeast* **13**, 1099 (1997).

(34) A. F. Straight, A. S. Belmont, C. C. Robinett, and A. W. Murray, *Curr. Biol.* **6**, 1599 (1996).

(35) T. Tanaka, J. Fuchs, J. Loidl, and K. Nasmyth, *Nat. Cell Biol.* **2**, 492 (2000).

(36) G. Goshima and M. Yanagida, *Cell* **100**, 619 (2000).

(37) R. Rothstein, C. Helms, and N. Rosenberg, *Mol. Cell. Biol.* **7**, 1198 (1987).

(38) E. Shor, S. Gangloff, M. Wagner, J. Weinstein, G. Price, and R. Rothstein, *Genetics* **162**, 647 (2002).

(39) J. E. Haber, *Methods Enzymol.* **350**, 141 (2002).

(40) L. S. Symington, *Microbiol. Mol. Biol. Rev.* **66**, 630 (2002).

(41) A. V. Avrutskaya and S. A. Leadon, *Methods* **22**, 127 (2000).

(42) G. F. Crouse, *Methods* **22**, 116 (2000).

(43) M. Berrios, *Nuclear structure and function, in* "Methods in Cell Biology," Vol. 53, Academic Press, San Diego; p. xvii, (1998).

(44) T. H. Giddings, Jr., E. T. O'Toole, M. Morphew, D. N. Mastronarde, J. R. McIntosh, and M. Winey, *Methods Cell Biol.* **67**, 27 (2001).

(45) P. Conn (ed.), *Confocal Microscopy, in* "Methods in Enzymology," Vol. 307. (1999).

(46) F. Iborra, P. R. Cook, and D. A. Jackson, *Methods* **29**, 131 (2003).

(47) D. J. Lew and S. I. Reed, *Curr. Opin. Genet. Dev.* **5**, 17 (1995).

(48) R. J. Deshaies, J. H. Seol, W. H. McDonald, G. Cope, S. Lyapina, A. Shevchenko, R. Verma, and J. R. Yates, 3rd, *Mol. Cell. Proteomics* **1**, 3 (2002).

(49) W. Shou, R. Verma, R. S. Annan, M. J. Huddleston, S. L. Chen, S. A. Carr, and R. J. Deshaies, *Methods Enzymol.* **351**, 279 (2002).

(50) M. H. Kuo and C. D. Allis, *Methods* **19**, 425 (1999).

(51) C. E. Horak and M. Snyder, *Methods Enzymol.* **350**, 469 (2002).

(52) L. L. Sandell and V. A. Zakian, *Cell* **75**, 729 (1993).

(53) M. S. Singer and D. E. Gottschling, *Science* **266**, 404 (1994).

(54) K. M. Kramer and J. E. Haber, *Genes Dev.* **7**, 2345 (1993).

(55) A. J. Lustig and T. D. Petes, *Proc. Natl. Acad. Sci. USA* **83**, 1398 (1986).

(56) M. L. Dubois, Z. W. Haimberger, M. W. McIntosh, and D. E. Gottschling, *Genetics* **161**, 995 (2002).

(57) M. Bucholc, Y. Park, and A. J. Lustig, *Mol. Cell. Biol.* **21**, 6559 (2001).

in vitro. Furthermore, *hta1*-S129E mutants that mimic the constitutively phosphorylated form of H2A exhibited subtle defects in chromatin structure. Both plasmid superhelical density and micrococcal nuclease digestion assays showed relaxed, but not grossly altered, chromatin structure. These investigators[83] proposed that Mec1-dependent modification of H2A S129 is important for ds break DNA repair in the NHEJ pathway. DNA repair may be facilitated by phosphorylation of S129 to relax chromatin structure thereby allowing access of repair machinery to DNA.

The N-terminal tail of H2A has also been subjected to extensive mutational analysis and is important for transcriptional repression of certain SWI/SNF-dependent genes.[84,85] The SWI/SNF ATP-dependent chromatin remodeling complex induces nucleosomal alterations that alleviate repression of a number of genes in yeast including *SUC2, INO1, GAL1,* and *HO* (reviewed in Fry and Peterson[86]). Although SWI/SNF is known to participate with HAT complexes in gene activation,[87,88] the role of post-translational modifications in H2A-mediated transcriptional regulation is not yet clear. Deletion of H2A amino acids 5-21 or N-terminal mutations recovered from a PCR-based random mutagenesis screen caused a *s*witch *in*dependent or Sin$^-$ phenotype.[84] Two of the *hta1* mutants recovered from the screen, *hta*-S20F and *hta*-G30D, were also cold sensitive. Further analysis of these mutants revealed decreased viability, altered chromosomal ploidy and segregation properties, and a delay in the G2/M stage of the cell cycle at the nonpermissive temperature.[89]

An additional link between histone H2A and transcriptional regulation came from the observation that deletion of amino acids 4-20 in the N-terminus, or simultaneous substitution of lysines 4 and 7 to arginine caused a telomeric silencing defect.[90] Interestingly, the histone acetyltransferase Esa1p acetylates H2A K4 and K7 in the context of free histones or nucleosomes[60,62] and *ESA1* mutants also have a telomeric silencing defect (Clarke and Pillus, personal communication). The H2A C-terminal region also contributes to transcriptional regulation. Deletion of this region likewise causes a telomeric silencing defect[90] and transcriptional phenotypes characteristic of SWI/SNF-dependent gene activation defects.[84] C-terminal mutants exhibited a significant loss of a phosphorylated H2A isoform that by site-directed mutagenesis appeared to be distinct from loss of S129

[85] J. N. Hirschhorn, S. A. Brown, C. D. Clark, and F. Winston, *Genes Dev.* **6,** 2288 (1992).
[86] C. J. Fry and C. L. Peterson, *Curr. Biol.* **11,** R185 (2001).
[87] M. P. Cosma, T. Tanaka, and K. Nasmyth, *Cell* **97,** 299 (1999).
[88] J. E. Krebs, M. H. Kuo, C. D. Allis, and C. L. Peterson, *Genes Dev.* **13,** 1412 (1999).
[89] I. Pinto and F. Winston, *EMBO J.* **19,** 1598 (2000).
[90] H. R. Wyatt, H. Liaw, G. R. Green, and A. J. Lustig, *Genetics* **164,** 47 (2003).

phosphorylation.[90] Considering that telomeres are reservoirs of repair-associated proteins,[91] it will be interesting to determine the relative contributions of C-terminal modifications in the processes of transcription and repair.

Histone H2B

As with H2A, N- or C-terminal deletions of histone H2B cause transcriptional defects, although the role of modifications in these cases is unclear. A number of N-terminal alterations, including deletion of amino acids 3-22 suppress a *snf5Δ* growth defect.[92] Additional site-directed mutagenesis revealed that several regions of H2A involved in critical interhistone contacts also caused selective transcriptional defects when mutated. Alteration of Tyr 40, 43, or 45 in the alpha 1 helix at the H2A/H2B interface, as well as Tyr 86 in the alpha 2 helix abutting histone H4, caused Sin⁻ phenotypes. Residues H2B Y86 and H4 Y72/Y88 form a hydrophobic cluster at the dimer–tetramer interface[1,93,94] and are likely to be important for core particle integrity. The coordinate function of these residues is further supported by mutations in H4 Tyr 88 that also caused a Sin⁻ phenotype.[71]

Whereas the above examples underscore the importance of specific histone contacts in core particle function, another remarkable example of interhistone communication has recently been reported. The C-terminal lysine 123 in H2B is ubiquitinated by Rad6p,[95] a protein implicated in a number of chromatin-dependent events including DNA repair, sporulation, retrotransposition, and transcriptional silencing. Ubiquitination of H2B K123 by Rad6p is required for methylation of histone H3 at positions K4 and K79.[29,48,96] As described in the histone H3 section above, methylation of histone H3 K4 by Set1p and K79 by Dot1p are implicated in both transcriptional activation and repression. Deletion of *RAD6* caused loss of H3 K4 methylation and correlated with the telomeric silencing defect of an *htb1*-K123R mutant.[96]

Different experimental approaches were used to identify Rad6p as the indirect mediator of H3 K4- or K79-dependent silencing. Based on earlier observations that H3 K4 methylation was important for rDNA silencing[32,26] and that mutation of *CAC* genes,[97] *SIR2*,[98] or *RAD6*[99] caused

[91] J. E. Lowell and L. Pillus, *Cell Mol. Life Sci.* **54**, 32 (1998).
[92] J. Recht and M. A. Osley, *EMBO J.* **18**, 229 (1999).
[93] A. M. Kleinschmidt and H. G. Martinson, *J. Biol. Chem.* **259**, 497 (1984).
[94] G. Arents and E. N. Moudrianakis, *Proc. Natl. Acad. Sci. USA* **92**, 11170 (1995).
[95] K. Robzyk, J. Recht, and M. A. Osley, *Science* **287**, 501 (2000).
[96] Z. W. Sun and C. D. Allis, *Nature* **418**, 104 (2002).
[97] P. D. Kaufman, R. Kobayashi, and B. Stillman, *Genes Dev.* **11**, 345 (1997).

rDNA and/or telomeric silencing defects, loss of K4 methylation was evaluated in each of these mutants. An alternative approach was to screen all viable yeast gene deletions by immunoblotting to assay for loss of histone H3 K4 methylation.[27] Interestingly, Rad6-dependent ubiquitination was required for methylation of H3 K4 and K79, but not K36,[48] suggesting potential hierarchies of specificity in methylating histone H3.

Histone Variants

In multicellular eukaryotes, a significant number of histone variants exist that contribute to a range of chromatin functions. In yeast, only three variants have been recognized to date. These genes encode orthologs of histones H1, CENP-A, and H2A.Z (reviewed in Smith[100]). Posttranslational modifications of these variants are to be expected and as in multicellular eukaryotes, are likely to be important in fine-tuning chromatin structure. For example, it appears that the yeast variant Cse4p may be subject to regulated phosphorylation.[101] Although many questions remain unanswered, the relative absence of histone variants in yeast may make it an ideal experimental setting in which to genetically examine the function of these proteins from other species (see later).

Histone Deacetylases

A key feature of most histone modifications is their reversibility, which allows dynamic control of chromatin related processes. To date, greatest progress has been made in studies of the deacetylases (as described later), acting in concert with acetyltransferases. A critical area for future research will be in defining the reversibility of other modifications. In some cases, reversibility will be accomplished by direct removal of a modification. In other cases, modification states may be "re-set" through the action of chromatin remodeling complexes or erased when DNA is replicated during cell division.

The histone deacetylases that remove acetyl groups from lysine residues of histones have a significant impact on chromatin function in both a global and gene-specific manner. There are ten genes in *S. cerevisiae* that have either been demonstrated to have histone deacetylase activity *in vitro* or *in vivo*, or are closely related to known deacetylases by sequence

[98] J. S. Smith and J. D. Boeke, *Genes Dev.* **11**, 241 (1997).

[99] H. Huang, A. Kahana, D. E. Gottschling, L. Prakash, and S. W. Liebman, *Mol. Cell. Biol.* **17**, 6693 (1997).

[100] M. M. Smith, *Curr. Opin. Cell Biol.* **14**, 279 (2002).

[101] S. Buvelot, S. Y. Tatsutani, D. Vermaak, and S. Biggins, *J. Cell Biol.* **160**, 329 (2003).

comparison: *RPD3, HDA1, HOS1, HOS2, HOS3, SIR2, HST1, HST2, HST3,* and *HST4.*

At promoters where an Rpd3p-containing complex is targeted by the DNA binding factor Ume6p,[102] deletion of *RPD3* caused increased acetylation of all lysines tested in the core histones (except H4 K16[6]) in a genomic acetylation microarray experiment. However, the greatest increase in acetylation occurred primarily at histone H4 K5[103] or K5 and K12[6] and histone H3 K9 and/or K14[4,104] over the span of approximately two promoter-proximal nucleosomes.[104] This increased acetylation correlated with an increased level of basal transcription and more rapid gene induction of the *INO1* and *IME2* genes. A more global role for Rpd3p was inferred from several observations. First, deletion of *RPD3* caused hyperacetylation of histone H4 K12 at the *PHO5* promoter,[4] which is not known to be targeted by Rpd3p, but is dependent upon Gcn5p for activation.[105] In this case, deletion of *RPD3* increased acetylation over a 4.25 kb region surrounding *PHO5* which was restored to normal levels in an *rpd3 gcn5* double mutant.[4] Second, genome-wide analysis of histone H4 K12 acetylation in intergenic regions (IGRs) revealed that 815 IGRs had increased H4 K12 acetylation by a factor of 1.95 or more in an *rpd3* null strain.[8] This is consistent with the observed increase in bulk histone acetylation primarily at H4 K5 and K12 (and also H3 K9 and 14) in *rpd3* mutant cell extracts.[106]

Gene-specific and regional deacetylation have also been observed with Hda1p, which associates with the transcriptional repressor Tup1p.[107] Deletion of *HDA1* caused increased acetylation of histone H3 N-terminal lysines K9, 14, 18, 23, 27, and H2B lysines K11 and K16 at several promoter-proximal nucleosomes of the Tup1p-responsive gene *ENA1.*[107] Genome-wide analysis of an *hda1* mutant showed that 647 intergenic regions had elevated histone H3 K9 and K18, and H2B K16 acetylation levels, preferentially in subtelomeric regions.[8] Although it is difficult to assess the relative contribution of indirect acetylation and transcriptional effects in these studies, the acetylation and transcription profiles of the *hda1*Δ strains correlated well with those from a *tup1*Δ strain.[8]

Although Rpd3p and Hda1p functions are associated with transcriptional repression, roles for deacetylation in gene activation have also been

[102] D. Kadosh and K. Struhl, *Cell* **89,** 365 (1997).

[103] S. E. Rundlett, A. A. Carmen, N. Suka, B. Turner, and M. Grunstein, *Nature* **392,** 831 (1998).

[104] D. Kadosh and K. Struhl, *Mol. Cell. Biol.* **18,** 5121 (1998).

[105] P. D. Gregory, A. Schmid, M. Zavari, L. Lui, S. L. Berger, and W. Horz, *Mol. Cell* **1,** 495 (1998).

[106] S. E. Rundlett, A. A. Carmen, R. Kobayashi, S. Bavykin, B. M. Turner, and M. Grunstein, *Proc. Natl. Acad. Sci. USA* **93,** 14503 (1996).

[107] J. Wu, N. Suka, M. Carlson, and M. Grunstein, *Mol. Cell* **7,** 117 (2001).

proposed. Hos2p has a modest effect on the acetylation of histones H3 and H4 N-terminal lysines at activated *GAL1* genes *in vivo*.[108] Hos2p associates with Set3p *in vivo* and deletion of *HOS2* or *SET3* leads to comparable defects in transcriptional activation of the *GAL1* gene.[108] Analysis of Hos3p deacetylase substrate specificity *in vitro* indicated that it deacetylated histone H3 primarily at K14 and K23, H4 at K5 and K8, H2A at K7 and H2B at K11.[109] This was supported by *in vivo* ChIP analyses with antisera directed toward different acetylated H4 isoforms, where deletion of *HOS3* caused increased acetylation of histone H4 primarily at K5, K8, and K12.[109] Subsequent genome-wide profiling analysis combining enzyme binding assays to determine the genomic binding sites of proteins under investigation, acetylation arrays to determine acetylation state at a particular genomic region, and RNA expression arrays to determine gene expression, indicated that *HOS1, HOS2*, and *HOS3* are required for deacetylation of histone H4 K12 primarily in intergenic regions within the ribosomal DNA array.[8]

The NAD-dependent enzyme Sir2p also deacetylates histones in the rDNA array and is important for transcriptional silencing of reporter genes placed within the rDNA, at telomeres and at the silent mating-type loci (reviewed in Denu[110]). Sir2p preferentially deacetylates histone H4 K16 and histone H3 K9 and K14 *in vitro*. Mutational analysis has indicated that its deacetylase activity is important for its roles in silencing, recombination and extension of life span *in vivo* (reviewed in Denu[110]). The closely related Hst2p has robust deactylase activity *in vitro* where it also targets histone H4,[111] although its biological function is unclear. Hst1p is known to participate in the Set3 complex[112] where it associates with Hos2p and Sum1p to repress transcription of the middle sporulation genes.[113]

Many Histone Mutations Cause Transcriptional Silencing Defects

The first *in vivo* evidence that changes in chromatin structure could disrupt gene regulation came from studies altering core histone gene dosage or mutating their N-termini.[56,114,115] Particularly striking was the

[108] A. Wang, S. K. Kurdistani, and M. Grunstein, *Science* **298,** 1412 (2002).
[109] A. A. Carmen, P. R. Griffin, J. Calaycay, S. E. Rundlett, Y. Suka, and M. Grunstein, *Proc. Natl. Acad. Sci. USA* **96,** 12356 (1999).
[110] J. M. Denu, *Trends Biochem. Sci.* **28,** 41 (2003).
[111] J. Landry, A. Sutton, S. T. Tafrov, R. C. Heller, J. Stebbins, L. Pillus, and R. Sternglanz, *Proc. Natl. Acad. Sci. USA* **97,** 5807 (2000).
[112] W. W. Pijnappel, D. Schaft, A. Roguev, A. Shevchenko, H. Tekotte, M. Wilm, G. Rigaut, B. Seraphin, R. Aasland, and A. F. Stewart, *Genes Dev.* **15,** 2991 (2001).
[113] J. Xie, M. Pierce, V. Gailus-Durner, M. Wagner, E. Winter, and A. K. Vershon, *EMBO J.* **18,** 6448 (1999).

observation that deletion of the H4 terminus resulted in a loss of normal transcriptional silencing of the *HM* loci.[56] These studies have now been extended by genome wide mRNA profiling in strains depleted of histone H4, which show a marked increase in expression of genes in subtelomeric regions.[9]

Since these early studies, chromatin mediated silencing has been a productive focus of study for dissecting histone modifications and chromatin modifiers. Silencing is a form of transcriptional repression that is chromatin-dependent, but is largely gene-independent. Entire regions of the genome may be encompassed by silent chromatin, and in multicellular eukaryotes these regions assume a condensed chromatin structure, termed heterochromatin. Yeast has proven to be an excellent model of silencing because there are three well-characterized chromosomal domains in yeast that confer silencing upon reporter genes: the silent mating-type loci (*HML* and *HMR*), telomeres and the rDNA array (see Fig. 4A).

Efficient silencing depends on histones, enzymes that posttranslationally modify histones, and a number of factors of unknown function that constitute silent chromatin. The silent information regulator proteins, Sir1p, Sir2p, Sir3p, and Sir4p, differentially contribute to silencing. Sir1p is required at the *HM* loci, Sir3p, and Sir4p are required at the *HM* loci and at telomeres, whereas Sir2p is distinguished by its participation at all three domains (reviewed in Rusche *et al.*[116]).

Silencing phenotypes caused by mutations in the reversibly acetylated lysine residues of H3 and H4 tails supported the model that lysine acetylation state was important for silencing[117–119] and reviewed in Smith.[59] This model was further supported by biochemical studies showing that the tail lysines of histones H3 and H4 were indeed hypoacetylated in silenced regions.[6,120] The demonstration that Sir2p had NAD-dependent histone deacetylase activity (reviewed in Gottschling[121]) was key in understanding mechanistically how hypoacetylation correlated with known silencing factors. Sir3p and Sir4p physically interact with histone N-termini[122] and

[114] C. D. Clark-Adams, D. Norris, M. A. Osley, J. S. Fassler, and F. Winston, *Genes Dev.* **2**, 150 (1988).

[115] M. Han, U. J. Kim, P. Kayne, and M. Grunstein, *EMBO J.* **7**, 2221 (1988).

[116] L. N. Rusche, A. L. Kirchmaier, and J. Rine, *Ann. Rev. Biochem.* **72**, 481 (2003).

[117] E. C. Park and J. W. Szostak, *Mol. Cell. Biol.* **10**, 4932 (1990).

[118] L. M. Johnson, P. S. Kayne, E. S. Kahn, and M. Grunstein, *Proc. Natl. Acad. Sci. USA* **87**, 6286 (1990).

[119] P. C. Megee, B. A. Morgan, B. A. Mittman, and M. M. Smith, *Science* **247**, S41 (1990).

[120] M. Braunstein, R. E. Sobel, C. D. Allis, B. M. Turner, and J. R. Broach, *Mol. Cell. Biol.* **16**, 4349 (1996).

[121] D. E. Gottschling, *Curr. Biol.* **10**, R708 (2000).

[122] A. Hecht, T. Laroche, S. Strahl-Bolsinger, S. M. Gasser, and G. M., *Cell* **80**, 583 (1995).

are thought to propagate silencing through cooperative binding along neighboring nucleosomes.[54] Indeed, *in vitro* binding of an H4 N terminal peptide to a C-terminal fragment of Sir3 decreases with increasing acetylation of H4.[123] Recent work has also pointed to silencing roles for histone H3 lysines 4 and 79, which are methylated by Set1p and Dot1p, respectively (see earlier).

Transcriptional silencing may be studied by standard molecular assays for transcription. However, a number of sensitive bioassays exist that allow experiments to be performed under various physiological conditions. The assays developed to study silencing *in vivo* rely on the use of chromosomally engineered reporter genes and have been presented in detail in an excellent review.[124] A number of strains in several of the standard backgrounds are currently available that contain reporter genes at silenced loci (see Table VI). Because a variety of prototrophic markers are available, usually a single genetic cross can yield a strain with reporter genes at one or more silenced loci combined with a genomic deletion of a gene under investigation. Below is a protocol for a typical colony growth dilution assay to analyze silencing phenotypes. This *in vivo* assay is easy to perform and evaluate and can be used with combinations of mutant histones or histone modifying enzymes. Further, it can be used as a platform for searching for environmental, pharmacological or genetic modifiers, and can be performed in parallel with direct transcriptional assays.

Protocol[a]

1. Inoculate 2 ml of selective medium or YPD (rich) medium with a single colony of a strain containing the desired silencing reporter genes and either a deletion of *GENE X* or wild-type control strain.
2. Grow 48 h at 30° in a tube roller until A_{600} is about 2.0–3.0.
3. Prepare a sterile 96-well plate for 5-fold dilutions by using an 8-tip multipipetter to aliquot 160 μl of sterile water into five sequential rows of wells, leaving the left-most row of wells empty.
4. Normalize concentration of strains to A_{600} of 2.0.
5. Put 200 μl of sample into left-most well.
6. Serially transfer 40 μl from normalized sample well across the five rows using a multipipetter. Pipette up and down several times to mix well.

[123] A. A. Carmen, L. Milne, and M. Grunstein, *J. Biol. Chem.* **277,** 4778 (2002).
[124] F. van Leeuwen and D. E. Gottschling, *Methods Enzymol.* **350,** 165 (2002).
[a] See Fig. 4B for a sample dilution assay.

7. Use 48-prong metal pin applicator (previously ethanol dipped and flamed) to transfer uniformly about 10 μl of fluid from each well to a petri dish containing the appropriate solid medium.[b]
8. Let plate dry for 5–10 min.
9. Invert and incubate 48–72 h at 30° or desired temperature until single colonies are visible.
10. Photograph plates with lids removed and agar-side down on black velvet background using lighting projected from above.

Chromatin-Related Assays for Phenotypic Analysis of Histone and Histone-Modifier Mutants

Many mutations that impair histone or histone-modifier functions have defects in a variety of cellular chromatin-dependent functions including chromatin assembly, transcriptional repression or activation, DNA replication, DNA recombination and DNA repair, and chromosome segregation. Table V lists commonly used assays to investigate phenotypes arising from mutations in histones and histone-modifying factors. Many of these assays, along with detailed protocols, are presented in this volume.

Strategies for Studying Human Chromatin-Related Disease Genes in Yeast

From studies in yeast and in mammalian cells, it is clear that some genes implicated in chromatin-related diseases, such as the human *CHD3* and *SMARCA4* genes involved in chromatin remodeling, are highly conserved in yeast (see Table VII). However, other genes implicated in chromatin-related diseases, such as the genes involved in DNA methylation, do not exist in yeast. Yeast cells can be manipulated with such ease that they are useful for studying both highly conserved and nonconserved human chromatin-related disease genes. Although the approaches used to study either conserved or nonconserved genes differ, especially at the outset, both approaches can yield valuable insights into disease gene function.

Experimental Strategies When the Human Gene Shares a Yeast Homolog

If a chromatin-related human disease gene is clearly conserved between humans and yeast, cross-species complementation analysis can be performed (see Fig. 5). The analysis begins with a yeast strain carrying a

---

[b] Note that as presented in Table I, a variety of drugs and/or negative selectable reagents can be used in these assays. Because different strain backgrounds vary in their sensitivities, it is useful to establish in pilot assays a working range for both drugs and cell numbers. Pin applicator is available from Sigma.

TABLE VI
REPORTER CONSTRUCTS FOR SILENCING ASSAYS[a]

Locus	Reporter	Ref.[b]	Assay evaluation[c]
HML	*TRP1*	(1)	Ability to grow on –trp or 5-FAA[d] medium
	URA3	(2,3)	Ability to grow on –ura or 5-FOA medium
	P_{URA3}-*ADE2*	(4)	Ability to grow on –ade medium; colony sectoring/color on low ade medium
	ADE2	(5)	Ability to grow on –ade medium; colony sectoring/color on low ade medium
	SUP4-o	(2)	Ability to grow on –ade medium or colony sectoring/color on low ade medium in strain containing *ade2–1* allele
	P_{LEU2}-*lacZ*	(6)	β-gal activity; color on XGal medium[e]
HMR	*TRP1*	(7,8)	Ability to grow on –trp or 5-FAA[d] medium
	ADE2	(5,9,10)	Ability to grow on –ade medium; colony sectoring/color on low ade medium
	P_{URA3}-*ADE2*	(10)	Ability to grow on –ade medium; colony sectoring/color on low ade medium
	URA3	(11)	Ability to grow on –ura or 5-FOA medium
	P_{ADH1}-*GFP*	(12)	FACS analysis[f,g]
	SUP3-a	(13)	Ability to grow on YPD + 1.5 M ethylene glycol; invertase activity[h] in strain containing *suc2–215am* allele
Telomere	*URA3-VIIL*	(14)	Ability to grow on –ura or 5-FOA medium
	HIS3-VIIL	(15)	Ability to grow on –his medium or 3-AT
	URA3-ADE2-VIIL	(14)	Ability to grow on –ura or 5-FOA medium; ability to grow on –ade medium; colony sectoring/color on low ade medium
	URA3-TRP1-VIIL	(14)	Ability to grow on –ura or 5-FOA medium; ability to grow on –trp or 5-FAA[d] medium
	URA3-VR	(14)	Ability to grow on –ura or 5-FOA medium
	ADE2-VR	(16)	Ability to grow on –ade medium; colony sectoring/color on low ade medium
	SUP4-o-VIIL	(17)	Ability to grow on –ade medium or colony sectoring/color on low ade medium in strain with *ade2–101* allele
rDNA	*ADE2-CAN1*	(18)	Ability to grow on canavanine; ability to grow on –ade medium
	ADE2	(18,19)	Ability to grow on –ade medium
	Ty1-mURA3	(20)	Ability to grow on –ura medium
	Ty1-MET15	(20)	Colony color/sectoring on MLA medium
	mURA3-LEU2	(20)	Ability to grow on –leu medium
	mURA3/HIS3	(20,21)	Ability to grow on –ura medium

(continued)

TABLE VI *(continued)*

Locus	Reporter	Ref.[b]	Assay evaluation[c]
Ty1-his3A1	(22)	Transposition rates assessed by frequency of His$^+$ prototrophy
mURA3/HIS3 flanking the array	(23)	Ability to grow on –ura medium

[a] Many of these silencing reporters and several useful silencing strains are detailed in an excellent review[i].

[b] Key to References:

(1) S. Le, C. Davis, J. B. Konopka, and R. Sternglanz, *Yeast* **13**, 1029 (1997).

(2) D. J. Mahoney and J. R. Broach, *Mol. Cell. Biol.* **9**, 4621 (1989).

(3) X. Bi, M. Braunstein, G. J. Shei, and J. R. Broach, *Proc. Natl. Acad. Sci. USA* **96**, 11934 (1999).

(4) T. H. Cheng and M. R. Gartenberg, *Genes Dev.* **14**, 452 (2000).

(5) X. Bi and J. R. Broach, *Genes Dev* **13**, 1089 (1999).

(6) C. Boscheron, L. Maillet, S. Marcand, M. Tsai-Pflugfelder, S. M. Gasser, and E. Gilson, *EMBO J.* **15**, 2184 (1996).

(7) A. H. Brand, L. Breeden, J. Abraham, R. Sternglanz, and K. Nasmyth, *Cell* **41**, 41 (1985).

(8) L. Sussel and D. Shore, *Proc. Natl. Acad. Sci. USA* **88**, 7749 (1991).

(9) L. Sussel, D. Vannier, and D. Shore, *Mol. Cell. Biol.* **13**, 3919 (1993).

(10) D. H. Rivier, J. L. Ekena, and J. Rine, *Genetics* **151**, 521 (1999).

(11) M. S. Singer, A. Kahana, A. J. Wolf, L. L. Meisinger, S. E. Peterson, C. Goggin, M. Mahowald, and D. E. Gottschling, *Genetics* **150**, 613 (1998).

(12) P. U. Park, P. A. Defossez, and L. Guarente, *Mol. Cell. Biol.* **19**, 3848 (1999).

(13) R. Schnell and J. Rine, *Mol. Cell. Biol.* **6**, 494 (1986).

(14) D. E. Gottschling, O. M. Aparicio, B. L. Billington, and V. A. Zakian, *Cell* **63**, 751 (1990).

(15) B. D. Bourns, M. K. Alexander, A. M. Smith, and V. A. Zakian, *Mol. Cell. Biol.* **18**, 5600 (1998).

(16) M. S. Singer and D. E. Gottschling, *Science* **266**, 404 (1994).

(17) H. Huang, A. Kahana, D. E. Gottschling, L. Prakash, and S. W. Liebman, *Mol. Cell. Biol.* **17**, 6693 (1997).

(18) C. E. Fritze, K. Verschueren, R. Strich, and R. Easton Esposito, *EMBO J.* **16**, 6495 (1997).

(19) S. Imai, C. M. Armstrong, M. Kaeberlein, and L. Guarente, *Nature* **403**, 795 (2000).

(20) J. S. Smith and J. D. Boeke, *Genes Dev.* **11**, 241 (1997).

(21) J. S. Smith, E. Caputo, and J. D. Boeke, *Mol. Cell. Biol.* **19**, 3184 (1999).

(22) M. Bryk, M. Banerjee, M. Murphy, K. E. Knudsen, D. J. Garfinkel, and M. J. Curcio, *Genes Dev.* **11**, 255 (1997).

(23) S. W. Buck, J. J. Sandmeier, and J. S. Smith, *Cell* **111**, 1003 (2002).

[c] Unless noted, details of media preparation have been described[i].

[d] J. H. Toyn, P. L. Gunyuzlu, W. H. White, L. A. Thompson, and G. F. Hollis, *Yeast* **16**, 553 (2000).

[e] S. Rupp, *Methods Enzymol.* **350**, 112 (2002).

[f] C. Mateus and S. V. Avery, *Yeast* **16**, 1313 (2000).

[g] R. K. Niedenthal, L. Riles, M. Johnston, and J. H. Hegemann, *Yeast* **12**, 773 (1996).

[h] R. Schnell and J. Rine, *Mol. Cell. Biol.* **6**, 494 (1986).

[i] F. van Leeuwen and D. E. Gottschling, *Methods Enzymol.* **350**, 165 (2002).

TABLE VII

HUMAN CHROMATIN-RELATED DISEASE GENES

Category[a]	NCBI symbol[b]	Common name	Disease	Candidate yeast ortholog[c] (P value)
DNA methylation	MECP2		Rett Syndrome	None
	MBD2		Colon Cancer antigen	None
	MBD4		Tumors w/microsatellite instability	None
	DNMT3B		ICF Syndrome	None
Epigentic regulation of genes	FMR1		Fragile X associated mental retardation	None
	IGF2		Overgrowth disorder, cancers	None
Histone acetylation	CREBBP	CBP	Rubinstein-Taybi syndrome	BDF2 (4.6×10^{-12})
	EP300	p300	Gastric cancer, colon cancer, brain tumor	GCN5 (9.1×10^{-12})
	ZNF220-CREBBP	MOZ-CBP	Acute myelocytic leukemia	ZNF220: ESA1 (1.2×10^{-31}); CREBBP: BDF2 (4.6×10^{-12})
	ZNF220-EP300	MOZ-p300	Leukemia	ZNF220: ESA1 (1.2×10^{-31}); EP300: GCN5 (9.1×10^{-12})
	ZNF220-NCOA2	MOZ-TIF-2	Leukemia	ZNF220: ESA1 (1.2×10^{-31}); NCOA2: none
	MLL-CREBBP	MLL-CBP	Leukemia	MLL: SET1 (7.7×10^{-31}); CREBBP: BDF2 (4.6×10^{-12})
	MLL-EP300	MLL-p300	Leukemia	MLL: SET1 (7.7×10^{-31}); EP300: GCN5 (9.1×10^{-12})
Histone modification	RPS6KA3	RSK2	Coffin-Lowry syndrome	SCH9 (3.9×10^{-66}) + 74 ORFs[d]
Chromatin remodeling system	CHD3/4	Mi2α/Mi2β	Autoantibody in dermatomyositis; neoplasia	CHD1 (1.8×10^{-128})/(1.6×10^{-126}) + 15/15 ORFs[d]
	MTA1		Metastatic potential of carcinoma	None
	SMARCB1	hSNF5/INI1	Rhabdoid tumor	SFH1 (5.9×10^{-24}) + 1 ORF
	SMARCA4	BRG1	Tumors	SNF2 (3.4×10^{-224}) + 15 ORFs[d]

	ATRX	α-Thalassemia/mental retardation syndrome, X-linked	*RAD54* $(5 \times 10^{-54}) + 15$ ORFs^d
Transcriptional control	*SMARCA2*	Tumors	*SNF2* $(1.2 \times 10^{-221}) + 15$ ORFs^d
	TP53	Neoplasia	None
	RB1	Neoplasia	None
	BCL6	Lymphoma as fusion protein	*ZAP1* $(6.3 \times 10^{-24}) + 4$ ORFS
	BRCA1	Breast and ovarian cancer	None
	GCCR	Cortisol resistance	None
	E1A (adenoviral gene)	Transformation	None
	ZNF145-RARA	Leukemia	*RARA*: none; *ZNF145*: *ZAP1* $(1.1 \times 10^{-30}) + 3$ ORFs
	PML-RARA	Leukemia	*PML*: none; *RARA*: none
	RUNX1-CBFA2T1	Leukemia	None
	RUNX1-MDS1	Leukemia	None
	RUNX1-EVI1	Leukemia	*RUNX1*: none; *EVI1*: *ZAP1* (2.5×10^{-22})
	RUNX1-CBFA2T3	Leukemia	None
	ETV6-RUNX1	Leukemia	None
	MLL-MLLT1	Leukemia	*MLL*: *SET1* (7.7×10^{-31}); *MLLT1*: *YAF9* (3.5×10^{-18})
	MLL-MLLT7	Leukemia	*MLL*: *SET1* (7.7×10^{-31}); *MLLT7*: *HCM1* (5.3×10^{-11})
	MLL-MLLT3	Leukemia	*MLL*: *SET1* (7.7×10^{-31}); *MLLT3*: *YAF9* (3.4×10^{-19})
	ING1	Putative tumor suppressor	*YNG1* (1.5×10^{-18})
	ING3	Candidate tumor suppressor	*PHO23* $(6.1 \times 10^{-27}) + 2$ ORFs
Chromatin assembly	*SET*	Leukemia as fusion protein	*NAPI* (1.5×10^{-13})

^{a} Genes have been grouped to simplify organization. Note that some genes may have overlapping functions, and chromosomal translocation partners may logically fit into more than one category. Because of space constraints, the large number of human disease genes involved in DNA repair^{e-j} are not included.

(*continued*)

TABLE VII (continued)

[b] Refers to the human "approved gene symbol" name at URL http://www.gene.ucl.ac.uk/nomenclature/Chromatin-related human disease genes have been reviewed[k–o]. Inclusion of a gene as a "human disease gene" relied on currated summaries in Online Mendelian Inheritance in Man, OMIM™, McKusick-Nathans Institute for Genetic Medicine, Johns Hopkins University (Baltimore, MD) and National Center for Biotechnology Information, National Library of Medicine (Bethesda, MD), 2000. URL: http://www.ncbi.nlm.nih.gov/omim/.

[c] BLAST searches were performed between the human predicted protein product and the $Saccharomyces\ cerevisiae$ predicted ORFs. The program WU BLAST 2.0 was used [Gish, W., (1996–2003) http://blast.wustl.edu]. Yeast genes homologous to the human disease gene at p values of $\leq 1 \times 10^{-10}$ were included. A p value of $<1 \times 10^{-40}$ has often been used as a predictor of orthology[p,q], but as discussed[p–r], exceptions occur.

[d] The large number of yeast genes predicted to be orthologous to the human disease gene reflects the high degree of conservation of certain protein catalytic domains.

[e] M. A. Resnick and B. S. Cox, $Mutat.\ Res.$ $\mathbf{451}$, 1 (2000).

[f] A. R. Venkitaraman, $N.\ Engl.\ J.\ Med.$ $\mathbf{348}$, 1917 (2003).

[g] L. H. Thompson and D. Schild, $Mutat.\ Res.$ $\mathbf{509}$, 49 (2002).

[h] P. Mohaghegh and I. D. Hickson, $Int.\ J.\ Biochem.\ Cell\ Biol.$ $\mathbf{34}$ 1496 (2002).

[i] A. Duval and R. Hamelin, $Ann.\ Genet.$ $\mathbf{45}$, 71 (2002).

[j] R. E. Moses, $Annu.\ Rev.\ Genomics\ Hum.\ Genet.$ $\mathbf{2}$, 41 (2001).

[k] C. A. Johnson, $J.\ Med.\ Genet.$ $\mathbf{37}$, 905 (2000).

[l] S. Timmermann, H. Lehrmann, A. Polesskaya, and A. Harel-Bellan, $Cell\ Mol.\ Life\ Sci.$ $\mathbf{58}$, 728 (2001).

[m] B. Hendrich and W. Bickmore, $Hum.\ Mol.\ Genet.$ $\mathbf{10}$, 2233 (2001).

[n] M. Nakao, $Gene$ $\mathbf{278}$, 25 (2001).

[o] M. Esteller and J. G. Herman, $J.\ Pathol.$ $\mathbf{196}$, 1 (2002).

[p] D. E. Bassett, Jr., M. S. Boguski, F. Spencer, R. Reeves, S. Kim, T. Weaver, and P. Hieter, $Nat.\ Genet.$ $\mathbf{15}$, 339 (1997).

[q] F. Foury, $Gene$ $\mathbf{195}$, 1 (1997).

[r] N. Zhang, M. Osborn, P. Gitsham, K. Yen, J. R. Miller, and S. G. Oliver, $Gene$ $\mathbf{303}$, 121 (2003).

recessive mutation in the most closely related gene homolog. Complementation is tested using an assay that distinguishes between the mutant and wild-type phenotype. If the yeast strain has no known mutant phenotype, a number of assays can be explored that may reveal a mutant phenotype (listed in Table V and discussed in more detail in Hampsey[125]). The yeast strain is transformed with a plasmid containing the human cDNA under the control of a yeast promoter, in parallel with control plasmids. Multiple independent transformants are tested by immunoblot analysis to verify expression of the human protein and then the transformants are tested to determine if they have a mutant, wild-type or intermediate phenotype under conditions that select for maintenance of the plasmid. If the transformants expressing the human cDNA have a wild-type phenotype, then the human gene is able to complement the yeast mutation, providing strong evidence for *in vivo* conservation of function.

Plasmids allow a rapid assessment of complementation and are reliable, as long as multiple independent transformants are assayed for complementation and control plasmids are tested (see Fig. 5). In order to confirm that the human cDNA complements, the plasmid-dependence of the rescued phenotype is tested by screening or selecting for colonies that have lost the plasmid and then demonstrating that they have also lost the rescued phenotype. An additional confirmation can come from the demonstration that a mutant version of the human cDNA is expressed but fails to complement.

When performing cross-species complementation analysis, many plasmid vectors are available.[126–128] Researchers can use vectors maintained at 1–2 copies per cell (CEN vectors[129]) or 10–40 copies per cell (2 μ vectors[129]). In addition, researchers can choose plasmids containing constitutive or inducible promoters, different selectable markers, and different epitope tags. It is often prudent to make several constructs and test them in parallel, as complementation might be seen only at specific levels of protein expression, and epitope tags might interfere with the function of the protein. Once complementation by the human gene has been established using plasmids, it is preferable to use a chromosomally integrated version of the human cDNA for further genetic analyses. An integrated version can be more useful when performing screens or genetic crosses due to its inherent mitotic stability. Plasmid copy number variation and plasmid loss can cause phenotypic variability and complicate genetic crosses.

[125] M. Hampsey, *Yeast* **13,** 1099 (1997).
[126] M. Funk, R. Niedenthal, D. Mumberg, K. Brinkmann, V. Ronicke, and T. Henkel, *Methods Enzymol.* **350,** 248 (2002).
[127] D. Mumberg, R. Muller, and M. Funk, *Gene* **156,** 119 (1995).
[128] J. P. Brunelli and M. L. Pall, *Yeast* **9,** 1299 (1993).
[129] M. D. Rose and J. R. Broach, *Methods Enzymol.* **194,** 195 (1991).

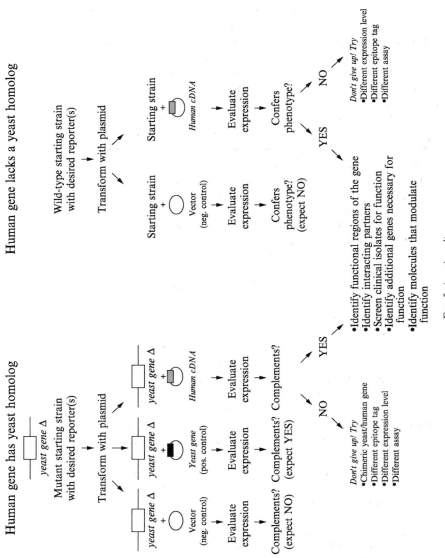

Fig. 5. (continued)

Testing Complementation When the Yeast Homolog is Essential

In cases where the yeast homolog is essential for viability, complementation analysis must be performed using different methods. In one method, the complementation analysis can be performed by expressing the yeast homolog under control of a repressible promoter. A plasmid containing the human cDNA, or even a human cDNA library, is then transformed into that strain and tested for viability under repressing conditions.[130] In a second method, complementation can be tested using a variation on the plasmid shuffle technique introduced earlier (Fig. 6 and Ref. 13).

In order to construct the strain for the plasmid shuffle, a diploid strain heterozygous for a deletion of the yeast homolog is transformed with a plasmid containing the yeast homolog and a yeast counterselectable marker such as $URA3$ (Table I). Transformants are selected on medium lacking uracil and independent transformants are sporulated and dissected. In tetrads in which all four spores are viable, two of the four spores will contain a chromosomal wild-type copy of the yeast homolog and two of the four spores will contain a chromosomal deletion of the yeast homolog and be Ura^+ (contain the plasmid). Mitotic descendents of spores with the latter genotype are then tested directly for complementation (see Fig. 6).

In order to perform the plasmid shuffle step, the segregants are transformed with the plasmid containing the human cDNA as well as control plasmids (vector alone and a wild-type copy of the yeast homolog). After verifying expression of the human protein, the transformants are tested for growth on 5-FOA. The strains transformed with the vector control should fail to grow on 5-FOA, the strains transformed with the wild-type yeast plasmid control should grow on 5-FOA, and the strains transformed with the experimental plasmid should grow if the human protein can functionally substitute for the yeast protein.

If the human gene appears not to substitute for the yeast gene, several options can be explored. Levels of protein expression can be manipulated by using different promoter or plasmid constructs. The protein can be

---

[130] N. Zhang, M. Osborn, P. Gitsham, K. Yen, J. R. Miller, and S. G. Oliver, *Gene* **303,** 121 (2003).

FIG. 5. Expression of human genes in yeast. This flowchart outlines two strategies for testing human gene function in yeast, depending on whether there is a yeast homolog. Genes are depicted as boxes on linear chromosomes or circular yeast plasmids. Open boxes indicate that the yeast gene is deleted; shaded boxes indicate that a gene is present. The notation + indicates that the strain bears a yeast plasmid. The *Human cDNA* gene construct contains a human cDNA whose expression is controlled by a yeast promoter.

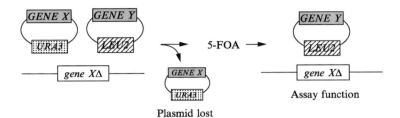

FIG. 6. Plasmid shuffle. Genes are depicted as boxes on linear chromosomes or circular yeast plasmids. Open boxes indicate that the yeast gene is deleted; shaded boxes indicate that a gene is present. Different selectable markers (*URA3* or *LEU2*) are present in plasmids containing *GENE X* or *GENE Y* and both plasmids are maintained by growth in a medium lacking these supplements. 5-FOA refers to the counterselectable medium used to identify isolates that have lost the *URA3*-containing plasmid.

tagged with different epitopes, or the epitopes may be added to different regions of the protein. It may be necessary to develop new phenotypic assays, or to expand the scope of phenotypes tested in order to detect complementation. Alternatively, a chimeric protein containing the most conserved portions of the human protein combined with the nonconserved portions of the yeast protein may complement.

A number of these approaches were taken to demonstrate cross-species complementation of the yeast *SIR2* gene by the human *SIRT2* gene. The human *SIRT2* gene failed to complement *sir2Δ* silencing defects at *HML, HMR*, a telomere or the rDNA when expressed at low levels from the *SIR2* promoter or at high levels from a *GAL* promoter.[131] However, a chimeric yeast/human *SIR2-hSIRT2* gene complemented the *sir2Δ* silencing defect at *HMR* when expressed from the *SIR2* promoter (see Fig. 4B). This complementation was locus-specific in that the chimeric yeast/human *SIR2-hSIRT2* gene was unable to complement the *sir2Δ* silencing defect of a telomeric reporter gene. Thus, the choice of assay as well as the use of a chimeric protein can allow the detection of complementation, particularly if it is partial or locus-specific.

Using Functional Conservation to Explore the Cellular Function of a Protein

If a human gene or portions of a human gene can complement a yeast mutation, this can be used as a foundation for experiments that delve into the human gene function. An example of the power of this approach comes

---

[131] J. M. Sherman, E. M. Stone, L. L. Freeman-Cook, C. B. Brachmann, J. D. Boeke, and L. Pillus, *Mol. Biol. Cell* **10**, 3045 (1999).

from studies of the human chromatin-related gene *SMARCA4 (BRG1)* and its yeast homolog *SNF2 (SWI2)*. *SNF2* encodes the catalytic subunit of the SWI2/SNF2 ATP-dependent chromatin-remodeling complex (reviewed in Peterson[132]). The high degree of conservation between the *SNF2* gene family members motivated studies in which the human homologs were expressed in yeast.[133–135] One study showed that although SMARCA4 failed to complement a yeast *snf2* mutation, a chimeric gene encoding a SMARCA4 ATPase domain-Snf2 protein fusion complemented the growth and transcriptional defects caused by a yeast *snf2* mutation.[135]

Based on these results, Khavari and co-workers created a point mutation in *SNF2* that mutated the ATPase domain to cause a dominant negative transcriptional phenotype in yeast. The corresponding alteration in SMARCA4 also caused a dominant negative transcriptional phenotype, both in yeast and in human cells,[135] due to change of a residue necessary for the DNA-stimulated ATPase activity of Snf2p.[133] Thus, the catalytic activity of Snf2p and by analogy of SMARCA4, was required for transcriptional regulation. In a later study, this dominant mutation was used as a tool to identify *c-fos* as a SMARCA4-modulated gene.[136] The functional conservation between *SMARCA4* and *SNF2* also led to the discovery of additional genes involved in chromatin remodeling, such as *SMARCB1 (hSNF5)*,[137] which is mutated in certain cancers.[138] Thus, complementation assays provided the basis for studies to elucidate the function of SMARCA4 in transcription, to identify genes regulated by *SMARCA4*, and to identify other members of SWI/SNF complexes.

Experimental Strategies When the Human Gene Lacks a Yeast Homolog

In situations where the function of a chromatin-related human disease gene is not conserved in yeast, it may still be feasible to study aspects of the human gene function to different ends. In these cases, the experimental starting point is again expression of the human gene in yeast (see Fig. 5).

---

[132] C. L. Peterson, *Curr. Biol.* **13**, R195 (2003).

[133] B. C. Laurent, I. Treich, and M. Carlson, *Genes Dev.* **7**, 583 (1993).

[134] I. Okabe, L. C. Bailey, O. Attree, S. Srinivasan, J. M. Perkel, B. C. Laurent, M. Carlson, D. L. Nelson, and R. L. Nussbaum, *Nucleic Acids Res.* **20**, 4649 (1992).

[135] P. A. Khavari, C. L. Peterson, J. W. Tamkun, D. B. Mendel, and G. R. Crabtree, *Nature* **366**, 170 (1993).

[136] D. J. Murphy, S. Hardy, and D. A. Engel, *Mol. Cell. Biol.* **19**, 2724 (1999).

[137] C. Muchardt, C. Sardet, B. Bourachot, C. Onufryk, and M. Yaniv, *Nucleic Acids Res.* **23**, 1127 (1995).

[138] I. Versteege, N. Sevenet, J. Lange, M. F. Rousseau-Merck, P. Ambros, R. Handgretinger, A. Aurias, and O. Delattre, *Nature* **394**, 203 (1998).

However, the goal is to produce an expression-dependent effect on yeast growth or function, which can indicate that some aspect of the normal human protein function is being reconstituted in yeast.

When human chromatin-related disease genes are expressed ectopically in yeast, they are typically tested in transcription-based assays and growth inhibition assays. In some transcriptional assays, the human protein is expressed and targeted to one or more reporter constructs containing the human protein's DNA binding site. Transcriptional output from the reporter gene is then used as a measure of the human protein function. Proteins such as nuclear hormone receptors (reviewed in McEwan[139]) and p53 (see later) have been studied productively in this way. In other transcriptional assays, the human protein is expressed as a Gal4p DNA binding domain fusion and targeted to a reporter gene containing Gal4p DNA binding sites. Again, transcriptional output from the reporter gene is used as a measure of the human protein function. Examples include $BRCA1$,[140] RB,[141,142] and tat.[143]

Many variations of these strategies have been used. For example, in one study, adenoviral early region 1A (E1A)-Gal4 DNA binding domain fusions or E1A-VP16 fusions were coexpressed with the thyroid hormone receptor (TR); the E1A fusions caused increased transcription of a reporter gene under the control of the TR responsive element.[144] In this example, the Gal4p DNA-binding domain or VP16 increased, but was not required for, the enhancement of transcription by E1A. Similar experimental approaches may be used in studying human gene function in other chromatin-related processes.

Alternatively, human disease genes can be tested for growth inhibitory, growth enhancing or other phenotypes (see Table V and Hampsey[145]) when expressed in yeast. For example, expression of distinct regions of E1A causes growth inhibition (Shuen et al.[146] and references therein) or pseudohyphal growth under low nitrogen conditions.[147] Expression of

[139] I. J. McEwan, *Trends Genet.* **17,** 239 (2001).

[140] A. N. Monteiro, A. August, and H. Hanafusa, *Proc. Natl. Acad. Sci. USA* **93,** 13595 (1996).

[141] M. Arneric, A. Traven, L. Staresincic, and M. Sopta, *J. Biol. Chem.* **277,** 8797 (2002).

[142] B. K. Kennedy, O. W. Liu, F. A. Dick, N. Dyson, E. Harlow, and M. Vidal, *Proc. Natl. Acad. Sci. USA* **98,** 8720 (2001).

[143] T. Subramanian, C. D'Sa-Eipper, B. Elangovan, and G. Chinnadurai, *Nucleic Acids Res.* **22,** 1496 (1994).

[144] X. Meng, Y. F. Yang, X. Cao, M. V. Govindan, M. Shuen, A. N. Hollenberg, J. S. Mymryk, and P. G. Walfish, *Mol. Endocrinol.* **17,** 1095 (2003).

[145] M. Hampsey, *Yeast* **13,** 1099 (1997).

[146] M. Shuen, N. Avvakumov, P. G. Walfish, C. J. Brandl, and J. S. Mymryk, *J. Biol. Chem.* **277,** 30844 (2002).

[147] Z. Zhang, M. M. Smith, and J. S. Mymryk, *Mol. Biol. Cell* **12,** 699 (2001).

BRCA1[148] and p53[149] in yeast also cause growth defects. Indeed, a study using an efficient method to overexpress human genes in yeast and screen for growth defects found that approximately 30% of the tested human genes caused growth inhibition.[150]

The techniques used to study a human gene lacking a yeast homolog are similar to those used for cross-species complementation analysis (see Fig. 5). Briefly, the human cDNA is expressed on a plasmid under the control of a yeast promoter and transformed into a yeast strain in parallel with a negative control plasmid. After verifying ectopic protein expression, the transformants are subjected to phenotypic analysis. If possible, the relevance of any phenotypes observed in yeast should be assessed in human cells.

The p53 Paradigm: How to Gain Insightful Information through Ectopic Expression in Yeast

An excellent example of a chromatin-related human disease gene that has been ectopically expressed and studied in yeast is *TP53* (encoding p53). In initial studies, critical transcription-based assays for p53 function in yeast were developed.[151–153] In the most commonly adopted assay, p53 was expressed in yeast and shown to *trans*-activate a reporter gene containing a p53 responsive element.[152] This assay has been exploited in numerous ways to study the function of p53 in yeast and four different methodological approaches are presented here to illustrate the range of studies possible once a phenotype has been established.

One approach used a yeast transcription assay to screen clinical samples for mutations in *TP53*.[154,155] The approach exploited the efficient homologous recombination system of yeast to recover RT-PCR amplified from clinical samples in a *TP53* expression plasmid (principles and methods

[148] J. S. Humphrey, A. Salim, M. R. Erdos, F. S. Collins, L. C. Brody, and R. D. Klausner, *Proc. Natl. Acad. Sci. USA* **94**, 5820 (1997).

[149] J. M. Nigro, R. Sikorski, S. I. Reed, and B. Vogelstein, *Mol. Cell. Biol.* **12**, 1357 (1992).

[150] S. Tugendreich, E. Perkins, J. Couto, P. Barthmaier, D. Sun, S. Tang, S. Tulac, A. Nguyen, E. Yeh, A. Mays, E. Wallace, T. Lila, D. Shivak, M. Prichard, L. Andrejka, R. Kim, and T. Mélèse, *Genome Res.* **11**, 1899 (2001).

[151] S. Fields and S. K. Jang, *Science* **249**, 1046 (1990).

[152] E. Scharer and R. Iggo, *Nucleic Acids Res.* **20**, 1539 (1992).

[153] S. E. Kern, J. A. Pietenpol, S. Thiagalingam, A. Seymour, K. W. Kinzler, and B. Vogelstein, *Science* **256**, 827 (1992).

[154] C. Ishioka, T. Frebourg, Y. X. Yan, M. Vidal, S. H. Friend, S. Schmidt, and R. Iggo, *Nat. Genet.* **5**, 124 (1993).

[155] J. M. Flaman, V. Robert, S. Lenglet, V. Moreau, R. Iggo, and T. Frebourg, *Oncogene* **16**, 1369 (1998).

detailed in Cormack and Castano[156]). Then the function of the *TP53* allele was tested by assaying expression of a reporter gene under the control of a p53-responsive element. This approach, which was initially pursued for clinical applications,[157–160] has subsequently proven to be useful for characterizing mutant *TP53* alleles. Such clinically derived alleles have been tested for temperature-sensitivity, for their ability to activate reporters under the control of different p53-responsive elements, and for dominance (e.g., Refs. 155,161–164).

A more pharmacological approach was taken in two studies that used the yeast transcription assay for genetic screens. In one study, intragenic suppressors of *TP53* missense mutants were isolated, with the goal of identifying regions of the p53 protein that could be targeted by small molecules to restore p53 function.[165] In a second study, several reagents were screened for their ability to improve reporter activation in strains containing missense mutations in *TP53*.[166]

A third approach, more epidemiological in nature, has tried to determine causal factors for particular types of p53 mutations. In those studies, a yeast transcription assay was used as a tool to screen for *TP53* mutations induced by a particular mutagenic treatment. Then the spectrum of mutations obtained in yeast was compared with that seen in clinical samples (reviewed in Fronza et al.[167]).

Finally, a fourth approach has used the yeast transcription-based assay to identify and characterize yeast genes (e.g., Refs. 168–170) as well as human genes[171] that modulate p53's transcriptional activation activity.

[156] B. Cormack and I. Castano, *Methods Enzymol.* **350,** 199 (2002).

[157] T. Soussi and C. Beroud, *Nat. Rev. Cancer* **1,** 233 (2001).

[158] R. S. Camplejohn and J. Rutherford, *Cell Prolif.* **34,** 1 (2001).

[159] R. W. deVere White, A. D. Deitch, P. H. Gumerlock, and X. B. Shi, *Prostate* **41,** 134 (1999).

[160] P. M. Duddy, A. M. Hanby, D. M. Barnes, and R. S. Camplejohn, *J. Mol. Diagn.* **2,** 139 (2000).

[161] A. Inga, F. Storici, T. A. Darden, and M. A. Resnick, *Mol. Cell. Biol.* **22,** 8612 (2002).

[162] C. J. Di Como and C. Prives, *Oncogene* **16,** 2527 (1998).

[163] A. Inga, S. Cresta, P. Monti, A. Aprile, G. Scott, A. Abbondandolo, R. Iggo, and G. Fronza, *Carcinogenesis* **18,** 2019 (1997).

[164] P. Monti, P. Campomenosi, Y. Ciribilli, R. Iannone, A. Inga, A. Abbondandolo, M. A. Resnick, and G. Fronza, *Oncogene* **21,** 1641 (2002).

[165] R. K. Brachmann, K. Yu, Y. Eby, N. P. Pavletich, and J. D. Boeke, *EMBO J.* **17,** 1847 (1998).

[166] D. Maurici, P. Monti, P. Campomenosi, S. North, T. Frebourg, G. Fronza, and P. Hainaut, *Oncogene* **20,** 3533 (2001).

[167] G. Fronza, A. Inga, P. Monti, G. Scott, P. Campomenosi, P. Menichini, L. Ottaggio, S. Viaggi, P. A. Burns, B. Gold, and A. Abbondandolo, *Mutat. Res.* **462,** 293 (2000).

In summary, once a phenotypic assay is developed for expression of a human gene in yeast, a wide range of different types of studies is available. Those studies may yield insights at multiple levels, from the identification of interacting partners to mutational detail to pharmacologically useful reagents.

Genomics/Proteomics Resources: Doing A Lot with a Little (Organism)

The yeast community has historically been collegial, generous, and technology-driven. Even as the *S. cerevisiae* genome was being sequenced, considerable attention was devoted to developing yeast as a model for genomic and proteomic resources to facilitate rapid and efficient experimentation and information gathering. These efforts have been exceptionally successful and now provide, virtually with just a few keystrokes from the comfort of one's own computer, access to mutants, interaction maps, localization data, and expression profiling.

Genetic Resources

After the sequence itself, a second major effort by a consortium of international laboratories was to systematically delete every ORF in the yeast genome.[172] The deletions were constructed as 'start-to-stop' mutations with the precision and efficiency afforded by PCR and homologous recombination (reviewed in Johnston *et al.*[173]). The deleted locus is marked by the phenotypically neutral drug resistance gene cassette *kan*MX, which encodes resistance to kanamycin (G418, geneticin, see Table I) and the deletion site is marked by a gene-specific molecular "barcode" that uniquely identifies each particular deletion.

[168] S. Thiagalingam, K. W. Kinzler, and B. Vogelstein, *Proc. Natl. Acad. Sci. USA* **92,** 6062 (1995).

[169] R. Candau, D. M. Scolnick, P. Darpino, C. Y. Ying, T. D. Halazonetis, and S. L. Berger, *Oncogene* **15,** 807 (1997).

[170] A. Nourani, L. Howe, M. G. Pray-Grant, J. L. Workman, P. A. Grant, and J. Côté, *J. Biol. Chem.* **278,** 19171 (2003).

[171] T. Wang, T. Kobayashi, R. Takimoto, A. E. Denes, E. L. Snyder, W. S. el-Deiry, and R. K. Brachmann, *EMBO J.* **20,** 6404 (2001).

[172] E. A. Winzeler, D. D. Shoemaker, A. Astromoff, H. Liang, K. Anderson, B. Andre, R. Bangham, R. Benito, J. D. Boeke, H. Bussey, A. M. Chu, C. Connelly, K. Davis, F. Dietrich, S. W. Dow, M. El Bakkoury, F. Foury, S. H. Friend, E. Gentalen, G. Giaever, J. H. Hegemann, T. Jones, M. Laub, H. Liao, R. W. Davis *et al.*, *Science* **285,** 901 (1999).

[173] M. Johnston, L. Riles, and J. H. Hegemann, *Methods Enzymol.* **350,** 290 (2002).

At this point, more than 96% of the predicted ORFs longer than 100 amino acids have been deleted and are available individually or as complete sets. A total of more than 20,000 mutant strains exist in combinations of haploids of both mating types, along with the heterozygous and homozygous diploids. All versions are available for nonessential genes; the essential genes may be obtained as heterozygous diploids and used experimentally as described in sections above. The deletion strains are currently available from several sources as noted in Table VIII and may be ordered individually or as complete sets.

The mutant strains may be assayed directly for chromatin-related phenotypes under study, or the deletions may be reconstructed into specialized strains that have been developed for evaluating transcriptional silencing or activation of particular reporter genes (see Table VI), or interactions with other mutations. In these cases, it is simplest to move the deletion construct by amplifying the locus with complementary primers several hundred nucleotides up- and down-stream of the selectable drug marker, transforming the product into the strain of choice, and selecting for resistance to G418.[173] Alternatively, a number of the deletion constructs have been cloned into plasmids. These may be released as fragments after

TABLE VIII
SOURCES OF YEAST DELETION STRAINS AND COLLECTIONS[a]

Source	e-mail	URL
American Type Culture Collection	sales@atcc.org	http://www.atcc.org/cydac/cydac.cfm
EUROSCARF	Euroscarf@em.uni-frankfurt.de	http://www.rz.uni-frankfurt.de/FB/fb16/mikro/euroscarf/index.html
Open Biosystems	info@openbiosystems.com	http://www.openbiosystems.com/productPage.php?pageType = yeast.knockout&q = 0
Research Genetics	info@resgen.com	http:/www.resgen.com/products/YEASTD.php3

[a] The home page of the Saccharomyces Genome Deletion Project is: http://www.yeastgenome.org/ExternalContents.shtml#deletions. References for many functional profiling studies with the 'knockout collection' are presented there which extend the original report of the resource[b].

[b] E. A. Winzeler, D. D. Shoemaker, A. Astromoff, H. Liang, K. Anderson, B. Andre, R. Bangham, R. Benito, J. D. Boeke, H. Bussey, A. M. Chu, C. Connelly, K. Davis, F. Dietrich, S. W. Dow, M. El Bakkoury, F. Foury, S. H. Friend, E. Gentalen, G. Giaever, J. H. Hegemann, T. Jones, M. Laub, H. Liao, R. W. Davis et al., Science 285, 901 (1999).

restriction endonuclease digestion, for use directly in transformations, and are available from EUROSCARF. Each new deletion constructed by recombination should be confirmed by molecular analysis before being subject to extensive phenotypic studies. Detailed advice on primer selection, fragment preparation and deletion verification have been reviewed.[173]

In addition to the bar-coded deletion strains, another large collection of yeast mutants has been constructed via transposon insertional mutagenesis (for details see Ross-Macdonald et al.[174] and Kumar et al.[175]). Although the sites of insertion vary for each gene, a special feature of these mutants is that each also contains a lacZ gene and a triple-HA epitope tag that are expressed when insertion has occurred in-frame in the gene of interest. Individual strains and resources are available from the Yale Genome Analysis Center (http://ygac.med.yale.edu/default.stm) and the collection of mutants and other reagents are available through Open Biosystems (see Table VIII). Utility of these insertional mutants is described later.

Cell Biological and Biochemical Resources

Beyond the mutant collections themselves, several significant resources are available for cell biological and biochemical studies. The insertional mutant collection introduced above has been analyzed in two broad ways. In the first, a survey of readily scored phenotypes has been performed.[174] In the second, localization of the tagged mutant proteins to both subcellular compartments and spread chromosomes has been completed.[174,176] The data from these studies are available in a readily searchable format, the "TRIPLES" database (http://ygac.med.yale.edu) and provide significant information about many otherwise uncharacterized ORFs as well as genes that have been studied earlier.

The National Center for Research Resources' Yeast Resources Center (http://depts.washington.edu/~yeastrc/index.html) at the University of Washington, Seattle is an NIH-funded center created to facilitate collaborative study of protein complexes. Participation in the Center provides expertise and access to collaborative two-hybrid, fluorescence microscopy, mass spectroscopy and structural prediction analyses.

[174] P. Ross-Macdonald, P. S. Coelho, T. Roemer, S. Agarwal, A. Kumar, R. Jansen, K. H. Cheung, A. Sheehan, D. Symoniatis, L. Umansky, M. Heidtman, F. K. Nelson, H. Iwasaki, K. Hager, M. Gerstein, P. Miller, G. S. Roeder, and M. Snyder, *Nature* **402**, 413 (1999).

[175] A. Kumar, S. A. des Etages, P. S. Coelho, G. S. Roeder, and M. Snyder, *Methods Enzymol.* **328**, 550 (2000).

[176] A. Kumar, S. Agarwal, J. A. Heyman, S. Matson, M. Heidtman, S. Piccirillo, L. Umansky, A. Drawid, R. Jansen, Y. Liu, K. H. Cheung, P. Miller, M. Gerstein, G. S. Roeder, and M. Snyder, *Genes Dev.* **16**, 707 (2002).

Significant progress has been made in producing either pools or arrays of purified proteins generated from each ORF (reviewed in Phizicky et al.[177] and Phizicky et al.[178]). The potential utility for assaying these for new chromatin modifying activities or in substrate identification is great. Such protein analysis will yield powerful biochemical strategies for enhancing and extending genome-wide genetic and cell biological approaches.

Databases

The genome-scale resources noted above provide new impetus for embarking on projects and characterization of new genes and activities. One tremendous advantage of studying new proteins and processes in yeast is that many insights and clues may already be available. How to find the clues?

Concurrently with the sequencing of the yeast genome, much creativity, insight and analysis was dedicated to the creation of the extraordinarily useful and user-friendly Saccharomyces Genome Database, SGD (http://www.yeastgenome.org/[179,180]). Essentially structured around the genome sequence itself, the database is organized to provide concise and well-annotated information about each genetic locus, from the most well-characterized HAT, to near-anonymous ORFs that may appear as hits in your next mass spec run. Information about mutant phenotypes, biochemical properties, structure and *in vivo* functions is summarized on a single page, on which direct links to many other key features are accessible with but a mouse-click. Notable among these are data for genome-wide two hybrid and affinity purification studies, identification of homologs in other organisms, and individual and collated microarray expression databases.

Conclusions

Our goal has been to introduce methods of studying yeast to develop deeper understanding of the function of chromatin modifications. The strength of genetic analysis in combination with biochemical studies makes

[177] E. M. Phizicky, M. R. Martzen, S. M. McCraith, S. L. Spinelli, F. Xing, N. P. Shull, C. Van Slyke, R. K. Montagne, F. M. Torres, S. Fields, and E. J. Grayhack, *Methods Enzymol.* **350,** 546 (2002).

[178] E. Phizicky, P. I. Bastiaens, H. Zhu, M. Snyder, and S. Fields, *Nature* **422,** 208 (2003).

[179] L. Issel-Tarver, K. R. Christie, K. Dolinski, R. Andrada, R. Balakrishnan, C. A. Ball, G. Binkley, S. Dong, S. S. Dwight, D. G. Fisk, M. Harris, M. Schroeder, A. Sethuraman, K. Tse, S. Weng, D. Botstein, and J. M. Cherry, *Methods Enzymol.* **350,** 329 (2002).

[180] S. Weng, Q. Dong, R. Balakrishnan, K. Christie, M. Costanzo, K. Dolinski, S. S. Dwight, S. Engel, D. G. Fisk, E. Hong, L. Issel-Tarver, A. Sethuraman, C. Theesfeld, R. Andrada, G. Binkley, C. Lane, M. Schroeder, D. Botstein, and J. Michael Cherry, *Nucleic Acids Res.* **31,** 216 (2003).

it possible to validate the significance and contributions of a given modification or enzyme *in vivo*. Beyond studies of yeast enzymes themselves, we have summarized and encouraged the use of yeast as a living crucible for analysis of important genes from other organisms that have a more limited scope of genetic tricks. Building on these strategies, and the genomic resources of yeast, we hope that this chapter will be useful for those beginning to consider genetic approaches to chromatin function.

Acknowledgments

We appreciate the critical comments of M. Smith, S. Berger, W. S. Lo, and M. Ruault. Work in our laboratory has been supported by the National Institutes of Health.

[2] Genetic Analysis of Chromatin Remodeling Using Budding Yeast as a Model

By DAVID J. STEGER and ERIN K. O'SHEA

Genetic studies of eukaryotic transcription often yield mutants that are impaired in the regulation of chromatin remodeling. Yeast or fly mutants have been isolated that carry mutations in genes encoding the enzymatic activities that covalently modify histones through acetylation, methylation, and phosphorylation.[1-4] Mutants of SWI/SNF and related complexes that use the energy of ATP hydrolysis to alter histone-DNA contacts have also been isolated.[5-7] In most cases, genetics has produced a mutant of a chromatin modifier before its biochemical characterization. The physiological relevance of many biochemical discoveries were validated by preexisting mutants, rapidly advancing our understanding of chromatin structure and how it is reconfigured for transcription.

In principle, any protein that remodels the chromatin structure of a gene of interest can be genetically identified as long as mutants are viable and display a detectable phenotype. This approach has worked well for

[1] G. Reuter and P. Spierer, *Bioessays* **14,** 605 (1992).
[2] G. Thon and A. J. Klar, *Genetics* **131,** 287 (1992).
[3] S. L. Berger, B. Pina, N. Silverman, G. A. Marcus, J. Agapite, J. L. Reigier, S. J. Triezenberg, and L. Guarente, *Cell* **70,** 251 (1992).
[4] R. C. Allshire, J. P. Javerzat, N. J. Redhead, and G. Cranston, *Cell* **76,** 157 (1994).
[5] M. Stern, R. Jensen, and I. Herskowitz, *J. Mol. Biol.* **178,** 853 (1984).
[6] L. Neigeborn and M. Carlson, *Genetics* **108,** 845 (1984).
[7] V. Pirrotta, *Cell* **93,** 333 (1998).

genes that are highly dependent on chromatin modifiers for transcription because the mutant phenotypes are robust. For example, the screens in yeast that identified genes encoding subunits of the SWI/SNF complex examined expression of the inducible genes HO and $SUC2$, which both require SWI/SNF for full activation.[8]

For some genes, multiple chromatin modifiers may work together to facilitate transcription, but no single modifier is absolutely required. An example of such a gene in yeast is $PHO5$, and although genetic screens have identified many regulators of its expression,[9–11] none of these activate transcription by altering chromatin. Because most screens are not saturating, it is possible that chromatin modifiers were missed. However, a likely possibility is that the phenotypes of these mutants were too weak to be detected under the conditions of the screens.

In this chapter, we present two examples of genetic strategies designed to identify chromatin modifiers that regulate $PHO5$ transcription. In the first example, we describe a genetic selection recently performed to isolate mutants that cannot activate $PHO5$ expression.[12] In the second example, we describe a screen that is currently in progress to identify chromatin modifiers by searching the collection of 4847 viable haploid yeast deletion strains for those that exhibit a kinetic delay in $PHO5$ activation (S. Huang and E. K. O'Shea, unpublished data). This screen is based on the observation that in strains with mutations in chromatin-modifying genes, including $gcn5$,[13] $arp8$, and $swi2$ (E. S. Haswell and E. K. O'Shea, unpublished data), the steady-state levels of activated $PHO5$ transcription are similar to the wild-type level, but the time required to reach steady-state is longer than that for the wild-type strain. Although the screens described here focus on $PHO5$, their underlying principles can be applied to other genes.

Genetic Selection for $PHO5$ Uninducible Mutants

Transcription of $PHO5$, which encodes a secreted acid phosphatase is regulated in response to phosphate availability by the transcription factors Pho4 and Pho2.[14] The activity of Pho4 is regulated through phosphorylation by the cyclin-CDK (cyclin-dependent kinase) complex Pho80-Pho85.[15]

[8] F. Winston and M. Carlson, *Trends Genet.* **8**, 387 (1992).

[9] E. A. To, Y. Ueda, S. I. Kakimoto, and Y. Oshima, *J. Bacteriol.* **113**, 727 (1973).

[10] Y. Ueda, E. A. To, and Y. Oshima, *J. Bacteriol.* **122**, 911 (1975).

[11] W. W. Lau, K. R. Schneider, and E. K. O'Shea, *Genetics* **150**, 1349 (1998).

[12] D. J. Steger, E. S. Haswell, A. L. Miller, S. R. Wente, and E. K. O'Shea, *Science* **299**, 114 (2003).

[13] S. Barbaric, J. Walker, A. Schmid, J. Q. Svejstrup, and W. Hörz, *EMBO J.* **20**, 4944 (2001).

[14] Y. Oshima, *Genes Genet. Syst.* **72**, 323 (1997).

[15] A. Kaffman, I. Herskowitz, R. Tjian, and E. K. O'Shea, *Science* **263**, 1153 (1994).

When yeast cells are grown in phosphate-rich medium, Pho4 is phosphorylated and localized to the cytoplasm.[16] In addition, four positioned nucleosomes reside over the *PHO5* promoter, and *PHO5* transcription is repressed.[17] When cells are starved for phosphate, the CDK inhibitor Pho81 inactivates Pho80-Pho85,[18] Pho4 is unphosphorylated and localized to the nucleus,[16] the positioned nucleosomes are no longer detectable,[17] and *PHO5* transcription is induced. Remodeling of *PHO5* chromatin structure requires Pho4 and Pho2[19] and is a prerequisite for transcriptional induction of *PHO5*.[20] A Gcn5-containing histone acetyltransferase (HAT) complex and the SWI/SNF and INO80 complexes have been implicated in facilitating this process.[12,13,21,22]

We recently performed a genetic selection to identify candidate regulators of *PHO5* expression by mutagenizing a *pho80ts* strain and looking for mutants that failed to induce *PHO5* in high phosphate when raised to the nonpermissive temperature[12] (see Fig. 1). In addition to replacing the endogenous *PHO80* gene with a temperature sensitive *pho80* allele in the starting strain, the promoter of the *CAN1* gene was replaced with the *PHO5* promoter. Cells expressing *CAN1* die in the presence of canavanine, a toxic arginine analog, such that cells carrying the *pPHO5-CAN1* fusion are sensitive to canavanine in conditions inducing *PHO5* transcription. The *pho80ts* allele was used to induce *PHO5* transcription in high phosphate media. Its gene product is functional and *PHO5* transcription repressed at 25°, while it is nonfunctional and *PHO5* transcription activated at 37°. In arginine dropout media containing 5–10 μg/ml canavanine, the starting strain grows at 25° and dies at 37°.

The *pho80ts* allele was used to help identify factors involved in chromatin remodeling. It limited our search to genes acting downstream of *PHO80*. It also provided a sensitized background for the selection because the activation of *PHO5* transcription is more dependent on chromatin modifiers. The activation of *PHO5* transcription is decreased many fold in *gcn5Δ pho80ts*, *snf6Δ pho80ts*, and *arp8Δpho80ts* mutants compared to *pho80ts* cells when induced in high phosphate conditions by temperature shift,[12] whereas it is activated to near wild-type levels in the *gcn5*, *swi2*, and *arp8* mutants in *PHO80* cells starved for phosphate[13] (E. S. Haswell and E. K. O'Shea, unpublished data).

[16] A. Komeili and E. K. O'Shea, *Science* **284,** 977 (1999).
[17] A. Almer, H. Rudolph, A. Hinnen, and W. Hörz, *EMBO J.* **5,** 2689 (1986).
[18] K. R. Schneider, R. L. Smith, and E. K. O'Shea, *Science* **266,** 122 (1994).
[19] K. D. Fascher, J. Schmitz, and W. Hörz, *EMBO J.* **9,** 2523 (1990).
[20] A. Schmid, K.-D. Fasher, and W. Hörz, *Cell* **71,** 853 (1992).
[21] M. Vogelauer, J. Wu, N. Suka, and M. Grunstein, *Nature* **408,** 495 (2000).
[22] X. Shen, G. Mizuguchi, A. Hamiche, and C. Wu, *Nature* **406,** 541 (2000).

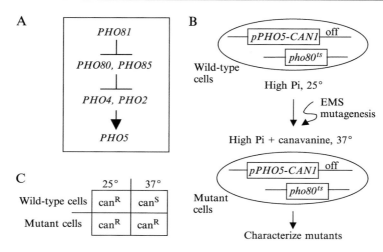

Fig. 1. Genetic selection to identify *PHO5* uninducible mutants. (A) A partial genetic view of *PHO5* regulation. (B) Schematic of the selection strategy. The starting strain (K699 *MATa pho80^{ts} can1 :: pPHO5-CAN1 pho81Δ :: TRP1B ade2-1 trp1-1 leu2-3, 112 his3-11, 15 ura3-1*) was treated with ethylmethane sulfonate (EMS) to approximately 50% viability, and plated on SD-arginine + 8 μg/ml canavanine medium (high phosphate [Pi]) at room temperature for 5 h. After recovery, the mutagenized cells were transferred to 37°. Cells unable to turn on *pPHO5-CAN1* upon inactivation of the *pho80^{ts}* allele grow at 37°. (C) Conditions under which the starting strain (wild-type cells) and mutant cells are either sensitive (can^S) or resistant (can^R) to canavanine.

To isolate mutant strains defective in the activation of *PHO5* transcription, the starting strain was treated with mutagen and plated for growth at 37° on high phosphate medium containing canavanine. We picked 964 canavanine-resistant mutants and analyzed these by secondary screens to eliminate undesirable classes of mutants. Mutations in the *pPHO5-CAN1* fusion gene were detected by examining expression of the endogenous *PHO5* gene by an acid phosphatase assay performed directly on cells on plates. We discarded 513 mutants that were able to induce expression of the endogenous *PHO5* gene at 37°. Mutations in *PHO4*, *PHO2*, or dominant mutations were detected by mating to a *pho2Δ pho4Δ* strain. Of the remaining 451 mutants, 431 failed to complement the *pho2Δ pho4Δ* strain, and these were discarded. Mutations causing improper localization of Pho4 were detected by examining the subcellular localization of Pho4-GFP. We found three mutants that failed to localize Pho4-GFP to the nucleus at 37°. They comprised a single complementation group, and by screening a yeast genomic library for plasmids that complement the *PHO5* uninducible phenotype, were determined to carry mutations in *PSE1*, the nuclear import receptor for Pho4.

Mutants surviving these secondary screens were thought to carry mutations in genes affecting *PHO5* transcription downstream of *PHO2* and *PHO4*. Of the remaining 17 mutants, only one displayed a defect in *PHO5* mRNA induction as measured by northern analysis. Mutants not impaired in *PHO5* mRNA induction were presumed to affect *PHO5* expression downstream of transcription and were not examined further. By examining the structure of *PHO5* promoter chromatin with a restriction enzyme accessibility assay, we found that the transcriptional defect in the remaining mutant resulted from a block in *PHO5* chromatin remodeling. We subsequently determined that the mutant carried a mutation in *ARG82/IPK2* by complementation analysis with the yeast genomic library and recovery of the mutant allele by gap repair.

Arg82 is a nuclear inositol polyphosphate kinase that functions in a pathway with Plc1 and Ipk1 to produce soluble inositol polyphosphates in the nucleus.[23] Its identification in the genetic selection was crucial to revealing a connection between inositol polyphosphate production and chromatin remodeling. Given that we obtained only one allele of *ARG82*, the selection was not saturating. By analyzing more mutants, additional unexpected activities may be discovered to regulate chromatin remodeling and transcription at *PHO5* and perhaps other genes.

Screening the Collection of 4847 Viable Haploid Yeast Deletion Strains for Those That Exhibit Impaired *PHO5* Expression

The kinetics of *PHO5* transcription are delayed in response to phosphate starvation in *gcn5Δ*, *snf6Δ*, and *arp8Δ* strains, but the steady-state levels of activated transcription are similar to the wild-type level[13] (E. S. Haswell and E. K. O'Shea, unpublished data). In contrast, *PHO5* transcription levels at steady state are significantly decreased in these mutants in the *pho80^ts* background when induced by temperature shift in high phosphate medium.[12] A simple interpretation of these observations is that an activity (or activities) is present in phosphate starvation conditions that functions redundantly with these chromatin modifiers to induce *PHO5*. Thus, it seems likely that there are factors participating in *PHO5* transcription and chromatin remodeling that remain unknown.

To identify these, we are currently screening the collection of 4847 viable haploid yeast deletion strains for strains that exhibit a kinetic delay in *PHO5* induction (S. Huang and E. K. O'Shea, unpublished data). Pho5

[23] A. R. Odom, A. Stahlberg, S. R. Wente, and J. D. York, *Science* **287,** 2026 (2000).

acid phosphatase activity is quantified as a function of time after cells are transferred to no phosphate medium. Each strain is grown in high phosphate medium to mid-log phase, then transferred to medium lacking phosphate. Aliquots of approximately equal cell numbers are removed after 0, 2, 4, and 6 h of growth in no phosphate medium and assayed for acid phosphatase activity. To expedite handling of the 4847 strains, all procedures are performed in 96-deep-well plates with the aid of a liquid handling robot.

Although the screen is still in progress, it is reassuring that we detect defects in Pho5 activity in strains carrying deletions in regulators of $PHO5$ expression. For example, the $pho80\Delta$ and $pho85\Delta$ strains display constitutively high levels of Pho5 activity. Furthermore, strains with deletions in subunits of the SAGA, SWI/SNF and INO80 complexes have less Pho5 activity at the early time points compared to the majority of the other deletion strains.

Several secondary screens will be performed for strains displaying a kinetic delay in Pho5 production. To determine that the $PHO5$ production phenotype is linked to the deletion, each mutant will be backcrossed to the wild-type strain and the phenotypes of progeny analyzed. Mutants whose phenotype is linked to the deletion will be transformed with a plasmid overexpressing a constitutively active allele of $PHO81$ and Pho5 activity examined. Mutants that remain impaired for $PHO5$ expression will carry mutations in genes functioning downstream of $PHO81$, and will be examined further. Pho4 localization, $PHO5$ mRNA production, and $PHO5$ chromatin remodeling will be examined as described earlier.

A strength of this screen is that it is saturating for nonessential genes. Another advantage is that the mutation within each strain is known. Thus, for strains passing the secondary screens, we may have immediate insight into how a particular gene functions to regulate $PHO5$ expression.

Summary

Novel discoveries result from genetic analyses of transcription and chromatin remodeling because these methods identify activities in an unbiased manner. By describing our genetic approaches to identify regulators of $PHO5$ transcription and chromatin remodeling, we hope to encourage others to apply similar strategies to their genes of interest.

Acknowledgments

We are grateful to members of the O'Shea laboratory for many interesting and helpful discussions. This work was supported by the National Institutes of Health grant GM51377. D. J. S. was funded by the Leukemia and Lymphoma Society as a Special Fellow.

[3] Introduction to Trx-G and Pc-G Genes

By James A. Kennison

The genetic characterization of genes that regulate the determination of segmental identity in *Drosophila melanogaster* has provided a rich source of mutations in proteins involved in chromatin modifications, as well as many new ideas about their importance in transcriptional regulation. Among the transacting factors that regulate segment identity, two groups of genes, the trithorax group (Trx-G) and the polycomb group (Pc-G), are of particular interest in understanding how chromatin modifications affect gene expression.

The identities of the body segments of *Drosophila* are specified by the homeotic genes of the Antennapedia complex (ANTC)[1] and the bithorax complex (BXC).[2,3] The homeotic genes of the ANTC and BXC encode DNA-binding transcription factors that directly control the transcription of downstream target genes. The transcription of the homeotic genes is also highly regulated.[4,5] Each homeotic gene is expressed in a specific domain along the anterior–posterior body axis. Loss of function within the domain of expression alters segment identity. Outside of its domain of expression, transcription of a homeotic gene is silenced. Failure of silencing and the ectopic expression of homeotic proteins also alters segment identity. The requirements for both maintenance of expression in the normal domains and silencing outside of these domains have made the homeotic genes important models for studying the genetic basis of gene expression and silencing.

The initial domains of expression and silencing for the homeotic genes are set very early in development by the segmentation genes, the same genes that divide the body into segments.[4-6] Studies of the maintenance of these patterns of expression throughout embryonic and larval development have defined two groups of required genes, the Trx-G genes required for maintenance of the expression within the normal domains,[6,7] and the Pc-G genes required for the maintenance of silencing outside of the

[1] T. C. Kaufman, M. A. Seeger, and G. Olsen, *Adv. Genet.* **27,** 309 (1990).
[2] I. M. Duncan, *Annu. Rev. Genet.* **21,** 285 (1987).
[3] E. B. Lewis, *Nature* **276,** 565 (1978).
[4] J. A. Kennison, *Trends Genet.* **9,** 75 (1993).
[5] J. A. Simon and J. W. Tamkun, *Curr. Opin. Genet. Dev.* **12,** 210 (2002).
[6] J. A. Kennison and J. W. Tamkun, *Proc. Natl. Acad. Sci. USA* **85,** 8136 (1988).
[7] A. Shearn, *Genetics* **121,** 517 (1989).

domains of expression.[8,9] Trx-G mutants have phenotypes that mimic loss-of-function mutations in the homeotic genes. Pc-G mutants have phenotypes that mimic gain-of-function mutations in the homeotic genes. Molecular and biochemical characterizations of the Trx-G and Pc-G genes and their proteins have shown that many of these genes encode subunits of chromatin-modifying complexes.[5,9] This chapter focuses on the genetic assays used to identify and characterize Trx-G and Pc-G mutations.

ANTC and BXC Genes Used in Assays for Trx-G and Pc-G Mutations

Although most of the homeotic genes have been used as assays for the effects of Trx-G and Pc-G mutations, the effects on some homeotic genes are easier to interpret than others. This is due to the cross-regulatory interactions among the homeotic genes. For example, the *Ultrabithorax* homeotic gene is transcriptionally repressed by the *abdominal-A* and *Abdominal-B* proteins,[10,11] whereas the *Ultrabithorax* protein represses transcription of the *Antennapedia* gene.[12,13] I will explain later how this complicates interpretations of the effects of Trx-G and Pc-G mutations. The transcriptional activations and repressions of at least three of the homeotic genes (*Sex combs reduced, Abdominal-B*, and *Deformed*) do not appear to be affected by expressions of the other homeotic genes.[14,15] The lack of cross-regulatory interactions for the expression of these three homeotic genes makes the results of genetic experiments easier to understand.

The Sex Combs Reduced *Homeotic Gene and the "Extra Sex Combs" Phenotype*

The Sex combs reduced *(Scr)* homeotic gene specifies the identities of the labial and first thoracic segments in *Drosophila*.[1] The expression of the *Scr* gene in the larval cells that give rise to the adult structures has been particularly useful in identifying new Trx-G and Pc-G genes. *Scr* proteins are only expressed at high levels in the larval cells that give rise to the first pair of adult legs; the larval cells that form the second and third pairs of legs

[8] G. Jürgens, *Nature* **316,** 153 (1985).
[9] J. A. Kennison, *Annu. Rev. Genet.* **29,** 289 (1995).
[10] R. A. H. White and M. Wilcox, *Nature* **318,** 563 (1985).
[11] G. Struhl and R. A. H. White, *Cell* **43,** 507 (1985).
[12] E. Hafen, M. Levine, and W. J. Gehring, *Nature* **307,** 287 (1984).
[13] A. Boulet and M. P. Scott, *Genes Dev.* **2,** 1600 (1988).
[14] P. Riley, S. Carroll, and M. P. Scott, *Genes Dev.* **1,** 716 (1987).
[15] W. McGinnis, T. Jack, R. Chadwick, M. Regulski, C. Bergson, N. McGinnis, and M. A. Kuziora, *Adv. Genet.* **27,** 363 (1990).

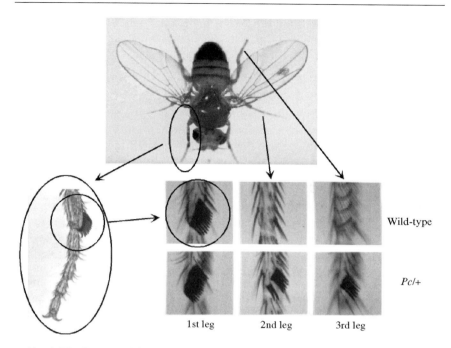

Wild-type

Pc/+

1st leg   2nd leg   3rd leg

FIG. 1. The *Scr* gene of the ANTC specifies the identity of the adult first leg. An adult male fly is shown at the top of the figure. Below are magnified views of the distal first leg (on the bottom left) and the first tarsal segments of all three pairs of legs (on the bottom right). First tarsal segments are shown for both wild-type and *Pc/+* heterozygous mutant males. The ovals indicate the first leg of the wild-type male with a complete sex comb (10 dark bristles) on the first tarsal segment. Partial ectopic sex combs can be seen on the first tarsal segments of the second and third legs of the *Pc/+* male.

express little or no *Scr* protein.[16] In the adult male, the first pair of legs has a row of very distinctive bristles on the first tarsal segment. This row of bristles is the "sex comb" and the individual bristles are "sex comb teeth." Sex comb teeth are not normally present on either the second or third pairs of legs. The sex comb of the adult male is shown in Fig. 1. The differentiation of sex comb teeth can be used to estimate the level of expression of the *Scr* gene. The higher the levels of expression, the higher the numbers of sex comb teeth. Mutations that affect silencing of the *Scr* gene in the larval cells that form the second and third legs cause ectopic expression of *Scr* proteins and the differentiation of ectopic sex comb teeth on the second

[16] A. M. Pattatucci and T. C. Kaufman, *Genetics* **129,** 443 (1991).

and third legs.[16] This is called the "extra sex combs" phenotype, and mutations outside the ANTC with an extra sex combs phenotype identify the Polycomb group of genes.[8,9] Figure 1 shows ectopic sex comb teeth on the first tarsal segments of the second and third legs of a fly heterozygous for a mutation in the Polycomb group gene *Polycomb (Pc)*. The extra sex combs phenotype has been used both to identify new Pc-G genes and in genetic assays to identify new Trx-G genes.

Three methods have been extensively used to identify Pc-G genes. A few Pc-G genes are sufficiently dosage-sensitive such that flies with only one wild-type copy (instead of the normal two copies) have some ectopic expression of *Scr* and ectopic sex comb teeth on the second and third legs of adult males. A good example is the *Pc* gene for which the group is named.[3,8] For other Pc-G genes, hypomorphic alleles have been isolated. Homozygotes for these hypomorphic alleles survive late enough in development to differentiate adult cuticle with ectopic sex comb teeth. *Pleiohomeotic (pho)*[17,18] and *super sex combs (sxc)*[19] are examples of Pc-G genes identified by the extra sex combs phenotype of homozygotes. In addition, mutations that enhance the homeotic phenotypes of known Pc-G mutations have also been isolated and characterized. The isolation of dominant enhancers takes advantage of the property that, for most genes in *Drosophila*, the amount of gene product is proportional to the number of wild-type copies of the gene.[20,21] Thus, flies that are heterozygous for a mutation or deletion (and which have only one wild-type copy of the mutated or deleted gene) have only half as much gene product as flies with two wild-type copies. Although reducing the expression of most genes in *Drosophila* does not alter the phenotype in wild-type flies, it often has an effect in a sensitized genetic background. For example, *Polycomblike (Pcl)* and *Enhancer of Pc (E[Pc])* mutations were both first identified as dominant enhancers of the weak homeotic phenotypes of heterozygous *Pc* mutant flies.[22,23] Several sensitized genetic backgrounds have been used to screen for mutations that enhance the phenotypes of known Pc-G mutations. Test flies heterozygous for mutations in *Pc* (*Pc³* and *Pc⁴* are the alleles generally used)[6,22] or hemizygous for mutations in one of the two X-linked *polyhomeotic (ph)*

[17] W. J. Gehring, *Drosophila Inform. Serv.* **45,** 103 (1970).
[18] J. L. Brown, D. Mucci, M. Whiteley, M. J. Dirksen, and J. A. Kassis, *Molec. Cell* **1,** 1057 (1998).
[19] P. W. Ingham, *Cell* **37,** 815 (1984).
[20] S. J. O'Brien and R. C. Gethmann, *Genetics* **75,** 155 (1973).
[21] J. A. Kennison and M. A. Russell, *Genetics* **116,** 75 (1987).
[22] I. M. Duncan, *Genetics* **102,** 49 (1982).
[23] T. Sato, M. A. Russell, and R. E. Denell, *Genetics* **105,** 357 (1983).

genes^{24} are crossed to candidate mutations or to mutagenized flies (to recover new mutations). If the candidate mutation is involved in homeotic gene silencing, an increase in the numbers of sex comb teeth on the adult second and third legs is expected. Expressivity (the average number of ectopic sex comb teeth on each second or third leg in a test genotype) is probably the more sensitive assay, but penetrance (the percentage of second and third legs that have at least one ectopic sex comb tooth) is easier to determine and has also been used.

The ectopic expression of *Scr* in Pc-G mutants has also been used to identify new Trx-G genes.^{6,24} In this case, Trx-G mutations are expected to reduce the numbers of ectopic sex comb teeth in the Pc-G mutant flies. This can be measured as a reduction in either the expressivity or penetrance of the extra sex combs phenotype. Again, the test genotypes have usually been heterozygous for *Pc* mutations or hemizygous for *ph* mutations. The Trx-G genes are more heterogeneous than the Pc-G genes (there are more ways to cause loss of gene function than there are ways to cause ectopic expression) and only a subset of the Trx-G genes encode proteins that modify chromatin. Among the Trx-G genes identified as genetic suppressors of Pc-G mutations, the *brahma (brm)*, *moira (mor)*, and *osa (osa)* genes encode subunits of the *Drosophila* BRM chromatin-remodeling complex.^{5,6}

The Abdominal-B *Homeotic Gene of the BXC*

A second homeotic gene that has been used in genetic assays to identify new Trx-G genes is the Abdominal-B *(Abd-B)* gene of the BXC. The *Abd-B* gene specifies the identities of the posterior abdominal segments, including the fifth abdominal segment of the adult.^{2,3} In the adult male, the dorsal cuticles of both the fifth and sixth segments are darkly pigmented throughout, whereas the dark pigmentation is restricted to the posterior edge of each of the more anterior segments (see panel A in Fig. 2). Loss of *Abd-B* function in the larval cells that form the fifth abdominal segment causes the cells to differentiate with a more anterior segment identity, resulting in the loss of dark pigmentation in the anterior portion of the dorsal cuticle (see panel B in Fig. 2). The *Abd-B* gene is dosage-sensitive, showing some loss of pigmentation in the fifth abdominal segment of mutant heterozygotes. Unfortunately, the *Abd-B* gene is also dosage-sensitive for fertility; the *Abd-B* mutant heterozygotes are usually sterile in both sexes.^{2,3} This makes *Abd-B* mutants more difficult to use in sensitized genetic backgrounds for

---

^{24} M. O. Fauvarque, P. Laurenti, A. Boivin, S. Bloyer, R. Griffin-Shea, H. M. Bourbon, and J. M. Dura, *Genet. Res. Camb.* **78**, 137 (2001).

FIG. 2. Phenotypes from loss-of-function for the *Abd-B* and *Antp* homeotic genes. Panel A shows the abdomen from a wild-type male with the fifth abdominal segment completely pigmented. Panel B shows the abdomen from a male with partial loss of *Abd-B* function. Part of the fifth abdominal segment is now lightly pigmented and resembles the fourth abdominal segment just above. Panel C shows a wild-type fly with the wings held parallel to the body. Panel D shows a fly with partial loss of function for the *Antp* P2 promoter, which causes the fly to holds its wings out from the body at an angle.

assaying effects of mutations in chromatin-modifying enzymes. To circumvent this problem, sensitized genetic backgrounds with mutations in Trx-G genes have also been developed. One such test genotype is heterozygous for mutations in both the *brm* and *trithorax (trx)* genes.[25,26] Flies heterozygous for the $brm^2\ trx^{E2}$ double mutant chromosome often have patches of transformed tissue in the fifth abdominal segment. When heterozygous for

[25] J. W. Tamkun, R. Deuring, M. P. Scott, M. Kissinger, A. M. Pattatucci, T. C. Kaufman, and J. A. Kennison, *Cell* **68,** 561 (1992).
[26] A. K. Dingwall, S. J. Beek, C. M. McCallum, J. W. Tamkun, G. V. Kalpana, S. P. Goff, and M. P. Scott, *Mol. Biol. Cell* **6,** 777 (1995).

mutations in chromatin-modifying enzymes, the expressivity or penetrance of the phenotype in the fifth abdominal segment can be either increased (for chromatin-modifying enzymes that affect transcriptional activation) or decreased (for chromatin-modifying enzymes that affect transcriptional silencing).

The Deformed (Dfd) *Homeotic Gene of the ANTC*

The third homeotic gene without cross-regulatory interactions that has been used to identify and test putative Trx-G mutations is the *Dfd* gene. The *Dfd* homeotic gene specifies identities of part of the head in *Drosophila* and does not appear to be transcriptionally regulated by other homeotic genes.[15] William McGinnis and co-workers developed a genetic assay to characterize mutations that cause a loss of *Dfd* function based on the reduced viability of a genotype with only partial *Dfd* function.[27–29] Flies transheterozygous for two different alleles of *Dfd* (Dfd^3/Dfd^{13}) are only about 50% as viable as expected at 29°. This provides a very sensitive assay for *Dfd* function. The assay again utilizes the gene-dosage dependence of Trx-G genes. Putative Trx-G mutations are tested as heterozygotes in the *Dfd* mutant background and identified by a reduction in viability of the test genotype. There are two advantages to the *Dfd* genetic assay. The first advantage is the ability to quantitate the effect by determining the viability of the test genotype compared to the test genotype that has two wild-type copies of the Trx-G gene. The second advantage is that the genetic assay depends on both the embryonic and larval functions of *Dfd*, whereas the *Scr* and *Abd-B* assays described earlier measure primarily the larval functions. The disadvantage of the *Dfd* genetic assay is the difficulty in constructing the test genotypes. Mutations to be tested must be introduced into the Dfd^3/Dfd^{13} mutant background. For mutations on the same chromosome as *Dfd*, this involves making a recombinant that carries both the test mutation and one of the *Dfd* mutations. The *Dfd* gene is located proximally on the third chromosome where meiotic recombination is reduced. About one quarter of the genes on the third chromosome (or about 10% of the total genome) lie within one map unit to either side of the *Dfd* gene. This makes the generation of the test genotype extremely difficult to construct in some cases. For this reason, the *Dfd* genetic assay has been

[27] K. W. Harding, G. Gellon, N. McGinnis, and W. McGinnis, *Genetics* **140,** 1339 (1995).
[28] G. Gellon, K. W. Harding, N. McGinnis, M. M. Martin, and W. McGinnis, *Development* **124,** 3321 (1997).
[29] B. Florence and W. McGinnis, *Genetics* **150,** 1497 (1998).

used mostly for identifying new candidate Trx-G mutations recovered after mutagenesis of the Dfd mutant flies.

The Antennapedia (Antp) and Ultrabithorax (Ubx) Homeotic Genes

Genetic assays for Trx-G mutations that involve other homeotic genes in $Drosophila$ have also been used, however, the cross-regulatory interactions among the homeotic genes make these assays difficult to interpret without additional experiments. For example, partial loss of function for the P2 promoter of the $Antp$ gene causes the adult flies to hold their wings out from the body at an angle (wild-type flies usually rest with the wings parallel to the body).[30] This phenotype is illustrated in panels C and D of Fig. 2. As a sensitized genetic background, the flies can be heterozygous for $Antp$ mutations that specifically reduce P2 promoter expression ($Antp^1$, $Antp^{23}$, or $Antp^B$),[30] they can be heterozygous for a Trx-G mutation that reduces P2 expression (such as brm^2 or osa^2),[30] or they can be heterozygous for a $Cbx^1 Ubx^1$ double mutant chromosome.[31] Flies heterozygous for the $Cbx^1 Ubx^1$ double mutant chromosome hold their wings out from the body because of homeotic cross-regulatory interactions. The $Cbx^1 Ubx^1$ double mutant causes ectopic Ubx expression in the cells that differentiate the adult wing. The ectopic Ubx expression represses transcription from the $Antp$ P2 promoter (a known target for Ubx transcriptional repression),[13] causing the wings to be held out at an angle from the body. In a sensitized genetic background in which $Antp$ P2 expression is reduced, the held-out wing phenotype can be altered by direct affects on $Antp$ expression, but also by changes in the silencing of the Ubx gene. Thus, although a Trx-G mutation could enhance loss of expression from the $Antp$ P2 promoter, a Pc-G mutation could also indirectly enhance the loss of expression from the $Antp$ P2 promoter by interfering with silencing of the Ubx gene. In contrast, while a Pc-G mutation could suppress the held-out wing phenotype by directly derepressing the $Antp$ P2 promoter, a Trx-G mutation could indirectly suppress the phenotype of the $Cbx^1 Ubx^1$ double mutant by decreasing the ectopic expression of Ubx. Thus, although the expression from the $Antp$ P2 promoter is a good assay, the effects should be interpreted only in conjunction with data on changes in the expression of other homeotic genes. For example, mutations in the $tonalli$ (tna) and $taranis$ $(tara)$ genes enhance the held-out wing phenotypes in the $Antp$ P2–sensitized genetic backgrounds.[32,33] That these genes are Trx-G genes

[30] M. Vázquez, L. Moore, and J. A. Kennison, $Development$ **126,** 733 (1999).
[31] E. B. Lewis, $Amer.$ $Nat.$ **90,** 73 (1955).
[32] S. Calgaro, M. Boube, D. L. Cribbs, and H. Bourbon, $Genetics$ **160,** 547 (2002).

(and not Pc-G genes) is inferred from their homozygous mutant phenotypes. Both genes have mutant phenotypes that mimic loss-of-function for the homeotic genes.[32,33]

The ETP Group of Homeotic Gene Regulators

I believe that the failure to take into account the homeotic cross-regulatory interactions for the Ubx gene may have lead to mistaken conclusions about the functions of a subset of Pc-G genes, the enhancers of trithorax and Polycomb mutations (ETP) group genes.[34] Using sensitized genetic backgrounds (flies heterozygous for one or more Trx-G mutations) that reduce Ubx expression, Gildea et $al.$[34] identified mutations and deletions that further enhanced the phenotypes caused by loss of Ubx expression. Surprisingly, mutations in several known Pc-G genes were among the mutations identified. These Pc-G mutations had been previously shown to cause ectopic Ubx expression,[35–37] and therefore, loss of Ubx function was unexpected. The authors concluded that this subset of Pc-G mutations identify proteins that are involved in both activation and repression of the homeotic genes and named them the ETP group. Whereas proteins involved in both homeotic gene activation and repression are expected to exist, I am not convinced that the ETP group identify such proteins. Ubx is not only silenced by the Pc-G, but it is also transcriptionally repressed by the $abd-A$ and $Abd-B$ proteins through cross-regulatory interactions.[10,11] In flies heterozygous for various Trx-G and Pc-G mutations, expressions of the homeotic genes is not uniformly affected, but individual Trx-G and Pc-G mutations have strikingly differential effects. For example, in brm mutant heterozygotes, ectopic expressions of $Antp$ and Scr are greatly reduced, whereas expression of Ubx is only mildly affected. In contrast, in $ash1$ mutant heterozygotes, Ubx expression is reduced to a much larger extent than $Antp$ or Scr expressions. I believe that the difference between the Pc-G and the ETP group mutations is their relative derepression of different homeotic gene promoters. The ETP group mutations may derepress the $abd-A$ and/or $Abd-B$ promoters to a much greater level than they derepress the Ubx promoter. The ectopic expressions of $abd-A$ and/or $Abd-B$ would be able to repress Ubx expression to an even greater extent in a genetic background in which Ubx expression is weakened (e.g., in flies heterozygous for Trx-G mutations that compromise Ubx expression

---

[33] L. Gutiérrez, M. Zurita, J. A. Kennison, and M. Vázquez, $Development$ 130, 343 (2003).

[34] J. J. Gildea, R. Lopez, and A. Shearn, $Genetics$ 156, 645 (2000).

[35] J. McKeon and H. W. Brock, $Roux's$ $Arch.$ $Dev.$ $Biol.$ 199, 387 (1991).

[36] M. C. Soto, T.-B. Chou, and W. Bender, $Genetics$ 140, 231 (1995).

[37] D. Beuchle, G. Struhl, and J. Müller, $Development$ 128, 993 (2001).

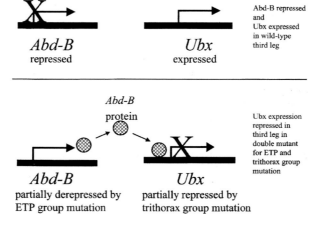

FIG. 3. Model for homeotic cross-regulatory interactions and the ETP group assay. At the top are shown the *Abd-B* (repressed) and *Ubx* (expressed) genes in the wild-type third thoracic segment. In the lower half are shown the effects of partial derepression of *Abd-B* by an ETP group mutation and partial loss of *Ubx* function by a Trx-G mutation. The ectopic *Abd-B* protein (shown as a cross-hatched circle) binds to sequences in the *Ubx* gene and represses *Ubx* transcription, further enhancing the loss of function caused by the Trx-G mutation.

to a greater extent than they compromise *abd-A* and/or *Abd-B* expression). A simplified version of this model to account for the ETP group effects is shown in Fig. 3. I am currently trying to test this model by independently altering expressions of the *abd-A* and *Abd-B* genes in the genetic assays that identified the ETP group. If true, the ETP group should remain a subset of the Pc-G.

[4] Genetic and Cytological Analysis of *Drosophila* Chromatin-Remodeling Factors

By DAVIDE F. V. CORONA, JENNIFER A. ARMSTRONG, and JOHN W. TAMKUN

Eukaryotic DNA is packaged into a highly compact, dynamic structure called chromatin. Although chromatin provides the cell with the obvious benefit of organizing a large and complex genome, it can also block the access of transcription factors and other proteins to DNA.[1] Two classes of enzymes, ATP-dependent chromatin-remodeling factors and

Copyright 2004, Elsevier Inc.
All rights reserved.
0076-6879/04 $35.00

histone-modifying enzymes, modulate chromatin accessibility by directly modifying the structure and/or position of nucleosomes.[2–4] These enzymes function in a variety of nuclear processes including gene expression.[5,6] Both classes of enzymes typically function in the context of multi-subunit complexes. Current efforts are directed towards understanding how the activities of these complexes are targeted, regulated, and coordinated *in vivo.*[7,8]

Drosophila melanogaster is a particularly useful model system for the study of chromatin remodeling and modifying enzymes since it permits both the genetic and biochemical analysis of these factors in a developing organism. We use *Drosophila* to examine the biological roles of several chromatin-remodeling factors, including Brahma (BRM) and Imitation-SWI (ISWI). *brm* was identified as a suppressor of *Polycomb*[9] and encodes the ATPase subunit of the BRM chromatin-remodeling complex: a 2-MDa complex highly related to the *S. cerevisiae* SWI/SNF and RSC complexes.[10–12] BRM maintains the transcription of homeotic genes during *Drosophila* development[13] and functions as a global activator of transcription.[14] The ISWI ATPase functions as the catalytic subunit of three distinct chromatin-remodeling complexes, NURF, CHRAC, and ACF, which catalyze nucleosome remodeling and spacing reactions *in vitro.*[15] ISWI has been implicated in both transcriptional activation and repression[16,17] as well as the maintenance of higher order chromatin structure.[16,18]

[1] G. Felsenfeld and M. Groudine, *Nature* **421**, 448 (2003).

[2] T. Jenuwein and C. D. Allis, *Science* **293**, 1074 (2001).

[3] P. B. Becker and W. Horz, *Ann. Rev. Biochem.* **71**, 247 (2002).

[4] B. M. Turner, *Cell* **111**, 285 (2002).

[5] J. A. Martens and F. Winston, *Curr. Opin. Genet. Dev.* **13**, 136 (2003).

[6] M. Iizuka and M. M. Smith, *Curr. Opin. Genet. Dev.* **13**, 154 (2003).

[7] S. L. Berger, *Curr. Opin. Genet. Dev.* **12**, 142 (2002).

[8] G. J. Narlikar, H. Y. Fan, and R. E. Kingston, *Cell* **108**, 475 (2002).

[9] J. A. Kennison and J. W. Tamkun, *Proc. Natl. Acad. Sci. USA* **85**, 8136 (1988).

[10] O. Papoulas, S. J. Beek, S. L. Moseley, C. M. McCallum, M. Sarte, A. Shearn, and J. W. Tamkun, *Development* **125**, 3955 (1998).

[11] R. T. Collins, T. Furukawa, N. Tanese, and J. E. Treisman, *EMBO J.* **18**, 7029 (1999).

[12] A. J. Kal, T. Mahmoudi, N. B. Zak, and C. P. Verrijzer, *Genes Dev.* **14**, 1058 (2000).

[13] J. A. Simon and J. W. Tamkun, *Curr. Opin. Genet. Dev.* **12**, 210 (2002).

[14] J. A. Armstrong, O. Papoulas, G. Daubresse, A. S. Sperling, J. T. Lis, M. P. Scott, and J. W. Tamkun, *EMBO J.* **21**, 5245 (2002).

[15] G. Langst and P. B. Becker, *J. Cell Sci.* **114**, 2561 (2001).

[16] R. Deuring, L. Fanti, J. A. Armstrong, M. Sarte, O. Papoulas, M. Prestel, G. Daubresse, M. Verardo, S. L. Moseley, M. Berloco, T. Tsukiyama, C. Wu, S. Pimpinelli, and J. W. Tamkun, *Mol. Cell* **5**, 355 (2000).

[17] P. Badenhorst, M. Voas, I. Rebay, and C. Wu, *Genes Dev.* **16**, 3186 (2002).

[18] D. F. Corona, C. R. Clapier, P. B. Becker, and J. W. Tamkun, *EMBO Rep.* **3**, 242 (2002).

In this chapter, we present methods that have proven useful in our studies of BRM and ISWI. We describe strategies for the generation and analysis of dominant-negative mutations in chromatin-remodeling factors and their use in genetic modifier screens. We also discuss the use of polytene chromosomes to study interactions between chromatin-remodeling factors and chromatin *in vivo*.

Use of Dominant-Negative Mutations for Studying Enzymes Involved in Chromatin-Remodeling

Due to their critical roles in gene expression and development, many *Drosophila* chromatin-remodeling factors are encoded by essential genes. Mutations in these genes can block oogenesis and cell proliferation[16,19,20]; as a result, it is difficult to generate cells or organisms that completely lack their activity. This problem can be circumvented using either conditional or dominant-negative alleles. Due to the ease with which they can be generated, dominant-negative mutations are particularly useful for studying the function of chromatin-remodeling factors. Replacing a conserved lysine in the ATP-binding site of a chromatin-remodeling factor with an arginine eliminates its catalytic activity without disrupting its ability to interact with other proteins.[20–24] These mutant proteins therefore have strong, dominant-negative effects when expressed at high levels *in vivo*. Similar mutations can be used to study the function of virtually any complex with chromatin-remodeling or modifying activity.

Generation and Characterization of Dominant-Negative Transgenes

A variety of inducible or tissue-specific promoters can be used to drive the expression of cDNA or genomic DNA fragments encoding dominant-negative chromatin-remodeling factors. Ideally, it is best to begin by expressing the dominant-negative transgene under the control of its normal promoter to ensure that it is expressed in a spatial and temporal pattern

[19] M. L. Ruhf, A. Braun, O. Papoulas, J. W. Tamkun, N. Randsholt, and M. Meister, *Development* **128,** 1429 (2001).

[20] L. K. Elfring, C. Daniel, O. Papoulas, R. Deuring, M. Sarte, S. Moseley, S. J. Beek, W. R. Waldrip, G. Daubresse, A. DePace, J. A. Kennison, and J. W. Tamkun, *Genetics* **148,** 251 (1998).

[21] E. Richmond and C. L. Peterson, *Nucleic Acids Res.* **24,** 3685 (1996).

[22] P. A. Khavari, C. L. Peterson, J. W. Tamkun, D. B. Mendel, and G. R. Crabtree, *Nature* **366,** 170 (1993).

[23] J. Cote, J. Quinn, J. L. Workman, and C. L. Peterson, *Science* **265,** 53 (1994).

[24] D. F. Corona, G. Langst, C. R. Clapier, E. J. Bonte, S. Ferrari, J. W. Tamkun, and P. B. Becker, *Mol. Cell* **3,** 239 (1999).

comparable to that of its wild-type counterpart. After transgenic *Drosophila* strains are generated, the ability of the transgene to rescue null mutations should be tested to verify that the mutation destroys the activity of the chromatin-remodeling factor. Western blotting and gel filtration chromatography should be used to ensure that the mutant protein is expressed at normal levels and incorporated into chromatin-remodeling complexes. We routinely place an epitope tag at the N- or C-terminus of the mutant protein so that it can be distinguished from the endogenous protein by immunofluorescence or Western blotting. Sufficient material for these biochemical assays can be obtained from relatively small numbers of embryos using the following protocol.

Preparation of Protein Extracts From Small Quantities of Drosophila *Embryos*

1. Place approximately 25 male and 25 female adults on grape juice-agar plates (2.25% agar, 0.625% sucrose, 25% concentrated grape juice, 0.015% methyl paraben) streaked with yeast paste. Change the plates every 12 h until the females begin laying sufficient numbers of eggs (approximately 2 days).

2. Collect 0–12 h embryos on grape juice-agar plates. One collection routinely yields about 50 to 100 μl of embryos. If necessary, embryos can usually be stored at $4°$ for 2 days before extracts are prepared.

3. Collect embryos into a small Nitex mesh sieve while rinsing with deionized water. Dechorionate embryos by incubation for 2 min in a freshly made solution of 50% bleach (2.6% sodium hypochlorite) with occasional gentle swirling.

4. Rinse embryos three times in deionized water and once in wash buffer (0.4% NaCl, 0.03% Triton X-100) for 1 min to remove bleach.

5. Gently scoop dechorionated embryos into 2 ml dounce homogenizer. Cover the embryos with wash buffer and let them settle on ice for 3 min.

6. Aspirate off the wash buffer and floating chorions. Remove excess liquid by placing the tip of a Pasteur pipette against the bottom of the dounce; capillary action will remove most of the excess liquid.

7. Add an equal volume of homogenization buffer (50 mM HEPES pH 7.6, 385 mM NaCl, 0.1% Tween, 0.1 mM EGTA, 1 mM MgCl$_2$, 10% glycerol, 1× complete protease inhibitors [Roche #1697498]) to the packed embryos. Homogenize the embryos with 10 strokes of a tight pestle. Let the homogenate sit on ice for 10 min to extract proteins from the damaged nuclei.

8. Centrifuge the homogenate at $4°$ for 30 min at 25,000 rpm in a TLA100.2 rotor. Remove the top lipid layer by touching it with the wide end of a 200 μl plastic pipette tip layer and quickly pulling it up.
9. Transfer the supernatant to an Eppendorf tube and determine its volume and protein concentration by standard Bradford assay.

The above protocol routinely yields 200 μl of extract with a protein concentration of 10–30 μg/μl. The extract can be used immediately for Western blotting, gel filtration chromatography, immunoprecipitation or other assays. Alternatively, the extract can be flash frozen in liquid nitrogen and stored at $-80°$ for later use.

Functional Characterization of Chromatin-Remodeling Factors Using Dominant-Negative Transgenes

Information concerning the function of chromatin-remodeling factors *in vivo* can often be obtained by characterizing phenotypes resulting from the expression of dominant-negative transgenes. For example, the expression of a dominant-negative form of the BRM protein (BRM^{K804R}) revealed an unanticipated role for this chromatin-remodeling factor in peripheral nervous system development.[20] The GAL4 system of Brand and Perrimon[25] is particularly useful for this purpose because it allows high levels of a dominant-negative transgene to be expressed in a wide variety of different cell or tissue types. To use the GAL4 system, a cDNA or genomic DNA fragment encoding a dominant-negative chromatin-remodeling factor is placed downstream of a minimal promoter and GAL4-responsive element in the transformation vector pUAST.[25] To induce expression of the dominant-negative protein, a strain bearing the transgene is crossed to a strain bearing a "driver" gene that expresses GAL4 in a temporally or spatially restricted pattern.[25–27] A current list of available driver lines and their expression patterns may be obtained from the Bloomington Stock Center (http://flystocks.bio.indiana.edu/gal4.htm).

Eye-based Screens for Identifying Genes that Interact with Chromatin-Remodeling Factors In vivo

Although *in vitro* systems have provided a wealth of information concerning the mechanism of action of chromatin-remodeling factors, they do not reflect the complexity of chromatin that exists *in vivo*. Furthermore,

[25] A. H. Brand and N. Perrimon, *Development* **118**, 401 (1993).
[26] P. P. D'Avino and C. S. Thummel, *Methods Enzymol.* **306**, 129 (1999).
[27] J. B. Duffy, *Genesis* **34**, 1 (2002).

proteins that directly or indirectly regulate the activity of chromatin-remodeling factors can be difficult to detect using biochemical assays. By contrast, genetic screens have the potential to identify novel components of chromatin and other proteins that functionally interact with chromatin-remodeling factors without bias concerning their mechanism of action.

The *Drosophila* eye provides an attractive system for the genetic analysis of complex biological processes, including chromatin-remodeling. Since the eye is not required for viability, it is relatively straightforward to study the function of essential genes in this tissue. Screens for modifiers of phenotypes resulting from the expression of wild-type or dominant-negative proteins in the developing eye have been used to analyze a wide variety of regulatory pathways.[28]

Phenotypes resulting from the expression of dominant-negative chromatin-remodeling factors using the GAL4 system are ideal for eye-based modifier screens. Many different GAL4 drivers can be used to drive the expression of dominant-negative transgenes in the eye-antennal disc. We routinely use a GAL4 driver under the control of the *eyeless* regulatory element *(ey-GAL4)*.[29] The expression of either BRM^{K804R} (see Fig. 1B) or $ISWI^{K159R}$ in the eye-antennal disc leads to the development of adults with rough, reduced, or missing eyes[16]; this phenotype can be quantitatively scored for enhancement or suppression, allowing the identification of genes that interact with chromatin-remodeling factors *in vivo*. Before initiating a dominant-modifier screen, it is important to verify that the eye defects result from the loss of function of the chromatin-remodeling factor of interest. For example, the rough eye phenotype resulting from expression of BRM^{K804R} is enhanced by reducing the level of wild-type BRM (see Fig. 1C). The sensitivity of the assay should be tested using mutations in candidate genes. For example, mutations in a gene encoding a subunit of the BRM complex, Moira (MOR),[30] strongly enhance eye defects resulting from the expression of BRM^{K804R} (see Fig. 2). To gain confidence in the selectivity of the screen, it is important to show that mutations in functionally unrelated chromatin-remodeling factors fail to interact in the eye assay. For example, *ISWI* mutations have no effect on eye defects resulting from BRM^{K804R} expression.

Eye-based assays can be used to investigate functional interactions between chromatin-remodeling factors and other proteins known to

[28] B. J. Thomas and D. A. Wassarman, *Trends Genet.* **15,** 184 (1999).

[29] D. J. Hazelett, M. Bourouis, U. Walldorf, and J. E. Treisman, *Development* **125,** 3741 (1998).

[30] M. A. Crosby, C. Miller, T. Alon, K. L. Watson, C. P. Verrijzer, R. Goldman-Levi, and N. B. Zak, *Mol. Cell. Biol.* **19,** 1159 (1999).

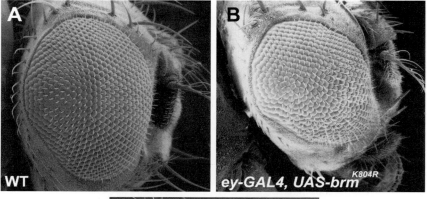

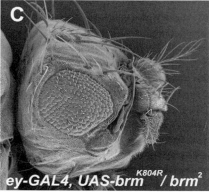

FIG. 1. Expression of dominant-negative brm (brm^{K804R}) in the developing eye results in a rough eye phenotype. (A) Scanning electron micrograph of wild-type eye. (B) Eye from an individual expressing brm^{K804R} under control of the ey-$GAL4$ driver. To facilitate the eye-based assay, the ey-$GAL4$ transgene was recombined onto the chromosome carrying the UAS-brm^{K804R} transgene. (C) A null mutation in the brm gene (brm^2) enhances the rough eye phenotype of ey-$GAL4$, UAS-brm^{K804R}.

modulate chromatin structure or gene expression. For example, using an eye-based assay we were able to demonstrate a functional antagonism between ISWI and the MOF histone acetyltransferase $in\ vivo$.[18] We were also able to confirm that the HMG-domain protein BAP111 is critical for the function of the BRM complex.[31]

Eye-based screens can also be used to identify potentially novel genes that interact with chromatin-remodeling factors. The GAL4 system is

[31] O. Papoulas, G. Daubresse, J. A. Armstrong, J. Jin, M. P. Scott, and J. W. Tamkun, $Proc.$ $Natl.\ Acad.\ Sci.\ USA$ **98**, 5728 (2001).

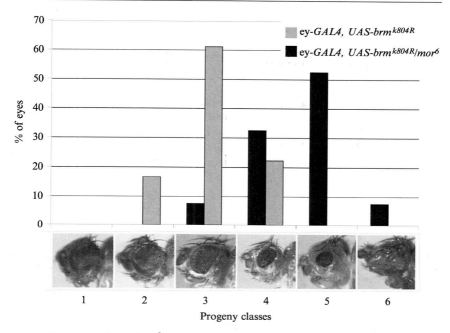

FIG. 2. A *mor* allele *(mor⁶)* enhances eye defects resulting from the expression of brm^{K804R}. *ey-GAL4, UAS-brm^{K804R}/TM3, Sb* males were crossed to *mor⁶/TM6B, Hu* virgins. Eyes of the progeny were scored using an arbitrary scale of 1 to 6: (1) normal eye; (2) 50% or less of the eye is rough (as determined by disordered ommatidia under the light microscope); (3) greater than 50% of the eye is rough; (4) the eye is rough and reduced in size by 50% or less; (5) the eye is rough and reduced in size by more than 50%; and (6) eye is absent. We use the Kolmogorov–Smirnov two-sample test to determine the statistical significance of the genetic interactions.

inherently temperature sensitive[25]; progeny with moderate eye defects can be generated by varying the temperature between 18° and 29°, thus allowing the recovery of both suppressor and enhancer mutations in an F_1 screen. An F_2 screen should be considered, however, if the number of progeny that display extremely strong or weak phenotypes is too high (>0.1%). Examples of both F_1 and F_2 screens for dominant modifiers of eye phenotypes are shown in Fig. 3. Screens for EMS-induced mutations have strong tendency to recover loss-of-function mutations. As a result, mutations in genes that are not expressed in limiting quantities may not be recovered in dominant-modifier screens. An alternative approach is to screen for genes that modify eye phenotypes when expressed at high levels. This approach has been greatly facilitated by the availability of a large

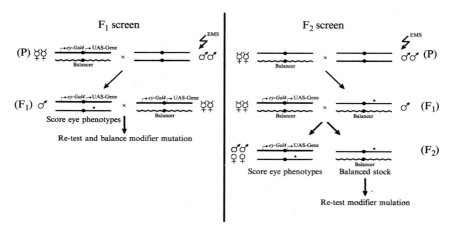

FIG. 3. Schemes for F_1 and F_2 eye-based modifier screens. In an F_1 screen, males are mutagenized and crossed *en masse* to virgins expressing the dominant-negative chromatin-remodeling factor in the eye. Single F_1 males are scored for enhancement or suppression of eye defects. Candidate modifier mutations are retested and stocks are established. In an F_2 screen, single F_1 progeny bearing mutagenized chromosomes are crossed to virgin females expressing the dominant-negative chromatin-remodeling factor in the eye. F_2 progeny are scored for enhancement or suppression of eye defects. Siblings carrying the mutation over a balancer are used to establish stocks and retest. Protocols for F_1 and F_2 screens for EMS-induced mutations have been published elsewhere.[32]

collection of inducible P insertions known as EPs (http://flystation.exelixis.com/).[33,34] Each insertion contains a GAL4-responsive promoter that has the potential to drive high levels of expression of a neighboring gene. These genes can be easily identified by their annotation in the Berkeley *Drosophila* Genome Project (http://www.fruitfly.org/index.html).

Once candidate mutations are identified, other criteria should be used to identify the most promising mutations for further analysis. For example, to test for specificity we ask whether a mutation that modifies a brm^{K804R} rough eye phenotype has any affect on an $ISWI^{K159R}$ rough eye phenotype, and vice versa. This allows us to exclude mutations that modify dominant-negative phenotypes for trivial reasons, such as altering the expression of the UAS transgenes. Standard techniques can then be used to map and characterize the genes identified in the modifier screens. Ultimately, the

[32] T. Grigliatti, *in* "Drosophila a Practical Approach" (D. B. Roberts, ed.), p. 39. IRL Press, Oxford, 1986.

[33] P. Rorth, K. Szabo, A. Bailey, T. Laverty, J. Rehm, G. M. Rubin, K. Weigmann, M. Milan, V. Benes, W. Ansorge, and S. M. Cohen, *Development* **125,** 1049 (1998).

[34] P. Rorth, *Proc. Natl. Acad. Sci. USA* **93,** 12418 (1996).

biochemical characterization of the genes products will be necessary to clarify the basis of their genetic interactions with chromatin-remodeling factors.

Using Salivary Gland Polytene Chromosomes to Study Chromatin-Remodeling Factors

One advantage of using *Drosophila* as a model system is the ability to directly visualize interphase chromosomes. The chromosomes of the salivary gland and many other larval tissues undergo multiple rounds of DNA replication in the absence of cell division. The resulting giant chromosomes display reproducible banding patterns that are easily visualized under the light microscope. Salivary gland polytene chromosomes can be stained with antibodies against chromatin-remodeling factors and other proteins to directly visualize their interactions with chromatin. Although this approach lacks the resolution of chromatin immunoprecipitation (ChIP), it provides a straightforward and inexpensive way to determine the genome-wide distribution of a protein of interest.

Many protocols for immunostaining polytene chromosomes have been published.[35-38] We have successfully used the following protocol to compare the distribution of several different proteins, including BRM, ISWI, the initiating and elongating forms of RNA polymerase II, and Polycomb.

Antibody Staining of Polytene Chromosomes

1. Dissect salivary glands of third-instar larvae in 0.7% NaCl. Remove as much of the fat as possible without damaging the glands. Detailed instructions on how to dissect and manipulate salivary glands have been published elsewhere.[35,39]

2. Transfer one set of glands to 12 μl of freshly prepared fixing solution (45% glacial acetic acid, 1.85% formaldehyde) on a silanized coverslip and fix for 10 min. To silanize coverslips, dip them in 1% dimethylsilane in chloroform (work in a hood), and let dry overnight. Silanized coverslips can be stored indefinitely in a dust free box.

---

[35] L. A. Pile and D. A. Wassarman, *Methods* **26**, 3 (2002).

[36] L. M. Silver and S. C. Elgin, *Proc. Natl. Acad. Sci. USA* **73**, 423 (1976).

[37] R. Paro, *in* "Drosophila Protocols" (W. Sullivan, M. Ashburner, and R. S. Hawley, eds.), p. 131. Cold Spring Harbor Press, Cold Spring Harbor, NY, 2000.

[38] D. J. Andrew and M. P. Scott, *Methods Cell Biol.* **44**, 353 (1994).

[39] J. A. Kennison, *in* "Drosophila Protocols" (W. Sullivan, M. Ashburner, and R. S. Hawley, eds.), p. 111. Cold Spring Harbor Press, Cold Spring Harbor, NY, 2000.

3. Pick up the coverslip by gently touching the drop of fixative with a clean slide. Place the slide (with the coverslip facing up) on a paper towel. Use the back end of a small paintbrush to disrupt the nuclei and spread the chromosomes. Start tapping in the center of the coverslip and gently tap out in a spiral, repeat once. Bracket the coverslip with gloved fingertips to keep it from sliding.

4. Being careful not to move the coverslip, cover the slide with a paper towel and squash with thumb. Check the quality of the chromosome squash by phase-contrast microscopy. A good squash will have well-spread chromosome arms and distinct banding patterns.

5. Mark the position of the coverslip on the slide with a diamond pen. Freeze slide in liquid nitrogen and remove the coverslip by wedging a razor blade under the corner and flipping off the coverslip. Immediately place slide in PBS (137 mM NaCl, 2.68 mM KCl, 10.14 mM Na$_2$HPO$_4$, 1.76 mM KH$_2$PO$_4$, adjust pH to 7.2 with 1 N NaOH). You can freeze slides as you go and accumulate them in PBS, otherwise they tend to dry out.

6. Wash slides in PBS for 5 min and in PBS-T (PBS containing 1% Triton X-100) for 10 min. Block slides in PBS-TB (PBS containing 1% BSA and 0.1% Triton X-100) for 30 min.

7. Dilute primary antibody in PBS-TB (when diluting antibodies, use RIA grade BSA, Sigma). In general, the concentration of primary antibody should be 10-fold higher than used for western blotting. Place 20 μl diluted primary antibody on a coverslip. Dry off as much of the slide as possible without disturbing the area containing the chromosome squash. Pick up the coverslip with the slide, avoiding bubbles. Place slide with the coverslip facing up in a humid chamber (a large Petri dish lined with wet kimwipes works well for this purpose). Incubate overnight at 4°.

8. Rinse slides in PBS to remove coverslips. Wash slides three times for 5 min each in PBS and twice for 15 min each in PBS-TB.

9. Dilute secondary antibody in PBS-TB according to the manufacturer's recommendations. For best results, the titer of the secondary antibody should be determined empirically. Place 20 μl diluted secondary antibody on a coverslip. Dry the slide and pick up the coverslip as before. Place in a humid chamber in the dark at room temperature for 1 h.

10. Rinse slides in PBS to remove the coverslips and wash three times for 5 min each in PBS.

11. If desired, place 20 μl of 0.05 μg/ml DAPI dissolved in 2× SSC (0.3 M NaCl, 0.03 M sodium citrate, 0.1 mM EDTA) on a coverslip

and pick up with the slide as before. Incubate 4 min at room temperature.

12. Wash slides three times for 5 min each in PBS. Place slides on 20 μl of PBS containing 80% glycerol and 2.5% *n*-propylgallate on a coverslip. Place slides face down on paper towels and gently press with fingers to remove excess mountant. Seal with nail polish if desired. Slides can be stored lying flat in the dark at 4°.

13. Visualize with a fluorescent microscope using appropriate filters.

Several controls can be used to confirm that the results obtained using the above protocol accurately reflect the chromosomal distribution of the protein of interest. Whenever possible, affinity-purified or monoclonal antibodies should be used and the specificity of the primary antibody should be confirmed by Western blotting. The specificity of the secondary antibody should be confirmed by omitting the primary antibody. These controls are critical because the quality of the results obtained using the above protocol is highly dependent on the quality of the primary and secondary antibodies. If possible, one should compare results obtained using antibodies against different epitopes of the same protein, or different subunits of the same complex. The ideal control is to show that the staining pattern is reduced or eliminated in larvae homozygous for a null allele of the gene of interest. However, this is not feasible for most chromatin-associated proteins, including those that are essential for early development.

Staining Polytene Chromosomes with Two Antibodies

The distribution of a chromatin-remodeling factor relative to RNA polymerase II (Pol II) and other proteins can be quite informative. For example, the chromosomal distribution of a chromatin-remodeling factor that plays a general role in facilitating transcription might be similar to that of Pol II, as has been observed for the BRM complex.[14] Antibodies specific to the paused, hypophosphorylated form of RNA Pol II, as well as the hyperphosphorylated, elongating forms of Pol II can be used to clarify whether a protein of interest facilitates a specific step of transcription. The comparison of chromatin-associated factors to specific forms of modified histones can also provide insights concerning their interactions *in vivo*.[40,41] Note that antibodies directed against modified histones (and other basic proteins) require specific staining protocols.[16,42,43]

[40] L. A. Pile and D. A. Wassarman, *EMBO J.* **19,** 6131 (2000).
[41] S. A. Jacobs, S. D. Taverna, Y. Zhang, S. D. Briggs, J. Li, J. C. Eissenberg, C. D. Allis, and S. Khorasanizadeh, *EMBO J.* **20,** 5232 (2001).

Assuming that the necessary immunological reagents are available, it is relatively straightforward to stain chromosomes with two or more antibodies. Polytene chromosomes can be stained with a mixture of primary antibodies raised in different species, followed by an appropriate combination of species-specific secondary antibodies. As controls, each of the primary antibodies must be omitted in turn to ensure that the secondary antibodies do not cross react with inappropriate primary antibodies.

Other approaches must be used in situations where the primary antibodies were generated in the same species. Primary antibodies can be covalently coupled to fluorochromes, thereby eliminating the need for secondary antibodies.[44] We have used this approach to compare the distributions of ISWI and GAGA factor on salivary gland polytene chromosomes.[16] One limitation of direct labeling is that it requires relatively large amounts of primary antibody. This technique often yields suboptimal results due to the loss of signal amplification resulting from the omission of secondary antibodies. Direct labeling is therefore best suited to robust antibodies against relatively abundant proteins.

Another technique utilizes Fab fragments to allow the comparison of the chromosomal distributions of two proteins, even when the relevant primary antibodies were raised in the same species and are present in limited quantities. In this technique, Fab fragments are used to both label and block the first primary antibody. We have used the following protocol (adapted from a method suggested by Jackson ImmunoResearch Laboratories, West Grove, PA) to stain chromosomes with rabbit antibodies directed against BRM and Polycomb.[14]

Using Fab Fragments to Stain Polytene Chromosomes with Two Primary Antibodies Raised in Rabbits

1. Prepare, wash, and block squashes of salivary gland chromosomes as described earlier.
2. Dilute first primary antibody with PBS-TB and place on slide as described earlier. It is best to choose the antibody with the fewest binding sites to be the first primary antibody. Incubate at room temperature for 1 h.

[42] J. Fang, Q. Feng, C. S. Ketel, H. Wang, R. Cao, L. Xia, H. Erdjument-Bromage, P. Tempst, J. A. Simon, and Y. Zhang, *Curr. Biol.* **12,** 1086 (2002).

[43] K. Nishioka, J. C. Rice, K. Sarma, H. Erdjument-Bromage, J. Werner, Y. Wang, S. Chuikov, P. Valenzuela, P. Tempst, R. Steward, J. T. Lis, C. D. Allis, and D. Reinberg, *Mol. Cell* **9,** 1201 (2002).

[44] E. Harlow and D. Lane, *in* "Antibodies a Laboratory Manual," p. 319. Cold Spring Harbor Laboratory, Cold Spring Harbor, NY, 1998.

3. Wash slides three times for 5 min each in PBS, two times for 15 min each in PBS-TB. Add biotin-labeled goat antirabbit Fab fragments (Jackson ImmunoResearch Laboratories, West Grove, PA) diluted 1:200 in PBS-TB for 1 h at room temperature. This step is necessary to both label the primary antibody and to block the binding of other antibodies to the primary antibody during subsequent steps.

4. Wash slides twice for 10 min each in PBS-TB. Incubate with fluorescently labeled streptavidin (Jackson ImmunoResearch Laboratories, West Grove, PA) diluted 1:400 in PBS-TB for 1 h at room temperature.

5. Wash slides twice for 10 min each in PBS-TB and stain with the second primary antibody and secondary antibody as described earlier.

The second secondary antibody is blocked from binding the first primary antibody by the Fab fragments. Therefore, this protocol will yield spurious results if the Fab fragments fail to completely mask all binding sites on the first primary antibody. As a control, the second primary antibody should be omitted to verify that the fluorescently labeled secondary antibody is unable to bind the first primary antibody following Fab blocking. If needed, unlabeled Fab fragments (Jackson ImmunoResearch Laboratories) can be used to further block the first primary antibody. Following incubation with the biotin-labeled Fab fragments, incubate slides with 20 μg/ml Fab fragments for 1 h at room temperature.

When directly comparing the distribution of two proteins in merged images of polytene chromosomes, high levels of one protein can mask low levels of another. As a result, the staining patterns of proteins with virtually identical distributions may appear quite different. This problem can be circumvented by generating "split" images of polytene chromosomes using Adobe Photoshop software as shown in Fig. 4.

Analysis of Chromosome Defects Resulting from the Loss of Chromatin-Remodeling Factor Function

The analysis of polytene chromosomes has revealed unanticipated roles for chromatin-remodeling factors *in vivo*. For example, the global architecture of the male X chromosome is dramatically altered in *ISWI* mutant larvae, suggesting that this chromatin-remodeling factor is required for the maintenance of higher order chromatin structure.[16] It is not always possible to examine polytene chromosomes of homozygous mutant larvae, since many chromatin-remodeling factors are essential for early development. The expression of dominant-negative forms of chromatin-remodeling factors can be used to circumvent this problem. For example, the expression

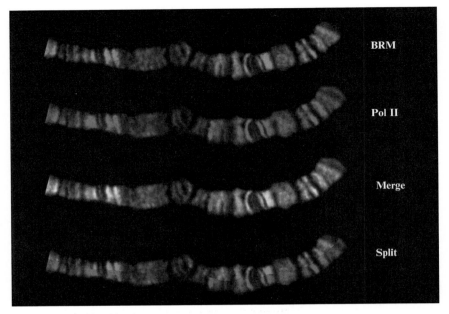

FIG. 4. "Split" chromosomes reveal that BRM and RNA Pol II colocalize on polytene chromosomes. Salivary glands from wild type larvae were stained with rabbit anti-BRM and goat anti-RNA Pol II (subunit IIc).[45] To generate a "split" image in Photoshop, individual images (BRM and IIc) were put in separate layers. The "lasso tool" was used to select half the chromosome arm in the upper layer and delete the selection to reveal the underlying layer. Displaying multiple staining patterns in a "split" format helps avoiding visual artifacts when high levels of one protein can mask low levels of another. (See color insert.)

of dominant-negative BRM protein in salivary gland nuclei decreases the levels of both paused and elongating forms of Pol II associated with salivary gland chromosomes, suggesting that the BRM complex plays a global role in gene expression.[14] To express dominant-negative proteins in salivary gland nuclei, we routinely use either the ey-$GAL4$ driver or the heat shock driver $P[w^{+mC} = GAL4\text{-}Hsp70.PB]89\text{-}2\text{-}1$, grown at $18°$ under nonheat shock conditions. When doing these types of experiments, it is critical to raise the flies at constant temperature in uncrowded conditions. UAS-$LACZ$ or UAS-GFP lines can be used as controls. The control and experimental chromosomes must always be stained in parallel, and the resulting images must be taken with the same camera settings.

[45] A. M. Skantar and A. L. Greenleaf, $Gene\ Expr.$ **5,** 49 (1995).

Although this chapter focuses on methods we have used to study the BRM and ISWI chromatin-remodeling factors, similar approaches can be used to study other chromatin-remodeling and modifying enzymes. Together with biochemical studies, these genetic and cytological methods should facilitate the investigation of the complex regulatory network of chromatin-modulating proteins.

Acknowledgments

We would like to thank Mary Kay Phillips from Jackson ImmunoResearch Laboratories (West Grove, PA), for advice concerning the Fab fragment-based polytene staining protocol. We also would like to thank Dr. Renate Deuring and Vidhya Srinivasan for their useful comments on this manuscript. Most of the stocks described in this article can be obtained from the Bloomington Stock Center or our laboratory. Work in our laboratory is supported by a Grant from the NIH (GM49883) to J. W. T. J. A. A. has been supported by the Damon Runyon Cancer Research Fund (Fellowship DRG-1556) and D. F. V. C. has been supported by EMBO and HFSP postdoctoral fellowships.

[5] Genetic Analysis of H1 Linker Histone Subtypes and Their Functions in Mice

By Yuhong Fan *and* Arthur I. Skoultchi

In most eukaryotic cells, the chromatin fiber consists of nearly one molecule of linker histone for each nucleosome core particle. Therefore, linker histones are expected to play a key role in the structure of the chromatin fiber. A large variety of *in vitro* experiments with chromatin, and other correlative findings, support this view. These studies indicate that two important functions of linker histones are to stabilize the DNA as it enters and exits the core particle and to facilitate the folding of nucleosome arrays into more compact structures. Linker histones also affect nucleosome core particle spacing and mobility *in vitro*.[1,2]

Surprisingly, however, elimination of the linker histone in Tetrahymena, yeast and fungi led to the conclusion that H1 is not essential in these unicellular eukaryotes.[3-7] But double stranded RNA-mediated interference

[1] K. E. van Holde, "Chromatin." Springer-Verlag, New York, 1989.
[2] A. P. Wolffe, "Chromatin: Structure and Function." Academic Press, San Diego, CA, 1998.
[3] X. Shen, L. Yu, J. W. Weir, and M. A. Gorovsky, *Cell* **82,** 47 (1995).
[4] S. C. Ushinsky, H. Bussey, A. A. Ahmed, Y. Wang, J. Friesen, B. A. Williams, and R. K. Storms, *Yeast* **13,** 151 (1997).

Copyright 2004, Elsevier Inc.
All rights reserved.
0076-6879/04 $35.00

(RNAi) of H1.1 in *C. elegans*[8] or antisense-mediated decrease of certain linker histone subtypes in tobacco[9] does lead to effects on germ line development. Delaying synthesis of somatic H1 in *Xenopus* embryos at the midblastula stage with specific ribozymes caused prolonged mesodermal competence.[10] These studies indicate that linker histones do play important roles in chromatin function during development in higher organisms.

In mice there are at least eight H1 subtypes, including the widely expressed somatic H1a through H1e, the testis-specific H1t, the oocyte-specific H1oo, and the replacement linker histone $H1^0$.[11,12] The genes for H1a–H1e and H1t are linked on mouse chromosome 13,[13] whereas the $H1^0$ gene is on mouse chromosome 15 and the H1oo gene is located on mouse chromosome 6. These H1 subtypes differ significantly in amino acid sequences; for example, $H1^0$ is only 30–38% identical to H1a–H1e and even amongst H1a–H1e sequence divergence approaches 40% between certain subtypes.[13] These subtypes also exhibit distinct patterns of expression during development.[11] Much of the differential regulation occurs at the transcription level. The accumulation of $H1^0$ is generally associated with terminal differentiation and terminal cell division.[14] The levels of certain subtypes during development can reach a major fraction of the linker histones present in specific cell types, for example $H1^0$ and H1e constitute 28 and 42%, respectively, of H1 in adult mouse hepatocytes[15] and H1t constitutes 40–50% of H1 in pachytene spermatocytes.[16,17] Many of these features are conserved between mice and humans, including genomic organization, amino acid sequence of corresponding subtypes, developmental

[5] H. G. Patterton, C. C. Landel, D. Landsman, C. L. Peterson, and R. T. Simpson, *J. Biol. Chem.* **273,** 7268 (1998).

[6] J. L. Barra, L. Rhounim, J. L. Rossignol, and G. Faugeron, *Mol. Cell. Biol.* **20,** 61 (2000).

[7] A. Ramon, M. I. Muro-Pastor, C. Scazzocchio, and R. Gonzalez, *Mol. Microbiol.* **35,** 223 (2000).

[8] M. A. Jedrusik and E. Schulze, *Development* **128,** 1069 (2001).

[9] M. Prymakowska-Bosak, M. R. Przewloka, J. Slusarczyk, M. Kuras, J. Lichota, B. Kilianczyk, and A. Jerzmanowski, *Plant Cell* **11,** 2317 (1999).

[10] O. C. Steinbach, A. P. Wolffe, and R. A. Rupp, *Nature* **389,** 395 (1997).

[11] R. W. Lennox and L. H. Cohen, *J. Biol. Chem.* **258,** 262 (1983).

[12] M. Tanaka, J. D. Hennebold, J. Macfarlane, and E. Y. Adashi, *Development* **128,** 655 (2001).

[13] Z. F. Wang, T. Krasikov, M. R. Frey, J. Wang, A. G. Matera, and W. F. Marzluff, *Genome. Res.* **6,** 688 (1996).

[14] J. Zlatanova and D. Doenecke, *FASEB J.* **8,** 1260 (1994).

[15] A. M. Sirotkin, W. Edelmann, G. Cheng, A. Klein-Szanto, R. Kucherlapati, and A. I. Skoultchi, *Proc. Natl. Acad. Sci. USA* **92,** 6434 (1995).

[16] M. L. Meistrich, L. R. Bucci, P. K. Trostle-Weige, and W. A. Brock, *Dev. Biol.* **112,** 230 (1985).

[17] R. W. Lennox and L. H. Cohen, *Dev. Biol.* **103,** 80 (1984).

regulation where studied, and even DNA sequences in regions of the mouse and human corresponding genes, such as the 5' and 3' untranslated regions and the promoters, that differ even more amongst the genes for the subtypes in one species.[18] This high level of conservation in mammalian evolution also suggests important roles for these subtypes in fine tuning chromatin structure.

The development of techniques for inactivating genes in murine embryonic stem (ES) cells and for producing mice from such modified ES cells provides an approach for undertaking a genetic analysis of H1 linker histones in mammals. In the following sections we first discuss methods for inactivating the genes for individual members of the H1 linker histone gene family and the procedures for analyzing the composition of chromatin from mice null for individual H1 subtypes. Surprisingly, these studies show that, despite the abundance of certain subtypes in specific tissues, none of the six individual H1 subtypes eliminated to date is essential for proper mouse development. Chromatin analyses from single H1 null mice show that compensation by upregulation of other H1 subtypes is a likely explanation for the absence of a phenotype in these mice. Nevertheless, by introducing into some of these single H1 null strains, a globin transgene, it is possible to show that specific H1 subtypes do have differential effects on gene expression. Finally we describe methods for producing compound H1 null mice. In certain compound null strains, the compensation among H1 family members is disrupted. These strains have been used to show that: (1) H1 linker histones are essential for mouse development; (2) marked reductions in stoichiometry of total H1 in chromatin are tolerated in mice.[19]

Production of Mice Lacking a Single H1 Subtype

The general procedure for inactivating a gene in mice is gene targeting by homologous recombination in ES cells (see Figs. 1 and 2). First, genomic clones including portions of the gene of interest are isolated from a mouse genomic library. For a gene family consisting of highly related members, such as the H1 histone family, clones containing a specific family member(s) can be identified by hybridization with gene-specific flanking region probes or by gene-specific PCR. The use of a genomic library derived from a strain of mice congenic with the ES cell line used is recommended. The advantage of this practice is to increase the homologous recombination

---

[18] Z. F. Wang, A. M. Sirotkin, G. M. Buchold, A. I. Skoultchi, and W. F. Marzluff, *J. Mol. Biol.* **271,** 124 (1997).

[19] Y. Fan, T. Nikitina, E. M. Morin-Kensicki, J. Zhao, T. R. Magnuson, C. L. Woodcock, and A. I. Skoultchi, *Mol. Cell. Biol.* **23,** 4559 (2003).

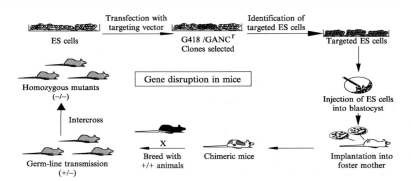

FIG. 1. Procedure of gene disruption in mice.

frequency during gene targeting. A variety of ES cell lines can be used. Many of these lines are derived from the 129J/sv mouse strain, for example, the E14-1 ES cell line,[20] or have a major portion of their genome derived from 129J/sv, for example, the WW6 ES cell line.[21] The use of genomic clones derived from the 129J/sv library increases the frequency of gene targeting by homologous recombination in these lines.[22] We have had particularly good success with the WW6 ES cell line which grows robustly, is easy to maintain, yields efficient germline transmission, and can be used for several rounds of sequenctial gene targetings with different selection markers (see later section). In addition, WW6 cells contain a 11-Mb β-globin transgene on chromosome 3 which can serve as an inert, nuclear-localized marker to identify donor ES-derived cells in chimeric mice using DNA in situ hybridization.[21]

Targeting vectors for gene inactivation are generally constructed containing both positive and negative selection marker genes for isolation of stably transfected ES cell clones.[22] Useful positive selection marker genes include those conferring resistance to G418 (Geneticin, Gibco-BRL), hygromycin (Boehringer Mannhem) or puromycin (Sigma). Transcription of these genes is usually driven by the strong, ubiquitously expressed phosphoglycerate kinase (PGK) promoter. The most widely used negative

[20] R. Fodde, W. Edelmann, K. Yang, C. van Leeuwen, C. Carlson, B. Renault, C. Breukel, E. Alt, M. Lipkin, P. M. Khan et al., Proc. Natl. Acad. Sci. USA 91, 8969 (1994).
[21] E. Ioffe, Y. Liu, M. Bhaumik, F. Poirier, S. M. Factor, and P. Stanley, Proc. Natl. Acad. Sci. USA 92, 7357 (1995).
[22] B. Hogan, R. Beddington, F. Constantini, and E. Lacy, "Manipulating the Mouse Embryo—A laboratory manual." Cold Spring Harbor Laboratory Press, Cold Spring Harbor, New York (1994).

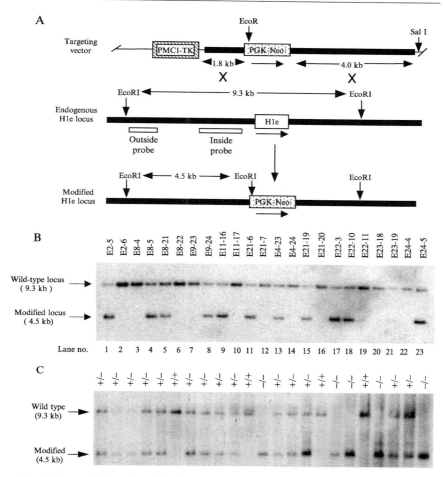

FIG. 2. Targeted disruption of the H1e gene in mouse ES cells and mice. (A) Homologous recombination strategy in ES cells. The H1e targeting vector (top) was constructed by replacing a 720 bp Msc I fragment containing the H1e coding region (open box) along with 49 bp of 5′ noncoding sequence and 11 bp of 3′ noncoding sequence with the PGK-Neo gene. A negative selection gene PMCI-TK was inserted 5′ of the 1.8 kb short arm of genomic DNA. A homologous recombination event (Xs) between the targeting vector and the endogenous H1e locus (middle) results in production of a modified H1e locus (bottom). (B) Identification and confirmation of ES cell clones containing the modified H1e allele. ES cell DNA (10 ug) was digested with EcoRI followed by Southern blot analysis using the 5′ outside probe shown in panel A. The expected position of the hybridizing fragments from the unmodified (wild-type) and modified H1e loci and their respective sizes are indicated. (C) Genotype analysis of offspring from parents heterozygous for the modified H1e allele. Siblings that were heterozygous for the modified H1e allele were bred and 15 ug of tail DNA from offspring were digested with EcoRI, blotted and hybridized with the 5′ inside probe. The deduced genotype of each animal is indicated above each lane. The wild-type and modified H1e loci and their corresponding sizes are indicated. Reproduced with permission from Fan et al.[23]

selection marker gene is Herpes Simplex thymidine kinase (HSV-tk) gene that, when expressed, causes cells to be sensitive to ganciclovir (Syntex). In the vector, the positive selectable marker gene is flanked by two arms of genomic DNA that are homologous to regions upstream and downstream of the gene to be inactivated. The two arms of genomic DNA in the vector may include none of the histone coding sequence, in which case, homologous recombination leads to complete deletion of the gene at the modified locus, or it may include a portion of the gene in which case an incomplete gene remains at the modified locus. Design of the targeting vectors for most H1 histone genes is facilitated by the fact that, except for the H1oo gene, none of the mouse H1 genes contain introns and their coding regions are small, ranging in size from 579 (H1^0) to 660 bp (H1d).[12,18] The negative selection marker gene (PMCI-TK in Fig. 2A) is placed either upstream or downstream of the regions of homology. Selection for loss of expression of this marker (with ganciclovir) increases the frequency of homologous recombinants among antibiotic resistant clones.[24]

Figure 2 shows an example of a gene targeting experiment at the H1e locus. To construct the H1e targeting vector, a positive selection marker gene, PGK-Neo, was flanked by two arms of genomic DNA that are homologous to the regions of the 5′ and 3′ region of H1e gene. The HSV-tk gene was inserted upstream of the 5′ homology region for negative selection. Homologous recombination between the targeting vector and the endogenous H1e locus will result in a modified locus in which a segment from 49 bp 5′ of the ATG codon to 11 bp 3′ of the stop codon in the H1e gene is replaced with the PGK-Neo gene (see Fig. 2A).

In the later sections we provide the protocol we have used to inactivate several H1 linker histone genes. Readers are referred to gene targeting manuals for even more detailed procedures.[22,24,25]

Methods for ES Cell Transfection and Selection

1. ES cells are grown on mitotically inactive feeder layers of G418^r SNL fibroblasts and cultured with ES cell medium supplemented with leukemia inhibitory factor (LIF) (ESGRO, Chemicon) at 1000 units/ml. (Note: In the sequential gene targeting experiments [see later], G418^r Hygro^r SNLH or G418^r Puro^r SNLP fibroblasts were used in subsequent targeting steps accordingly.)

[23] Y. Fan, A. Sirotkin, R. G. Russell, J. Ayala, and A. I. Skoultchi, *Mol. Cell. Biol.* **21,** 7933 (2001).

[24] R. Ramirez-Solis, A. C. Davis, and A. Bradley, *Methods Enzymol.* **225,** 855 (1993).

[25] E. J. Robertson, "Teratocarcinomas and Embryonic Stem Cells." IRL, Oxford, 1987.

2. Feed ES cells with fresh ES cell medium 4 h before harvesting them for electroporation.

3. Twenty-five micrograms of SalI linearized targeting vector is electroporated into 2×10^7 ES cells at 400 V and 250 μF using a Bio-Rad Gene Pulser. The electroporated ES cells are subsequently plated onto five 10-cm plates that were coated with inactivated SNL cells.

4. Twenty-four hours after electroporation, change the ES cell medium to ES cell medium containing G418 (200 μg/ml). Forty-eight hours after application of G418, change the culture medium to ES medium containing both G418 (200 μg/ml) and ganciclovir (2 μM). Refeed cells when the medium begins to become acidic, usually daily for the first five days and every two days afterwards.

5. Ten days after transfection, ES cell colonies resistant to G418 and ganciclovir are picked, trypsinized and cultured in feeder cell coated 24-well plates. After three days, cells from each well are trypsinized and split into two 24-well plates (one plate is treated with gelatin, and the other plate is coated with inactivated feeder cells). The cells grown on the gelatin-coated plate can be grown to confluence and used for extraction of DNA. The absence of feeder cells allows for detection of targeting events in the ES cells without interference by feeder cell DNA. After 2–3 days, the feeder cell-coated plate is frozen by changing the medium to ES cell freezing medium (ES medium with 10% v/v DMSO). The plate is then wrapped with foil, placed in a styrofoam box and deposited into a $-80°$ freezer. After determining which ES cell clones have undergone gene targeting, these clones can be recovered by thawing the plate and culturing cells from the appropriate wells.

Preparation of ES Cell DNA for Screening Clones

This method is modified from Laird et al.[26]

1. Allow the cells on the gelatinized plate to grow to confluence, which usually takes four days. During this period, change the medium daily with G418 selection only.

2. Aspirate the medium from the wells, rinse the well with PBS once, and fully aspirate the PBS at room temperature. There is no need to change tips during aspiration from one well to the other.

3. Add 500 μl of ES cell lysis buffer (100 mM Tris pH 8.5, 5 mM EDTA, 0.2% SDS, 200 mM NaCl, 100 μg/ml Proteinase K). Incubate at $37°$ for 3 h with agitation.

[26] P. W. Laird, A. Zijderveld, K. Linders, M. A. Rudnicki, R. Jaenisch, and A. Berns, *Nucleic Acids Res.* **19,** 4293 (1991).

4. Add one volume (500 μl) of isopropanol to each lysate.

5. Swirl a glass capillary sealed at one end in the well until the genomic DNA precipitates and wraps around the capillary. Dip the capillary in a beaker containing 75% ethanol and place it with the sealed end up in an Eppendorf tube to air dry.

6. After isolating DNA from one plate of clones, reverse the capillary so that now the end containing the DNA is inside the Eppendorf tube. Use a diamond-tipped pen to cut the capillary near the mouth of the tube, add 200 μl of 10 mM Tris pH 8.0 and allow the DNA to dissolve overnight with agitation at 55°.

In the gene targeting experiment described in Fig. 2, genomic DNA from 240 individual G418/ganciclovir-resistant colonies was prepared, pooled (2 colonies/pool), digested with EcoRI, and screened for homologous recombination events by Southern blot analysis (see Fig. 2B) with a 0.9 kb 5′ flanking region probe (Outside Probe—Fig. 2A). 11 clones gave rise to a 4.5-kb EcoRI hybridizing fragment expected from the modified allele (see Fig. 2B).

Four clones containing the modified H1e locus were injected into C57BL/6 recipient blastocysts, and the blastocysts were transferred to CD-1 pseudopregnant females to generate chimeric mice (these steps are usually carried out by highly trained personnel in a gene targeting core facility). Two cell lines generated chimeras ranging in chimerism between 70 and 100% based on coat color. Male chimeras from these two cell lines were mated with C57B1/6 mice. Both cell lines successfully transmitted the modified H1e allele through the germline. Male and female chimeras from each line were then mated with C57B1/6 mice. ES cells derived from the 129J/sv strain of mice contain the dominant agouti coat color gene, and the recipient C57B16 mice has a black coat color. Therefore, if the injected ES cells are able to contribute to the germline, the F1 progeny of the chimeric mice will have an agouti coat color. Generally, both agouti mice and black mice are present in the F1 progeny. The agouti mice among the F1 progeny were selected and their tail DNAs were extracted as described earlier.[22] Southern blot analyses were used to verify the transmission of the modified allele through the germline of chimeric mice produced from correctly targeted ES cell clones.

Mice heterozygous for H1e gene mutation were phenotypically normal.[23] These heterozygous mice were interbred to generate H1e homozygous mutant mice. Southern blot analyses (see Fig. 2C) or PCR assays were used to genotype F2 progeny using mouse tail DNA. Among 48 F2 progeny, 10 (21%) carried only the wild-type allele, 24 (50%) carried one copy of the modified allele and 14 (29%) carried two copies of the modified allele. The ratio of the three classes of animals, is not significantly different from the expected values for Mendelian transmission of the two alleles.

To confirm that the H1e gene targeting produced a null allele, total histone extracts from livers of F2 mice were analyzed by reverse-phase HPLC (see procedures later) (see Fig. 3A).[15,27–29] As shown in the figure, this method resolves $H1^0$ and the five somatic H1s. H1d and H1e elute as a single peak in the chromatogram, but by collecting this peak and subjecting it to TOF-MS analysis, these two subtypes also can be quantified separately. Whereas H1e was readily detected in extracts of wild-type animals, it was not detectable in extracts from homozygous mutant animals (see Fig. 3A).

Method for Preparation of Total Histones

All operations are done at $4°$, except as indicated.

1. Dissect tissue, rinse in PBS, mince into small pieces with a razor blade, suspend in sucrose buffer (0.3 M Sucrose, 15 mM NaCl, 10 mM HEPES [pH 7.9], 2 mM EDTA, 0.5 mM PMSF [added fresh]) at 1 g tissue per 10 ml sucrose buffer.

2. Homogenize in a Dounce homogenizer (B pestle), 10–15 strokes.

3. Transfer the homogenated tissue to 15 ml conical tubes, centrifuge at 500 rpm for 30 s. (Centrifuge model: Eppendorf 5810R)

4. Carefully transfer the supernatant to a fresh tube (discard the pellet) and centrifuge at 2000 rpm, 5 min.

5. Discard the supernatant; resuspend the pellet in 10 ml sucrose buffer + 0.5% NP-40 and homogenize with a B pestle, 10 strokes.

6. Centrifuge 2000 rpm for 5 min and discard supernatant.

7. Resuspend pellet together in high salt buffer at 3 ml per 1 g tissue, and homogenize in a small Dounce homogenizer, 10 strokes [high salt buffer: 0.35 M KCl, 10 mM Tris (7.2), 5 mM $MgCl_2$, 0.5 mM PMSF (added fresh)].

8. Transfer to a 1.5 ml microcentrifuge tube and let stand on ice for 20 min.

9. Centrifuge at 14,000 rpm for 10 min and discard supernatant.

10. Resuspend each pellet in 0.8 ml 0.2 N H_2SO_4 by using an Eppendorf tube pestle to grind the pellet well, and let stand at $4°$ for overnight.

[27] D. T. Brown and D. B. Sittman, *J. Biol. Chem.* **268,** 713 (1993).

[28] W. Helliger, H. Lindner, O. Grubl-Knosp, and B. Puschendorf, *Biochem. J.* **288,** 747 (1992).

[29] H. Lindner, W. Helliger, A. Dirschlmayer, M. Jaquemar, and B. Puschendorf, *Biochem. J.* **283,** 467 (1992).

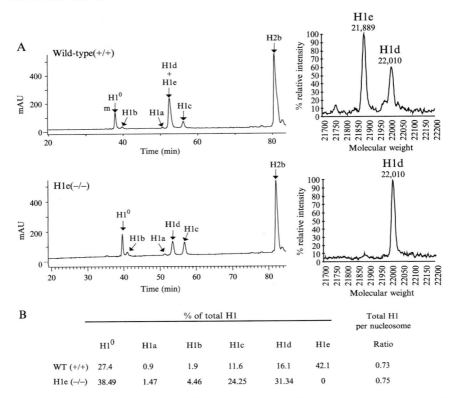

FIG. 3. Analysis of chromatin from H1e knockout mice. (A) Analysis of histones extracted from livers of wild-type and H1e homozygous mutant mice. Panels on the left show reverse phase HPLC analyses of approximately 100 μg of total liver histone extracts from a 20-week-old wild-type (upper panel) and a H1e homozygous mutant (lower panel). The abscissa represents the elution time and the ordinate represents absorbency at 214 Å. The identity of the histone subtype(s) in each peak is indicated. Panels on the right show TOF-MS analysis of a fraction eluting between 52 and 54 min (corresponding to the peak marked H1d + H1e). The identity of the H1d and H1e subtypes detected in this analysis was shown earlier.[18] (B) H1 subtype composition of liver chromatin from 5-month-old wild-type and H1e mutant mice. Data were calculated from HPLC analyses of wild-type and H1e mutant strains like that shown in (B). The percentage of total H1 was determined by the ratio of the $\overline{A}_{214}$ of the indicated H1 peak to the total $\overline{A}_{214}$ of all H1 peaks. Total H1 per nucleosome was determined by the ratio of the total $\overline{A}_{214}$ of all H1 peaks to half of the $\overline{A}_{214}$ of the H2b peak. The $\overline{A}_{214}$ values of the individual H1 peaks and the H2b peak were adjusted to account for the differences in the number of peptide bonds in each H1 subtype and H2b. (Fig. 3A is adapted with permission from Fan et al.[23])

11. Centrifuge at 14,000 rpm for 10 min, transfer the supernatant to two 1.5 ml microcentrifuge tubes (400 μl each).

12. Add 2.5 volumes of ice cold ethanol directly to each tube and leave at $-20°$ overnight.

13. Centrifuge at 14,000 rpm for 10 min and wash the pellet 3 times with 70% EtOH. Air dry about 20–30 min or longer as necessary or dry for 3–5 min in a speed-vac centrifuge. The dried protein can be stored at $-20°$ for up to 12 months.

Methods for HPLC Analysis of Histone Proteins

1. 50–100 μg of dried total histone preparations are resuspended in 1 ml H_2O/0.1% TFA and injected into a Vydac 300A, 5 mm C18 reverse phase column (0.46 cm × 25 cm) on a Hewlett-Packard 1090 HPLC system.

2. Total histone proteins are fractionated by an increasing acetonitrile gradient as follows:

Time (min)	H_2O/0.1%TFA (%)	Acetonitrile/0.1%TFA (%)
0.01	100.0	0.0
1.00	95.0	5.0
11.0	75.0	25.0
26.0	70.0	30.0
45.0	65.0	35.0
66.0	60.0	40.0
75.0	57.0	43.0
126.0	45.0	55.0
131.0	10.0	90.0
136.0	95.0	5.0
151.0	100.0	0.0

3. The effluent from the HPLC column is monitored at 214 nm and the peaks are recorded using the Hewlett-Packard 1090 system. Peak areas are determined with a Hewlett-Packard peak integrator program.

Compensation Among Linker Histone Subtypes in Single H1 Null Mice

Using the procedures described in the preceding section, strains of mice that are null for each of six individual H1 subtypes were generated, including the somatic H1s, H1a, H1c, H1d, and H1e, the testis-specific H1t, and the highly variant, replacement linker histone $H1^0$.[15,23,30,31] Each of the six types of single H1 null mice were found to be indistinguishable from their heterozygous and wild-type littermates. Homozygous null mice were fertile and they produced litters of normal size and their progeny appeared normal. Examination of hematoxylin-and-eosin-stained sections from each null strain did not show any pathology or histologic abnormality. No differences were found even in testis or liver in which particular subtypes are especially abundant (H1t 40–50% in pachytene spermatocytes; $H1^0$ and H1e 27% and 42% in liver).

In the absence of an obvious phenotype in knockout mice, it is often assumed that other proteins with similar functions compensate for the protein that has been eliminated. This hypothesis may be examined by generating compound mutants (see a later section). For H1 histones, another approach to investigating this possibility is to examine the content and stoichiometry of the remaining H1 subtypes by reverse phase HPLC. As shown in Fig. 3, HPLC, in combination with TOF-MS, allows quantitation of the relative amounts of $H1^0$ and each of the somatic H1 subtypes, H1a through H1e. H1t, which migrates at about 64 min in the chromatogram also can be quantitated.[30] Furthermore, the method also separates the H1s from the nucleosomal core histones. This allows one to estimate the ratio of linker histones to nucleosome core particles by measuring the total H1 amount compared to one of the core histones, for example, H2b.

The results of this type of analysis carried out with each of the six strains of single H1 null mice showed that the stoichiometry of total linker histones compared to nucleosomes is not altered in the knockout mice. For example, in H1e null mice, despite the fact that H1e accounts for 42% of total H1 in adult liver, the elimination of H1e did not result in a decrease in the H1/nucleosome ratio (see Fig. 3B). Instead, other H1 subtypes, especially $H1^0$, H1c, and H1d, are upregulated in chromatin so as to maintain the ratio in the null mice at the same level as in wild-type mice. This is also true in mice null for highly specialized H1 subtypes, such as the differentiation specific $H1^0$ which account for 27% of H1 in adult liver and the testis specific H1t which is 28% of H1 in testis.[15,23,30]

The mechanism(s) that serve to maintain a normal linker histone to nucleosome stoichiometry in single H1 null mice is not fully understood. One possibility is that excess linker histone molecules are synthesized normally and when the gene for one subtype is inactivated, some of the excess of the remaining subtypes is deposited in chromatin in place of the missing subtype. This explanation is supported by the finding that in several types of single H1 null mice, the relative amounts of the remaining subtypes found

[30] Q. Lin, A. Sirotkin, and A. I. Skoultchi, *Mol. Cell. Biol.* **20,** 2122 (2000).

[31] Q. Lin, Ph.D. thesis: Albert Einstein College of Medicine, Bronx, New York (1998).

in chromatin are nearly the same as their relative amounts in wild-type mice. However, observations made with compound H1 null mice, generated as described later, suggest that increased mRNA synthesis of specific H1 subtypes may also be involved in the feedback or cross-regulation of transcription within the H1 family. Interestingly, this cross-regulation may include the highly variant $H1^0$ gene (YF and AS unpublished observations) which is not located at the histone locus on mouse chromosome 13 and which, unlike most H1 genes, produces polyA + mRNA.

Deletion of Specific H1 Linker Histone Subtypes Affects Expression of Transgenes in Mice

Although compensation within the H1 gene family apparently leads to the absence of a phenotype in single H1 null mice, these strains of mice can be used to examine the effects of specific H1 subtypes on gene regulation *in vivo*. Many studies have shown that expression of transgenes in mice (and other eukaryotes) are subject to position effects. Although the basis for position effects on transgene expression are mostly not understood, they are thought to reflect aspects of chromatin structure. Transgenes are thought to be very sensitive indicators of perturbations in local chromatin structure. To study the effect of H1 subtypes on transgene expression, in collaboration with Dr. Eric Bouhassira and colleagues (Albert Einstein College of Medicine), we bred several different single H1 null mice with a transgenic strain carrying a human β-globin transgene that is subject to age-dependent silencing (see Fig. 4A, B). The 4.4 kb human β-globin transgene contains the entire coding sequence of the β-globin gene plus 2.1 kb of upstream flanking sequences and 0.8 kb of downstream coding sequences. Analysis of transgene expression at the single cell level with flow cytometry using a FITC-labelled monoclonal anti-human β-globin antibody[32] showed that the transgene is silenced due to a gradual decrease in the proportion of expressing cells as the transgenic mice age (see Fig. 4B). This transgene is therefore subject to an age-dependent variegating position effect.

To determine if the silencing of the transgene is modulated by linker histones, we produced by breeding mice hemizygous for the globin transgene and homozygous or heterozygous for deletion of the H1(0), H1a, H1c, H1d, or H1e genes. We then periodically monitored the fraction of red blood cells expressing the transgene and found that the absence of linker histones H1e or H1d strongly attenuated the silencing (see Fig. 4C). Loss of one allele of the H1d gene has less of an effect on attenuating

[32] R. Alami, J. M. Greally, K. Tanimoto, S. Hwang, Y. Q. Feng, J. D. Engel, S. Fiering, and E. E. Bouhassira, *Hum. Mol. Genet.* **9,** 631 (2000).

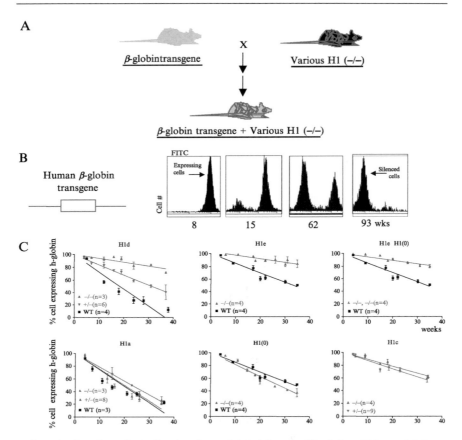

FIG. 4. Age-dependent silencing is modulated by specific linker histone deletions. (A) Scheme of experimental strategy for generating mice containing the human β-globin transgene and deletion of a single H1 subtype. (B) The 4.4 kb human β-globin gene transgene is subject to an age-dependent variegating position effect. Permeabilized RBCs were stained with FITC-labelled anti human-β-globin antibodies and periodically analysed by flow cytometry. The proportion of RBCs expressing the human transgene decreases with age. (C) Linker histone subtypes differentially affect transgene expression. Silencing of the human-β-globin transgene in mice homozygous or heterozygous for deletions of linker histones was monitored over time by flow cytometry and compared to controls. Deletion of H1d, or H1e or (H1e and H1^0) dramatically attenuates the rate of silencing. Deletions of H1a, H1c, and H1^0 has no effect on the rate of silencing. n is the number of mice analyzed. (Fig. 4B, C are reproduced from Alami et al.[33] Copyright National Academy of Sciences, USA.) (See color insert.)

[33] R. Alami, Y. Fan, S. Pack, T. M. Sonbuchner, A. Besse, Q. Lin, J. M. Greally, A. I. Skoultchi, and E. E. Bouhassira, Proc. Natl. Acad. Sci. USA 28, 28 (2003).

transgene silencing than homozygous H1d gene deletion, suggesting that even partial reduction in H1d amount can affect the rate of silencing. Absence of H1a, H1c, or $H1^0$ had no effect on the rate of silencing (see Fig. 4C). Mice lacking both H1e and $H1^0$ silenced at the same rate as mice lacking only H1e, confirming the results obtained with the H1e $-/-$ and $H1^0$ $-/-$ mice.[34]

The results of these studies show that, whereas mice null for individual H1 subtypes do not have apparent phenotypic changes, the absence of specific subtypes in such mice changes the properties of chromatin in subtle ways that lead to quantitative effects on expression of certain genes. These studies illustrate the utility of such H1 null mice for analysing the functions of specific H1 subtypes. In the future, single H1 null mice should be useful to study the role of specific H1s in regulation of other genes and other cellular processes that may be affected or controlled by specific subtypes.

Generation of Compound H1 Knockout Mice

As discussed in the earlier sections, despite the extensive sequence divergence within the H1 family and despite the abundance of certain subtypes in specific tissues, elimination of any one of six different H1 subtypes by gene inactivation does not appear to affect mouse development. The absence of an effect on development is most likely due to compensation by other H1 subtypes. HPLC analysis showed that the chromatin of mice lacking any one of the H1 subtypes has a normal ratio of total H1 linker histones to core histones, and that increased amounts of the remaining H1 subtypes were deposited in chromatin and led to maintainence of a normal linker-to-core histone stoichiometry. When such compensation is present in knockout mice, one way to proceed is to generate compound null mice by combining two or more null mutations in a single strain. As described below, we have used this approach successfully for the H1 gene family.

Histone genes are often located in clusters at specific chromosomal loci. For example, the genes for H1a through H1e and H1t are tightly linked on mouse chromosome 13 interspersed with core histone genes.[18] On the other hand, the solitary $H1^0$ gene is located on mouse chromosome 15 and the H1oo gene is located on mouse chromosome 6.

There are three approaches to making compound knockout mice in which two or more genes have been inactivated. One approach is simply to breed single gene knockout mice and subsequently to intercross the double heterozygous mutants to generate double homozygous mutants.

---

[34] A. Puech, B. Saint-Jore, S. Merscher, R. G. Russell, D. Cherif, H. Sirotkin, H. Xu, S. Factor, R. Kucherlapati, and A. I. Skoultchi, *Proc. Natl. Acad. Sci. USA* **97,** 10090 (2000).

This approach is suitable for generating compound mutant mice for genes that are not linked, such as for the $H1^0$ gene in combination with any of the other H1 genes. However, we found that this approach is not practical for most of the H1 genes on chromosome 13 because they are too tightly linked and thus the recombination frequency is very low. In these cases, a second, but more arduous approach of sequential gene targeting has been successful, as described later. A third approach, that we and others have used at nonhistone loci, is to generate a chromosomal deletion encompassing multiple genes.[34,35] Such deletions can be produced in ES cells or mice by engineering through gene targeting a chromosomal region flanked by loxP sites. The intervening sequences between loxP sites can then be removed by Cre-mediated recombination. Application of this approach to the H1 locus on mouse chromosome 13 is limited, however, because many core histone genes and possibly other genes[13,36] are interspersed among the six H1 genes at the locus. Nevertheless, the deletion approach has been used to attempt to address linker histone functions in chicken DT40 cells.[37–39]

Taking advantage of the fact that $H1^0$ is not located on chromosome 13 where the somatic H1 genes reside, we have used the first approach (breeding strategy) to generate three types of compound H1 mutant mice lacking $H1^0$ and either H1c or H1d or H1e. First, doubly heterozygous mutant mice were produced by breeding $H1^0-/-$ mice with each of strains null for one of the three somatic subtypes. Double mutant heterozygotes of each specific type were then interbred to produce double null mice, namely $H1^0H1c$, $H1^0H1d$, and $H1^0H1e$ homozygous mutants. All three types of double H1 null mice are fertile and appear to be normal.[23] Just as in the analysis of single H1 null mice, here also the chromatin from tissues of these mice had a normal linker histone to nucleosome ratio.[23] This is the case even in the livers of $H1^0H1e$ null mice in which 70% of the H1 normally present has been eliminated. Thus, it seems that there is sufficient excess capacity within the H1 gene family to maintain normal total H1 amounts even when the genes for two abundant subtypes are inactivated. In this circumstance, it is necessary to produce mice with three (or even more) H1 genes inactivated.

[35] B. Zheng, M. Sage, E. A. Sheppeard, V. Jurecic, and A. Bradley, *Mol. Cell. Biol.* **20,** 648 (2000).

[36] W. Albig, B. Drabent, N. Burmester, C. Bode, and D. Doenecke, *J. Cell. Biochem.* **69,** 117 (1998).

[37] Y. Takami and T. Nakayama, *Genes Cells* **2,** 711 (1997).

[38] Y. Takami and T. Nakayama, *Biochim. Biophys. Acta* **1354,** 105 (1997).

[39] Y. Takami, S. Takeda, and T. Nakayama, *J. Mol. Biol.* **265,** 394 (1997).

As mentioned earlier, to inactivate two or more of the six H1 genes on chromosome 13, the second approach of sequential gene targeting strategy is needed (see Fig. 5). The first step in the sequential strategy was the generation of ES cells in which one allele of the H1c gene is replaced by a PGK-Hygro cassette. The second step in the sequential gene inactivation strategy was the inactivation of the H1e gene in H1c +/− ES cells by replacing the H1e coding region with a PGK-Neo cassette (see Fig. 5).

To generate compound knockout mice by sequential gene targeting, it is essential to obtain ES cell clones in which the targeting events have

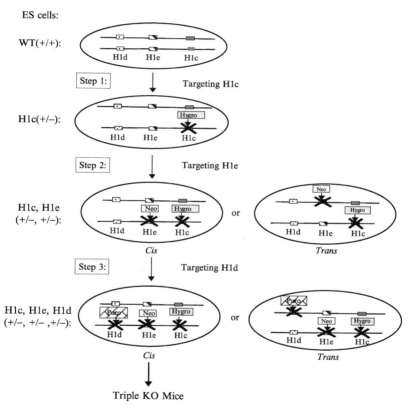

FIG. 5. Strategy for sequential inactivation of three H1 histone genes in mouse ES cells. The figure depicts the chromosome homologues containing the three, linked H1 genes to be targeted and the three steps used for targeting the genes, each with a different selectable marker gene. The figure also depicts the *cis* and *trans* configurations of gene targetings that can occur in Steps 2 and 3. Reproduced from Fan *et al.*[19] with permission from the American Society of Biology.

occurred in *cis* (see Fig. 5). There are two approaches for determining whether the two homologous recombination events occurred in the *cis* or the *trans* configuration. One approach is to use traditional genetic linkage analysis. For example, in the experiment shown in Fig. 5, several doubly targeted (H1c +/−, H1e +/−) ES cell clones were isolated, injected into blastocysts and chimeric mice were obtained from each cell line. The chimeric mice were mated with C57B16 mice and agouti progeny were genotyped by PCR analysis or Southern blot analysis of tail DNA to determine whether the modified H1c and H1e alleles cosegregated. Cosegregation of the H1c and H1e modified alleles indicates that the two targeting events occurred in *cis*, whereas independent segregation of the two modified alleles indicates that they occurred in *trans*.

An alternative approach to genetic linkage analysis is a new application of Expression FISH that we developed to distinguish between *cis* and *trans* gene targeting events in ES cells. The method is based on detection of transcripts from commonly used selectable marker genes inserted during homologous recombination. Nascent transcripts at the sites of transcription of these genes (e.g., PGKHygro and PGKNeo at Step 2 in Fig. 5) are detected in interphase nuclei making the preparation of mitotic cells unnecessary. The nascent transcripts are detected by *in situ* hybridization with specific, fluorescent oligonucleotides. In this way, one can determine the *cis* versus *trans* configuration of the two marker genes inserted during homologous recombination by examining whether or not the two signals colocalize. Compared to genetic linkage testing in chimeric mice, this method can greatly shorten the time required for determining the configuration of two gene targeting events in ES cells. Moreover, by providing such information about the modified ES cells prior to their injection into mouse blastocysts, the method allows for preselecting ES cell clones with a desired configuration. Readers who are interested in this approach are referred to the published procedure.[40]

Using the foregoing procedures, we developed mice null for the H1c and H1e subtypes. Surprisingly, these mice are also apparently normal. Once again, HPLC analysis showed that this strain had a normal H1 to nucleosome ratio in adult liver, indicating that other H1s (mainly $H1^0$ and H1d in liver) could compensate for loss of H1c and H1e. Accordingly, we carried out a third step of gene inactivation in *cis*-doubly targeted H1c, H1e (+, +/−, −) ES cells with a H1d targeting vector in which a PGK-Puro gene replaced the H1d coding region (see Fig. 5). Mice heterozygous for the three, linked mutant alleles were derived from several independent ES cell clones. These mice were intercrossed and their progeny were genotyped by

[40] Y. Fan, S. A. Braut, Q. Lin, R. H. Singer, and A. I. Skoultchi, *Genomics* **91**, 66 (2001).

PCR assays. From of a total of 638 F2 progeny analyzed, no triple, homozygous mutant animals were found.[19] The absence of triple homozygous mutants in litters is not due solely to H1d deficiency, because H1d null mice obtained from three H1d-*trans* targeted ES cell lines develop normally.[23] These results imply that H1 histones are required for proper mouse development. Thus, unlike lower eukaryotes, H1 is essential in mammals.

Genotype analysis of embryos (see Fig. 6A) at various stages of gestation revealed that no homozygous mutants were recovered after E11.5. The homozygous mutant embryos recovered from E8.5 to E11.5 were increasingly underrepresented in litters. They exhibited a wide spectrum of abnormalities, ranging from slight growth retardation to severe growth retardation with additional disruptions in development.[19]

HPLC analysis of relatively normal appearing E10.5 triple mutant embryos showed that the H1-to-core histone ratio in these embryos is nearly 50% of that in wild-type embryos (see Fig. 6B, C). Thus inactivation of the three H1 genes indeed disrupts the compensation observed in single and double mutants.

It is interesting that reduction of total H1 levels in the embryos to about 50% of normal leads to lethality. Perhaps 50% is a critical amount for H1 protein, as it is for other proteins involved in haploinsufficiency syndromes. To explore this question further, we generated two different strains of compound H1 null mice. Mice null for H1c and H1e, but with one wild-type allele of H1d, (H1cH1dH1e [$-, +, -/-, -, -$], were produced by breeding H1cH1e double null animals with H1cH1dH1e triple heterozygous mutants. Mice null for $H1^0$ and H1c and H1e [$H1^0$ ($-/-$)H1cH1e($-, -/-, -$]) were produced by intercrossing $H1^0$($+/-$)H1cH1e($+, +/-, -$) animals. Both types of mutants were very significantly underrepresented in litters, and in both cases, surviving mutants were much smaller shortly after birth and as adults. HPLC analyses of chromatin extracts from these mutants showed that their tissues have 20–50% reductions in the total H1 to nucleosome core ratio (see Table I). In both strains, we found that total H1 stoichiometry is most severely depressed in the thymus, in which the H1 to nucleosome ratio approaches 50% of wild-type levels (see Table I). Interestingly, both strains also have significantly smaller thymi than their wild-type littermates, again suggesting that having more than 50% normal total H1 levels is critical.

Future Prospects

The procedures described above have allowed the generation of both single H1 null mice which have a normal H1 to nucleosome ratio as well as compound H1 mutant mice and embryos, some of which have reduced

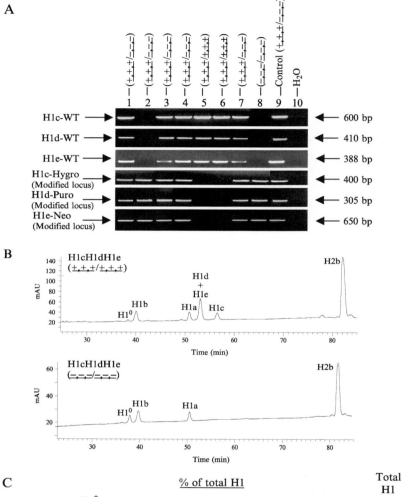

total amounts of linker histones in chromatin, in some cases to levels approaching 50% of normal. It is possible to prepare various types of cultured cells from such animals, for example, ES cells, mouse embryonic fibroblasts, etc., that also can be used to study the role of H1 in processes that are more readily assayed in cell culture. The mice and cells lacking one of the H1 genes should be very useful to further understand how specific subtypes affect specific gene regulation. On the other hand, the compound mutants and cells derived from them can further our understanding of how H1 stoichiometry in general affects gene expression. The first step in this direction will be to use genome wide assays, such as microarray technology, to analyze the gene expression profile changes associated with the loss of H1 or specific H1 variants.

The compound mutants with reduced amounts of H1 also are useful for studying how H1 stoichiometry affects various aspects of global chromosome structure and functions. Much of our knowledge about the role of H1 in the structure of the chromatin fiber is derived from studies in which chromatin has been reconstituted *in vitro* with or without H1. Some of these studies have shown that other basic molecules (e.g., polyamines) can substitute for certain H1 functions.[41] It will be interesting to determine whether or not this is true *in vivo*. Analysis of nucleosome repeat lengths in some of the compound H1 null mice, carried out in collaboration with Dr. Christopher Woodcock (University of Massachusetts, Amherst), shows that reduced H1 content does lead to shortened nucleosome spacing, in agreement with *in vitro* findings.[19]

[41] T. A. Blank and P. B. Becker, *J. Mol. Biol.* **252,** 305 (1995).

FIG. 6. Analysis of chromatin from H1c, H1d, H1e triple null mouse embryos. (A) PCR genotype analysis of E7.5 embryos from intercrosses of H1cH1dH1e $(+, +, +/-, -, -)$ mice. Embryo DNA was prepared and analyzed by PCR assays for H1c, H1d, and H1e wild-type and modified loci.[19] The deduced genotype of each embryo is indicated above each lane. The position of the PCR products from the wild-type and modified alleles are indicated. Control reactions contained tail DNA from a H1c, H1d, H1e triple heterozygous mutant mouse. (B) Reverse-phase HPLC analysis of histones in extracts from E10.5 wild-type and homozygous H1c, H1d, H1e mutant embryos. Approximately 20 ug of total histone extract of chromatin from wild-type (upper panel) and homozygous triple H1c,H1d,H1e mutant (lower panel) 10.5 d.p.c embryos were fractionated by reverse phase HPLC. Other details are as in the legend to Fig. 2B. (C) H1 subtype composition of chromatin from wild-type and H1cH1dH1e $(-, -, -/ -, -, -)$ 10.5 dpc embryos. Data were calculated from HPLC analyses of wild-type and H1c, H1d, H1e triple mutant embryos like that shown in (B). Reproduced from Fan *et al.*[19] with permission from the American Society of Biology.

TABLE I

H1 Subtype Composition and Total H1 Stoichiometry per Nucleosome in Wild-type and H1 Compound Knockout Mice[@]

Tissue		Genotype	% of Total H1[*]							Total H1 per Nucleosome[#]
			$H1^0$	H1a	H1b	H1c	H1d	H1e		
Thymus		Wild-type (+/+)	1.9 ± 1.6	10.7 ± 1.3	16.5 ± 0.9	26.7 ± 3.5	35.4 ± 2.7	9.6 ± 1.5		$\mathbf{0.83 \pm 0.10, \mathit{n} = 12}$
		$H1^0(-/-)$H1cH1e $(-,-/-,-)$	0	17.0 ± 3.5	28.3 ± 3.0	0	54.7 ± 5.5	0		$\mathbf{0.47 \pm 0.26, \mathit{n} = 8}$
		H1cH1dH1e $(-,+,-/(-,-,-)$	5.9 ± 4.6	25.0 ± 3.0	35.4 ± 3.1	0	31.7 ± 2.6	0		$\mathbf{0.41 \pm 0.16, \mathit{n} = 6}$
Liver	Adult	Wild-type (+/+)	29.0 ± 0.8	1.4 ± 0.6	15.8 ± 3.5	12.5 ± 1.0	14.3 ± 0.6	39.9 ± 2.5		$\mathbf{0.79 \pm 0.07, \mathit{n} = 5}$
	Liver	$H1^0(-/-)$H1cH1e $(-,-/-,-)$	0	2.2 ± 0.3	6.3 ± 1.3	0	91.5 ± 1.6	0		$\mathbf{0.64 \pm 0.01, \mathit{n} = 4}$
	Neonatal	Wild-type (+/+)	9.5 ± 4.9	6.1 ± 2.4	14.2 ± 1.7	20.9 ± 4.5	30.3 ± 2.3	19.0 ± 1.5		$\mathbf{0.76 \pm 0.07, \mathit{n} = 5}$
	Liver	H1cH1dH1e $(-,+,-/-,-,-)$	26.7 ± 6.5	9.9 ± 2.2	29.3 ± 4.6	0	32.7 ± 5.2	0		$\mathbf{0.50 \pm 0.15, \mathit{n} = 6}$

Reproduced from Y. Fan, T. Nikitina, E. M. Morin-Kensicki, J. Zhao, T. R. Magnuson, C. L. Woodcock, and A. I. Skoultchi, *Mol. Cell. Biol.*[19] with permission from the American Society of Biology.

[@] Data are from HPLC analyses of wild-type and H1 compound knockout mice. Values are means ± standard deviations of individual determinations made on each of the indicated genotypes.

[*] Determined by ratio of A214 of the indicated H1 peak to total A214 of all H1 peaks. The A214 values of the individual H1 peaks were adjusted to account for the differences in the number of peptide bonds in each H1 subtype.

[#] Determined by the ratio of the total A214 of all H1 peaks to half of the A214 of the H2b peak. The A214 values of the H1 and H2b peaks were adjusted to account for the differences in the number of peptide bonds in each H1 subtype and H2b.

The techniques of histone gene manipulation described in the earlier sections also open the way for the generation of other types of interesting mutant mice. For example, homologous recombination at histone loci can be used not only to delete the gene, but also to replace it with a mutant version of the same gene or different histone variant, for example, another H1 subtype. Providing that important control elements are not present in the replaced coding sequences, this should allow for normal regulation of the substituted histone protein. In this way, for example, it may be possible to study the consequences of expressing highly variant H1 subtypes, such as $H1^0$, H1t, and H1oo, at times or places during development when they are not ordinarily expressed.

As discussed briefly above, methods for creating defined chromosomal deletions are also now available and they could be used to eliminate all of the H1 histone genes on mouse chromosome 13. Based on our studies of the triple H1 knockout embryos, it is expected that homozygous deletion of all of the H1 genes on chromosome 13 will be lethal for the embryo. However, it is not known whether ES cells with such a deletion are viable. Even if they are not, it might be possible to rescue them with a conditional expressed H1 gene. This approach might allow for production of cells and possibly mice expressing only a single H1 subtype, and then mutants of that subtype. Alternative approaches, such as RNA interference, are much more limited when dealing with a multigene family with members that diverge in sequence.

Section II

Histone Modifying Enzymes

[6] The Use of Mass Spectrometry for the Analysis of Histone Modifications

By Tiziana Bonaldi, Jörg T. Regula, and Axel Imhof

Histone Modifications

Traditionally acid urea gels with or without Triton,[1] reversed phase HPLC[2] or specific antibodies were used[3,4] to determine the state of histone modifications in chromatin. All three methods have serious limitations when studying specific modifications. Triton acid urea or acid urea gels and reversed phase HPLC can only provide information on the number of modifications but not on the site. In addition, those techniques cannot be used to study histone methylation, as there is no difference in charge between methylated and unmethylated histone. Specific antibodies in theory could be used to overcome these problems but it turned out to be very difficult to generate antisera with sufficient specificity to distinguish between subtle differences such as mono-, di- or trimethylated lysines or epitopes with similar primary sequence. Furthermore, combinations of histone modifications at the same tail can severely affect antibody binding to its epitope and therefore mimic the loss of a particular modification. The recent developments of soft ionization techniques to study proteins and peptides by mass spectroscopy (MS)[5,6] allow it to study histone modifications directly without the need of generating antibodies against specific modifications.

Mass spectroscopy cannot only be used for the analysis of naturally occurring histone modifications but also for a more detailed analysis of the reaction products of histone modifying or demodifying enzymes. Especially in light of growing evidence for the effects of histone modifications on the generation of additional modifications,[7,8] a rapid and high-resolution technique to study the reaction products of modifying enzymes is needed.

[1] W. M. Bonner, M. H. West, and J. D. Stedman, *Eur. J. Biochem.* **109,** 17 (1980).
[2] K. W. Marvin, P. Yau, and E. M. Bradbury, *J. Biol. Chem.* **265,** 19839 (1990).
[3] T. R. Hebbes, C. H. Turner, A. W. Thorne, and C. Crane-Robinson, *Mol. Immunol.* **26,** 865 (1989).
[4] B. M. Turner, A. J. Birley, and J. Lavender, *Cell* **69,** 375 (1992).
[5] M. Wilm and M. Mann, *Anal. Chem.* **68,** 1 (1996).
[6] M. Wilm, A. Shevchenko, T. Houthaeve, S. Breit, L. Schweigerer, T. Fotsis, and M. Mann, *Nature* **379,** 466 (1996).
[7] W. S. Lo, R. C. Trievel, J. R. Rojas, L. Duggan, J. Y. Hsu, C. D. Allis, R. Marmorstein, and S. L. Berger, *Mol. Cell* **5,** 917 (2000).
[8] P. Cheung, K. G. Tanner, W. L. Cheung, P. Sassone-Corsi, J. M. Denu, and C. D. Allis, *Mol. Cell* **5,** 905 (2000).

Mass Spectrometry

For the analysis of posttranslational modifications (PTMs) of histones we use two different types of mass spectrometers, a MALDI-TOF (Voyager DE STR, Applied Biosystems; see Fig. 1A) and a tandem hybrid quadrupol time of flight (Q-STAR XL, Applied Biosystems; see Fig. 1B). The two mass spectrometers use two different ways to ionize peptides

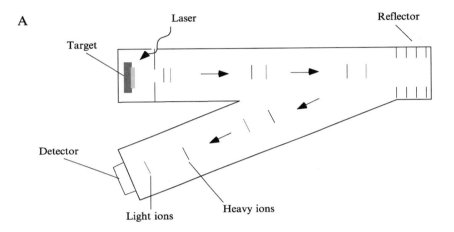

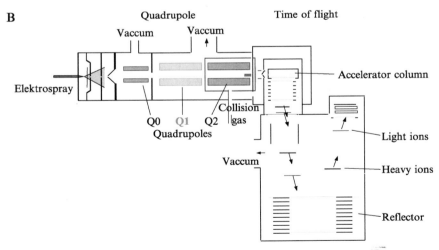

Fig. 1. Schematic drawing of the mass spectrometers used in this study. (A) Reflector MALDI-TOF and (B) ESI Quadrupol-TOF Hybrid. The path of heavy and light ions within the mass spectrometer is depicted by dark and light bars, respectively.

and are mostly supplementary when it comes to the analysis of PTMs. The MALDI-TOF uses a high-energy laser beam aimed at a mixture of small organic molecules (the matrix) and the analyte (the peptides) to ionize the peptides and transfer them into the gas phase. The generated ions are then accelerated by a strong electric field and are allowed to drift through a vaccum tube. The time it takes to reach the detector is proportional to the mass versus charge value of the corresponding peptide as described in Eq. (1)

$$\frac{m}{z} = \frac{2t^2 K}{L^2} \tag{1}$$

Where m is the mass of the ion, z is the charge of the ion, t is the time, K is the constant value representing e (1.602×10^{-19} C) $\times U$ (accelerating voltage), L is the length of the flight tube.

This technique is used to determine the mass of a given molecule with high sensitivity and resolution rapidly. With regards to histone modifications MALDI-TOF can be used to determine, which peptides carry a particular modification and of which nature the modification is. However, because of the limited capability of standard MALDI-TOF instruments to further fragment peptides generated by a proteolytic digest the exact site of modification within a particular peptide is difficult to discriminate.

A potential way to determine the site of modification is the use of electrospray ionization quadrupol-time of flight instruments in which particular peptide ions carrying a modification can be selected and further fragmented in a collision cell. These fragment ions can be analyzed in a similar way as we described easier. This tandem MS/MS analysis has recently been shown to be extremely useful to study posttranslational modifications even in very complex mixtures of peptides. Very often specific product ions can be identified, which are specific for amino acids carrying a particular modification.[9] These ions are then used as markers for posttranslationally modified peptides, thereby dramatically reducing the number of peptides which have to be analyzed. In the case of histones the primary sequence of the peptides and the potential sites of modifications are known and the site of modification can be identified (see Fig. 6).

It is important to mention that in both instruments the peptide coverage is not 100%. However, it is frequently observed that some peptides get more easily ionized by one method or the other and the two techniques are often complementary, resulting in a high percentage of peptide coverage.

[9] M. Wilm, G. Neubauer, and M. Mann, *Anal. Chem.* **68,** 527 (1996).

Isolation of Histones

To study histone modifications, it is useful to compare peptides resulting from a proteolytic digest of histones expressed in *E. coli*, which are not modified with the ones isolated from native sources such as *Drosophila* embryos. This enables the researcher to directly see which peptide is specifically found in the sample of the native sources and therefore potentially modified.

Over Expression of Drosophila Histones in Bacteria

A cDNA for each of the *Drosophila* histones H3, H2A H2B, and H4 has been cloned into a pET3 expression vector, which allow high-level expression in bacteria.[10] The expression plasmids are transformed into electro competent BL21 (pLysS) DE3 cells. The transformation mixture is then plated on LB plates containing ampicillin and chloramphenicol. Between 3 and 10 colonies are used to inoculate a 10 ml preculture containing antibiotics. It is essential to have chloramphenicol present at all media to prevent a leaky expression of histones, which leads to a rapid loss of the expression plasmid and a low protein yield. It is also advisable to grow the preculture for no more than 10 h to avoid overgrowth of the culture.

On the next day 8 ml of the preculture are used to inoculate 400 ml LB containing antibiotics. The culture is incubated for about 3 h with vigorous shaking at $37°$ (rpm > 200) until it reaches a optical density (OD) at 600 nm of 0.5–0.75.

Production of the histones is induced by the addition of IPTG to a final concentration of 1 mM and incubated for another 2.5 h at $37°$. The optical density of the culture is measured and bacteria are spun down for 20 min at 5000 rpm at $4°$. The supernatant is discarded and the bacterial pellet is resuspended in 10 ml of wash buffer (50 mM Tris, pH 7.5, 100 mM NaCl, 1 mM EDTA, 1 mM β-mercaptoethanol, 0.2 mM PMSF) and frozen.

Due to the high level expression of histones, they are present as insoluble particles (inclusion bodies) within the bacteria and can therefore be purified easily by a series of centrifugation steps.

To purify the inclusion bodies bacterial pellets are thawed in a $37°$ water bath. As the bacteria lyse, the solution will become very viscous and is sonicated two times for 1 min with a 1 min break. It is important to avoid frothing or a heating up as this could lead to a fragmentation of the histones. The bacterial lysate is centrifuged for 20 min at 20,000g and the supernatant is discarded. The pellet is then thoroughly resuspended in

---

[10] K. Luger, A. W. Mader, R. K. Richmond, D. F. Sargent, and T. J. Richmond, *Nature* **389,** 251 (1997).

10 ml of wash buffer containing 1% Triton and centrifuged again for 20 min at 20,000g. This wash step is repeated and the pellets are similarly washed two times with wash buffer without Triton. The so purified inclusion bodies have a whitish color and are resuspended in 4 ml unfolding buffer (8 M Urea, 20 mM Tris, pH 7.5; 10 mM DTT) for 60 min with constant stirring. Residual insoluble material is spun down for 20 min at 20,000g and the histones, which are now in the supernatant, are dialyzed over night against SAU200 buffer (7 M urea, 20 mM sodium acetate, pH 5.2, 200 mM NaCl, 5 mM β-mercaptoethanol, 1 mM Na-EDTA) with at least three changes of buffer. The insoluble material is removed by centrifugation for 30 min at 20,000g at 4° (Sorvall SS34). The soluble fraction is filtered through a 0.45μ syringe filter and loaded onto a 5 ml cation exchange column (Hitrap-SP-Sepharose, Amersham) using a ÄKTA FPLC system (Amersham).

Most of the nucleic acids contaminating the preparation will not bind to the column and will be in the flow through fraction whereas histone proteins bind to the resin and can be eluted using a linear gradient from SAU200 to SAU600 (7 M urea, 20 mM sodium acetate, pH 5.2, 600 mM NaCl, 5 mM β-mercaptoethanol, 1 mM Na-EDTA) in six column volumes.

Eluted proteins are checked on 15% SDS-PAGE. Peak fractions containing the histones are pooled and dialyzed O/N against three changes of 11 H_2O. Histone concentration is then checked by Bradford assay and proteins are flash frozen in liquid nitrogen.

Preparation of Bulk Histones from Drosophila Embryos

All buffers are prepared the day before, cooled down and kept at 4°. At the day of use 0.2 mM PMSF and 1 mM DTT are added to all buffers. About 50 g of 0–12 h old *Drosophila* embryos are washed off apple juice agar plates using 0.7% NaCl and dechorionated using 13% bleach. Dechorioneted embryos can be either processed immediately or kept frozen at −80° until processing the sample.

Embryos are resuspended in 40 ml hypotonic buffer (15 mM HEPES-KOH pH 7.6, 10 mM KCl, 5 mM MgCl$_2$, 0.05 mM EDTA, 0.25 mM EGTA, 1 mM DTT, 0.2 mM PMSF, 10% glycerol), homogenized using a Yamamoto homogenisator (1000 rpm; 6 strokes; 4°) and filtered through a single layer of Mira cloth. The embryo lysate is centrifuged for 10 min at 8000 rpm in a HB-4 rotor at 4°. This results in a biphasic pellet with a loose nuclear pellet on top and a tight yellow pellet at the bottom. The supernatant is discarded and the nuclear pellet is carefully resuspended in a final volume of 50 ml of SUC buffer (15 mM HEPES-KOH pH 7.6, 10 mM KCl, 5 mM MgCl$_2$, 0.05 mM EDTA, 0.25 mM EGTA, 1 mM

DTT, 0.2 mM PMSF, 350 mM sucrose). The nuclear pellet is washed twice in 50 ml of SUC buffer and finally resuspended in 30 ml of SUC buffer.

Ninety micro-liters of a 1 M CaCl$_2$ solution is added to the resuspended nuclei and the solution is brought to 26° for 5 min. To release chromatin, 125 μl Micrococcus nuclease (Roche 50 U/μl) is added and incubated for 10 min at 26°. The reaction is then stopped by the addition of 600 μl of 0.5 M EDTA. The nuclease digestion is the critical step in the procedure and the amount of nuclease added as well as the time needed for digestion should be tested in preliminary experiments. To test the efficiency of chromatin release, UV readings of the extracted material (see later) at 254 nm should be taken.

After the digestion nuclei are centrifuged in a HB-4 rotor (Sorvall) for 10 min at 8000 rpm, the nuclear pellet is resuspended in 6 ml TE pH 7.6, 1 mM DTT, 0.2 mM PMSF and rotated for 30–45 min in the cold room to fully extract the chromatin fragments.

After another centrifugation step (30 min; 12,000 rpm; HB-4 rotor; 4°) the released chromatin can be found in the supernatant and is adjusted to a final concentration of 0.63 M NaCl by adding HP elution buffer [2 M NaCl/ 100 mM K-PO$_4$ (pH 7.2)].

The chromatin is then loaded on a 30 ml hydroxyl apatite column (BIO RAD) preequilibrated in column buffer [0.63 M NaCl/100 mM K-PO$_4$ (pH 7.2)]. A 30 ml hydroxyl apatite column is enough to bind chromatin from up to 100 g of dechorionated embryos, which in turn will result in about 100 mg of pure core histones. After binding, the unbound material is washed off the column until no more material is eluted from the column (at least 10 column volumes). The flow through fraction contains the linker histone H1 and can be stored at −20°.

The intact histone octamer is eluted from the column using 2 M NaCl/ 100 mM K-PO$_4$ (pH 7.2). Fractions containing histones are concentrated with Filtron Macrosep ultrafiltration units with an exclusion limit of 6000–8000 to a concentration of at least 2 mg/ml as measured by Bradford. 100% glycerol is added to a final concentration of 50% and the fractions are stored at −20°.

Mass Spectrometry

To analyze histone modifications in bulk chromatin or within histones modified *in vitro*, it is possible to either compare the molecular mass of intact recombinant histones with the ones found in histones isolated from native sources or to use proteolytic digestion of the histones and then to look at the mass of the resulting peptides. In general we found this comparative approach more reliable than a predictive method, which is more difficult to interpret and leaves more ambiguities when interpreting the results. Intact histones can be analyzed by MALDI-TOF using sinapinic acid as a matrix (see Fig. 3). The information gathered from whole histone analysis from bulk histones is somewhat limited as there is no information about the exact site of modification and in many cases there is a certain uncertainty associated with the interpretation of the data, which does increase with number of modified sites. So is the mass difference between a trimethylated lysine and an unmethylated lysine 42 Da, which is the same for an acetylated, or three monomethylated lysines. The advantage of the measurement of intact histones is the fact that it is easier to distinguish between closely related histone variants such as H3.1 or H3.3, which only differ in a few amino acids and generate very similar peptide mass fingerprints when digested with trypsin.

A better way of identifying and characterizing histone modifications is to proteolytically digest the histones and then analyze the resulting peptides. This procedure is commonly used to identify unknown proteins in so called peptide mass fingerprints (PMF). The traditional way to perform those experiments is by tryptic digestion of the proteins within polyacrylamide gels, the extraction of the peptides and their analysis by MALDI-TOF. The masses of the peptides are then compared with virtual tryptic digests of protein sequences deduced from databases. Depending on the molecular weight of the protein a peptide coverage of 10–30% is usually enough to unambiguously identify a protein. However, to study posttranslational modifications in histones it is essential to be able to detect and analyze the peptides carrying the modifications and to get information about their amino acid sequence. As the histones contain a high percentage of lysine and arginine residues cleavage with trypsin results in a mixture of very small fragments that are difficult to analyze by MS. In order to circumvent this problem, one could either use a different protease, such as V8 or Arg-C, or use condition for the tryptic digest in which not all residues are cleaved.[11,12] We routinely use the latter procedure, which is remarkably reproducible and uses a protease, which is very well characterized in gel digestion experiments (see Fig. 2).

Trypsin Digestion

Histone proteins are separated on 15% SDS-PAGE and stained with a colloidal comassie staining kit (Invitrogen) according to the manufacturers instructions. Small pieces of individual histone bands are then cut out of the

[11] K. Zhang, H. Tang, L. Huang, J. W. Blankenship, P. R. Jones, F. Xiang, P. M. Yau, and A. L. Burlingame, *Anal. Biochem.* **306,** 259 (2002).
[12] K. Zhang, K. E. Williams, L. Huang, P. Yau, J. S. Siino, E. M. Bradbury, P. R. Jones, M. J. Minch, and A. L. Burlingame, *Mol. Cell Proteomics* **1,** 500 (2002).

Fig. 2. (continued)

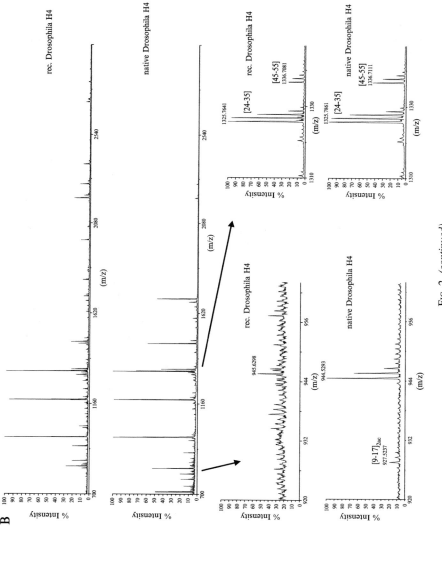

FIG. 2. (continued)

gel with a clean razor blade or a small pipette tip and transferred into a 0.2 ml reaction tube (Nunc). 150 μl of H_2O is added to the isolated gel piece after which it can be stored for several days at $-20°$. For all subsequent procedures it is important to store all solutions in Teflon bottles to prevent the leaking of small contaminants from plastic or glass bottles into the solutions.

After the removal of the H_2O supernatant from the gel piece another 150 μl of water is added and incubated for 30 min at $37°$ on a shaking incubator. Water is removed and the washing step is repeated.

The gel is then incubated in freshly prepared acetonitrile (ACN):H_2O (1:1; v/v) solution and incubated for 15 min at $37°$. If necessary the step of rehydrating/shrinking with ACN/H_2O is repeated until the gel pieces are completely destained.

The gel pieces are then fully dehydrated by the addition of 100 μl acetonitrile and another incubation of 15 min at $37°$. The acetonitrile is removed and the gel pieces are dried thoroughly in a speed vac for 30 min.

FIG. 2. MALDI-TOF spectra of tryptic-peptides form histone H3 and H4. (A) Upper panels: Total spectra form recombinant (top) and native (bottom) *Drosophila* histone H3; the peaks were detected in a mass/charge region between 700 and 3000. The spectra were two-point internally calibrated with auto-proteolytic peptides form trypsin (842.51 and 2211.10). About 5-6 spectra, each collecting 100 laser shots, were acquired at different spots. Database searches using MASCOT led to the identification of the histone H3 in both cases. The peaks present in the lower spectra (native H3) that are not present in the upper one, were further analysed to search for PT modifications; detailed examples are provided in the lower panels. Lower panels: Left—The peak at 1032.61 corresponds to unmodified peptide 41-49 and is present in both spectra (recombinant/native *Drosophila* H3) while peaks 1028.62 and 1070.62 are specific for the native H3 and are in good correspondence with the peptide 18-26 (986.60) carrying 1 and 2 acetylations ($\Delta M = 42.01$ Da) respectively. Right—The peak at 1335.68 corresponds to the unmodified peptide 73-83 with one missed cleavage at Lys79. Two additional peaks, 1349.70 and 1363.72, can be found in the lower spectra (native H3), which show an increased mass of 14 and 28 from the original peak 1335.68, respectively. This mass difference is suggestive of mono- and dimethylation of Lysine 79. The methylation of this lysine has already been detected as a major modification site in histones from many other eukaryotes. Reproduced by permission from F. van Leeuwen, P. R. Gafken, and D. E. Gottschling, *Cell* **109**, 745 (2002); Q. Feng, H. Wang, H. H. Ng, H. Erdjument-Bromage, P. Tempst, K. Struhl, and Y. Zhang, *Curr. Biol.* **12**, 1052 (2002).

(B) Upper panels: Total spectra form recombinant (top) and native (bottom) *Drosophila* histone H4. Spectra were recorded as described above and MASCOT database searches were done to unambiguously identify both proteins as H4. In the *lower right panel* peaks 1325.76 and 1336.71 were identified as unmodified peptides (24-35) and (45-55) of H4. Additional peaks in the spectrum from native H4 were investigated for PT modifications. The *lower left panel* shows the expanded region region of m/z values between 920 and 960: the peak 927.52 corresponds to the H4 peptide (9-17) (calculated mass of 843.54) of H4 carrying two acetyl groups ($2 \times 42.01 = 84.02$).

For the tryptic digestion we use a methylated version of trypsin (Promega) that is largely protected from auto proteolysis. One vial of trypsin (20 μg) is resuspended in 100 μl trypsin reconstitution buffer (Promega), which leads to a final concentration of 200 ng/μl. Reconstituted trypsin is stored in small aliquots at $-20°$. For limited protease digestion the enzyme was further diluted 1:10 in 40 mM Ammonium Bicarbonate to a concentration of 20 ng/μl. The approximate substrate enzyme ratio for the digestion was around 400:1.

To digest the histones 1 μl of Trypsin solution is added directly to the dry gel piece and incubated for 1 min to allow diffusion of the enzyme solution into the gel. Another 5 μl of ice-cold 40 mM ammonium Bicarbonate solution is added to the gel piece and incubated on ice until the it is completely rehydrated (and the concentrated trypsin is fully absorbed by the gel). Another 35 μl of room temperature ammonium bicarbonate is added and the mixture is incubated for 1 h in an orbital shaker at $37°$ at a shaking speed of 180 rpm. To stop the reaction reactions are either frozen at $-20°$ or acidified with 4 μl formic acid. Five microliters of this supernatant can be used directly for the preparation of the MALDI target (see later).

In some cases (like for MS/MS analyses or in cases where the histone concentration is very low) it is necessary to extract peptides more efficiently. Here the entire supernatant is removed, 30 μl ammonium bicarbonate is added to the remaining gel piece and incubated for 15 min on a shaking incubator at a shaking speed of 250 rpm. After centrifugation the supernatant is collected and transferred to a new tube. The gel pieces are then dehydrated with 50 μl ACN for 15 min at $37°$. If necessary this extraction and dehydration step can be repeated once or twice. All supernatants are pooled, dried in a speed vac and kept dry at $-20°$ until usage in nano-ESI or for MALDI target preparation.

Target Preparation

In order to purify the peptides from contaminating salts or acryl amide the peptide solution is passed over a pipette tip containing small amounts of C18 reversed phase material (ZipTip, Millipore). To do this, the ZipTip is first hydrated by repeated pipetting of 10 μl of 50% ACN and then equilibrated with 10 μl of 0.1% trifluoroacetic acid (TFA). Ten microliters of the sample (see above) are then pipetted three times through the ZipTip to bind the peptides to the C18 material. After three 10 μl wash steps with 0.1% TFA, the bound peptides are eluted with 1 μl of prepared matrix solution directly onto the target plate. Samples are air dried to allow cocrystallization of the peptides and the matrix, the target plate is loaded in the MALDI-TOF spectrometer and analyzed.

For the analysis of intact proteins, an alternative, dry droplet protocol is used. The sample is acidified using TFA to a final concentration of about 0.1–0.5% and mixed with the matrix (sinapinic acid, see below) at a 1:1 ratio. One microliter of this mixture is loaded onto the target and air-dried.

Matrix Preparations

Matrices are prepared freshly every day

1. α–cyano-hydroxy-cinnamic-acid (for classical pmf)

One spatula of α-cyano-hydroxy-cinnamic-acid (Sigma) is dissolved in 50% ACN/0.3% TFA, vortexed for 30 s and incubated for 10 min at room temperature. Excess matrix is then centrifuged in a bench top centrifuge at maximum speed for 1 min and the supernatant is used for the samples.

2. sinapinic acid (for total histones MALDI)

Six to ten milligrams of sinapinic acid powder (Fluka) is dissolved in 1 ml of 30% ACN/0.45% TFA. The matrix solution is then vortexed for 30 s and incubated for 10 min at room temperature. Excess matrix is centrifuged in a bench top centrifuge at maximum speed for 1 min and the supernatant is used for the samples

Measurements Using the MALDI-TOF

Mass spectra were recorded on a Applied Biosystems Voyager DE STR mass spectrometer (Applied Biosystems) with delayed extraction operated either in the reflector mode for peptide mass fingerprints and peptides (see Figs. 2 and 5) or the linear mode for whole histones (see Fig. 3) with the following settings:

Accelerating voltage	20 kV (reflector); 25 kV (linear)
Shots/spectrum	100–300
Mass range	700–3500 (reflector); 3000-open (linear)
Low mass gate	500 (reflector); 2500 (linear)

All spectra were two-point internally calibrated with auto-proteolytic peptides from trypsin (842.51 and 2211.10) in the reflector mode or using the predicted mass of the measured recombinant proteins in the linear mode (see Table I).

Sample Preparation for Nano-ESI

Before any desalting step all the samples were acidified with formic acid to a final concentration of 0.2–5%. The sample were desalted using a Zip-Tip (Millipore) approach, rehydrating the ZipTip with 50% ACN/H_2O and

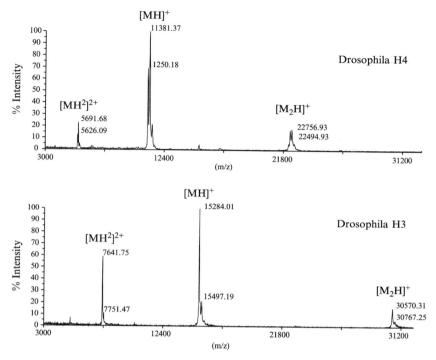

Fig. 3. Maldi-TOF analysis of total histones H3 and H4. Total histone proteins were cocrystallized with sinapinic acid matrix and spectra were acquired in the linear mode. Five spectra were acquired at different spots (each spectrum of 100 shots). The spectra were internally calibrated using the predicted mass of the recombinant proteins (11250.18 and 15284.01 for H4 and H3, respectively). For both *Drosophila* histone H4 (top spectrum) and *Drosophila* histone H3 (bottom spectrum) we detect the singly charged molecule $[MH^+]$, the doubly charged $[MH^2]^{2+}$ and a singly charged dimer of the the molecules $[M_2H]^+$. In addition to the peak corresponding to the unmodified proteins both spectra contain one (H3) or two additional peaks. In naturally occurring histones the N-terminal methionine is cleaved, however this cleavage does not seem to occur on all H4 molecules produced in bacteria. In the histone H4 spectrum, one of the additional peaks observed has an additional mass of 131, which corresponds to a methionine residue.

equilibrating it 5% formic acid before loading the samples. After three washes with 5% formic acid, the peptides are eluted with 5% formic acid/50% methanol directly into a gold-covered needle (Protana), by assembling on a special needle-holder (see Fig. 4) provided by the EMBL. The needle is then connected to the Q-STAR and MS-TOF and MS/MS spectra are acquired using a nano spray voltage of 900 V.

TABLE I

PEPTIDE SEQUENCES IDENTIFIED BY MALDI-TOF MASS SPECTRA[a]

H3 *peptides*

Observed mass (Da)	Calculated mass (Da) (MH+)	ΔM (Da)	Identified peptide (start-end)	Sequence
715.41	715.41	0.00	123-128	DILQR
788.50	788.48	+0.02	64-69	KLPFQR
831.49	831.49	0.00	57-63	STELLIR
900.51	$858.52 + 42.01$ $= 900.53$	−0.02	19-26	$QLATK_{(ac)}AAR$
943.53	$901.52 + 42.01$ $= 943.53$	0.00	9-17	$KSTGGK_{(ac)}APR$
943.53	$901.52 + 42.01$ $= 943.53$	0.00	10-18	$STGGK_{(ac)}APRK$
1028.63*	1028.64	−0.01	65-72	LPFQRLVR
1028.63*	$986.61 + 42.01$ $= 986.62$	+0.01	18-26	$KQLATK_{(ac)}AAR$
1070.62	$986.61 + 2(42.01)$ $= 1070.63$	−0.01	18-26	$K_{(ac)}QLATK_{(ac)}AAR$
1032.61	1032.60	+0.01	41-49	YRPGTVALR
1335.68	1335.69	−0.01	73-83	EIAQDFKTDLR
1349.70	$1335.69 + 14.01$ $= 1349.69$	+0.01	73-83	$EIAQDFK_{(me)}TDLR$
1363.71	$1335.69 + 2(14.01)$ $= 1363.71$	0.00	73-83	$EIAQDFK_{(me)2}TDLR$
1461.85	$1433.83 + 2(14.01)$ $= 1461.85$	0.00	27-40	KSAPATGGVKKPHR + 2me
1489.88	$1433.83 + 4(14.01)$ $= 1489.87$	+0.01	27-40	KSAPATGGVKKPHR + 4me
1503.89	$1433.83 + 5(14.01)$ $= 1503.88$	+0.01	27-40	KSAPATGGVKKPHR + 5me

H4 *peptides*

Observed mass (Da)	Calculated mass (Da) (MH+)	ΔM (Da)	Identified peptide (start-end)	Sequence
927.54	$843.54 + 2(42.01)$ $= 927.54$	0.00	9-17	$GLGK_{(ac)}GGAK_{(ac)}R$
1180.63	1180.62	+0.01	46-55	ISGLIYEETR
1325.76	1325.75	+0.01	24-35	DNGQGITKPAIR
1336.71	1336.72	−0.01	45-55	RISLGIYEETR

[a] Comparison of native bulk histone H3 from *Drosophila* embryos with recombinant unmodified H3 form. Peaks, which could be derived from two different peptides, are underlined and the two possibilities are shown.

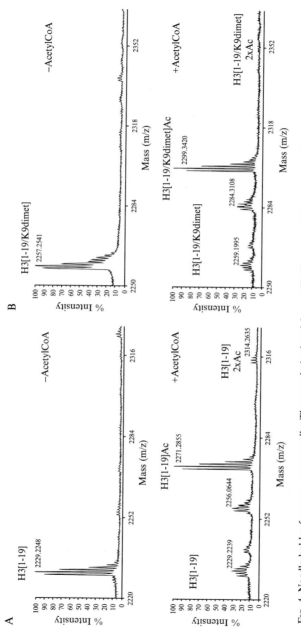

FIG. 4. Needle holder for nano spray needle. The sample is eluted from a ZipTip into a GELloader tip (Eppendorf), which is inserted in a nanospray needle. The assembly is centrifuged for 10 s in a conventional benchtop centrifuge to load the nano spray needle.

In Vitro Modification of Histones to Generate Specifically Modified Histones

In addition to the analysis of naturally occurring histone modifications, MS can also be used to better characterize histone-modifying enzymes with regards to their efficiency and specificity *in vitro* (see Figs. 5 and 6). Many histone acetyltransferases, methyltransferases and kinases can be expressed as active enzymes in heterologous systems. Most of those recombinantly expressed proteins are able to modify chemically synthesized peptides as well as intact proteins with similar specificities.

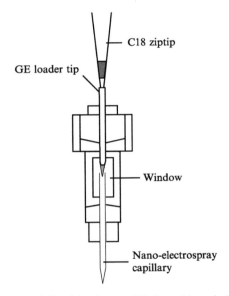

Fig. 5. MALDI-TOF analysis of *in vitro* modified peptides mimicking unmodified and modified histone H3 tails. Peptides (1-19) and (1-19/K9 di-methylated) from histone H3 were *in vitro* acetylated by recombinant purified GCN5. Spectra were acquired in the reflector mode and internally one-point calibrated using the exact mass of the reduced/alkylated peptide as reference for calibration (m/z 2229.22 and 2257.25 for the wt and the dimethylated peptide, respectively). Beween 100 and 400 shots were collected for each spectrum. Acetylation of the peptides was investigated by comparing the upper spectra (negative controls without AcCoA) with the lower ones (products of acetylation reaction), searching for increment of 42.01 (and multiples) in the m/z values of the substrate peaks. Although in both peptides the monoacetylated product is abundant (2271.29 and 2299.34 for wt and dimethylated, respectively), the diacetylated peptide can only be detected in the reaction with the wt peptide (lower left spectrum, m/z = 2213.30), while no obvious diacetylation is detectable in the methylated one (lower right spectrum).

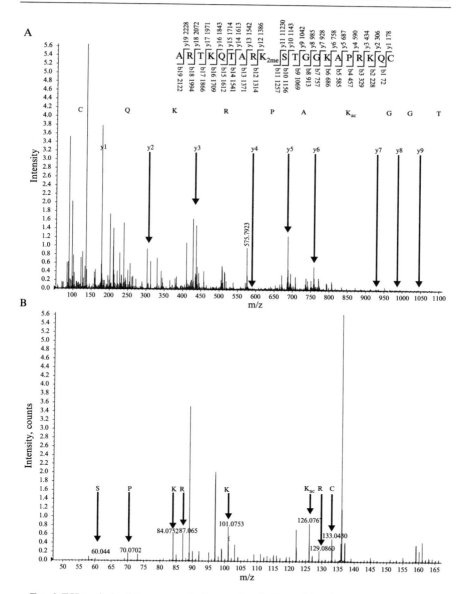

FIG. 6. ESI-analysis of the enzymatically acetylated H3 peptide. The H3 peptide carrying a dimethyl group at lysine 9 was *in vitro* acetylated by gcn5 (Fig. 4B) and then further analyzed by electrospray mass spectrometry. (A) A quadruply charged peptide with a mass/charge value of 575.8 has been fragmented using a collision energy of 47%. The most commonly fragmented bond is the peptide bond and the resulting ions are named with regards to the

Expression of GST-GCN5

An cDNA for yeast GCN5 was cloned into a bacterial expression vector (pGEX-4T-2 [Amersham])[13] allowing it to be expressed in *E. coli* as a fusion protein with glutathione-S-transferase. The expression plasmids are transformed into electro competent BL21 (pLysS) DE3 cells. The transformation mixture is then plated on LB plates containing ampicillin and chloramphenicol. A single colony is used to inoculate a 10 ml preculture containing ampicillin and chloramphenicol. After incubation for 12 h at $37°$ 5 ml of the preculture are used to inoculate 500 ml of LB containing antibiotics. The culture is grown for 3 h at $37°$ until the optical density at 600 nm reaches a value of 0.5 or 0.7. Expression of the protein is induced by the addition of 1 M IPTG to a final concentration of 1 mM and the culture is then incubated for another 3 h at $37°$ and harvested by centrifugation in a Sorvall centrifuge for 20 min at 5000 rpm. The bacterial pellet is resuspended in 10 ml of PBS (1.4 mM NaCl, 27 mM KCl, 100 mM Na_2HPO_4, 18 mM KH_2PO_4, pH 7.3) and flash frozen in liquid nitrogen. After thawing the solution is sonicated two times for 1 min with a 1 min break to lyse the cells and shear the DNA. To the sonified solution Triton X-100 is added to a final concentration of 1%. To completely solubilize the GST-Gcn5 protein the solution is rotated for 30 min on a rotating wheel at $4°$. The insoluble material is centrifuged for 30 min with 1500 rpm at $4°$ in a Sorvall centrifuge and the soluble supernatant is loaded onto a 500 μl Glutathione agarose column. After extensive washing of the column (at least 5 column volumes) bound proteins are eluted in 500 μl elution buffer (50 mM Tris–HCl, pH 8.0, 10 mM reduced glutathione). Eluted proteins are checked for activity and protein content, dialyzed against 31 of BC100 (25 mM HEPES-KOH pH 7.6; 100 mM KCl, 1 mM $MgCl_2$, 0.5 mM EGTA, 1 mM DTT, 0.2 mM PMSF, 10% glycerol) and stored at $-80°$.

---

[13] A. Imhof, X. J. Yang, V. V. Ogryzko, Y. Nakatani, A. P. Wolffe, and H. Ge, *Curr. Biol.* **7,** 689 (1997).

---

termini of the peptide as b-ions (containing the N-terminus) or y-ions (containing the C-terminus). The masses of the complete y and b series for the alkylated peptide is shown inserted into the spectrum. Only the peaks corresponding to the singly charged y-ions are marked in the figure. Other peaks can be assigned to either b, a or z ions or doubly charged fragments ions derived from the same parent ion (575.79; marked in the spectrum). A continous y- and b-series could only be found for an acetylated lysine 14 but not lysine 4 or 18 suggesting that lysine 14 is the main product of the acetylation reaction. (B) Expanded region of the immonium ions produced after fragmentation of the quadruply charged peptides. Note the appearence of a peak at a m/z value of 126, which is diagnostic for an acetylated lysine residue. Reproduced by permission from K. Zhang, H. Tang, L. Huang, J. W. Blankenship, P. R. Jones, F. Xiang, P. M. Yau, and A. L. Burlingame, *Anal. Biochem.* **306,** 259 (2002).

In Vitro *Modification of Peptides*

To analyze the reaction products of histone modifying enzymes it is advisable to first determine the kinetic parameters of the particular enzyme one is interested in. To do this, the easiest and quickest assay is a filter binding assay that was first established by Horiuchi and Fujimoto.[14] In our experience most histone modifying enzymes bind peptides and histones with a high affinity (K_m in the nM range) and have a K_m in the μM range for the small cofactors Acetyl-CoA, S-Adenosyl Methionine and ATP. As the commercially available radioactive precursor molecules usually are of low concentration we add nonradioactive precursors to vary the concentration of the cofactors. Another important aspect to consider when measuring the initial velocity of the reaction in an end-point assay is the fact that product inhibition is an often-observed phenomenon in enzymatic reactions. Therefore, we measure the incorporated radioactivity after 1 min of reaction time to ensure that product inhibition is low. For generation of *in vitro* modified histones and peptides, we use a concentration of at least 5 times the K_m for the corresponding precursor molecule. The salt and pH dependency of histone modifying enzymes can also vary over a large range and should be determined before using MS to analyze the reaction products. In general histone-modifying enzymes work at neutral pH (with the exception of several histone methyltransferases, which have their pH optimum at high pH value[15]) and at a low salt concentration. One should bear in mind, however, that this is an *in vitro* assay and does not necessarily reflect modifications observed *in vivo*.

A standard reaction for the enzyme reaction consists of 50–100 ng of enzyme, 0.1–1 pmol of peptide or histone substrate at a final concentration of 20–100 μM of acetyl-CoA or SAM or ATP (with a specific activity of 0.6 Ci per mmol) in a volume of 41.5 μl. Reactions are incubated for various times at 30° and then either spotted onto P81 paper (when radioactivity was used) or acidified with 4 μl of acetic acid (for MS analysis). The spotted samples are washed in a large volume of 50 mM sodium carbonate buffer (pH 9.2), rinsed with acetone and incorporated radioactivity is measured by scintillation counting.

For MS analysis (see Fig. 5), the acidified samples are loaded directly onto a ZipTip, purified and spotted onto the target plate (see procedure above). Because our peptides contain an additional cystein residue at the C-terminus to enable coupling to a carrier it was necessary to reduce and alkylate the peptides to prevent dimer formation. This was especially

[14] K. Horiuchi and D. Fujimoto, *Anal. Biochem.* **69**, 491 (1975).

[15] X. Zhang, H. Tamaru, S. I. Khan, J. R. Horton, L. J. Keefe, E. U. Selker, and X. Cheng, *Cell* **111**, 117 (2002).

important for elctrospray measurements as this technique somehow favors dimer formation. To do this, peptides are incubated for 15 min at $37°$ in reduction buffer (final concentration of 2 mM DTT) and reduced cystein residues are subsequently alkylated using 20 mM iodine Acetamide in 40 mM ammonium bicarbonate.

For MS/MS analysis triply and quadruply charged ions of the *in vitro* acetylated peptides are selected (see Fig. 6), fragmented and the resulting spectra were analyzed using the BioAnalyst™ software (Applied Biosystems).

Concluding Remarks

Over the last two years mass spectroscopic techniques have become a very versatile tool to identify proteins and to study posttranslational modifications. There is growing evidence that an elaborate code has evolved, which makes use of the combinatorial power of various histone modifications to determine the availability of the genetic information for the cell through its packaging into chromatin. Mass spectrometry will allow us to read this epigenetic code and therefore to better understand the generation and the interpretation of the multitude of posttranslational histone modifications.

Acknowledgments

The authors thank Peter Becker, Gernot Langst, and Alexander Brehm for their support and critical comments on the work. We would also like to thank Gabrielle Mengus for providing the expression constructs for the *Drosophila* histones and Volker Seitz, Ragnhild Eskeland and Birgit Czermin for generating essential reagents. Work in A.I.s lab is funded by grants from the Deutsche Forschungsgemeinschaft (IM23/3-2; IM23/4-1) and the FöFoLe program of the University of Munich.

[7] Histone Modification Patterns During Gene Activation

By Wan-Sheng Lo, Karl W. Henry, Marc F. Schwartz, and Shelley L. Berger

An increasing number of core histone covalent modifications are being identified, including acetylation (Ac), methylation (Me), phosphorylation (P), and ubiquitylation (Ub). The occurrence and role of these modifications in gene-specific activation appears in many cases to be evolutionarily conserved, in organisms ranging from *S. cerevisiae* to humans.

Copyright 2004, Elsevier Inc.
All rights reserved.
0076-6879/04 $35.00

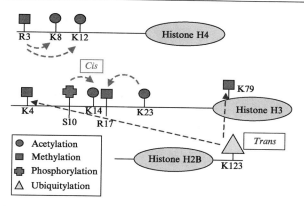

Fig. 1. Schematic display of histone modification patterns. *Cis* relationships are modifications on a single histone tail, that include histone H3 and H4. *Trans* relationships are modification patterns between two tails, that include H2B and H3. The modifications include acetylation, methylation, phosphorylation and ubiquitylation. (See color insert.)

Well-characterized modifications on histone H3 having a role in transcription include Me-Lys-4, P-Ser-10, Ac-Lys-14, and Me-Arg-17.[1]

An important question that has emerged concerns the reason for this plethora of modifications. One answer is that certain modifications occur in a temporal sequence such that an initial modification promotes a second one, leading to a specific pattern. Such patterns support the "histone code" hypothesis,[2,3] in that the sequence and combination of modifications may yield an ordered series of events and more specific outcomes than single modifications. The outcomes may be specific structural changes to the chromatin or to promote interaction with effector proteins. The effector proteins possess specialized domains for interaction with modified tails, such as bromodomain binding to acetyl-lysine.[4] A modification pattern is postulated to provide increased specificity and affinity in such binding, or more dramatic structural changes in the chromatin, compared to single modifications.

Only a few such sequences/patterns have been recognized and they consist of two broad types, namely modifications that occur in *cis* (on the same histone tail) or those that occur in *trans* (on different histone tails) (see Fig. 1).[5] There are three examples of *cis* modifications. Sequences that

[1] S. L. Berger, *Curr. Opin. Genet. Dev.* **12**, 142 (2002).
[2] T. Jenuwein and C. D. Allis, *Science* **293**, 1074 (2001).
[3] B. M. Turner, *Bioessays* **22**, 836 (2000).
[4] R. Marmorstein and S. L. Berger, *Gene* **272**, 1 (2001).
[5] K. W. Henry and S. L. Berger, *Nat. Struct. Biol.* **9**, 565 (2002).

occur on histone H3 include P-Ser-10/Ac-Lys-14,[6,7] and Ac-Lys-18,23/Me-Arg-17.[8] The former relationship has been characterized in both *S. cerevisiae* and in mammalian cells, while the latter has been detected only in mammalian cells. A third *cis* modification sequence occurs on the histone H4 tail in mammalian cells and consists of Me-Arg-3/Ac-Lys-8,12.[9] In each of these cases the sequence of modifications has been revealed initially using *in vitro* tests using purified enzymes that modify the sites and peptides bearing specific modifications. For the H3 tail sequences, the *in vitro* data have been strongly bolstered by analysis *in vivo*, using various techniques to explore the relationship, including the development and use of antibodies that specifically recognize the dual modification.[7] In general, in both *S. cerevisiae* and mammalian cells the timing and pattern can be investigated using the antibodies to detect modifications, as well as altering the amount and activity of the enzymes that catalyze the modifications.[8,10] These examinations are most efficient and revealing in *S. cerevisiae*, where it is possible to delete the endogenous histone genes, providing a source of the genes on plasmids where specific substitution mutations can be installed in the sites to be examined (see Chapter 1) and where the single or dual substitutions can be tested for functional effects.[10]

Trans-tail modification has been found in *S. cerevisiae* occurring between ubiquitylation of the histone H2B tail at Lys-123 and methylation of histone H3, at either Lys-4,[11] in the tail or Lys-79[12] in the globular domain (one of the few currently known modifications outside of the amino or carboxyl tails). This relationship was initially revealed by examination of bulk histones from *S. cerevisiae*, isolated from strains bearing specific substitutions mutations in the histones.

For P-Ser-10 and Ac-Lys-14 there is a good deal of genetic, biochemical, and structural data supporting an interrelationship. This discussion will encompass methods *in vitro* and *in vivo* to investigate a possible functional

[6] W.-S. Lo, R. C. Trievel, J. R. Rojas, L. Duggan, J.-Y. Hsu, C. D. Allis, R. Marmorstein, and L. Berger, *Mol. Cell* **5,** 917 (2000).

[7] P. Cheung, K. G. Tanner, W. L. Cheung, P. Sassone-Corsi, J. M. Denu, and C. D. Allis, *Mol. Cell* **5,** 905 (2000).

[8] S. Daujat, U. M. Bauer, V. Shah, B. Turner, S. Berger, and T. Kouzarides, *Curr. Biol.* **12,** 2090 (2002).

[9] H. Wang, Z. Q. Huang, L. Xia, Q. Feng, H. Erdjument-Bromage, B. D. Strahl, S. D. Briggs, C. D. Allis, J. Wong, P. Tempst, and Y. Zhang, *Science* **293,** 853 (2001).

[10] W.-S. Lo, L. Duggan, N. C. T. Emre, R. Belotserkovskya, W. S. Lane, R. Shiekhattar, and S. L. Berger, *Science* **293,** 1142 (2001).

[11] Z. W. Sun and C. D. Allis, *Nature* **418,** 104 (2002).

[12] S. D. Briggs, T. Xiao, Z. W. Sun, J. A. Caldwell, J. Shabanowitz, D. F. Hunt, C. D. Allis, and B. D. Strahl, *Nature* **418,** 498 (2002).

relationship between two modifications, using P-Ser-10 and Ac-Lys-14 as the principal model for detailed methods, but diverging where appropriate to use other examples of methods that are useful in these analyses.

Alteration of Enzymatic Activity by Prior Modification

Upon discovering that there are multiple modifications associated with a particular process, such as transcription, an initial question is whether these modifications mutually influence enzymatic activity. To investigate this question *in vitro*, it is necessary to prepare three synthetic peptides corresponding to the unmodified sequence and to each modification separately.[6,7] The concentration of these peptides must be measured accurately, in order to directly compare enzymatic activity on each one. Purified enzymes that carry out each modification must be obtained in recombinant form, or preferably, purified in native complexes from the relevant source, since enzyme activity has been found to differ in protein complexes. Once the peptides and enzymes are prepared, then quantitative assays are carried out to determine whether an initial modification alters the activity of the second enzyme.

Synthesis and Maintenance of H3 Peptides

To test whether there is a mutual influence of phosphorylation of Ser-10 and acetylation of Lys-14 *in vitro*, 26 residue peptides were generated, either nonmodified or bearing single modifications at Ser-10, Lys-14 or both modifications (see Fig. 2). Lysines are often used for attachment to activated beads (such as Affi-Gel, Bio-Rad Laboratories), but because lysines are substrate residues for acetylation, cysteine was incorporated at the C-terminus of each peptide to provide means of attachment to beads and other manipulations, as well as their quantitation (see later).

The histone H3 peptides were synthesized and mass spectrometry quality tested by an outside vendor (Protein Chemistry Core Facility, Baylor College of Medicine). The peptides were hydrated in water with trace amounts of TFA as needed for solubility, aliquoted, and stored at $-80°$. Repeated freeze/thaw cycles should be avoided. As the low pH of the peptide solutions may interfere with downstream applications, the peptides were adjusted to pH $\sim$7.5–8 using appropriate amounts of a 1.5 M Tris pH 8.8 stock solution prior to use.

Measurement of Peptide Concentration Using Ellman's Reagent

As described earlier, cysteine was incorporated at the C-terminus of each peptide. The single cysteine can be used for quantitation using DTNB (5,5′-dithio-bis-[2-nitrobenzoic acid]), a versatile water-soluble compound

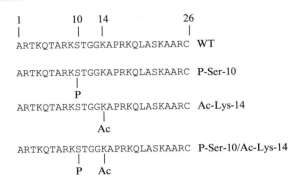

Fig. 2. Histone H3 peptides used in the *in vitro* enzyme assays to investigate relationships between modifications. The peptides encompass residues 1-36, with a cysteine at the C-terminus.

for quantitating free sulfhydryl groups in solution.[13] DTNB reacts with a free sulfhydryl group to yield a mixed disulfide and TNB (2-nitro-5-thobenzoic acid). TNB is yellow and has a high molar extinction coefficient in the visible range.

Methods

1. To generate a standard curve, prepare a serial set of dilutions (from 0.25 mM to 2 mM) of cysteine hydrochloride monohydrate (MW = 175.6) by dissolving into reaction buffer (0.1 M sodium phosphate, pH 8.0). The standards should be prepared immediately before use.
2. Prepare a set of tubes, each containing 2.5 ml of reaction buffer and 50 μl Ellman's Reagent solution (dissolve 4 mg Ellman's Reagent [Pierce] in 1 ml reaction buffer).
3. Mix with 250 μl of cysteine standard or testing peptide and incubate at room temperature for 15 min.
4. Measure absorbance at 412 nm. Calculate the peptide concentrations from the standard curve.

Preparation of Enzymes for Testing Activity In Vitro

There are multiple histone acetyltransferases (HAT) and histone kinases (HK) that modify Lys-14 and Ser-10, respectively, in the H3 amino-terminal tail in the context of transcription.[14,15] HATs include

[13] G. L. Ellman, *Arch Biochem. Biophys.* **82,** 70 (1969).
[14] D. E. Sterner and S. L. Berger, *Microbiol. Mol. Biol. Rev.* **64,** 435 (2000).

members of the Gcn5 family, the CBP/p300 family and some members of the MYST family, such as Esa1.[16] HKs include Rsk2,[17] Msk1,[18] and Snf1.[10]

Modification enzymes can be prepared from recombinant or native sources. The preparation of recombinant HATs, such as Gcn5, is described elsewhere (see Chapter 7 in volume 376) and we have prepared GST-Snf1 by standard GST purification.[10] HATs and HKs as components of native complexes often display different catalytic activities than recombinant enzymes.[10,19] Here we describe our method for preparation of the SAGA and Snf1 complexes.

HAT and HK Complex Purification

Certain HAT and HK complexes in *S. cerevisiae* can be purified based upon fortuitous association with Ni^{2+} beads, followed by MonoQ ion exchange chromatography to separate the individual complexes.[10,19] Further purification has been characterized elsewhere for SAGA[20] and Snf1.[10] The following is an adaptation of this protocol for partial purification of several HAT and HK complexes from yeast, including ADA, NuA3, NuA4, SAGA, and Snf1 complexes.

Materials

Lysis buffer (LyB): 40 mM HEPES, pH 7.5; 350 mM NaCl; 0.1% Tween 20; 10% glycerol.

Imidazole buffer (IB20): 40 mM HEPES, pH 7.5; 20 mM imidazole-HCl, pH 7.0; 100 mM NaCl; 0.1% Tween 20; 10% glycerol. IB300: same as above but 300 mM imidazole-HCl, pH 7.0.

MonoQ A (MQA) 50 mM Tris, pH 8.0, 0.1% Tween 20, 10% glycerol, 0.5 mM DTT, 10 mM $ZnCl_2$, 100 mM NaCl; MQB: same as above, but 1 M NaCl.

Protease inhibitors: 1 mM PMSF, 2 μg/ml leupeptin, 2 μg/ml pepstatin A, 5 μg/ml protinin.

0.5 mm acid washed glass beads and Bead Beater (Biospec Products).

[15] P. Cheung, C. D. Allis, and P. Sassone-Corsi, *Cell* **103**, 263 (2000).

[16] S. Y. Roth, J. M. Denu, and C. D. Allis, *Annu. Rev. Biochem.* **70**, 81 (2001).

[17] P. Sassone-Corsi *et al.*, *Science* **285**, 886 (1999).

[18] S. Thomson *et al.*, *EMBO J.* **18**, 4779 (1999).

[19] P. A. Grant, L. Duggan, J. Côté, S. M. Roberts, J. E. Brownell, R. Candau, R. Ohba, T. Owen-Hughes, C. D. Allis, F. Winston, S. L. Berger, and J. L. Workman, *Genes Dev.* **11**, 1640 (1997).

[20] P. A. Grant, D. Schieltz, M. G. Pray-Grant, D. J. Steger, J. C. Reese, J. R. Yates, III, and J. L. Workman, *Cell* **94**, 45 (1998).

Methods

1. Yeast cells are grown in 6 L of 1% yeast extract, 2% peptone and 2% dextrose extract (YPD) to an optical density of 1.0 to 2.0 at 600 nm. Centrifuge at 5000 $\times g$ for 10 min at $4°$. Wash cell pellet once with distilled water and then LyB. Store frozen cell paste at $-80°$.

2. Resuspend cells in 30 ml LyB and add fresh protease inhibitors. Transfer cells to a chilled Bead Beater (Biospec; 80 ml Bead Beater chamber). Fill with glass beads to the top of chamber and break cells at $4°$ using 15–20 cycles of 30 s bead beating alternating with 1 min rest. The lysate is removed and the beads are rinsed twice with 50 ml LyB and combined with lysate.

3. Crude extract is centrifuged at 17,000 rpm for 30 min in SS34 rotor. Avoid the pellet and pipette the supernatant into a chilled ultracentrifuge tube (Beckmann Ti-60). Spin at 40,000 rpm in Ti-60 for 1 h. Transfer the extract into chilled tube and measure the protein concentration by Bradford assay.

4. Incubate extract with pre-equilibrated Ni-NTA-agarose (Qiagen, 1 ml agarose for 1L yeast culture) 6 h to overnight while rotating at $4°$. Pour the mixture into a column. Collect the flowthrough and rerun over the column once. Wash column with 50 ml of LyB followed by 50 ml of IB20. Elute with 25 ml of IB300.

5. Fractionate complexes over a 1 ml Mono Q 5/5 column (Amersham Pharmacia) using an AKTA FPLC (Amersham Pharmacia). The column is pre-equilibrated with 10 ml MQA. Sample is loaded and unbound material is washed with 10 ml of MQA. Protein is eluted with 100 to 500 mM NaCl gradient (MQA to MQB), collecting 0.5 ml fractions.

6. Fractions corresponding to ADA, NuA4, Snf1, NuA3, and SAGA are determined by HAT or HK assays (described below) on free core histones or nucleosomes. Fractions containing Gcn5 (ADA and SAGA) and Snf1 are confirmed by Western blot. Fractions are aliquoted and frozen at $-80°$.

HAT and HK Assays

Once sources of recombinant or native enzymes have been obtained, then assays are performed on unmodified peptides or peptides bearing modifications. The determination of altered activity requires comparative assays of enzyme activities on different substrates, thus it is essential that the assays are done under conditions of excess substrates where activity is in the linear range with respect to enzyme concentration.

Effect of Phosphorylation on HAT Activity

The enzymatic activity of purified HAT enzymes can be monitored as follows.

$$\text{histone H3} + {}^3\text{H-AcCoA} \rightarrow {}^3\text{H-Ac-histone H3} + \text{CoA}$$

Detailed kinetic analyses of HAT enzymes show that Gcn5 HAT domain binds more tightly to the P-Ser-10 H3 peptide compared to the unmodified H3 peptide.[7] Here we describe a liquid HAT assay followed by filter binding to quantitate overall activity. This relatively simple assay, adapted to different enzymatic activities, has been used to investigate differences in activity on modified substrates.[6–9] Reaction components typically include HAT enzyme sample ($\sim$33 nM), substrates ($\sim$1.5 nmol peptide, 1 mg free histone, or 2 μg of oligonucleosomes), 0.25 μCi of ^{3}H-labeled acetyl coenzyme A cofactor (75 pmol, 250 μCi/ml, 3.5 Ci/mmole [NEN Life Science Products]) and reaction buffer (20 mM Tris, pH 7.6; 50 mM NaCl; 5% glycerol, 10 mM sodium butyrate, 10 μg purified BSA, and 1 mM PMSF).

Materials

Histone H3 peptides dissolved in PBS to 50 mM.

Purified HAT enzymes diluted in PBS, 0.01% BSA (BSA used to stabilize the enzyme). Histone H3 peptides, purified HAT enzymes and enzyme complexes are stored in aliquots at $-80°$ and can be used for several years. However, repeated freeze-thaw cycles should be avoided to prevent precipitation of essential components and loss of activity.

5× HAT buffer (100 mM Tris, pH 7.6; 250 mM NaCl; 25% glycerol).

0.5 M NaHCO$_3$/Na$_2$CO$_3$, pH 8.5.

100% Acetone.

PE81 filter paper (Whatman).

Methods

1. Assemble reagents on ice. A 30 μl HAT reaction contains 6 ml 5× HAT buffer, 3 μl 100 mM sodium butyrate, 3 μl 100 mM PMSF, 1 μl purified 10 mg/ml BSA, 1 μl acetyl [^{3}H]CoA, 250 μCi/ml, 1.5 μl 50 μM peptide stocks (or 1 ml 10 mg/ml free histone). Add appropriate amount enzyme before incubation.
2. Incubate at 30° for 30 min then put on ice to quench the reaction. Stop reactions by spotting onto a square of PE81 filter paper (Whatman).

3. Wash filters four to five times in 50 mM NaHCO$_3$/Na$_2$CO$_3$, pH 8.5 with gentle shaking for 10 min, then wash briefly in acetone and air dry.

4. Place each filter into a scintillation vial with ~4 ml scintillation fluid and count. The nonenzymatic activity is subtracted from each sample.

The assay should be done using a range of enzyme concentrations to be certain that substrate is in excess. Initially, nonmodified peptide is used to establish the linear range. Incorporation should increase commensurately with the amount of enzyme. Time courses are then set up at one enzyme concentration within the linear range, to compare activity on unmodified and modified peptides. Volumes are scaled up into two tubes and only the peptides differ between the reactions. An appropriate amount of enzyme is added and time points are removed at short intervals (usually 30 s to 2 min) and spotted onto filter paper. An example of results of this assay for unmodified H3 tail peptide compared to P-Ser-10-modified peptide are shown in Fig. 3.

If it is observed that a modification is indeed improved by a prior modification, then substitution mutations can be introduced into the second enzyme to try to selectively reduce the stimulation. Predictions can be made using detailed structural data, if available. For example, in the case

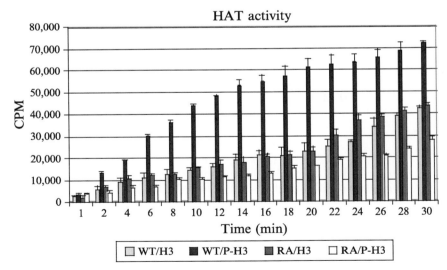

Fig. 3. Gcn5 HAT activity over a time course on unmodified or P-Ser-10 modified histone H3 peptides. WT, HAT domain of Gcn5; RA, Gcn5 HAT domain bearing a substitution mutation at Arg-164. (See color insert.)

of Gcn5, the HAT domain X-ray crystal structure was solved in complex with the H3 tail peptide, and the structure suggested that Arg-164 in the yeast enzyme might contact the phosphorylated form of H3. *In vitro*, Arg-164-Ala (RA) mutations caused a reduction of acetylation on the phosphorylated tail without lowering acetylation on the nonmodified tail (see Fig. 3).[6,7] The mutation was tested *in vivo* and was found, similar to the S10A mutation in H3, to lower transcriptional activity of certain Gcn5-regulated promoters.[6]

Effect of Acetylation on HK Activity

Similar activity assays are also carried out to determine whether the inverse relationship occurs, that is, in this example, whether prior acetylation has an effect on kinase activity. Again, the nonmodified peptide must be tested initially at a range of kinase concentrations, to be certain the reaction is within linear range. In the case of Ser-10 and Lys-14, phosphorylation of the nonmodified peptide is compared to the Ac-Lys-14 peptide. Our examination of this relationship indicated that, while phosphorylation promotes acetylation, the reverse does not occur.[6] Here we describe a solution kinase assay.

Materials

5× Assay Buffer (AB): 250 mM HEPES, pH 7.5; 50 mM MgCl$_2$, 50 mM MnCl$_2$; 10 mM DTT; 5 mM sodium orthovanadate, 250 mM sodium fluoride, 25 mM β-glycerophosphate.
5× ATP cocktail: 100 μM nonradioactive ATP, 1 μCi/μl [γ-^{32}P] ATP.
Substrates: histone H3 peptides; prepare a 250 μM solution in AB.
5 × BSA (no substrate): 2.5 mg/ml BSA.
Enzyme diluting buffer (EDB): 20 mM HEPES, pH 7.5; 5 mM MgCl$_2$, and 5 mM MnCl$_2$, 10% glycerol.

Methods

1. Prepare reaction buffer mixture on ice: 10 μl of each component (AB, ATP cocktail) times the number of assay samples to a single large tube.
2. Dispense the appropriate amount (30 μl) of reaction buffer mixture to each tube (on ice) containing 10 μl substrates (histone H3 peptides or the negative control + BSA).
3. Dilute the enzyme complexes with EDB to appropriate amount, then add the enzyme to corresponding reaction mixtures. Incubate at 30° for 30 min.
4. Terminate the reaction similar to above, by spotting on filters for scintillation counting.

It may also be desirable to examine the reaction products by polyacrylamide gel electrophoresis and autoradiography, in order to determine whether there is autocatalysis by the enzymes or other contaminating substrates that alter the interpretation of the filter assays. A protocol for examining small peptides by electrophoresis is described.

Methods. Boil samples for 5 min and load onto a 20% SDS-PAGE mini gel (8 cm × 10 cm × 1 mm). Electrophorese samples until the loading dye has migrated one-third the length of the gel to sharpen resolution of the small peptides. Tris–tricine rather than Tris–glycine buffers are superior for resolution. The 26-mer histone H3 peptide migrates approximately at 3 kDa. Fix gel with 10% acetic acid and 10% methanol for ~15 min. For kinase assays, wash gel with 0.1% tetrasodium pyrophosphate to remove unbound radiolabel. Dry gel for one hour and audioradiograph with X-OMAT AR film (Kodak).

Synthesis, Purification, and Testing of Modification-Specific
   Antibodies

A powerful method to investigate relationships between modifications is to prepare polyclonal antibodies that specifically recognize the histone modifications, including either the single modifications, or combinations of adjacent modifications. These antibodies can be used both to determine global relationships between modifications through bulk histone analysis, and to identify gene-specific relationships between modifications via chromatin immunoprcipitation (ChIP) experiments. In either approach, epistasis analysis is effective to tease apart sequential modificatins and related patterns, as described in detail later on. Purification of these polyclonal antibodies is critical to eliminate potential contaminating background antibodies. These contaminants include antibodies that recognize the non-modified histone residues, and, in the case of multiple adjacent modifications, antibodies that recognize only a single modification.

The example we use here is to the H3 P-Ser-10 antibody, describing how to design and purify the antibody. Similar methods would be used for the dual specificity antibody, which are described later.

Design and Purification of Single and Dual Modification Antibodies

Antibodies should be made to small epitopes, to limit background antibodies that detect backbone. Normally, nine to ten residue peptides are used, and, if possible, the modification should be in the center. The peptide is conjugated to BSA or KLM as carrier and injected into rabbits to raise polyclonal antibodies. The peptide synthesis and antibody production is routinely done by commercial vendors.

We prepared a histone H3 P-Ser-10 peptide antibody using the sequence TARK-S-(P)-TGGK-C, which contained a C-terminal cysteine for conjugation to beads and other manipulations.[10] A similar nonmodified peptide was synthesized for use as a negative control. In general, an ELISA test is performed at 10 weeks to determine the extent of serum reactivity to nonmodified and modified peptides. Positive results are followed by antibody purification and further specificity testing.

This method is used for affinity purification of anti-peptide antibodies from polyclonal serum. The protocol involves two steps: negative selection on an nonmodified peptide column, and positive selection on a modified peptide column.

Materials

SulfoLink beads (Pierce).
Coupling buffer (CB): 50 mM Tris, pH 8.5; 5 mM EDTA.
Wash buffer (WB): 1 M NaCl, 0.05% NaN$_3$.
50 mM cysteine solution in CB.
TBS-azide: 20 mM Tris, pH 7.5; 150 mM NaCl; 0.05% NaN$_3$.
PBS-azide: 4.3 mM Na$_2$HPO$_4$; 1.4 mM KH$_2$PO$_4$; 137 mM Nacl; 2.7 mM KCl; 0.05% NaN$_3$.
1 M Tris, pH 8.
0.1 M glycine, pH 2.5.
peptides ($\sim$1 mg/ml).

Preparation of Columns

1. The peptides are coupled to the beads through the C-terminal cysteine. To insure availability of this residue, peptides were first reduced using bead-immobilized TCEP (Pierce). The TCEP beads were prepared by washing with excess CB, combined with the pH 7–8 peptide solutions, and rotated at room temperature for one hour. Peptides were seperated from the beads by centrifugation, and passed through a 0.22 μm spin filter (Millipore).
2. SulfoLink beads (at most 1 ml bead bed volume per 1 mg peptide) are washed (6× bed volume) with CB and mixed with either nonmodified peptide (for negative selection) or phosphorylated peptide (for positive selection).
3. Bead/peptide mixtures are rotated at room temperature for 15 min, poured into 10 ml columns and allowed to sit for 30 min.
4. Flowthrough is collected and column washed (6× bed volume) with CB, and combined with 1× bed volume of 0.05 M cysteine to block nonspecific binding. Beads are rotated for 15 min and allowed to sit for 30 min. Peptide/beads are washed with 10 bed volumes of WB, followed by wash with excess PBS-azide and stored at 4°.

Negative Selection

1. Sequentially wash nonmodified peptide column with 10 ml PBS-azide, 10 ml 0.1 M glycine, pH 2.5, 30 ml PBS-azide before loading serum.
2. Dilute 1.5 ml serum to 4.5 ml in PBS-azide, and filter through a 0.22 μm syringe filter. Allow the diluted serum to flow over the equilibrated column 3 times. Collect flowthrough. Add 300 μl PBS-azide to the column (approximately the void volume for the column) and collect the flowthrough. Combine with the previous flowthrough and save at 4°.
3. Recycle the nonmodified peptide column by sequential washes with 30 ml PBS-azide, 20 ml 0.1 M glycine (pH 2.5), 30 ml PBS-azide. Leave 2–3 ml PBS-azide on top of the beads. Seal the column applying the lower plug before applying the upper plug. Wrap both plugs with parafilm. Store column at 4°.

Positive Selection

1. Prepare the modified peptide column as described above. Apply the flowthrough (from negative selection) to the column. Collect the flowthrough and store at 4° to test later.
2. Wash the column with excess PBS-azide until the A_{280} of the flowthrough is <0.03.
3. Prepare collection tubes for elutions, apply excess 0.1 M glycine, and collect 3 ml fractions into tubes containing 1 M, Tris, pH 8.0. After each fraction is collected, invert a few times to mix well, and measure the A_{280} against the glycine + Tris blank. Stop collecting the elutions after the A_{280} drops to <0.03. To recycle the column, continue to wash with at least 10 ml of glycine, then wash with excess PBS-azide and store at 4° as described earlier.
4. Combine elutions and concentrate with a Centriprep-30 (Millipore) at 4°. Dialyze into an appropriate storage buffer (such as TBS), estimate antibody concentration by A_{280} ($A_{280} = 1$ is approximately 0.74 mg/ml rabbit IgG), add glycerol to ~50% and sodium azide to 0.05%, and store at−20°.

Similar methods are used for the dual specificity antibody, except the antibody is negatively selected over two columns containing each single modification peptide; this will eliminate antibodies to both non-modified peptide and to either single modified peptide. The last step is a positive selection over the column bearing the dual modification peptide. The dual antibody (P-Ser-10/Ac-Lys-14) is commercially available (Upstate).

Testing Antibodies for Specificity

It is crucial to examine the specificity of the purified antibodies, prior to using them for functional studies. First, the antibodies are tested *in vitro* in dot blot format. In general, the four peptides (unmodified, single modifications and dual modification) are tested, using single and dual specificity antibodies. An example is shown (see Fig. 4A), where the P-Ser-10 antibody detected both the corresponding single modification peptide and the dual modification peptide, but not the others. Further, the dual specificity antibody (P-Ser-10/Ac-Lys-14) detected only the dual modified peptide. Specificity was further checked using peptides to block the detection (see Fig. 4A, right). The dot blot method is described here.

Methods

1. Prepare a dilution series for the peptides from 0 to 1000 ng/200 μl in PBS, from a fresh 1 ml stock solution.
2. Presoak a nitrocellulose membrane (NC, 0.2 μm) in PBS.
3. Place two sheets of 3 MM paper (prewetted in PBS buffer) on the filter support plate of the dot blot apparatus. Place the pre-wetted NC paper on the top of the 3 MM paper, and clamp the sample well plate into place. The membrane and 3 MM paper should be the size of the support plate.
4. Apply low vacuum ($\sim$1 ml/min) and wash individual wells with 500 ul of PBS. Check the system for leaks.
5. While vacuum is applied, add 200 μl of histone peptide to the well and filter. Wash with 500 μl PBS.
6. Remove membrane when wells are washed. Perform Western blot.

Note that the dot blot method will not indicate whether there is background detection in a more physiological setting, for example, of bulk histones isolated from cells, or in ChIP assays. One useful method to check an antibody for specificity is to prepare bulk histones from cells that bear substitutions in the sites to be tested. In *S. cerevisiae* this is particularly useful since the mutant histone H3 can be the only source. For example, histone substitutions at Ser-10 (S10A) or Lys-14 (K14A) or at both sites (S10A/K14A) are tested for immunoreactivity using single and double specificity antibodies. Using the method described in the next section, bulk histones are isolated from cells and probed with the P-Ser-10 antibody. Detection by the single specificity antibody, P-Ser-10, is eliminated in the S10A strain, but not the K14A strain (see Fig. 4B) and further, should be abrogated in the double mutant strain. Detection by the dual specificity antibody should be eliminated in the single or double mutant strains.

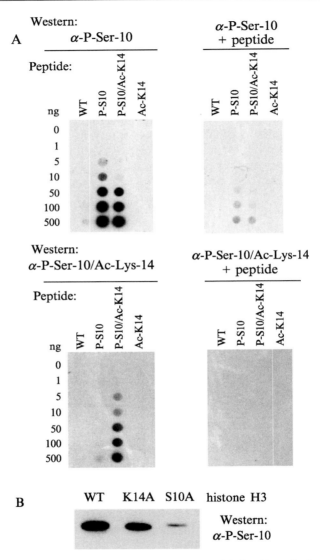

FIG. 4. Western blot assays measuring specificity of P-Ser-10 antibody and double specificity P-Ser-10/Ac-Lys-14 antibody. Panel A is a dot blot assay. The dilutions (in ng) of each of the four peptides (shown across the top of each panel) is indicated. Peptide, the specific single or double modification peptide corresponding to the antibody. Panel B is a Western blot of bulk histones obtained from *S. cerevisiae* strains either wild type or bearing substitutions K14A or S10A on histone H3. The antibody is P-Ser-10.

Examining Global and Local Histone Modification Patterns *In Vivo*

Once antibodies are obtained and tested, then they are used for analysis *in vivo*, to examine patterns, sequences, and dependencies (see Fig. 5A). There are two general *in vivo* approaches. The first is to isolate bulk histones to investigate global changes in modifications. The second approach is ChIP analysis to examine specific promoters or DNA sequences (described in the next section). The bulk histone approach reveals changes on a large cellular scale, while the ChIP approach reveals limited changes at specific localized positions.

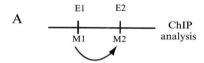

B	S.c. or mamm?	How? Westerns/ChIP and...	Reference
1. Characterization of single and dual modifications			
a. Do single M1 and M2 exist and does M1/M2 exist?	Both	No manipulation	(7,10)
b. Does M1, M2 and M1/M2 change in amount when a specific pathway is induced?	Both	No manipulation	(7,10)
c. Does M1 and M2 occur in a specific sequence during a time course after induction, and does M1/M2 appear when M2 is detected?	Both	No manipulation	(7,8)
2. Alter amount of modification enzymes			
a. Does disruption of E1 reduce M1 and M2, but disruption of E2 reduce M2 but not M1?	Both	S.c.: gene disruption mamm: siRNA or DN	(10)
b. Does increased expression of E1, but not E2, speed the appearance of M1 and M2 ?	Both	Both: OE	(8)
3. Disrupt modification sites			
a. Does substitution of M1 lower M2, but not the reverse order?	S.c.	S.c.: plasmid shuffle	(10,11,12)
b. Does substitution of M1 or M2 lower the M1/M2 pattern?	S.c.	S.c.: plasmid shuffle	(10)

FIG. 5. Global and ChIP analysis of histone modifications in *S. cerevisiae* (S.c.) and mammalian cells (mamm). Panel A is a schematic of a hypothetical sequence and pattern of modifications. Panel B shows a series of experimental approaches described in the text to investigate a potential sequence and pattern of modifications. Antibodies are used to modification 1 (M1), modification 2 (M2) and double modification (M1/M2). The enzyme that catalyzes M1 is E1 and M2 is E2. siRNA, small inhibitory RNA; DN, dominant negative; OE, overexpression.

The general logical flow of experiments and published examples of the approaches are described in Fig. 5, to test the model shown in Panel A. These can be done either by bulk histone examination or by ChIP analysis. The most straightforward experiment is to determine whether the single modifications and the pattern increases in a certain pathway (and at a certain gene) (see Fig. 5B.1). If so, then time courses can be examined, to determine whether the modifications arise in sequence and whether the double modification is found at the time the second modification is detected. Second, dependencies can be investigated by manipulating enzyme levels to determine whether altering the first enzyme affects the second modification, but not the reverse (see Fig. 5B.2). The first and second types of approaches can be done in both *S. cerevisiae* (gene disruption of the enzymes) and mammalian cells (siRNA or dominant negative mutations in the enzymes) and by over expression of the enzymes in both organisms. The third approach is to disrupt the modifications sites themselves (see Fig. 5B.3), which can be done in *S. cerevisiae* by the "plasmid shuffle" technique (see Chapter 1). In this case substitution of the first modification site should lower the second modification, but not the reverse.

Here we describe two key protocols for these analyses. The first is bulk histone purification from *S. cerevisiae*. The second is quantitative PCR in real-time, which greatly facilitates ChIP analysis from multiple samples. Quantitative PCR provides a much faster and more reproducible signal than older electrophoretic methods using ethidium bromide staining or radiolabeling to visualize reaction products.

Bulk Histone Purification from Yeast

Examination of changes in histone modifications in bulk histones reveals large-scale variations and patterns. Bulk histone preparation is followed by gel electrophoresis and Western blot analysis to monitor specific modification sites and patterns. An example of results from this approach using an antibody that detects a single modification site is shown in Fig. 4B.

Materials

Buffer A (BA): 50 mM Tris, pH 7.5; 30 mM DTT.
Buffer S (BS): 1.2 M Sorbitol, 20 mM HEPES, 7.4.
Buffer B (BB): 1.2 M sorbitol; 20 mM PIPES, pH 6.8; 1 mM MgCl$_2$.
Buffer NIB (BNIB): 0.25 M sucrose; 60 mM KCl; 140 mM NaCl; 5 mM MgCl$_2$; 1 mM CaCl; 15 mM MES, pH 6.6; 0.8% Triton X-100; 1 mM PMSF.

Wash buffer A (WBA): 10 mM Tris, pH 8; 0.5% NP-40; 75 mM NaCl; 30 mM sodium butyrate; 1 mM sodium flouride; and 1 mM PMSF.
Wash buffer B (WBB): same as WBA except 0.4 M NaCl and no NP-40.

Cell Preparation

Day 1. Inoculate yeast into 10 ml YPD and grow at 30° with shaking.
Day 2. Late in the evening, dilute the culture to $OD_{600} = 0.025-0.04$ in 100 ml YPD, and incubate as above to $OD_{600} = \sim 1.0$.
Day 3. Spin the culture in SS34 rotor at 5000 rpm for 5 min, wash 1× sterile water ($\sim 3 \times 10^8$ cells) in 15 ml round bottom snap cap Falcon tubes. Weigh pellet.

Methods

1. Wash cell pellet once in water, and resuspend in 5 ml of BA. Incubate at 30° for 15 min, with gentle shaking.
2. Centrifuge at 5000 rpm for 5 min. Wash pellet with 10 ml BS and centrifuge again.
3. Resuspend pellet in 5 ml BS with 0.2 ml of 10 mg/ml Zymolyase (2 mg of Zymolyase 100T per gram of cells). Incubate at 30° with gentle shaking, ~ 1 h (inspect to make certain that spheroplasts have formed).
4. Add 10 ml ice-cold BB, and harvest cells at 5000 rpm in SS34 rotor for 5 min, 4°.
5. Resuspend pellet in 5 ml ice-cold BNIB. Very important point: for efficient resuspension of pellets from this point, first use ap1000 tip with end cut off for rough resuspending, and then another uncut tip for complete resuspension. Hold 20 min and centrifuge again. Repeat twice.
6. Wash the pellet three times in 5 ml WBA. Hold on ice-water 15 min and centrifuge again. Repeat the wash procedure twice.
7. Wash the pellet in WBB for 10 min and centrifuge again. Repeat the wash procedure once.
8. Extract histones by resuspending the pellet in cold 0.4 N H_2SO_4 (100 ml cells = 1 ml). Incubate 1 h with occasional vortex.
9. Spin at 10,000 rpm in an SS34 rotor for 10 min and save the supernatant, which contains the extracted histone.
10. Precipitate the histone by adding 100% TCA to a final concentration of 20%. After 1 h incubation, collect the protein pellet by centrifugation for 30 min at 12,000 rpm in SS34.
11. Wash pellet once in chilled acidified acetone (acetone + 0.1% HCl) and then chilled 100% acetone. Air-dry the pellet and resuspend histones in ~ 1 ml of 10 mM Tris, pH 8, and store at $-20°$.

ChIP Analysis Using Quantitative PCR in Real Time

Preparation of whole cell extracts and chromatin immunoprecipitation has been reviewed extensively. Here we briefly discuss the use of quantitative PCR in real time to evaluate ChIP samples. We have found this method to be highly advantageous for quantitative questions, such as comparing multiple antibodies and other experimental manipulations. Examples are, for time courses using many antibodies, primer combinations, and for epistasis analysis in *S. cerevisiae* involving many different strains.

Thermocycler systems, such as those from Applied Biosystems (ABI PRISM), Roche (Lightcycler) or Stratagene (M × 3000P), allow facile quantitative analysis of multiple ChIP samples in real time. The software packages for primer design and data analysis greatly speed up generation and analysis of results.

In general, the amplification products produced using real time PCR systems are much smaller (50 to 100 bp) than those in traditional PCR followed by ethidium bromide staining or incorporation of radiolabeled primers or nucleotides. Real time systems allow two methods of nucleic acid detection: detection of a specific PCR product using the degradation of an internal, labeled oligonucleotide, or detection of a nonspecific dye that binds to double stranded DNA (e.g., SYBR green). The first method is optimal for analysis of a small number of genomic regions in repetitive samples. In the TaqMan (Applied Biosystems) system, a fluorescent label is located at the $5'$ end of the probe with a quencher at the $3'$ end. Amplification with primers outside of the probe results in its digestion via Taq's nuclease activity and the release of the fluor, resulting in fluorescence directly related to amplification. Specificity is high since the signal depends on binding of both amplification primers and probe. However, each PCR requires a unique and expensive fluorescent probe.

More suitable for analysis of multiple genomic regions in ChIP analysis is the second, nonspecific detection of amplification products. Specificity is lower than the fluorogenic probe because total double stranded DNA is detected. However, two methods exist to determine whether the signal is due to one or more amplification products. First, PCR products can be directly visualized by electrophoresis. In addition, for each new set of primers, a dissociation curve can be included in the real time assay. In this test, the block temperature is increased stepwise (generally from $60°$ to $95°$) causing dissociation of double stranded DNA at a characteristic temperature (T_m) which is displayed graphically as a peak along a temperature gradient. The presence of multiple peaks indicates non-specific amplification. Optimization of PCR or synthesis of new primers is then required.

Materials

ChIP samples (initial *S. cerevisiae* volume 50 ml; final elution volume 25 μl)
ABI PRISM 7000 Sequence Detection System (Applied Biosystems)
Primer Express Software (Applied Biosystems)
SYBR Green PCR Master Mix (Applied Biosystems)
Forward and reverse primers, 10 μM
Ultrapure DNA-free water
MicroAmp 96-Well Splash-Free Support Base (Applied Biosystems)
ABI PRISM 96-Well Optical Reaction Plates (Applied Biosystems)
ABI PRISM Optical Adhesive Covers (Applied Biosystems)
ABI PRISM Optical Cover Compression Pad (Applied Biosystems)

Methods

The default settings of the ABI PRISM Sequence Detection System work well for PCR of ChIP samples. Primers are designed from selected regions of DNA (following BLAST analysis) using the Primer Express software package. The PCR reaction volume is 25 μl. This is half of the recommended volume, but is equal in amplification and fluorescence detection and reduces the expense. Three independent PCR reactions are done and each of these is split into two duplicate wells to test reproducibility.

1. Dilute ChIP input and immunoprecipitated samples. Input samples (which are generated from 10% of the immunoprecipitation input) are 1:10,000 and 1:50,000 and elutions are 1:5, 1:10 and 1:25.
2. Prepare PCR master mix. For each PCR reaction, add the following:
   12.5 μl 2× SYBR Green PCR Master Mix
   7.5 μl ultrapure water
   0.125 μl each 10 μM primer
3. Transfer 40 μl of the PCR master mix to a separate tube for each ChIP sample. Add 10 μl of diluted input and ChIP templates, mix by pipeting and centrifuge (3000 rpm, 10 s). Prepare an additional set of reactions that contain dilutions of DNA and/or ChIP samples (undiluted, 1:10, 1:100, 1:1000) to test the efficiency of the PCR reaction. Also include a reaction without template to control for DNA contamination.
4. Place a 96-Well Optical Reaction Plate into a 96-Well Splash-Free Support Base. Carefully transfer 25 μl of each reaction mix into the bottom of duplicate wells. Be careful to avoid air bubbles in the reaction mixture, as they may interfere with accurate signal detection.

5. Cover the PCR plate with optical adhesive cover and compression pad.
6. Define reactions within the ABI PRISM 7000 SDS program and save settings.

Default cycling conditions are as follows:

$1\times$, $50°$, 2 min

$1\times$, $95°$, 10 min

$40\times$, $95°$, 15 s and $60°$, 1 min

Place the reaction plate into the machine and start program.

Note: When setting up the program, change the default volume to 25 μl and check that a dissociation protocol will be performed at the end to determine if the signal is from one or multiple PCR products.

Investigating the Role of the Modification Pattern *In Vivo*

Finally, it is crucial to test the role of the modifications *in vivo*. The general logical flow of experiments, and published examples of the approaches, is described in Fig. 6, to test the model shown in Panel A. Again, in *S. cerevisiae* RNA levels are examined in strains bearing substitution mutations in the modification sites (see Fig. 5B.1). If the modifications are in a linear, dependent sequence then the single modification substitutions and the dual modification substitutions will result in a similar reduction in transcription.[10] However, this simple relationship may be complicated by one modification having a second role in the same process, and then substitution of that modification would have a more severe effect. However, in any case, the double modification is expected to be no worse than the most deleterious single modification, if the modifications are working together. A second approach is to alter the amount of the enzymes and to examine RNA levels (see Fig. 6B.2; and correlate with modification site changes, as in Fig. 5). In this case, a dependent sequence implies that increased expression of the enzyme that catalyzes the first modification (but not the second modification) should speed up the timing of RNA synthesis. Note that interesting predictions can also be made about the effect of increasing the first enzyme on facilitating recruitment of the second enzyme, which can be monitored by ChIP.[8]

RT-PCR of RNA Extracted from Cells

Primer design is performed using the RT-PCR primer design setting on the Primer Express software package from the vendor. Initially, BLAST analysis determines the region of the coding sequence to be used for primer design to minimize nonspecific amplification. As described earlier,

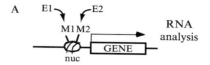

B	S.c. or mamm?	How? RT-PCR and...	Reference
1. Disrupt modification sites			
a. Does mutation of M1 or M2 reduce level of RNA synthesis to the same extent?	S.c.	Plasmid shuffle	(10)
b. Does double M1/M2 mutation reduce level to the same extent as single mutations?	S.c.	Plasmid shuffle	(10)
2. Alter amount of modification enzymes			
a. Does increased expression of E1, but not E2, speed up timing of RNA expression?	Both	Both: OE	(8)
b. Does decreased expression of E1 or E2 lower RNA synthesis to the same extent?	Both	S.c.: gene disruption mamm: siRNA or DN	(8,10)

Fig. 6. RNA analysis of the effect of altering histone modifications in *S. cerevisiae* (S.c.) and mammalian cells (mamm). Panel A is a schematic of a hypothetical pattern of histone modifications in regulating gene transcription. Panel B shows a series of experimental approaches described in the text to investigate the effect of altering histone modifications on RNA levels. Abbreviations are the same as Fig. 5.

qRT-PCR primers amplify shorter regions compared to conventional RT-PCR.

We have had excellent results using the default settings of the ABI PRISM 7000 Sequence Detection System. Amplification of cDNA is performed in triplicate in three independent reactions (a total of nine amplifications for each RT+ or RT− template) to be confident of the relative expression levels. We have reduced the costs of the assay by reducing the reaction volume to half of the recommended volume (50 μl) and have obtained equivalent results.

Materials

RNA samples (e.g., initial *S. cerevisiae* volume 1.7 ml; final RNA 40 μl)
RNase-free water
DNA-*free* (Ambion)
TaqMan Reverse Transcription Reagents (Applied Biosystems)
qPCR materials and hardware (same as for ChIP, above).

DNase Treatment

1. Quantitate RNA (extracted using Zymo Research YeaStar RNA kit) by measuring OD260/280. Transfer 2 μg of RNA to a fresh tube and add RNase-free water to 43 μl.
2. Add 5 μl of 10× DNase I buffer and 2 μl DNase I (DNA-*free*, Ambion), centrifuge (3000 rpm, 10 s), mix and spin again.
3. Incubate for 30 min at 37°.
4. Add 5 μl of DNase I Inactivation Reagent (DNA-*free*, Ambion) and mix, incubate at room temperature for 2 min. Centrifuge at 14,000 rpm for 10 s and transfer supernatant to a fresh tube.

Reverse Transcription

1. Add the following to a separate tube
   2 μl 10× TaqMan RT Buffer (Applied Biosystems)
   4.4 μl 25 mM $MgCl_2$
   1.0 μl 10 mM dNTP mixture
   1.0 μl 50 mM random hexamers (Applied Biosystems)
   0.4 μl RNase Inhibitor (20 U/μl) (Applied Biosystems)
   0.5 μl Multiscribe Reverse Transcriptase (Applied Biosystems)
   10 μl of the DNase I reaction supernatant
   RNase-free water to 20 μl final volume
   Centrifuge (3000 rpm, 10 s), mix and repeat.
   Prepare a similar reaction mixture but do not include reverse transcriptase (RT$^-$)
2. Incubate at room temperature for 10 min, followed by 30 min at 48° and 5 min at 95°.

Quantitative PCR Analysis of cDNA

1. Dilute reverse transcription reaction samples 1:10, 1:50, and 1:100 in ultrapure DNA-free water.
2. Prepare a master mix for real time PCR reaction as for ChIP analysis above.
3. Transfer 60 μl of the PCR master mix to a separate tube for each cDNA sample. Add 15 μl of appropriate RT$^+$ and RT$^-$ reaction to tubes, mix by pipeting and spin quickly (3000 rpm, 10 s). Dilution controls are the same as for ChIP analysis.
4. Follow directions 4–6 as in ChIP analysis.

Notes: (1) We generally perform triplicate reactions for each sample dilution with primers for the gene of interest and for a housekeeping/control gene *(ACT1, ADH1, GAPDH)* on the same reaction plate to ensure that the PCR conditions are similar. (2) Baseline levels are recorded between cycles 6 and 15 to determine the level of background fluorescence. If template levels are too high, background fluorescence will be recorded as high and sensitivity may be affected. Dilute the samples to resolve this issue. Also, the number of cycles used for background detection can be lowered.

Further Considerations Concerning the Role of Histone Modification Patterns

The presence and importance of a modification pattern begs the question of its specific mechanism. One obvious question is whether the pattern promotes binding of effector proteins beyond that achieved by single modifications. In the case of Ac-Lys-14, it is known that bromodomains within Gcn5 and Taf250 or Bdf1 in *S. cerevisiae*, bind with higher affinity to acetylated peptides compared to non-acetylated peptides.[21–23] Thus, it may be desirable to investigate improved binding to the dual modified peptide. Two general methods can be used *in vitro*: one is to use simple peptide binding approaches. For this method it is necessary to synthesize the peptides in a manner to provide for linkage to beads, such as a C-terminal cysteine (as shown in Fig. 2) or a biotin-conjugated terminal residue. Each peptide can be linked to beads and incubated with the protein domain to be queried for binding association. A second more quantitative method is to use surface plasmon resonance (using a BIAcor instrument), where the peptides are immobilized on a grid, again using a linkage (such as cysteine or biotin) that must be incorporated into the peptide and the protein domain is passed over the grid. Both of these methods have been extensively used to assay single modifications and binding, such as for acetyllysine and methyl-lysine (e.g., Bannister *et al.*[24]) but, to date, no pattern has been shown to provide increased binding capacity relative to single modifications.

Acknowledgments

Research in the laboratory is supported by the National Institutes of Health and the National Science Foundation. W.-S. Lo thanks the Leukemia and Lymphoma Foundation for fellowship support. K.W.H. is supported by a National Institutes of Health training grant to The Wistar Institute. We also wish to thank members of the Berger lab and L. Pillus for valuable discussions.

[21] C. Dhalluin, J. E. Carlson, L. Zeng, C. He, A. K. Aggarwal, and M. M. Zhou, *Nature* **399**, 491 (1999).

[22] R. H. Jacobson, A. G. Ladurner, D. S. King, and R. Tjian, *Science* **288**, 1422 (2000).

[23] O. Matangkasombut and S. Buratowski, *Mol. Cell* **11**, 353 (2003).

[24] A. J. Bannister, P. Zegerman, J. F. Partridge, E. A. Miska, J. O. Thomas, R. C. Allshire, and T. Kouzarides, *Nature* **410**, 120 (2001).

[8]　Identification and Analysis of Native HAT Complexes

By STACEY J. MCMAHON, YANNICK DOYON, JACQUES CÔTÉ, and PATRICK A. GRANT

Acetylation of the amino-terminal tails of core histone proteins has been linked to many cellular processes, including cell cycle progression, DNA replication, chromatin assembly, and the regulation of gene expression.[1] Therefore, characterization of histone acetyltransferases (HATs), the enzymes responsible for posttranslational acetylation of histones, is necessary to better understand the regulation of these processes. Two classes of HATs have been described: nuclear type A HATs, which mediate transcription related acetylation,[2] and cytoplasmic type B HATs, which mediate histone acetylation related to subsequent chromatin assembly.[1]

Upon identification and purification of the first nuclear HAT from *Tetrahymena*, a homologue of the yeast transcriptional activator Gcn5, a direct link between histone acetylation and gene activation was established.[2] It has since become apparent that substrate specificity is different between recombinant HAT enzymes and HATs as they typically exist *in vivo* in high molecular weight complexes.[3] Therefore, it is necessary to purify native histone acetyltransferase complexes in order to fully understand the functions of the HAT enzymes themselves. Here, we will describe methods commonly used to purify and analyze native HAT complexes from the yeast *Saccharomyces cerevisiae* and human cells. These approaches have also been adapted for the purification of histone kinase and methytransferase complexes.[4,5]

Purification of Native Yeast HAT Complexes

The first method of purification relies on the binding of HATs to nickel agarose resin in the absence of any tags. It was found that numerous native yeast HAT complexes fortuitously bind to nickel.[6] Whole cell extracts from yeast are prepared in the following manner.

[1] S. Y. Roth, J. M. Denu, and C. D. Allis, *Annu. Rev. Biochem.* **70,** 81 (2001).

[2] J. E. Brownell, J. Zhou, T. Ranalli, R. Kobayashi, D. G. Edmondson, S. Y. Roth, and C. D. Allis, *Cell* **84,** 843 (1996).

[3] P. A. Grant and S. L. Berger, *Semin. Cell Dev. Biol.* **10,** 169 (1999).

[4] B. D. Strahl, P. A. Grant, S. D. Briggs, Z. W. Sun, J. R. Bone, J. A. Caldwell, S. Mollah, R. G. Cook, J. Shabanowitz, D. F. Hunt, and C. D. Allis, *Mol. Cell. Biol.* **22,** 1298 (2002).

Copyright 2004, Elsevier Inc.
All rights reserved.
0076-6879/04 $35.00

1. Pick a single yeast colony and inoculate 5–10 ml YPD (1% yeast extract, 2% peptone, 2% dextrose) medium. Grow over night at $30°$ in shaking incubator.

2. Inoculate 4–6 L YPD with the 5–10 ml preculture. Grow yeast at $30°$ shaking at 300 rpm. Grow cells to mid log phase, or an OD_{600} of 1.0.

3. Cool cells on ice. Pellet cells at 4500 rpm for 15 min at $4°$.

4. Resuspend cells in 200 ml cold extraction buffer (40 mM HEPES-KOH, pH 7.5, 350 mM NaCl, 0.1% Tween 20, 10% glycerol, 2 μg/ml pepstatin A, 2 μg/ml leupeptin, 1 mM PMSF), repeat step 3.

5. Resuspend pellet in 25 ml extraction buffer.

6. Fill 100 ml bead-beater polycarbonate chamber half full with 0.5-mm glass beads, washed in extraction buffer. Pour in yeast suspension and fill container the remainder of the way with extraction buffer.

7. Homogenize yeast 10×30 s pulse followed by a 1 min pause.

8. Pour glass bead/yeast suspension into a 50-ml conical tube. Spin at 1000 rpm at $4°$ to remove glass beads and cell debris.

9. Clarify extract by ultracentrifugation at 25,000 rpm for 1 h at $4°$.

10. Wash 5 ml 100% Ni-NTA (Qiagen, Valencia, CA) slurry with 40 ml extraction buffer and spin at 1000 rpm for 5 min. Repeat.

11. Incubate yeast extract with Ni-NTA agarose for at least 3 h at $4°$ on a rotating wheel.

12. Pour the mixture into a disposable 10 ml Poly-Prep chromatography column fitted with a 200 ml funnel (Biorad). Collect the flow through.

13. Wash the resin with 12.5 ml extraction buffer at $4°$. Collect 5 ml with the flow through and discard the rest.

14. Wash the resin with 12.5 ml wash buffer (20 mM Imidazole, pH 7.0, 100 mM NaCl, 0.1% Tween 20, 10% glycerol, 2 μg/ml pepstatin A, 2 μg/ml leupeptin, 1 mM PMSF) at $4°$.

15. Elute the bound proteins with 12.5 ml elution buffer (300 mM Imidazole, pH 7.0, 100 mM NaCl, 0.1% Tween 20, 10% glycerol, 2 μg/ml pepstatin A, 2 μg/ml leupeptin, 1 mM PMSF) at $4°$.

The eluates and/or flow throughs can now be used for further chromatographic analysis and subsequent purification to separate the many different yeast HAT complexes from each other. This method has been used to isolate the SAGA, NuA4, ADA, and NuA3 HAT complexes.[6]

[5] W. S. Lo, L. Duggan, N. C. Tolga, Emre, R. Belotserkovskya, W. S. Lane, R. Shiekhattar, and S. L. Berger, *Science* **293**, 1142 (2001).

[6] P. A. Grant, L. Duggan, J. Cote, S. M. Roberts, J. E. Brownell, R. Candau, R. Ohba, T. Owen-Hughes, C. D. Allis, F. Winston, S. L. Berger, and J. L. Workman, *Genes Dev.* **11**, 1640 (1997).

Native complexes can also be purified from yeast as a result of their association with a tagged component of that complex. Tagging a protein provides a manner by which to follow the presence of that protein within a complex by using commercially available antisera that recognizes the tag. This method saves time, expense, and uncertainty associated with developing unique antibodies against new, uncharacterized proteins. Several tags are used frequently, including GST, Myc, HA, and TAP, which allows purification of the tagged component and its associated proteins by selecting those that bind specific resins. Here, we will describe the creation of a TAP-tagged protein, and purification of it and the proteins that associate with it in $vivo$.

This method relies on the highly efficient homologous recombination that takes place in the yeast genome to fuse the TAP-tag downstream from a gene of interest.[7] The TAP-tag consists of protein A and CBP (calmodulin binding protein) separated by a TEV cleavage site, digestion of which releases the protein A part of the tag. First, forward and reverse oligonucleotides are designed to be complementary to a TAP-tag marker cassette at their 3′ end and to contain a region of homology with the yeast genome to allow in frame fusion of the TAP-tag downstream of the gene of interest.[8] The PCR reaction mix used to amplify the TAP-tag marker cassette contains the following: 4 μl plasmid template (1 ng/μl),[9] 12 μl forward oligo (10 μM), 12 μl reverse oligo (10 μM), 12 μl dNTPs mix (2.5 mM each), 20 μl Taq 10× buffer, 139.2 μl water, and 1 μl Taq. Divide the mix into 5 × 40 μl aliquots in small PCR tubes. The PCR reaction is as follows:

1. Denature at 94° for 5 min.
2. Thirty-five cycles of 94° for 30 s denaturation, 48° for 30 s hybridization and 72° extension.
3. Fill in at 72° for 5 min.

After PCR is complete, pool the fractions and extract with phenol/chloroform/isoamyalcohol. Precipitate DNA with 2.5 volumes of salt/ethanol (1 volume 3 M KoAc : 24 volumes EtOH) for at least 2 h at −20°. Spin down DNA at maximum speed, wash with 70% ethanol, dry DNA. Resuspend pellet in 10 μl TE buffer. This DNA can then be transformed into the desired yeast strain and the TAP-tag will integrate near the gene of interest

[7] O. Puig, B. Rutz, B. G. Luukkonen, S. Kandels-Lewis, E. Bragado-Nilsson, and B. Seraphin, $Yeast$ $14,$ 1139 (1998).
[8] G. Rigaut, A. Shevchenko, B. Rutz, M. Wilm, M. Mann, and B. Seraphin, $Nat.$ $Biotechnol.$ $17,$ 1030 (1999).
[9] O. Puig, F. Caspary, G. Rigaut, B. Rutz, E. Bouveret, E. Bragado-Nilsson, M. Wilm, and B. Seraphin, $Methods$ $24,$ 218 (2001).

via homologous recombination. Transformants are selected by plating yeast on drop out media selecting for the appropriate marker. PCR is used to screen genomic DNA from yeast colonies for correct integration events. A single transformant is then used to inoculate a 2–5 ml preculture, followed by inoculation of a larger culture with the preculture and preparation of whole cell extracts as described earlier. The TAP-tag adds approximately 20 KD to the size of a protein.[8]

Once whole cell extracts have been prepared from a 100 ml yeast culture, dilute 6 ml of whole cell extract to 150 mM NaCl with 0 M calmodulin binding buffer (10 mM Beta MercaptoEthanol [BME], 10 mM Tris–Cl pH 8.0, 1 mM Mg-Acetate, 1 mM Imidazole, 2 mM CaCl$_2$, and 0.1% NP40). Wash 200 μl calmodulin sepharose bead slurry with 5 ml 150 mM calmodulin binding buffer (150 mM NaCl, 10 mM BME, 10 mM Tris–Cl pH 8.0, 1 mM Mg-Acetate, 1 mM Imidazole, 2 mM CaCl$_2$, and 0.1% NP-40). Add diluted sample to resin and rotate on a wheel at least 1 h at 4°. Run through a Poly-Prep chromatography column and collect flow through. Wash with 30 ml 150 mM calmodulin binding buffer. Elute 7 × 200 μl with calmodulin elution buffer (150 mM NaCl, 10 mM BME, 10 mM Tris–Cl pH 8.0, 1 mM Mg-Acetate, 1 mM Imidazole, 2 mM EDTA, and 0.1% NP-40). Elution fractions can be analyzed by Western blotting (see Fig. 1) and/or pooled for further chromatographic analysis. The TAP-tagged protein can be "followed" over successive steps by using antisera against the protein A portion of the tag. Classical use of the TAP-tag for purification involves successive IgG-sepharose and calmodulin-sepharose binding steps.[8]

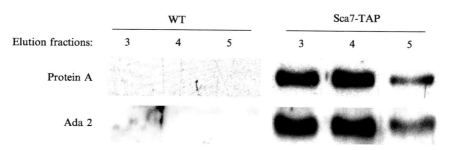

FIG. 1. Example of the first purification step of a complex by association with its tagged component. Whole cell extracts from wild type and TAP-Sca7/Sgf73 yeast strains were bound to and eluted from calmodulin beads. The elution fractions were run on SDS-PAGE gels and transferred to nitrocellulose. Western blot analysis with α-Protein A and α-Ada2 antibodies shows coprecipitation of TAP-tagged Sca7/Sgf73, and Ada2, a component of both SAGA and SLIK, two HAT complexes from *S. cerevisiae*.

Different tags may be chosen depending on what experiments are going to be done with the tagged protein. The complexes they associate with may be purified in the manner described earlier altering certain conditions to allow binding of a tag to its specific resin. Some things to be considered when choosing a tag are: size of the tag, whether it would be better to integrate your tag at the $5'$ or $3'$ end of your gene, and what charge a tag has when considering what further purification steps will follow. It is also important to remember that no matter what tag is chosen, once the tagged protein is incorporated into a complex, the tag may no longer be accessible. An example of purification of a complex using a tagged protein is Ngg1p (Ada3), which is a component of the SAGA and SLIK HAT complexes.[10] NGG1 was epitope tagged by introducing a restriction site into a plasmid upstream of the TAA stop codon of the NGG1 gene. An oligonucleotide encoding a triple hemagglutinin (HA) epitope was cloned into that site.[11] Ngg1 was fractionated by ion exchange and gel filtration chromatography, and the presence of HA-NGG1 was assayed by immunoblotting with an anti-HA antibody.[11] HAT activity was monitored by HAT assays followed by fluorograms.

Assembly of Recombinant HAT Complexes

It is also important to note the usefulness of generating recombinant protein complexes. For example, several proteins can be reconstituted together *in vitro* using polycistronic coexpression vectors. Polycistronic expression plasmids are capable of coexpressing up to four individual genes.[12] By this method, it is possible to observe minimal components of a complex that may exist as a subcomplex, and to determine whether or not the subcomplex carries out all the functions associated with a particular native complex in its entirety, or simply a subset of functions. Balasubramanian *et al.*[13] amplified the Ada2, Ada3, and Gcn5 genes (which are known to interact as components of the ADA, HAT A2, SAGA, and SLIK HAT/coactivator complexes)[6,10,11] from yeast genomic DNA. Temperature and length of induction time were optimized to obtain maximum expression of soluble proteins, the proteins were expressed from the polycistronic vector in BL21 bacteria and purified for use in further

[10] M. G. Pray-Grant, D. Schieltz, S. J. McMahon, J. M. Wood, E. L. Kennedy, R. G. Cook, J. L. Workman, J. R. Yates, III, and P. A. Grant, *Mol. Cell. Biol.* **22,** 8774 (2002).

[11] A. Saleh, V. Lang, R. Cook, and C. J. Brandl, *J. Biol. Chem.* **272,** 5571 (1997).

[12] S. Tan, *Protein Expr. Purif.* **21,** 224 (2001).

[13] R. Balasubramanian, M. G. Pray-Grant, W. Selleck, P. A. Grant, and S. Tan, *J. Biol. Chem.* **277,** 7989 (2002).

experiments. In this manner, it was possible to more precisely determine what functions of the SAGA complex these 3 proteins contribute to.[13] This was also successfully used to characterize the core nucleosomal HAT module of the yeast NuA4 complex, formed by the Esa1, Ep11, and Yng2 proteins.[14]

Analysis of Purified Complexes

Here, we will describe a simple method used to assay for HAT activity. The use of free histones or nucleosomes as substrates in this assay provides a means by which to distinguish between type A and type B HATs.[15] Type A HATs are able to acetylate both histone and nucleosome substrates, whereas, type B HATs are generally only able to acetylate free histone substrates. Liquid and gel fluorography HAT assays, described later, can be used to help separate and purify distinct HAT activities.

Liquid Assay

1. In 30 μl reactions, add 1 μl of a column fraction you would like to check for activity, 6 μl of 5× HAT buffer (250 mM Tris–HCl, pH 8.0, 25% glycerol, 0.5 mM EDTA, pH 8.0, 250 mM KCl, 5 mM DTT, 5 mM PMSF, 50 mM sodium butyrate), 0.25 μCi of tritiated acetyl CoA, and 2 μg of nucleosomes or free histones.

2. Incubate at 30° for 30 min.

3. Spot 15 μl of each reaction on P81 phosphocellulose filter paper circles (Whatman) and allow to air dry. The other 15 μl of each reaction can be used for gel fluorography (described later).

4. Wash the filters with 50 ml of 50 mM NaHCO$_3$–NaCO$_3$ buffer, pH 9.2, on a shaking platform for 5 min. Repeat two more times.

5. Rinse the filters briefly in acetone and allow to air dry. Place the filters in scintillation vials with 4 ml scintillation fluid and count in a scintillation counter for at least 2 min per sample.

Liquid assays allow for the determination of any HAT activity. Fluorography allows for the identification of the particular modified histone within a mixture by virtue of their separation by SDS-PAGE.

[14] A. A. Boudreault, D. Cronier, W. Selleck, N. Lacoste, R. T. Utley, S. Allard, J. Savard, W. S. Lane, S. Tan, and J. Cote, *Genes Dev.* **17,** 1415 (2003).

[15] P. A. Grant, S. L. Berger, and J. L. Workman, *Methods Mol. Biol.* **119,** 311 (1999).

Gel Fluorography

1. Pour 15% SDS-PAGE mini protein gels with 10-well, 0.75-mm combs.

2. Add 5 μl 4× SDS loading dye to the remaining 15 μl of each HAT reaction.

3. Boil samples for 5 min, spin briefly in a microfuge, and load samples onto the SDS-polyacrylamide gel. Load 5 μl of prestained molecular weight standard into a spare well.

4. Electrophorese samples at 150 V for about 2 h. Histones run between 10 and 20 KD.

5. Carefully remove gel from between plates and stain with approximately 50 ml Coomassie staining solution for at least 30 min with gentle shaking.

6. Destain gel with destaining solution (10% acetic acid, 45% H_2O, 45% methanol) until histones are visible.

7. Incubate gel in EN^{3}HANCE autoradiography enhancer (NEN Research Products) in hood with gentle shaking for 30–60 min.

8. Rinse gel several times with deionized water, then incubate with deionized water with gentle shaking for 30 min.

9. Place gel on a wet piece of blotting paper, cover with plastic wrap, and dry under vacuum for 90 min at 60°.

10. Expose dried gel to X-ray film at −80°. A fluorogram is usually readable within 24–48 h (see Fig. 2).

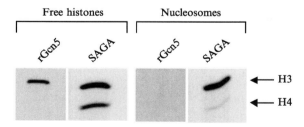

Fig. 2. Example of a fluorogram of HAT assays using HeLa free histones and nucleosomes as substrates. HAT assays using recombinant Gcn5 (rGcn5) or SAGA complex with either free histone or nucleosome substrates demonstrates a broader acetylation ability of Gcn5 when in complex with several other proteins. rGcn5 is unable to acetylate nucleosomes and only acetylates histone H3 when given free histones as a substrate. However, in the context of the SAGA complex, Gcn5 is able to acetylate histones H3 and H4 from both free histones and nucleosomes.

Activity Gel Assay

Activity gel, or "in gel," assays can be used to identify catalytic polypeptides according to their approximate molecular weight. This is another method used to monitor HAT activities during the course of their purification[16] in which samples containing HATs are resolved on SDS-PAGE gels that contain the substrate (i.e., free histones, nucleosomes, etc.) polymerized within them. We will describe the conditions used for mini gels, however, parameters may be adjusted to fit the size of the gel that is being used. Prepare a 10% polyacrylamide resolving portion of the gel with 1 mg/ml histones and allow it to polymerize at room temperature for at least 1 h. A control gel containing 1 mg/ml BSA can also be prepared in parallel (because it is known not to be acetylated *in vivo*). Conventional stacking gels without proteins are polymerized on top of the substrate containing gels. Electrophorese samples using standard SDS-PAGE conditions at room temperature. Substrate proteins may be depleted from the gel during the course of electrophoresis, therefore 0.1 mg/ml histones (or BSA, etc.) can be added to the upper buffer reservoir. The following steps, adapted from Mizzen *et al.*[17] are carried out at room temperature with gentle agitation unless otherwise noted.

1. Remove SDS by washing gels in Buffer 1 (50 mM Tris–HCl, pH 8.0, 20% 2-propanol, 0.1 mM EDTA, 1 mM DTT). Wash 4 × 15 min with about 100 ml Buffer 1.

2. Denature proteins by washing gels 4 × 15 min with 100 ml Buffer 2 (50 mM Tris–HCl, pH 8.0, 8 M urea, 0.1 mM EDTA, 1 mM DTT).

3. Renature proteins slowly by incubating gels with Buffer 3 (50 mM Tris–HCl, pH 8.0, 0.005% Tween 40, 0.1 mM EDTA, 1 mM DTT) at 4° without agitation.

4. Rinse gels once in 100 ml Buffer 3 for 15 min at 4° followed by overnight (12 h) incubation in 100 ml Buffer 3 at 4°. The next day, wash gels again with 100 ml Buffer 3 at 4°.

5. To allow the gel and buffer to equilibrate at room temperature, do a final 30 min wash with 100 ml Buffer 3 at room temperature with no agitation.

6. Equilibrate gel in 100 ml Buffer 4 (50 mM Tris–HCl, pH 8.0, 10% glycerol, 0.1 mM EDTA, 1 mM DTT) at room temperature with gentle agitation for 15 min.

[16] J. E. Brownell and C. D. Allis, *Proc. Natl. Acad. Sci. USA* **92,** 6364 (1995).
[17] C. A. Mizzen, J. E. Brownell, R. G. Cook, and C. D. Allis, *Methods Enzymol.* **304,** 675 (1999).

7. Place gel in a heat sealable plastic bag that is not much larger than the gel itself. Add 3 ml fresh Buffer 4 containing 5 μCi [^{3}H]acetyl-CoA. Seal the bag being careful to get as much air out as possible.

8. Make sure the buffer is distributed about the entire gel. Incubate for 1 h at 30°. Optimal reaction times must be determined based on the amount of enzyme activity recovered.

9. To terminate the acetylation reaction, place the gel in Coomassie blue staining solution for about 30 min.

10. Destain with several changes of the destaining solution to reduce background staining. Gels can then be processed for fluorography as described earlier and exposed to film.

Mapping Acetylation Sites

After HAT activity has been established, there are several methods that can be employed to identify sites acetylated by specific HATs *in vitro*. The first method depends on the ability to generate short peptides representing sequences which exist, for example the amino acid sequence of the histone H3 N-terminal tail, with only certain lysines available as acetylation sites. The peptides are synthesized with certain lysines preacetylated so that they are no longer available to act as receptors for ^{3}H-Ac in HAT reactions. Therefore, all of the HAT activity obtained in the reaction will be a result of acetylation at the unacetylated available sites. These peptides are then used in place of histones or nucleosomes as the substrates for HAT assays as described earlier. Assays are performed using 300 ng of peptides and equivalent amounts of enzyme for each reaction.[18]

Acetylation sites can also be mapped by microsequencing. HAT assays are performed as described earlier, subjected to SDS-PAGE on 15% gels, and then transferred to Immobilin P^{SQ} membrane (Millipore, Bedford, MA). Prior to setting up the transfer, the membrane is soaked in 100% methanol, followed by water, and is finally equilibrated in transfer buffer (48 mM Tris, 39 mM glycine, 20% methanol) + 0.0375% SDS. The SDS gel itself is equilibrated in transfer buffer + 0.05% SDS. 0.0375% SDS is added to the buffer during the transfer as well.[19] Determination of radioactivity at each position in the sequence defines which residues are acetylated since the ^{3}H is incorporated directly into the acetyl modification itself. Therefore, labeled histones of interest can be excised from the membrane, sequenced as described,[20] and the radioactivity at each position

[18] P. A. Grant, A. Eberharter, S. John, R. G. Cook, B. M. Turner, and J. L. Workman, *J. Biol. Chem.* **274**, 5895 (1999).

[19] M. Vettese-Dadey, P. A. Grant, T. R. Hebbes, C. Crane-Robinson, C. D. Allis, and J. L. Workman, *EMBO J.* **15**, 2508 (1996).

can be determined by liquid scintillation counting. In the case of histone H4 an extra deblocking step is required since its N-terminus is naturally blocked *in vivo*.

Finally, acetylation sites can be mapped by Western blotting with antiacetyl antibodies. Purified histones can be used in HAT assays *in vitro* or native histones can be extracted in the following manner from logarithmically growing yeast cells. The histones are then run on SDS-PAGE gels, transferred to Immobilin P^{SQ}, and analyzed for the presence of specific acetyl-lysines *in vivo* via Western blotting with antibodies against specifically acetylated lysines (e.g., commercially available from Upstate Inc).

1. Grow a 50 ml culture of your desired yeast strain in YPD or selective drop out media to an OD_{600} of about 1, to give you approximately 10×10^8 cells—Pellet.

2. Resuspend the cells in 400 μl H_2SO_4, add 0.45 g glass beads (SIGMA G-8772). Homogenize 5 min at maximum speed.

3. Centrifuge at 13,000 rpm for 10 min at 4°.

4. Collect 300 μl of the supernatant and add TCA (trichloroacetic acid) to a final concentration of 25%. Put on ice for at least 20 min.

5. Repeat step #3.

6. Remove supernatant and add 200 μl 100% ethanol. Vortex and leave at $-20°$ overnight.

7. Repeat step #3. Remove supernatant and dry pellet.

8. Add 20 μl 1× loading buffer (100 μl 10% SDS, pH 7.4, 100 μl 0.625 M Tris–Cl, pH 6.8, 100 μl glycerol, 50 μl Bromophenol blue [saturated solution], 50 μl 2-mercaptoethanol, 600 μl H_2O), vortex at least 20 s.

9. Boil 5 min, put on ice 10 min. Centrifuge at 13,000 rpm for 10 min at 4°.

10. Run the supernatants out on 15% SDS-PAGE gels, transfer to Immobilin P^{SQ} as described before, and analyze by Western blotting with antiacetyl antibodies.

These protocols for mapping acetylation sites can also be adapted to other types of histone modifications (e.g., methylation, phosphorylation).

Purification of Native Human HAT Complexes

Protocols for purification of native chromatin modifying complexes from mammalian cells have improved in efficiency and yields in recent years. A popular approach uses a combination of classical chromatographic steps (e.g., phosphocellulose and DEAE-cellulose) followed by a final

[20] R. E. Sobel, R. G. Cook, and C. D. Allis, *J. Biol. Chem.* **269,** 18576 (1994).

immunoaffinity purification using polyclonal antibodies or an epitope-tagged subunit.[21] More recently direct immunoaffinity purification from nuclear extracts using multiple tags has become popular (e.g., FLAG and HA tags added to the same protein).[22] The method presented here relies on the use of multiple affinity tags in order to purify to near homogeneity native multiprotein HAT complexes from human cells. Although eliminating the need for multiple classical and technically challenging chromatographic steps, it requires the engineering of cell lines expressing functional target fusion proteins. We took advantage of the powerful tandem affinity purification procedure (TAP) developed in yeast[8] and combined it with the popular FLAG epitope to achieve the purification of diverse low abundance HAT complexes from HeLa cells. We focused on human members of the MYST family of HATs.[23] For example, we successfully purified the TIP60-associated complex to near homogeneity in an amount sufficient for subunit identification by mass spectrometry.[24] TIP60 is highly related to the essential Esal protein from *S. cerevisiae*, is a transcriptional coactivator for steroid nuclear receptors and NFκB and it has been implicated in the cellular response to DNA damage and apoptosis.[25] Moreover, TIP60 is in low abundance in the cell since it is highly regulated by a Mdm2-dependent degradation pathway,[26] which emphasizes the efficiency of our protocol. The purification scheme consists of a standard TAP purification procedure[9] followed by immunopurification on α-FLAG-crosslinked beads as depicted in Fig. 3A. In our experience, depending on the abundance of the protein to be purified, it is possible to skip the final α-FLAG step. Alternatively, the α-FLAG immunoaffinity can be performed as the first step of the purification, though it should be considered that more FLAG peptide is required at this stage of the purification, which significantly increases the cost of the experiment.

Generation of Cell Lines

A critical factor for the choice of vector used to drive the expression of the transgene is the amount of endogenous target protein. The expression of the fusion protein must be sufficient to compete with the endogenous

[21] Y. Xue, J. Wong, G. T. Moreno, M. K. Young, J. Cote, and W. Wang, *Mol. Cell* **2,** 851 (1998).

[22] T. Ikura, V. V. Ogryzko, M. Grigoriev, R. Groisman, J. Wang, M. Horikoshi, R. Scully, J. Qin, and Y. Nakatani, *Cell* **102,** 463 (2000).

[23] R. T. Utley and J. Cote, *Curr. Top Microbiol. Immunol.* **274,** 203 (2003).

[24] Y. Doyon, W. Selleck, W. S. Lane, S. Tan, and J. Cote, *Mol. Cell. Biol.* in press (2004).

[25] M. Carrozza, R. T. Utley, J. L. Workman, and J. Cote, *Trends Genet.* **19,** 321 (2003).

[26] G. Legube, L. K. Linares, C. Lemercier, M. Scheffner, S. Khochbin, and D. Trouche, *EMBO J.* **21,** 1704 (2002).

protein for stable multisubunit complex formation *in vivo*, but not overexpressed as this will lead to the isolation of partial complexes due to subunit titration. Moreover, the use of an inducible system of protein expression allows one to bypass potential problems with protein toxicity (e.g., when using mutant forms of an essential protein). We modified the retroviral Tetracyclin-regulated system from Clontech in order to control the expression of an N-terminal FLAG-tagged and C-terminal TAP-tagged TIP60 fusion protein in HeLa S3 cells. Transduction by retrovirus vectors also limits the risk of overexpression versus the level of endogenous protein because of the very low number of integration events per cell. Figure 3B illustrates the retroviral vector we constructed for the TIP60 HAT protein which was used to transduce the HeLa cells (expressing the Tet-OFF regulator, Clontech). Following production of infectious retroviral particles by transfection of the PA317 packaging cell line, HeLa S3 cells are infected and hygromycin selection is applied for 2 weeks at a concentration of 500 μg/ml. Resistant colonies are screened for protein expression by western analysis and amplified for the purification.

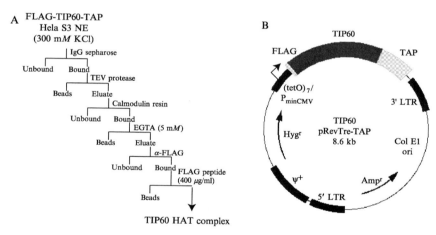

FIG. 3. Purification protocol for human multisubunit HAT complexes. (A) The native complex purification scheme with three affinity purification steps. The last step is necessary only for low abundance protein complexes like the TIP60 complex. (B) The modified retroviral expression vector containing the tandem affinity purification (TAP) cassette. The tetracyclin response vector pRevTre (Clontech) was modified to produce C-terminal TAP-tagged fusion proteins by subcloning a BamH1/HindIII fragment of pBS1479[9] into the respective sites of pRevTre. The FLAG-tag is carried over from prior cloning of TIP60 in pCDNA3 (Invitrogen) followed by PCR amplification and subcloning of FLAG-TIP60 into the BamH1 site of pRevTre. Following infection, Hela S3 cells can be selected using the hygromycin resistance (Hyg^r) gene and protein expression is controlled using the tetracycline derivative doxycycline.

HAT Complex Purification

Nuclear extracts are prepared following standard procedures (adjusted to 0.1% NP-40[27]) from transduced Hela S3 cells (3–6 l at 1–1.5 × 10^6 cells/ml) grown in suspension in Minimum Essential Medium Joklik's Modified supplemented with 10% tetracyclin-free fetal calf serum. Approximately 100 mg of nuclear extract is precleared with 100 μl of Sepharose CL-6B (Sigma) equilibrated in IgG-sepharose wash buffer (20 mM HEPES pH 7.9, 10% glycerol, 300 mM KCl, 0.1% NP-40, 1 mM DTT, 1 mM PMSF, 2 μg/μl pepstatin, 2 μg/μl leupeptin, and 5 μg/μl aprotinin) for 45 min at 4°. Five hundred microliters of IgG-sepharose beads (Amersham Bioscience, Piscataway, NJ) are then added to the extract and rotated for 2 h at 4°. The beads are washed with 10 column volumes of IgG-sepharose wash buffer and equilibrated with 10 column volumes of TEV cleavage buffer (20 mM HEPES pH 7.9, 10% glycerol, 150 mM KCl, 0.1% NP-40, 0.5 mM EDTA, and 1 mM DTT). Bound proteins are eluted with 25 units of TEV protease (Invitrogen) in 1.25 ml of TEV cleavage buffer for 2 h at 16°. The beads are washed once with 1.25 ml of TEV cleavage buffer and this fraction is pooled with the first eluate. The pool is then incubated with 25 μl of protein A-sepharose (Amersham Bioscience) equilibrated in TEV cleavage buffer for 15 min at 4° in order to eliminate IgGs leaking from the IgG-sepharose resin. The supernatant is then diluted with three volumes of calmodulin binding buffer (20 mM HEPES pH 7.9, 10% glycerol, 150 mM KCl, 0.1% NP-40, 1 mM Imidazole, 1 mM Mg-acetate, 1 mM DTT, 2 mM CaCl$_2$ supplemented with protease inhibitors) and 7.5 μl of CaCl$_2$ 1 M is added. Five hundred microliters of calmodulin resin (Stratagene) is added and the suspension is incubated on a wheel O/N at 4°. The beads are then transferred to a polyprep column (Bio-Rad) and washed with 10 column volumes of calmodulin binding buffer. Proteins are eluted with calmodulin elution buffer (20 mM HEPES pH 7.9, 10% glycerol, 150 mM KCl, 0.1% NP-40, 10 mM β-mercaptoethanol, 1 mM Imidazole, 1 mM Mg-acetate, 5 mM EGTA, 1 mM DTT, supplemented with protease inhibitors). Five fractions of 500 μl are collected and tested for HAT activity. Active fractions (usually second and third) are pooled and incubated with 25 μl of α-FLAG M2 resin (Sigma) for 4 h at 4°. The beads are then washed with FLAG elution buffer (20 mM Tris–Cl pH 7.5, 20% glycerol, 0.1% NP-40, 100 mM KCl, 1 mM DTT, supplemented with protease inhibitors) and eluted in batch with 25 μl of FLAG elution buffer containing 400 μg/ml 3× FLAG peptide (Sigma) for 2 h. This step is repeated four times. The

[27] J. L. Workman, I. C. Taylor, R. E. Kingston, and R. G. Roeder, *Methods Cell Biol.* **35,** 419 (1991).

fractions are tested for activity and pooled. For gel analysis the samples are precipitated at $-80°$ with 9 volumes of acetone:triethylamine:acetic acid in a ratio of 90:5:5, washed with cold acetone, resuspended in $1\times$ SDS loading dye, boiled for 5 min and loaded on a 10% Tris-glycine gel followed by silver staining. Preparative gels for tandem mass spectrometry are stained with the sensitive dye Sypro ruby (Bio-Rad) and bands are excised. In our hands this protocol is very efficient for purification of native human multisubunit complexes and can be applied to other types of chromatin modifying complexes.

Together, these assays describe methods to purify and analyze native HAT complexes from yeast and human cells. We have described different methods that can be carried out with or without antibodies against specific proteins.

Acknowledgments

P.A.G. is the recipient of a Burroughs Wellcome Fund Career Award in Biomedical Sciences. This work is supported by a NIDDK Grant DK58646 to P.A.G. Y.D. is supported by a Canadian Institute of Health Research (CIHR) studentship and J.C. is supported by an operating Grant from the Cancer Research Society.

[9] Histone Deacetylases: Purification of the Enzymes, Substrates, and Assay Conditions

By NATALIE REZAI-ZADEH, SHIH-CHANG TSAI, YU-DER WEN, YA-LI YAO, WEN-MING YANG, and EDWARD SETO

Histone deacetylase (HDAC) enzymes catalyze the removal of acetyl groups from the epsilon amino group of conserved lysine residues in the amino terminal tail domains of histones. This modification strengthens histone-DNA interactions and potentially generates specific docking surfaces for proteins that regulate chromatin folding and/or transcription. In addition to their crucial roles in chromatin regulation, HDACs are involved in many different cellular processes such as cell cycle control and differentiation.

Although the discovery of HDACs dates back to 1969,[1] research in this area of histone modification was stagnant for a long time because attempts

---

[1] A. Inoue and D. Fujimoto, *Biochem. Biophys. Res. Commun.* **36,** 146 (1969).

to purify HDACs to homogeneity by conventional chromatography were largely unsuccessful. In 1996 the field advanced significantly with the isolation and cloning of the first bona fide histone deacetylase, HDAC1.[2] Since that time, merely 7 years ago, close to 2000 papers have been published on this subject. Now, most discussions in eukaryotic transcriptional repression mention some aspect of histone deacetylation. In addition to increasing our understanding of the regulation of transcription in eukaryotic cells, the study of HDACs is important because there is a direct connection between HDAC function and a number of human diseases. HDACs play a role in the function of a number of cellular oncogenes and tumor suppressor genes. Several drugs that inhibit HDAC activity are currently in clinical trials for treatment of cancer. A thorough understanding of HDACs, therefore, will not only augment our knowledge of chromatin structure and transcription, but will directly contribute to the overall understanding of normal and abnormal cellular processes.

In humans, HDACs are divided into three categories[3–7]: the class I RPD3-like proteins (HDAC1, HDAC2, HDAC3, and HDAC8); the class II HDA1-like proteins (HDAC4, HDAC5, HDAC6, HDAC7, HDAC9, and HDAC10); and the class III SIR2-like proteins (SIRT1-7). The class II enzymes can further be divided into two subclasses: IIa (HDAC4, HDAC5, HDAC7, and HDAC9) and IIb (HDAC6 and HDAC10). HDAC11, the most recently discovered human HDAC, uniquely shares sequence homology to the catalytic regions of both class I and class II HDAC enzymes. The class III proteins do not show any sequence resemblance to other HDACs and differ from other HDACs in that their activity requires the cofactor NAD. This chapter focuses on methods and approaches for studying human class I and class II HDACs. Readers are referred to two excellent reviews on protocols for studies of the NAD-dependent class III enzymes.[8,9]

[2] J. Taunton, C. A. Hassig, and S. L. Schreiber, *Science* **272,** 408 (1996).
[3] X. J. Yang and E. Seto, *Curr. Opin. Genet. Dev.* **13,** 143 (2003).
[4] A. J. De Ruijter, A. H. Van Gennip, H. N. Caron, S. Kemp, and A. B. Van Kuilenburg, *Biochem. J.* **370,** 737 (2003).
[5] W. D. Cress and E. Seto, *J. Cell Physiol.* **184,** 1 (2000).
[6] S. Khochbin and H. Kao, *FEBS Lett.* **494,** 141 (2001).
[7] S. Khochbin, A. Verdel, C. Lemercier, and D. Seigneurin-Berny, *Curr. Opin. Genet. Dev.* **11,** 162 (2001).
[8] J. S. Smith, J. Avalos, I. Celic, S. Muhammad, C. Wolberger, and J. D. Boeke, *Methods Enzymol.* **353,** 282 (2002).
[9] J. Landry and R. Sternglanz, *Methods* **31,** 33 (2003).

Expression and Purification of Recombinant HDACs

All class I and many of the class II HDACs can easily be expressed as recombinant proteins in bacteria, thereby serving as an excellent source for antibody production. However, with the exception of HDAC8, we have found that none of the proteins produced in *E. coli* are enzymatically active. In contrast, many of the HDACs expressed in insect cells possess deacetylase activity (albeit sometimes at a very low level) and can be useful for most functional experiments. For examples, see Hassig *et al.*[10] and Tsai *et al.*[11] Here, we provide a protocol for the expression and purification of HDAC2 using the baculovirus system in Sf9 insect cells. We have successfully used this protocol to express and purify all other class I enzymes. It is important to note that other methods for successful expression and purification of HDACs exist, and each laboratory may have their own preferred method of expressing recombinant HDAC proteins.

Generation of Recombinant Baculoviruses and Expression and Purification of HDAC2

1. Subclone HDAC2 cDNA into the pFastBac-HTa vector (Invitrogen). This vector introduces a six histidine tag downstream of the polyhedrin promoter at the N-terminus of the resultant protein.

2. Transform recombinant plasmid into DH10Bac competent *E. coli* cells (Invitrogen). Briefly, aliquot 100 μl of competent cells into a chilled polypropylene tube and mix with DNA. Incubate for 30 min on ice then heat-shock for 45 s at 42°. Add 0.9 ml of SOC medium, incubate with shaking for 4 h, and then plate on selective medium containing X-Gal and IPTG.

3. Select white colonies, grow overnight in 1.5 ml of media, and verify recombinant bacmid with HDAC2 DNA using polymerase chain reaction (PCR) with M13 primers. The PCR consists of one cycle of initial denaturation at 94° for 2 min, followed by 30 cycles of denaturation at 94° for 45 s, annealing at 55° for 45 s, and extension at 72° for 4 min, plus the final extension at 72° for 10 min. The expected size of PCR product is approximately 4 Kb.

4. Prepare high molecular weight mini-prep DNA from selected *E. coli* clones containing the recombinant bacmid using standard alkaline lysis method.

---

[10] C. A. Hassig, J. K. Tong, T. C. Fleischer, T. Owa, P. G. Grable, D. E. Ayer, and S. L. Schreiber, *Proc. Natl. Acad. Sci. USA* **95,** 3519 (1998).

[11] S. C. Tsai, N. Valkov, W. M. Yang, J. Gump, D. Sullivan, and E. Seto, *Nat. Genet.* **26,** 349 (2000).

5. Transfect recombinant bacmid DNA into Sf9 insect cells using CellFECTIN reagent (Invitrogen) according to manufacturer's protocol. Harvest recombinant baculovirus from cell culture medium 72 h after transfection.

6. Amplify recombinant virus to produce a high titer stock: add virus supernatant to Sf9 cells at a Multiplicity of Infection (MOI) of 0.01 to 0.1. Once 15–20% cell survival is achieved, harvest virus supernatant by centrifugation at $2000g$ for 3 min at room temperature.

7. Infect insect Sf9 cells with viral particles (virus/cell ratio of 10:1) and collect cells after 2 days by centrifugation at $500g$ for 2 min at room temperature.

8. Resuspend 5×10^7 cells in 4 ml of ice-cold lysis buffer (50 mM Tris–HCl [pH 7.5], 100 mM NaCl, 1 mM MgCl$_2$, 5 mM β-mercaptoethanol, 10% glycerol, and 0.5% Triton X-100) containing 1 mM phenylmethane-sulfonyl fluoride and 1$\times$ protease inhibitor mix without EDTA (Roche). Centrifuge at $10,000g$ for 10 min.

9. Incubate supernatant with Ni-NTA (nitrilo-triacetic acid) resin (Qiagen) or the Talon affinity resin (Clontech) on a rotator for 1 h at 4°.

10. Wash resin successively with 10 volumes of buffer A (20 mM Tris–HCl, 1 M KCl, 20 mM Imidazole, 10% [v/v] glycerol, 5 mM 2-mercap-toethanol, pH 8.5) and 2 volumes of buffer B (20 mM Tris–HCl, 1 M KCl, 10% [v/v] glycerol, 5 mM 2-mercaptoethanol, pH 8.5).

11. Elute protein with buffer C (20 mM Tris–HCl, 100 mM KCl, 200 mM imidazole, 10% [v/v] glycerol, 5 mM 2-mercaptoethanol, pH 8.5).

12. Determine protein concentration by the Bradford method using the Bio-Rad dye reagent with BSA as a standard. Analyze purified HDAC2 protein by SDS-PAGE and Coomassie Blue staining (see Fig. 1).

Purification of HDACs from Mammalian Cells

After the discovery of the first HDAC, it immediately became apparent that many HDACs are members of protein complexes and function properly only together with other factors. For example, the metastasis-associated protein 2 (MTA2) modulates the enzymatic activity of HDAC1/2[12] and the activity of HDAC3 is exquisitely regulated by nuclear hormone receptor corepressor (N-CoR) and the closely related silencing mediator for retinoic acid and thyroid hormone receptors (SMRT) protein.[13,14] Here, we

[12] Y. Zhang, H. H. Ng, H. Erdjument-Bromage, P. Tempst, A. Bird, and D. Reinberg, *Genes Dev.* **13**, 1924 (1999).
[13] Y. D. Wen, V. Perissi, L. M. Staszewski, W. M. Yang, A. Krones, C. K. Glass, M. G. Rosenfeld, and E. Seto, *Proc. Natl. Acad. Sci. USA* **97**, 7202 (2000).
[14] M. G. Guenther, O. Barak, and M. A. Lazar, *Mol. Cell. Biol.* **21**, 6091 (2001).

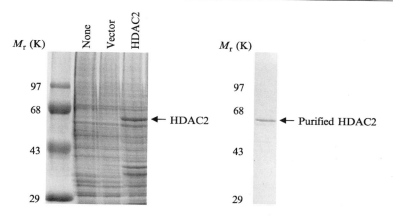

FIG. 1. Expression of recombinant HDAC2 protein in insect cells. A full-length HDAC2 cDNA was subcloned into the pFASTBacHTa vector (Invitrogen) in-frame with respect to the six histidine residues using linker/adaptors. Transformation of DH10Bac cells, transfection, and subsequent infection of insect cells, and viral amplification were performed as described. Proteins were resolved on a 10% SDS-polyacrylamide gel followed by Coomassie blue staining. Left panel, whole cell lysates showing His-HDAC2 fusion proteins over-expressed in Sf9 cells. Right panel, purified His-HDAC2.

provide a protocol for immunoaffinity purification of an endogenous HDAC3 complex that preserves the activating cofactors. Because the success of this type of purification relies heavily on the quality of the antibodies, it is imperative that one first optimize immunoprecipitation conditions for the antibody of interest.[15] In principle, this method can be applied to purification of any HDAC complex provided the necessary antibodies are available.

Immunoaffinity Purification of an Endogenous HDAC3 Complex from HeLa Cells

1. For purification of anti-HDAC3 peptide antibody, incubate 1 ml Affi-gel 10 resin (Bio-Rad) with 1 mg of peptide antigen in 10 ml of PBS at $4°$ for 3 h. Wash and resuspend resin in 2 ml of PBS. Add $1\,M$ ethanolamine/HCl (61 μl of ethanolamine in 939 μl of $1\,M$ HCl) to a final concentration of 0.1 M. Incubate peptide-bound resins at $4°$ for another hour. Load resin into a disposable column (Bio-Rad) and wash with 30 ml PBS to remove unbound peptide.

---

[15] E. Harlow and D. Lane, "Using Antibodies: A Laboratory Manual." Cold Spring Harbor Laboratory Press, Cold Spring Harbor, 1999.

2. Dilute 10 ml of anti-HDAC3 serum into 50 ml of PBS and slowly load the resulting mixture onto the column. Wash column with 30 ml of PBS. Elute antibodies with 0.2 M glycine (pH 2.2). Collect approximately 700 μl of eluate in microcentrifuge tubes containing 300 μl of 1 M K_2HPO_4. Combine fractions containing antibodies and dialyze in a buffer containing 20 mM Tris–HCl (pH 7.9), 0.1 mM EDTA, 0.1 mM DTT, 0.05% NP40, 10% glycerol, and 100 mM NaCl. Store antibody at -20° in small aliquots.

3. Prepare an anti-HDAC3 affinity column as described[13] by adding 2 mg of purified anti-HDAC3 antibody to 1 ml of protein A–agarose beads. For control, prepare a separate anti-HDAC3 column identically but preincubate with HDAC3 peptide (200 μg/ml) at 4° for 2 h before mixing with cell extracts.

4. Culture 12 L of HeLa cells (0.5 $\times$ 10^6 cells/ml). Harvest cells by centrifugation for 10 min at 2500g and wash twice with PBS. Resuspend cell pellet in 10 ml of buffer A (50 mM Tris–HCl [pH 7.9], 0.1 mM EDTA, 0.2 mM PMSF, 0.5% Triton X-100, and 10% glycerol) containing 150 mM KCl. Incubate lysates at 4° for 20 min with inversion and centrifuge for 30 min at 15,000g.

5. Load supernatants onto the anti-HDAC3 antibody affinity column and wash extensively with buffer B (50 mM Tris–HCl [pH 7.9], 0.1 mM EDTA, 0.2 mM PMSF, 2 mM DTT, 150 mM KCl, and 10% glycerol).

6. Elute bound proteins stepwise with HDAC3 peptide antigen (200 μg/ml) and collect approximately 200 μl of eluate per microcentrifuge tube.

7. Determine which fractions contain HDAC3 by Western blotting with anti-HDAC3 antibody. It is also helpful to determine HDAC activity in each fraction at this time.

8. Combine fractions containing HDAC3 protein and concentrate to approximately 200 μl using Centron 10 (Millipore). Resolve an aliquot of sample on an SDS gel and analyze by silver stain (see Fig. 2).

A major obstacle in the purification of HDACs or HDAC complexes is that, unfortunately, most anti-HDAC antibodies are not suitable for immunoaffinity purifications. To circumvent this problem, we have routinely used recombinant adenoviruses to over-express recombinant epitope-tagged HDACs in mammalian cells and have purified the tagged-HDAC proteins with immunoaffinity columns. We have successfully used this system to express and purify nearly all of the class I and class II HDACs, and as an example, we provide a protocol here for the expression and purification of HDAC5 in HeLa cells.

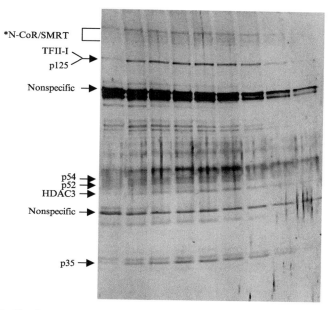

FIG. 2. Purification of an endogenous HDAC3 complex. Silver-stained SDS-polyacrylamide gel of the HDAC3 complex is shown. Modified from Wen et al.,[13] with permission from the National Academy of Sciences, USA.

Expression of Epitope-Tagged HDAC5 Using Recombinant Adenoviruses and Purification of an HDAC5 Complex with Immunoaffinity Chromatography

1. Use the AdEasy system[16] to generate a recombinant adenovirus that expresses GFP and an HA/FLAG-tagged HDAC5. For convenience, we engineered the HA epitope tag at the N-terminus in-frame with HDAC5 and the FLAG tag fused to the C-terminal end. Store purified virus in PBS containing 10% glycerol in cryotubes at $-80°$.

2. Culture HeLa cells to 90% confluency in several six well plates. Using filtered pipet tips, dispense different amounts of virus into each of the wells. A good start would be 0.5 μl of virus stock for one well, 1 μl for the second well, 1.5 μl for the third well, 2 μl for the fourth well, etc.

---

[16] T. C. He, S. Zhou, L. T. da Costa, J. Yu, K. W. Kinzler, and B. Vogelstein, *Proc. Natl. Acad. Sci. USA* **95**, 2509 (1998).

Gently rock the plates a few times to mix virus with culture medium. Incubate overnight at 37°.

3. Examine GFP expression under an inverted fluorescence microscope. Visually determine which concentration of virus stock results in the maximum number of GFP-expressing cells without causing cell death.

4. Add the appropriate amount of virus stock to two 150-mm plates of HeLa cells grown to 90% confluence. For example, if 1 μl of virus stock was optimal for cells grown in a 6-well plate, then use 14 μl of virus stock to infect cells grown in each 150-mm dish.

5. After 2 days, remove the medium from cells, wash cells twice with 10 ml of PBS, replace with 10 ml of PBS and scrape cells into a 50 ml conical tube. Collect cells by centrifugation for 5 min at 3500 rpm at 4°. A Beckman J6-HC centrifuge and a JS-4.2 rotor can be used for this purpose.

6. Remove PBS and lyse cells by adding 5 ml (2.5 ml per 150-mm plate) of PBS with 10% glycerol and 0.1% NP40 plus a complete protease inhibitor cocktail (Roche).

7. Sonicate sample for 10 s with a sonic dismembrator. Then, centrifuge for 10 min at 10,000 rpm with a Beckman JA2550 rotor at 4°. Transfer supernatant into a fresh tube.

8. Analyze a small aliquot for HDAC5 expression by Western blotting with anti-HA or anti-FLAG antibody. Save the remaining sample for further purification.

9. Working in the cold room, equilibrate a Poly-Prep® chromatography column (Bio-Rad, cat. no. 731-1550) with a few milliliters of buffer (PBS containing 10% glycerol).

10. Add 200 μl (100 μl per one 150-mm plate of cells) of Anti-Flag® M2-agarose affinity gel (Sigma A-2220) and let it settle in the column.

11. Wash column with 10 ml of buffer and then cap bottom of the column.

12. Add the total volume of lysate to the column, cap column top, and mix lysate with agarose beads by gentle rotation for 1 h in the cold room.

13. Uncap column and collect flow-through.

14. Wash column with buffer. Continue with wash until the absorbance of the wash samples is less than 0.01 at 280 nm.

15. Elute bound protein with 1 ml of 0.1 mg/ml FLAG peptide and collect at least five fractions of eluate (200 μl each fraction). Analyze a small aliquot from each fraction by silver staining.

16. Pool fractions containing HA-HDAC5-FLAG and reload into a second column prepared with monoclonal anti-HA agarose conjugate (Sigma A-2095).

17. Wash and eluate identically with HA peptide. Collect fractions and analyze a small aliquot again from each fraction by silver staining (see Fig. 3).

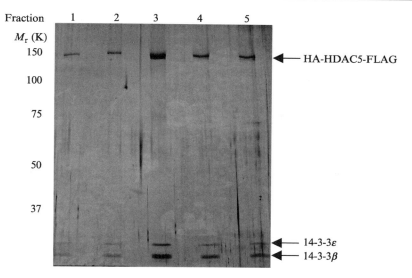

FIG. 3. Purification of an HDAC5 complex using over-expression of an epitope-tagged HDAC5. Silver stain of purified proteins after resolving by SDS-PAGE is shown. The identity of 14-3-3 was determined by Western blot with anti-14-3-3 specific antibodies.

Protocols Related to Histone Deacetylase Assay

A large number of protocols have been described for preparation of substrates for HDAC activity assay. The first, and perhaps still most commonly used, procedure involves incubation of immature chicken erythrocytes, calf thymus nuclei, or HeLa cells with ^{3}H- or ^{14}C-acetate follow by isolation of radiolabeled, hyperacetylated histones.[1,17–19] A chief advantage in using these biologically acetylated substrates is that they contain physiologically relevant, naturally acetylated sites. An alternative to *in vivo* labeling of histones is to use purified recombinant histone acetyltransferase to deposit ^{3}H-acetate onto purified histones *in vitro*.[20] A benefit to using this method is that the histone substrates can be labeled with very high specific activity. However, this method requires the preparation of high quality histone acetyltransferase. A third method involves chemical acetylation of histones with ^{3}H- or ^{14}C-acetic anhydride. While this last method offers the

[17] M. J. Hendzel, G. P. Delcuve, and J. R. Davie, *J. Biol. Chem.* **266,** 21936 (1991).
[18] A. A. Carmen, S. E. Rundlett, and M. Grunstein, *J. Biol. Chem.* **271,** 15837 (1996).
[19] J. M. Sun, V. A. Spencer, H. Y. Chen, L. Li, and J. R. Davie, *Methods* **31,** 12 (2003).
[20] P. A. Wade, P. L. Jones, D. Vermaak, and A. P. Wolffe, *Methods Enzymol.* **304,** 715 (1999).

advantage of obtaining very high specific activity substrates, it does present the problem of introducing nonspecific acetyl groups onto lysine residues.

Using unlabeled, acetylated peptides corresponding to the N-termini of histones H3 or H4, Guarente and co-workers analyzed Sir2 deacetylase activity by high-pressure liquid chromatography (HPLC).[21] Recently, a fluorogenic HDAC assay was developed that is well suited for high-throughput activity screening.[22] Nonradioactive HDAC fluorescent activity assay and HDAC colorimetric assay kits are commercially available from BIOMOL®.

In this section, we provide a basic protocol that uses a labeled peptide corresponding to the N-terminal of histone H4 as a substrate. Although this procedure does not generate substrates with naturally acetylated sites, it offers a quick and very convenient way to measure HDAC activity. The original procedure was pioneered in the Schreiber lab[2] and has been modified in our laboratory. It is important to note that many histone deacetylases also are capable of deacetylation of nonhistone cellular proteins. Therefore, this method can be applied to studies of nonhistone proteins by synthesizing and labeling peptides corresponding to the nonhistone substrate of interest.

Labeling of H4 Peptide with [³H] for Histone Deacetylase Assays

1. Synthesize a H4 peptide (SGRGKGGKGLGKGGAKRHRKVLR) corresponding to residues 2 to 24 of histone H4 with a free amine at the N-terminus and an amide at the C-terminus. Synthetic H4 peptide should be purified by HPLC to greater than 90% purity.

2. For radiolabeling of the synthetic H4 peptide, add 5 mCi of [³H]acetic acid (NEN cat. #NET-003H, 2–5 Ci/mmol in ethanol) to 0.4 mg of the H4 peptide. Add 10 μl of freshly prepared BOP solution (0.24 M BOP and 0.2 M triethylamine in acetonitrile) to the peptide/label mix and rock gently on a rocker at room temperature overnight.

3. Use a Microcon-SCX spin column (Millipore cat. #42460) to purify the labeled peptide. Prewash column with 500 μl of 10 mM HCl in methanol once and then with 500 μl of 10 mM HCl in 10% methanol. Spin down wash solutions from the column and discard.

4. Load 250 μl of the labeling mixture onto the prewashed column and spin column at 1200g for 1 min.

5. Wash column twice with 500 μl of 10 mM HCl in 10% methanol. Invert column and place in a new collection tube.

[21] S. Imai, C. M. Armstrong, M. Kaeberlein, and L. Guarente, *Nature* **403,** 795 (2000).
[22] D. Wegener, F. Wirsching, D. Riester, and A. Schwienhorst, *Chem. Biol.* **10,** 61 (2003).

6. To elute the labeled peptide, apply 50 μl of 3 N HCl in 50% isopropanol to the column and spin at 14,000g for 15 s. This step can be repeated once to ensure complete elution of the labeled peptide.

7. Dry labeled peptide in a fume hood with the cap of the collection tube open or in a SpeedVac. Add 500 μl of dd H_2O to dissolve the peptide. Aliquot dissolved radioactive peptide and store at $-80°$.

To estimate the purity of the radiolabeled peptide, the following equation is used:

$$\frac{\text{Purified CPM}}{\text{Ethylacetate} - \text{Extractable CPM}}$$

where purified CPM refers to the CPM of the final purified peptide and the ethyl acetate–extractable CPM refers to the CPM of the same volume of the purified peptide diluted in dd H_2O and extracted by ethyl acetate. A successful acetylation reaction should yield a value not smaller than 1000.

Histone Deacetylase Assay

1. For each reaction, the following reagents are mixed in a microcentrifuge tube: 40 μl of 5× HDAC buffer (50 mM Tris–HCl [pH 8.0], 750 mM NaCl, and 50% glycerol), 20,000 cpm [^{3}H]acetyl histone H4 peptide, the HDAC sample, and dd H_2O to a total volume of 200 μl.

2. Incubate reactions at room temperature overnight.

3. Stop reaction by the addition of 50 μl of stop solution (1 M HCl and 0.4 M acetic acid).

4. Extract released [^{3}H]acetate by adding 400 μl of ethyl acetate to the stop reaction. Vortex the mixture briefly, and centrifuge at 14,000g for 3 min at room temperature to separate phases.

5. Transfer 200 μl of the organic phase to a scintillation vial and measure CPM. Examples of reaction results are shown in Fig. 4.

A popular technique for analyzing histone deacetylation without radioisotopes uses triton-acid-urea (TAU) gels, first described by Panyim and Chalkley.[23] This method uses electrophoresis to separate various histone subtypes and allows monitoring of different forms of acetylated histones. We provide a very brief protocol here for using the TAU gel for assaying HDAC activity. For a comprehensive review on this technique, see Lennox and Cohen.[24]

[23] S. Panyim and R. Chalkley, *Arch. Biochem. Biophys.* **130,** 337 (1969).
[24] R. W. Lennox and L. H. Cohen, *Methods Enzymol.* **170,** 532 (1989).

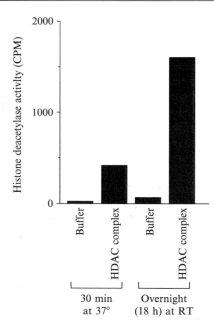

FIG. 4. Representative experiments to determine HDAC enzymatic activity. Histone deacetylase activity was assayed from a complex of proteins consisting of FLAG-HDAC1, His-HDAC2, RbAP46, and RbAP48 expressed and purified from Sf9 cells. Reactions were performed either at 37° for 30 min or overnight at room temperature. Each assay was carried out in duplicate from three independent samples, and the values shown are the averages.

Preparation and Electrophoresis of TAU Gels

1. To prepare separating gel mix and loading gel mix follow the combinations according to the table below.

	Separating gel (40 ml)	Loading gel (10 ml)
Urea (g)	18.0	4.5
40% acrylamide (ml)	14.4	1.8
30% Triton X-100 (ml)	0.5	0.123
Acetic acid (ml)	2	0.5
Add H_2O (ml)	40	10

2. Pour a TAU gel similar to standard SDS-polyacrylamide gels with the loading gel on top of the separating gel.

3. Prerun TAU gel as follows: use 5% acetic acid as the gel running buffer and wash away the urea from the well on the loading gel. Apply 30 μl of overloading solution (2.5 M urea, 5% acetic acid, 0.02% pyronin Y) on the surface of the well. Connect electrodes with the cathode ($-$) at the bottom and the anode ($+$) at the top of the gel. Apply 300 V to the gel.

4. When the dye front migrates to the bottom of the gel, wash the surface of the well and apply 10 μl of scavenge solution (2.5 M urea, 5% acetic acid, 0.02% pyronin Y, 2.5 M cystemine) to the well. Run gel at 300 V until the dye front migrates to the bottom.

5. Apply 10 μl of scavenge solution to the well again and run gel as before. Then replace the gel running buffer with a new batch and apply 10 μl of protamine sulfate solution (2.5 M urea, 5% acetic acid, 0.02% pyronin Y, 25 mg/ml) to the well. Run gel at 300 V until the dye front reaches the bottom.

6. When the protamine sulfate run is complete, insert a shark tooth comb into the well. Prepare a 2× protamine sulfate solution as the 2× loading buffer. Mix equal volumes of the sample and 2× loading buffer and apply to the well.

7. Use cytochrome C (30–40 μg per lane) as a marker. Carry out electrophoresis until cytochrome C migrates about 1.5 cm above the bottom of the gel.

8. Stain gel with Coomassie Brilliant Blue.

Concluding Remarks

The studies of HDACs, as with the studies of almost any area in histone modification, have experienced tremendous growth and rapid changes in the last several years. The analysis of HDACs is highly complex and will require a combination of many different approaches. The intent of this article is not to present a comprehensive approach to study HDAC enzymes. Rather, it is our desire to provide beginners a starting point to apply some of these techniques to their ongoing research. There is no reason to believe that the protocols presented here are superior to alternative methods currently employed by other laboratories studying HDACs. We do hope that some of the methods we presented will help in providing answers to many intriguing questions in the HDAC field.

Acknowledgments

We thank Nancy Olashaw for critical review of the manuscript. This work was supported by Grants from the NIH (GM64850, GM58486) to E.S.

[10] Measurement of Mammalian Histone Deacetylase Activity

By Eric Verdin, Franck Dequiedt, Wolfgang Fischle, Roy Frye, Brett Marshall, and Brian North

All core histone proteins are reversibly and dynamically acetylated at multiple sites in their amino-terminal tails. Hyperacetylated histones are generally found in transcriptionally active genes, and hypoacetylated histones in transcriptionally silent regions, such as heterochromatin. The level of histone acetylation at a particular locus in chromatin reflects the competing activities of histone acetyltransferases and histone deacetylases (HDACs). The identification and characterization of numerous transcriptional regulators possessing histone acetyltransferase or HDAC activities has validated the prediction that histone acetylation plays a critical role in transcriptional regulatory mechanisms (for general review see Cheung et al.[1]).

So far, 18 distinct human HDACs have been identified. They are grouped into three classes based on their primary homology to three *Saccharomyces cerevisiae* HDACs. Class I HDACs (HDAC1, 2, 3, 8, and 11) are homologous to yRPD3, share a compact structure, and are predominantly nuclear proteins expressed in most tissues and cell lines (see Table I) (reviewed in Fischle et al.[2]). Class II HDACs are homologous to yHDA1 and are subdivided in two subclasses, IIa (HDAC4, 5, 7, and 9 and its splice variant MITR) and IIb (HDAC6 and HDAC10), based on sequence homology and domain organization (for recent review, see Verdin et al.[3]). While class I and class II HDACs and their *S. cerevisiae* orthologs, yRPD3 and yHDA1, all share some degree of homology in their catalytic domain, class III HDACs are homologous to ySIR2 and show no homology to class I and class II proteins (see Table I). The class III enzymes are characterized by their dependence on nicotinamide adenine dinucleotide (NAD+). This cofactor serves as an acetyl-group acceptor in the deacetylation reaction generating acetylADP ribose.[4] The identity of proteins in each class and some of their salient properties are summarized in Table I.

[1] W. L. Cheung, S. D. Briggs, and C. D. Allis, *Curr. Opin. Cell Biol.* **12**, 326 (2000).

[2] W. Fischle, V. Kiermer, F. Dequiedt, and E. Verdin, *Biochem. Cell. Biol.* **79**, 337 (2001).

[3] E. Verdin, F. Dequiedt, and H. G. Kasler, *Trends Genet.* **19**, 286 (2003).

[4] M. T. Borra, F. J. O'Neill, M. D. Jackson, B. Marshall, E. Verdin, K. R. Foltz, and J. M. Denu, *J. Biol. Chem.* **277**, 12632 (2002).

METHODS IN ENZYMOLOGY, VOL. 377

TABLE I

CHARACTERISTICS OF HUMAN HISTONE DEACETYLASES

	Class I	Class IIa	Class IIb	Class III
Protein	HDAC1, 2, 3, 8, 11	HDAC4, 5, 7, 9	HDAC6, 10	SIRT1, 2, 3, 4, 5, 6, 7
Yeast ortholog	Rpd3	Hda1	Hda1	Sir2
Subcellular distribution	Predominantly nuclear	Nucleo-cytoplasmic shuttling	Nucleo-cytoplasmic shuttling	Nuclear (SIRT1), cytoplasmic (SIRT2), mitochondrial (SIRT3)
Target(s)	Histones p53 (HDAC1), NFκB (HDAC3)	Histones	Histones tubulin	Histones tubulin (SIRT2), p53 (SIRT1), TAF(I)68 (Sir2α/SIRT1)
Tissue expression	Ubiquitous	Tissue-specific (brain, muscle, thymus)	Tissue-specific (Testis, . . .)	Unknown
Protein cofactor necessary for enzymatic activity	SMRT/N-CoR (HDAC3)	HDAC3 + SMRT/N-CoR	?	?
Expression system				
Mammalian (stable and transient)	+	+	+	+
E. coli	− (+ for HDAC8)	−	−	+ (SIRT1, 2, 3)
In vitro T&T	− (+ for HDAC3)	−	−	+ (SIRT3)
Enzyme cofactor	Zinc	Zinc	Zinc	NAD+
Inhibition by:				
Trichostatin A, SAHA	+	+	+	−
Trapoxin, Na+ Butyrate	+	+	−	−
Nicotinamide (Vitamin B3)	−	−	−	+
Tubacin	−	−	+	−

The properties of these enzymes have been described in detail in several recent, extensive reviews.[3,5,6]

This review focuses on the experimental protocols that are used to detect the enzymatic activity associated with these proteins.

Purification of Enzymatically Active Class I and II HDACs

All class I and class II HDACs have proven relatively difficult to purify in an enzymatically active form. With a couple of exceptions, all proteins need to be immunoprecipitated after transient or stable transfection in mammalian cells. While activity can be detected after immunoprecipitation of endogenous proteins, transient or stable transfections of epitope-tagged proteins allow for the standardization of immunoprecipitation protocols and for the purification of larger amounts of enzymatically active enzymes. Of note, the enzymatic activity of class I enzymes, HDAC1 and HDAC3, can be inhibited when the epitope tag is placed at the N-terminus of the protein (W. Fischle and E. Verdin, unpublished observations). For this reason, fusion proteins incorporating a C-terminal epitope have been favored.

Mammalian Cell Culture Systems

Plasmids. Plasmid constructs for the expression of different human HDACs are based on the pcDNA3.1 vector(s). In a wide range of mammalian cells, the human cytomegalovirus immediate-early promoter of this vector provides high levels of expression. In addition, the vector will replicate episomally in cell lines latently infected with SV40 or expressing the SV40 large T antigen (e.g., 293T, COS7). A neomycin (G418) resistance gene cassette allows for the selection of stable transfected cells. High-level stable and nonreplicative transient expression can therefore be carried out in most mammalian cells. To generate C-terminal epitope-tagged constructs of different HDACs, we used polymerase chain reaction (Pfu polymerase) amplification with a reverse primer containing the sequence for the FLAG peptide.[7] First, the two primers used to amplify HDAC3 were: forward = 5'-CCGGATCCGAATTC ACCATGGCCAAGACCGTGGCC-3'; backward = 5'-GCTCTAGATTA CTTGTCATCGTCGTCCTTGTA GTCTCCTCC GAATTCAATCTCCACATCGCTTTCC-3'. (Restriction sites are underlined; the sequence encoding the FLAG epitope is in italics,

[5] J. M. Denu, *Trends Biochem. Sci.* **28,** 41 (2003).
[6] P. Marks, R. A. Rifkind, V. M. Richon, R. Breslow, T. Miller, and W. K. Kelly, *Nat. Rev. Cancer* **1,** 194 (2001).
[7] W. Fischle, F. Dequiedt, M. Fillion, M. J. Hendzel, W. Voelter, and E. Verdin, *J. Biol. Chem.* **276,** 35826 (2001).

and the stop codon is in bold.) The PCR product was digested with the restriction enzymes $BamHI$ and $XbaI$ and inserted into the corresponding sites of the pcDNA3.1(+) vector by standard protocols. The cDNA sequence encoding HDAC3 was then replaced with PCR fragments encoding each of the other HDACs by subcloning into newly generated $EcoRI$ sites engineered into the PCR primers. Other vectors with different promoters for the expression in mammalian cells can be used. Besides the FLAG epitope, HA and Myc tags have also been used successfully with different human HDACs (W. Fischle and E. Verdin, unpublished observations).

Transient Transfection. Constructs based on pcDNA3.1 are used to transiently express different human HDACs in a variety of mammalian cell lines, including: 293, HeLa, Jurkat, NIH3T3, COS7, and CHO cells. Depending on the cell type, 5×10^6 to 1×10^7 cells (corresponding to a 10 cm culture dish) are transfected according to standard procedures either by the $CaPO_4$ coprecipitation method or by lipofection. Cells are maintained in growth medium for 24–48 h before harvesting to allow for accumulation of HDAC proteins. For further experimental details, see mammalian cell culture system section.

Note. When the transfection efficiency ranges from 50–90%, 5×10^6 to 1×10^7 cells are sufficient for expression and immunoprecipitation of enough HDAC protein for up to four radioactive enzymatic HDAC assays.

Generation of Stably Transfected HDAC–FLAG Cell Lines. pcDNA3.1 vectors encoding human HDACs fused to the FLAG-epitope are linearized by digestion with the restriction enzymes $SspI$ (HDACs1–5 and HDAC7) or $SacI$ (HDAC6) and purified by agarose gel electrophoresis. 293 HEK cells of a low passage number (1×10^5 in a 6-well dish) are transfected with the linearized pcDNA3.1 HDAC-FLAG expression plasmids by lipofection according to common procedures. After 48 h, the cells are split into 15-cm tissue culture dishes at 3×10^4, 1×10^5, and 3×10^5 cells/dish. For selection of stable expressing cell clones, cultures are incubated in standard culture medium containing 700 μg/ml G418 sulfate until visible colonies appear. This takes approximately 3 weeks, and the medium should be changed every 3–4 days. Nontransfected cells (1×10^6 cells/dish as a control) die over this selection period. Plates are rinsed with phosphate-buffered saline (PBS), and single colonies are isolated by trypsinization using cloning cylinders. For further culturing of the stable transfected cell clones, the G418 sulfate concentration is reduced to 450 μg/ml. As soon as the clones reach more than 1×10^6 cells, they are frozen in liquid nitrogen. Expression of HDAC-FLAG is analyzed by western blotting of total cellular extracts (see later) with anti-FLAG antibodies, and small-scale immunoprecipitations are tested for HDAC activity in enzymatic assays (see later).[8,9] Indirect immunofluorescence is used to verify protein expression in the majority of cells. Single clones are expanded for further use if they test positive for expression of exogenous proteins, have high expression levels of HDAC-FLAG, and high deacetylase activity in immunoprecipitation-mediated HDAC activity assays. Detailed analysis of extracts from 293 HEK cells stably expressing HDAC1-FLAG by gradient centrifugation and gel filtration chromatography verified that the majority of the fusion protein is incorporated in high molecular weight multiprotein complexes similar to the endogenous HDAC1 protein. In addition, coimmunoprecipitation verified the interaction of HDAC1-FLAG with Sin3 and Mi2, components of different endogenous HDAC1 complexes (not shown).

Several aliquots of the expanded clonal cell lines are frozen in liquid N_2 (90% fetal calf serum, 10% dimethyl sulfoxide) at low passage numbers (10–15 passages after initial selection). Cells are routinely grown in medium containing 450 μg/ml G418. However, omission of the antibiotic for 6 weeks (approximately 10 passages) did not result in a significant reduction in the expression level of HDAC1-FLAG. In addition, large-scale spinner cultures have been grown for several weeks without addition of G418 sulfate. However, the viability of the clonal HDAC-FLAG cell lines is significantly lower than that of the parental cell line, and on a number of occasions, expression of HDAC has been lost during continuous culture. Therefore, cultures must be monitored for expression of the HDAC fusion protein, particularly after new cultures are started from frozen stocks.

Notes. The same protocol is suitable for generation of stable transfected clones derived from other cell lines. 293 HEK cells were chosen because they show high levels of HDAC expression in transient transfections and can easily be adapted for growth in suspension cultures. After maintaining clones in culture for more than 40 passages, an average of 70–80% of cells expressed the fusion constructs, as judged by immunofluorescence staining with anti-FLAG antibodies. Therefore, a second round of clonal selection by limited dilution might be necessary if a totally homogeneous cell population with optimized expression characteristics is desired.

If large-scale cell cultures of stable transfected cell lines are needed, the cells are transferred to small spinner flasks (250 ml, approximately 60 rpm, no additional aeration, 37° incubator) and grown in OptiMEM containing 2% fetal calf serum, 50 units/ml penicillin, 50 mg/ml streptomycin, 2 mM glutamine. After adaptation to suspension growth, cells are expanded to

[8] S. Emiliani, W. Fischle, C. Van Lint, Y. Al-Abed, and E. Verdin, *Proc. Natl. Acad. Sci. USA* **95,** 2795 (1998).

[9] W. Fischle, S. Emiliani, M. J. Hendzel, T. Nagase, N. Nomura, W. Voelter, and E. Verdin, *J. Biol. Chem.* **274,** 11713 (1999).

4 L in spinner flasks. Cells are harvested by centrifugation at a density of 1 $\times 10^6$ cells/ml and washed three times in ice-cold PBS buffer. Cell pellets can either be processed directly or flash-frozen in liquid N_2 and stored at $-80°$ for at least 6 months without significant loss of HDAC activity.

Immunoprecipitation of HDACs from Mammalian Cell Culture Systems

Buffers

1. Buffer A: 50 mM Tris–HCl, pH 7.5, 120 mM NaCl, 0.5 mM EDTA, 0.5% NP-40, and 10% glycerol.
2. Buffer B: 50 mM Tris–HCl, pH 7.5, 1 M NaCl, 0.5 mM EDTA, 0.5% NP-40, and 10% glycerol.
3. Buffer C: 10 mM Tris–HCl, pH 8.0, 10 mM NaCl, and 10% glycerol.
4. Buffer D: 10 mM HEPES (NaOH), pH 7.9, 10 mM KCl, 1.5 mM $MgCl_2$, and 0.5 mM dithiothreitol (DTT).
5. Buffer E: 20 mM HEPES (NaOH), pH 7.9, 450 mM NaCl, 1.5 mM $MgCl_2$, 1% NP-40, 0.5 mM DTT, 0.2 mM EDTA, and 25% glycerol.
6. Buffer F: 50 mM Tris–HCl, pH 7.5, 120 mM NaCl, 0.5 mM EDTA, 0.1% NP-40, and 10% glycerol.

All buffers are filtered (0.4 μm) and stored at 4°. DTT and protease inhibitors are added immediately before use as needed.

Procedure. Fresh or frozen cell pellets are resuspended in buffer A containing protease inhibitors at a ratio of 5 volumes of buffer to 1 volume of wet cell mass. Lysis is allowed for 30 min at 4° with rotation. Large volumes of cell lysates (>5 ml) are transferred to a Dounce homogenizer and homogenized on ice by 30 strokes with pestle B. Lysates are cleared by centrifugation at 14,000g for 30 min at 4°.

Alternatively, fresh cell pellets are resuspended at a ratio of 5 volumes of buffer D containing protease inhibitors to 1 volume of wet cell mass and incubated on ice for 10 min. Swollen cells are recovered by centrifugation at 1000g for 10 min at 4° and resuspended in 2 volumes of buffer D containing protease inhibitors. Cell membranes are broken by homogenization in a Dounce homogenizer (pestle B, 30 strokes, on ice). Nuclei are pelleted by centrifugation at 2000g for 10 min at 4°. After removal of the supernatant representing the cytoplasmic extract, nuclei are resuspended in 1 volume buffer E containing protease inhibitors. Nuclei are extracted by homogenization in a Dounce homogenizer (pestle B, 4°, 15 strokes every 5 min for 40 min). Extracts are cleared by centrifugation at 25,000g for 30 min at 4° and dialyzed three times against 20 volumes of buffer F containing 1 mM phenylmethylsulfonyl fluoride at 4°. Nuclear extracts are finally cleared by centrifugation at 25,000g for 30 min at 4°. Extracts can be flash-frozen in liquid N_2 and stored at $-80°$ for at least 6 months without significant loss of HDAC activity.

Exogenous HDACs from mammalian cell lines transiently or stably overexpressing HDACs can be immunoprecipitated on a small- or large-scale depending on the need. For small-scale immunoprecipitation, whole cell lysates are incubated with M2 agarose beads at 10 ml/ml (4 mg M2 antibody/ml) overnight with rocking at 4°. HDAC-FLAG immunocomplexes are recovered and washed as described for the immunoprecipitation of endogenous HDACs. For large-scale immunoprecipitations, cell lysates or nuclear extracts are incubated with M2 agarose beads (4 mg M2 antibody/ml) using 5 μl/ml resin for 4 h with rocking at 4°. Immunocomplexes bound to the agarose beads are recovered by centrifugation at 3000g for 10 min at 4°. The supernatant is incubated with M2 agarose for a second time (5 μl/ml, 4 h). M2 agarose beads from the two immunoprecipitations are combined and packed into a small disposable column. The column is washed with 100 volumes of buffer A, 100 volumes of buffer B, and 30 volumes of buffer C. HDAC-FLAG containing complexes are eluted in five column volumes of buffer C containing 0.25 mg/ml FLAG peptide at a low flow rate. At this step, the partially purified material can be aliquoted and stored at −80° for several months without loss of enzymatic activity. Two to three freeze—thaw cycles do not seem to interfere with the enzymatic activity.

To immunoprecipitate endogenous HDACs, cell lysates or nuclear extracts are incubated with antibodies (IgG fraction) specific for human HDACs at 1 μg/ml for overnight at 4°. Immunocomplexes are recovered by adding 20 μl/ml Protein G-Sepharose beads with rocking at 4° for 3 h. Agarose beads are recovered by centrifugation at 3000g for 5 min at 4° and washed three times with buffer A, three times with buffer B, and two times with buffer C (about 1 ml of buffer per 10 ml of beads, each wash for 5–10 min with rocking at 4°, careful aspiration of the supernatant using gel loader tips). After the final wash, immune complexes are either directly used for enzymatic HDAC activity assays or stored in the residual volume of buffer C at −80° (stable for several months).

A representative experiment with transfected HDAC1, 3, and 4 is shown below (see Fig. 1A), including an inhibition study of HDAC1 and 3 by trichostatin A and trapoxin (see Fig. 1B).

Notes. Only nuclear class II HDACs 4, 5, and 7 were found to be enzymatically active. Unlike the enzymatically inactive cytoplasmic fractions, these are incorporated into large multiprotein complexes.[10] A significant fraction of nuclear HDAC activity seems to be stably associated with the nuclear matrix and must be solubilized with detergents.

---

[10] W. Fischle, F. Dequiedt, M. J. Hendzel, M. G. Guenther, M. A. Lazar, W. Voelter, and E. Verdin, *Mol. Cell* **9**, 45 (2002).

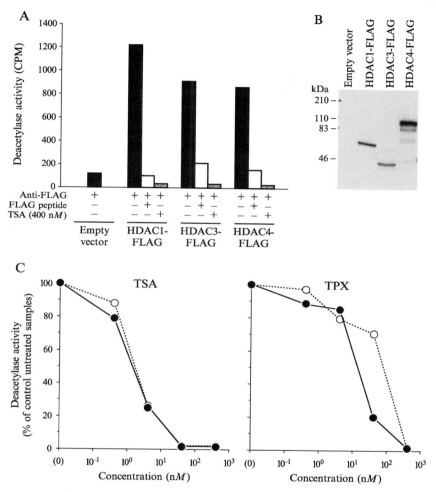

FIG. 1. HDAC activity associated with HDAC1, 3, and 4. (A) Cell extracts prepared from 293 HEK cells transfected with plasmids encoding the different HDACs were immunoprecipitated with M2 antibody in the presence or absence of the FLAG peptide (100-fold molar excess over antibody). Immunoprecipitated material was assayed in the presence or absence of trichostatin A (TSA, 400 nM) on a ^{3}H-acetylated histone H4 peptide (see later). The deacetylase activity (measured as cpm released) from a representative experiment is shown. (B) Cell lysates were prepared from cells transfected with the different HDACs, and HDAC expression was analyzed by western blot with anti-FLAG antiserum. (C) Immunoprecipitated HDAC1 (open circles) and 3 (black circles) were tested for inhibition by TSA and trapoxin (TPX) over a range of concentrations. Results are expressed as percentage of the activity obtained in the absence of inhibitors. Adapted with modifications from Refs. 8 and 9.

For further purification, the eluate can be loaded directly onto a gel filtration column (Superose 6 HR 10/30) equilibrated in buffer C. Elution of HDAC-FLAG protein is monitored by Western blotting and enzymatic HDAC assays. For example, HDAC1-FLAG elutes in a broad peak centered at about 1 MDa with some protein detected in lower molecular mass fractions. High molecular mass HDAC1-FLAG fractions are pooled and stored at $-80°$.

In vitro *Translated Protein*

Recent experiments have provided evidence that, while HDAC3 is inactive when expressed alone, it can be activated enzymatically when incubated with its corepressor, SMRT or N-CoR, *in vitro*.[10,11] These experiments are consistent with the hypothesis that other class I and class II HDACs also depend on unique cofactors that regulate their enzymatic activity. Such a model might provide a plausible explanation for our failure to purify recombinant enzymatically active proteins when these proteins are overexpressed alone. We provide below a protocol for the expression of enzymatically active HDAC3 using proteins translated *in vitro*.

Recombinant FLAG-tagged N-CoR and Myc-tagged human HDAC3, HDAC4, HDAC5, and HDAC7, all cloned in pcDNA3.1 (see mammalian cell culture systems section), are translated independently *in vitro* using the TNT T7 Quick coupled transcription-translation kit (Promega), according to the manufacturer's protocol using 200 μl of reticulocyte lysate. Increasing amounts (0, 2, 10, 50, or 200 μl) of each recombinant HDAC are mixed with 100 μl of unprogrammed reticulocyte lysate ($-$N-CoR) or with 100 μl of reticulocyte lysate programmed for FLAG-N-CoR. Mixed proteins are diluted three times with immunoprecipitation buffer (20 mM HEPES, pH 7.9, 300 mM KCl, 0.25 mM EDTA, 10% glycerol, 0.1% Tween 20) supplemented with protease inhibitors (Boehringer Mannheim, complete cocktail). Complexes are immunoprecipitated with anti-Myc-beads (Santa Cruz Biotechnology) overnight at $4°$. Beads are washed three times with immunoprecipitation buffer and twice in 20 mM Tris–HCl, pH 8.0, 150 mM NaCl, and 10% glycerol. One half of the immunoprecipitated material is tested for HDAC activity with the histone H4 peptide acetylated *in vitro* (see Fig. 2). HDAC3, but not the other HDACs, is activated upon binding to N-CoR. The second half of the immunoprecipitated material is analyzed for N-CoR by Western blotting with an anti-Flag antiserum (not shown). Western blots show that N-CoR coimmunoprecipitates readily with HDAC3, 4, 5, and 7 (data not shown).

---

[11] M. G. Guenther, O. Barak, and M. A. Lazar, *Mol. Cell. Biol.* **21**, 6091 (2001).

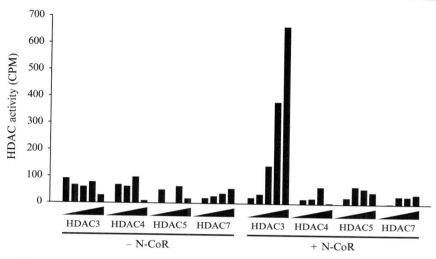

FIG. 2. HDAC3, but not class II HDACs (HDAC4, 5, and 7) is activated by binding to SMRT. Myc-tagged HDAC3, 4, 5, and 7 were translated *in vitro*. Increasing amounts of each translation reaction (0, 2, 10, 50, or 200 μl of reticulocyte extract for each HDAC) were added to unprogrammed reticulocyte lysates (−N-CoR) or to translation reactions containing N-CoR (+ N-CoR). The resulting mixtures were immunoprecipitated with an anti-Myc epitope antiserum and tested for HDAC activity. The immunoprecipitation step is required in this procedure since unprogrammed reticulocyte lysates contain significant amounts of endogenous HDAC activity. Adapted with modifications from Ref. 10.

Notes

1. Decreasing the immunoprecipitation incubation time to 2 h leads to a very significant reduction in the recovery of enzymatically active HDAC3.
2. The same experiment can be performed by cotranslating each HDAC individually together with N-CoR. The immunoprecipitated HDAC3-N-CoR complex works equally well as an enzymatically active HDAC as when the proteins are translated independently, but not better.
3. SMRT can be used in place of N-CoR with similar efficiency.

Purification of Enzymatically Active Class III HDACs

Class III enzymes have generally proven easier to manipulate than class I and class II HDACs. While not all class III HDACs are known to have histone deacetylase activity (see later), proteins with high enzymatic activity can be purified by immunoprecipitation after transient or stable transfection in mammalian cells. Recombinant proteins can also be expressed in *E. coli* in an enzymatically active form.

Mammalian Cell Culture Systems

Each of the class III human HDACs has been cloned in the pcDNA3.1 vector using the same strategy as described under mammalian cell culture systems section for class I and class II HDACs.

To purify active enzyme, HEK 293T cells are transiently transfected with the expression vector by the calcium phosphate DNA precipitation method. The DNA precipitate is left on the cells for 8–16 h followed by a medium change. Cells are washed twice in PBS by centrifugation and resuspended in fresh PBS. After the last centrifugation, the cell pellet is lysed in five packed cell pellet volumes of lysis buffer (buffer A see earlier) in the presence of protease inhibitor cocktail (Complete; Roche Molecular Biochemicals, Indianapolis, IN), for more than 30 min at $4°$ with agitation. After lysis, cellular debris are cleared by centrifugation at $14,000g$ for 10 min at $4°$, and the supernatant is transferred to a new precooled tube.

The resulting lysates are subjected to immunoprecipitation with an anti-FLAG peptide antiserum. We favor an anti-FLAG antibody already conjugated to agarose beads (anti-FLAG M2 agarose affinity gel; Sigma). Protein concentration is measured in cell lysates with a detergent-compatible protein assay kit (DC protein assay kit, Biorad). Anti-FLAG M2 affinity gel should be added at 10 μl/ml of lysate and incubated for at least 2 h at $4°$ with constant agitation. After incubation, immune complexes are pelleted with the agarose beads by centrifugation at $6000g$ for 5 min at $4°$. The supernatant should be carefully removed from beads so as to not disturb the pellet. The immune complexes are washed three times in buffer A for 15 min each at $4°$ with agitation. After each wash, the beads are pelleted and the supernatant is replaced with buffer A. The immune complexes are then washed twice for 15 min each in the buffer used for the enzymatic assay. The beads containing the immunoprecipitated material are now ready to use in the assay.

Using this protocol, significant histone deacetylase activity is detected with SIRT1, 2, 3, and 5 (see Fig. 3) while SIRT2 only shows tubulin deacetylation activity (see Fig. 3). These experiments suggest that the other SIRT proteins, SIRT4, 6, and 7, might target other substrates for deacetylation or that other cofactors, proteins or small molecules, are needed for their enzymatic activity.

Notes. HEK 293T cells do not attach tightly to the plastic of the culture dish and tend to come off during the washes. We therefore favor removing the cells from the plate by pipetting and transfer into a conical tube.

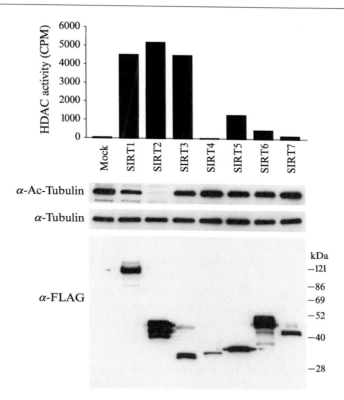

Fig. 3. Measurement of histone and tubulin deacetylase activity associated with human SIRT proteins. HEK 293T cells were transfected with expression vectors for each of the human SIRT proteins. One half of the immunoprecipitated material was subjected to HDAC activity assay with a [^{3}H] acetylated histone H4 peptide. The other half of the immunoprecipitated protein corresponding to each SIRT-FLAG protein was incubated with total cellular lysate. The reaction products were separated by SDS-PAGE and visualized by western blotting with antisera specific for acetylated tubulin, tubulin, and FLAG. Adapted with modifications from North et al.[12]

This lysis buffer does not give complete nuclear lysis in all cell types and should be supplemented with sonication if nuclear proteins are to be efficiently released into the lysate. Determination if nuclear lysis has occurred can be monitored by Trypan blue to visualize both cell lysis and the presence of intact nuclei.

[12] B. J. North, B. L. Marshall, M. T. Borra, J. M. Denu, and E. Verdin, Mol. Cell 11, 437 (2003).

Expression in Escherichia coli

In contrast to class I and class II enzymes, class III HDACs can be readily purified as enzymatically active protein after overexpression in *E. coli*.

Plasmid. Full-length human SIRT2 cDNA was cloned into pHEX, a modified version of pGEX-2T (Pharmacia) in which the GST-encoding sequence was replaced with a hexahistidine-encoding sequence (6xHis). The resulting protein is an N-tagged 6xHis-SIRT2 fusion protein (see Frye[13] for details). This vector is transformed in DH5αF'IQ bacteria (Gibco) for expression.

Buffers

1. Buffer G: 50 mM NaH$_2$PO$_4$, pH 8.0, 300 mM NaCl, 10 mM imidazole.
2. Buffer H: 50 mM NaH$_2$PO$_4$, pH 8.0, 300 mM NaCl, 20 mM imidazole.
3. Buffer I: 50 mM NaH$_2$PO$_4$, pH 8.0, 300 mM NaCl, 250 mM imidazole.
4. Buffer J: 50 mM Tris–HCl, pH 8.0, 0.2 mM DTT, 10% glycerol, 50 mM NaCl.
5. Buffer K: 50 mM Tris–HCl, pH 8.0, 0.2 mM DTT, 10% glycerol.
6. Buffer L: 50 mM Tris–HCl, pH 8.0, 265 mM NaCl, 0.2 mM DTT, 10% glycerol.

Procedure

1. One liter of transformed bacterial culture is grown in LB-Amp to an optical density of 0.6 (A$_{600}$), induced with 0.1 mM IPTG at 37° for 2 h, and pelleted.
2. The pellet is resuspended in buffer G (2 ml/g wet weight) and incubated on ice for 30 min in the presence of 1 mg/ml lysozyme. This mixture is sonicated on ice (four 10–15 s bursts at 40–60% power) and centrifuged at 4° at 14,000g for 30 min.
3. Supernatant (cleared lysate) is bound to Ni-NTA resin (1 ml of 50% Ni-NTA slurry for 4 ml of cleared lysate, batch method, Qiagen) on a rotary mixer at 4° for 60 min.
4. The batch mixture is passed through a commercial column (Polyprep, BioRad), and the flow-through is saved. The resin bed is washed twice with 4 ml of buffer H.
5. Bound proteins are eluted four times with 0.5 ml of elution buffer (buffer I). SDS-PAGE analysis revealed that the second eluting fraction contains a majority of recombinant protein along with contaminating proteins.

[13] R. A. Frye, *Biochem. Biophys. Res. Commun.* **260,** 273 (1999).

6. Ion-exchange chromatography is used to further purify the recombinant protein. The peak eluted fraction is shifted to ion exchange buffer (buffer J) with a HiPrep26/10 desalting column (Pharmacia, 2.0 ml/min flow rate, 1.0 ml fractions).
7. The peak fractions (as determined by UV monitoring or SDS-PAGE) are pooled (5–7, 1 ml fractions) and loaded on a Sepharose Q column and washed with buffer A.
8. Recombinant SIRT2 is eluted with a linear gradient of 50–1000 mM NaCl in buffer K. Under these conditions, SIRT2 elutes at ~265 mM NaCl.
9. Peak fractions are pooled and concentrated using centrifuge concentrating spin columns (Centricon, MWCO 30 KD). Recombinant protein is aliquoted and stored at $-20°$ in buffer.

Notes

1. This protocol yields a recombinant SIRT2 protein with high enzymatic activity and >90% purity as determined by SDS-PAGE.
2. Using a linear gradient of imidazole (20–500 mM) for elution of SIRT2 bound to the Ni-NTA resin has yielded a significantly purer recombinant preparation (J. Denu, personal communication).
3. The above protocol has also been used in our laboratory to purify an N-terminally truncated form of SIRT3 with high enzymatic activity (B. Marshall, B. Schwer, and E. Verdin, unpublished observations).
4. Recombinant purified SIRT1 enzyme can be obtained using a similar protocol and is commercially available (BIOMOL, Plymouth Meeting, PA).
5. Examples of enzymatic reactions performed with SIRT2 proteins expressed in *E. coli* are shown later (see Fig. 4). The activity of the protein purified from *E. coli* is compared to that of the protein expressed in mammalian cells and its inhibition by nicotinamide and pH dependency are illustrated.

Enzymatic Substrates

A large variety of substrates and assay systems have been used to detect the deacetylase activity of classes I, II, and III enzymes. A recent method uses purified histone acetyltransferase to acetylate isolated nucleosomes *in vitro* with high specific activity.[14] This method offers the very significant advantage of a physiologically relevant substrate (chromatin) and should prove useful for examining the activity of HDACs on their natural substrates. A variety of novel nonradioactive assays, based on fluorescent

[14] P. A. Wade, P. L. Jones, D. Vermaak, and A. P. Wolffe, *Methods Enzymol.* **304,** 715 (1999).

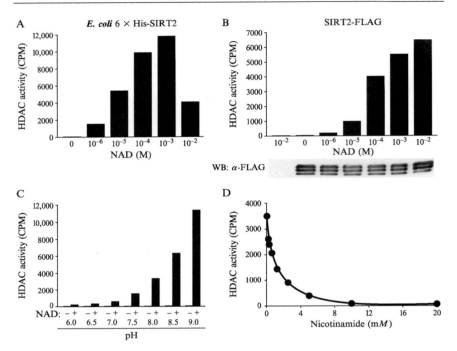

FIG. 4. Measurement of HDAC activity associated with SIRT2 expressed in *E. coli* and in mammalian cells. (A) The enzymatic activity of recombinant $6 \times$ His-SIRT2 on a [^{3}H] acetylated histone H4 peptide was measured in the presence of increasing concentrations of NAD (0, 0.001, 0.01, 0.1, 1.0, and 10 mM). Released acetate was extracted and measured by scintillation counting. (B) The enzymatic activity of immunoprecipitated protein from mock– or SIRT2-FLAG–transfected 293T cells on a [^{3}H] acetylated histone H4 peptide was measured in the presence of increasing concentrations of NAD (1 mM for mock and 0, 0.001, 0.01, 0.1, 1.0, and 10 mM for SIRT2-FLAG). Ten percent of immunoprecipitated input into the enzymatic reaction was analyzed by SDS-PAGE and by Western blotting with an antiserum specific for the FLAG peptide. (C) The activity of $6 \times$ His-SIRT2 on a [^{3}H] acetylated histone H4 peptide was measured in a range of pH (6–9) in the presence of 1 mM NAD. (D) Immunoprecipitations and reactions similar to those described in (B) of SIRT2-FLAG $\pm$ 1 mM NAD were tested in the presence of increasing concentrations of nicotinamide (0, 0.1, 0.3, 0.6, 1.2, 2.5, 5, 10, and 20 mM).

oligopeptides[15,16] or on an acetylated lysine derivative[17,18] have been developed recently. Since an extensive description of these substrates is beyond the scope of this review, we describe a sensitive assay using a peptide corresponding to the N-terminal extremity of histone H4 acetylated *in vitro*.

[15] K. Hoffmann, R. M. Soll, A. G. Beck-Sickinger, and M. Jung, *Bioconjug. Chem.* **12,** 51 (2001).
[16] D. Wegener, F. Wirsching, D. Riester, and A. Schwienhorst, *Chem. Biol.* **10,** 61 (2003).
[17] B. Heltweg and M. Jung, *Anal. Biochem.* **302,** 175 (2002).
[18] B. Heltweg and M. Jung, *Arch. Pharm. (Weinheim)* **335,** 296 (2002).

Peracetylation of Histone Peptides using [^{3}H] Acetate

Synthetic peptides corresponding to the amino-terminal tails of either histone H3 (amino acids 1–24: ARTKQTARKSTGGKAPRKQLATKA) or histone H4 (amino acids 1–24: SGRGKGGKGLGKGGAKRHRKV LRD) can be chemically acetylated *in vitro* using the protocol below. The advantages of this technique are the large amount of substrate obtained and its high specific activity. However, the peracetylation reaction leads to the modification of nonlysine residues in the peptide and requires the manipulation of volatile radioactive material and an HPLC purification step (modified from Taunton *et al.*).[19]

Procedure

1. Lyophilize 200 μg of peptide (stored at 2 mg/ml in water).
2. Resuspend peptide in 250 ml of [^{3}H] acetic acid (NEN NET003H, 10 Ci/mmol, 50 mCi/ml in ethanol) in a 5-ml scintillation vial using a small magnetic bar.
3. Add 5 ml of BOP (Aldrich #22,608-4, 0.24 M, 0.12 mmol) and triethylamine (0.2 M, 2 mmol) in acetonitrile (this solution should be freshly prepared) and stored overnight at room temperature.
4. Dry under vacuum (Speedvac, no heat, do not over dry).
5. Dissolve residue in 500 ml 10% MetOH/H_2O, 0.1% trifluoroacetic acid.
6. Separate peracetylated peptide from free acetate by HPLC on a reverse-flow column (RPC18 column, flow rate 1 ml/min, gradient from 0 to 40% acetonitrile, 0.1% trifluoroacetic acid in 40 min, Vydac #218TP54). Several acetylated forms of the peptide are detected, with the largest peak corresponding to pentaacetylated H4 peptide.
7. Peak fractions from different HPLC runs are combined and counted. We generally obtain from 2×10^8 to 4×10^8 cpm of acetylated peptide per reaction. The purified peptide is further characterized by analytical SDS-PAGE (20% Tris/Tricine).
8. Test radiolabeled peptide in HDAC assay. We typically engage from 2×10^5 to 10×10^5 counts/min cpm per enzymatic reaction. A background of $\sim$200 cpm or lower is expected.
9. The peptide is stored in small aliquots at $-70°$. A progressive increase in background counts is observed over time during extended storage ($>$3 months).

[19] J. Taunton, C. A. Hassig, and S. L. Schreiber, *Science* **272,** 408 (1996).

Enzymatic Assays

Class I and Class II HDACs

Complexes immobilized on beads from immunoprecipitations are incubated in 100 μl of HDAC buffer (10 mM Tris–HCl, pH 8.0, 10 mM NaCl, and 10% glycerol) containing 100,000 cpm of ^{3}H-histone H4 peptide substrate for 2 h at 37° with agitation. Reactions are stopped by adding 25 μl of 100 mM HCl and 160 mM acetic acid. Released acetate is extracted in 500 μl of ethyl acetate by vortexing for 15 s. After centrifugation at 14,000g for 5 min, 400 μl of the ethyl acetate fraction is mixed with 5 ml of scintillation fluid and counted.

Class III HDACs

Immunoprecipitated material and recombinant SIRT2 are resuspended in 100 μl of SIRT2 deacetylase buffer containing NAD (Sigma) and 100,000 cpm of ^{3}H-histone H4 peptide substrate. Nicotinamide (resuspended in water, Sigma) is added to reactions at the indicated concentration. Reactions containing inhibitors are preincubated with all components of the reaction in the absence of NAD for 10 min at room temperature. The enzymatic reactions are started by adding NAD. Reactions are incubated for 2 h at room temperature with agitation and stopped by adding 25 μl of 100 mM HCl and 160 mM acetic acid. Released acetate is extracted in 500 μl of ethyl acetate by vortexing for 15 s. After centrifugation at 14,000g for 5 min, 400 μl of the ethyl acetate fraction is mixed with 5 ml of scintillation fluid and counted.

Notes. The product of the enzymatic reaction catalyzed by class III HDAC is not acetate but acetylADPribose.[20] It is therefore not entirely clear why the acetate release assay described earlier functions with these enzymes. It is possible that free acetate is released from acetylADPribose under these acidic experimental conditions.

Acknowledgments

We thank Sarah Sande for help with the preparation of this manuscript, John Carroll and Jack Hull for graphics, Stephen Ordway and Gary Howard for editorial assistance. W.F. is a Robert Black fellow of the Damon Runyon Cancer Research Foundation. We apologize to colleagues whose work could not be cited due to space constraints. Our research is supported by the NIH and the J. David Gladstone Institutes.

[20] K. G. Tanner, J. Landry, R. Sternglanz, and J. M. Denu, *Proc. Natl. Acad. Sci. USA* **97,** 14178 (2000).

[11]   Analysis of Histone Phosphorylation: Coupling Intracellular Signaling to Chromatin Remodeling

By ROMAIN LOURY and PAOLO SASSONE-CORSI

The N-terminal domains of histones are subjected to several types of covalent modifications, such as acetylation,[1,2] phosphorylation,[3,4] methylation,[5] but also—although not as well investigated—ADP-ribosylation[6] and ubiquitination.[7] A number of these modifications have been associated with distinct chromatin-based outputs. In particular, position-specific modifications of the histone H3 N-terminal tail have been coupled to transcriptional regulation (Lys9/Lys14 acetylation, Ser10 phosphorylation), transcriptional silencing (Lys9 methylation), histone deposition (Lys9 acetylation), and chromosome condensation/segregation (Ser10/Ser28 phosphorylation).[8,9] It is generally thought that various combinations of histone modifications may elicit differential regulation of the chromatin condensation state which correspond to distinct biological responses.[10]

Histone phosphorylation seems to be involved in a wide range of cellular processes. One essential feature of phosphorylation versus all other histone modifications is that it can be directly linked to the activation of distinct signaling pathways, and thereby to unique physiological responses. All core histones contain phosphoacceptor sites in their N-terminal domains[11]: H2A is phosphorylated on Ser1, H2B on Ser14 and Ser32, H3 on Ser10 and Ser28, while H4 gets phosphorylated on Ser1. A significant number of studies on histone phosphorylation have focused on the importance of the Ser10 residue of histone H3. The privileged position of Ser10 in H3 is mainly due to its pivotal role in a number of cellular responses and the availability of a powerful antibody recognizing the phosphorylated

---

[1] P. D. Gregory, K. Wagner, and W. Horz, *Exp. Cell Res.* **265,** 195 (2001).
[2] S. Y. Roth, J. M. Denu, and C. D. Allis, *Annu. Rev. Biochem.* **70,** 81 (2001).
[3] L. R. Gurley, R. A. Walters, and R. A. Tobey, *Biochem. Biophys. Res. Commun.* **50,** 744 (1973).
[4] L. C. Mahadevan, A. C. Willis, and M. J. Barratt, *Cell* **65,** 775 (1991).
[5] A. J. Bannister, R. Scheider, and T. Kouzarides, *Cell* **109,** 801 (2002).
[6] A. Huletsky, C. Niedergang, A. Frechette, R. Aubin, A. Gaudreau, and G. G. Poirier, *Eur. J. Biochem.* **146,** 277 (1985).
[7] Z. W. Sun and C. D. Allis, *Nature* **418,** 104 (2002).
[8] T. Jenuwein and C. D. Allis, *Science* **293,** 1074 (2001).
[9] P. Cheung, C. D. Allis, and P. Sassone-Corsi, *Cell* **103,** 263 (2000).
[10] B. D. Strahl and C. D. Allis, *Nature* **403,** 41 (2000).
[11] S. C. Galasinski, D. F. Louie, K. K. Gloor, K. A. Resing, and N. G. Ahn, *J. Biol. Chem.* **277,** 2579 (2002).

Copyright 2004, Elsevier Inc.
All rights reserved.
0076-6879/04 $35.00

TABLE I
PHOSPHORYLATION OF HISTONES DURING DISTINCT CELLULAR PROCESSES

Cellular process	Histone	Residue	Kinase	Refs.
Gene activation	H3	Ser10	RSK2	a
		Ser10	MSK1	b
		Ser10	IKK-α	c
		Ser28	MSK1	d
Mitotic chromosome condensation	H3	Ser10	Aurora-B	e
		Ser28	Aurora-B	f
		Thr11	Dlk	g
Apoptosis	H2AX	Ser139	?	h
	H2B	Ser14	Mst 1	i
	H2B	Ser32	?	j
	H3	Ser10	?	k
Damage repair	H2AX	Ser139	ATM	l

^{a} P. Sassone-Corsi, C. A. Mizzen, P. Cheung, C. Crosio, L. Monaco, S. Jacquot, A. Hanauer, and C. D. Allis, *Science* **285,** 886 (1999).

^{b} S. Thomson, A. L. Clayton, C. A. Hazzalin, S. Rose, M. J. Barratt, and L. C. Mahadevan, *EMBO J.* **18,** 4779 (1999).

^{c} Y. Yamamoto, U. N. Verma, S. Prajapati, Y. T. Kwak, and R. B. Gaynor, *Nature* **423,** 655 (2003); V. Anest, J. L. Hanson, P. C. Cogswell, K. A. Steinbrecher, B. D. Strahl, and A. S. Baldwin, *Nature* **423,** 659 (2003).

^{d} S. Zhong, C. Jansen, Q. B. She, H. Goto, M. Inagaki, A. M. Bode, W. Y. Ma, and Z. Dong, *J. Biol. Chem.* **276,** 33213 (2001).

^{e} C. Crosio, G. M. Fimia, R. Loury, M. Kimura, Y. Okano, H. Zhou, S. Sen, C. D. Allis, and P. Sassone-Corsi, *Mol. Cell. Biol.* **22,** 874 (2002).

^{f} H. Goto, Y. Yasui, E. A. Nigg, and M. Inagaki, *Genes Cells* **7,** 11 (2002).

^{g} U. Preuss, G. Landsberg, and K. H. Scheidtmann, *Nucleic Acid Res.* **31,** 878 (2003).

^{h} E. P. Rogakou, W. Nieves-Neira, C. Boon, Y. Pommier, and W. M. Bonner, *J. Biol. Chem.* **275,** 9390 (2000).

^{i} W. L. Cheung, K. Ajiro, K. Samejima, M. Kloc, P. Cheung, C. A. Mizzen, A. Beeser, L. D. Etkin, J. Chernoff, W. C. Earnshaw, and C. D. Allis, *Cell* **113,** 507 (2003).

^{j} K. Ajiro, *J. Biol. Chem.* **275,** 439 (2000).

^{k} P. Waring, T. Khan, and A. Sjaarda, *J. Biol. Chem.* **272,** 17929 (1997).

^{l} E. P. Rogakou, D. R. Pilch, A. H. Orr, V. S. Ivanova, and W. M. Bonner, *J. Biol. Chem.* **273,** 5858 (1998).

site.[12] In more general terms, phosphorylation of histones has been coupled to various cellular processes, which are briefly summarized here and presented in Table I.

[12] M. J. Hendzel, Y. Wei, M. A. Mancini, A. Van Hooser, T. Ranalli, B. R. Brinkley, D. P. Bazett-Jones, and C. D. Allis, *Chromosoma* **106,** 348 (1997).

Mitosis

Histone phosphorylation during mitosis is a cell cycle-regulated chromatin modification.[3] H3 is phosphorylated at the Ser10 residue during mitosis, revealing the dual personality of this site.[13,14] Indeed, this modification tightly follows the process of chromosome condensation in a spatio-temporal manner,[12] while mutation of the Ser10 residue to Ala in *Tetrahymena* has been shown to lead to aberrant chromatin condensation.[13] In mammals, this modification has been shown to be involved in the initiation of the chromosome condensation process, but not for maintenance of the condensed state of the chromatin.[15] Phosphorylation at H3 Ser10 in mammals appears to be mediated by the Aurora-B kinase,[16] a protein belonging to the Ipl1/Aurora family of mitotic kinases, strongly implicated at diverse levels of chromosome segregation in yeast as well as in mammals.[17] Thus, the same phosphorylation event has a dual "personality." Indeed, histone H3 Ser10 phosphorylation appears to elicit opposite effects: chromatin opening in the case of mitogenically-induced gene activation, chromosome condensation during mitosis. This notion highlights in a clear manner the complexity of the histone code.

Additional histone phosphorylation events were reported to take place during mitosis: histone H3 Ser28 has also been reported to be phosphorylated, possibly also by Aurora-B,[18] although with slightly different kinetics than Ser10 phosphorylation.[19] The physiological significance of this event remains to be further clarified. Histone H1, the major linker histone, is also known to be phosphorylated during entry into mitosis. H1 phosphorylation depends on cdc2 kinase activity.[20] However, as recent studies have shown that chromosome condensation can occur without histone H1 phosphorylation,[21] or even without H1 itself,[22] the biological significance of H1 phosphorylation remains undefined.

[13] Y. Wei, L. Yu, J. Bowen, M. A. Gorovsky, and C. D. Allis, *Cell* **97**, 99 (1999).

[14] Y. Wei, C. A. Mizzen, R. G. Cook, M. A. Gorovsky, and C. D. Allis, *Proc. Natl. Acad. Sci. USA* **95**, 7480 (1998).

[15] A. Van Hooser, D. W. Goodrich, C. D. Allis, B. R. Brinkley, and M. A. Mancini, *J. Cell Sci.* **111**, 3497 (1998).

[16] C. Crosio, G. M. Fimia, R. Loury, M. Kimura, Y. Okano, H. Zhou, S. Sen, C. D. Allis, and P. Sassone-Corsi, *Mol. Cell. Biol.* **22**, 874 (2002).

[17] R. Giet and C. Prigent, *J. Cell Sci.* **112**, 3591 (1999).

[18] H. Goto, Y. Yasui, E. A. Nigg, and M. Inagaki, *Genes Cells* **7**, 11 (2002).

[19] H. Goto, Y. Tomono, K. Ajiro, H. Kosako, M. Fujita, M. Sakurai, K. Okawa, A. Iwamatsu, T. Okigaki, T. Takahashi, and M. Inagaki, *J. Biol. Chem.* **274**, 25543 (1999).

[20] T. A. Langan, J. Gautier, M. Lohka, R. Hollingsworth, S. Moreno, P. Nurse, J. Maller, and R. A. Sclafani, *Mol. Cell. Biol.* **9**, 3860 (1989).

[21] X. W. Guo, J. P. Th'ng, R. A. Swank, H. J. Anderson, C. Tudan, E. M. Bradbury, and M. Roberge, *EMBO J.* **14**, 976 (1995).

Activation of Gene Expression

The importance of histone N-termini in transcriptional induction is well recognized.[23,24] The first evidence for a specific kinase involved in the phosphorylation of a distinct site on a histone, came from the study of the Coffin-Lowry syndrome (CLS), an X-chromosome linked genetic disorder caused by the absence of a functional RSK2 kinase.[25] Cells derived from CLS patients display a strong impairment in the process of mitogenic stimulation. This kind of approach led to the discovery that RSK2 was responsible, after stimulation by EGF and through the Ras/MAPK cascade, for CREB phosphorylation,[26] but also histone H3 phosphorylation on Ser10,[27] in perfect correlation with the induction of immediate-early genes, as c-*fos*. Thus, RSK2 was the first physiological histone kinase to be directly linked to the activation of gene expression.

The role for this transduction-coupled modification remains to be further clarified, but it is tempting to think that histone H3 phosphorylation is restricted to the immediate-early genes, and that by a yet unknown recognition mechanism, the phosphorylated tail becomes a privileged substrate for subsequent acetylation by an histone acetyltranferase (HAT) such as CBP (CREB-binding protein) on Lys14, thus leading to promoter opening and transcription of the gene.[28] Interestingly, the enzymatic activities eliciting acetylation and phosphorylation are also coupled, as CBP and RSK2 have been found to be associated in a signaling-dependent fashion.[29]

Several other studies demonstrated the role of histone H3 Ser10 phosphorylation in response to other stimuli, as UV irradiation for example.[4] But in this case, it seems that the kinase responsible for histone H3 phosphorylation is MSK1, an effector kinase similar in structure to RSK2 that lies downstream of both the Ras/MAPK and SAPK cascades.[30] Moreover,

[22] K. Ohsumi, C. Katagiri, and T. Kishimoto, *Science* **262**, 2033 (1993).

[23] P. Allegra, R. Sterner, D. F. Clayton, and V. G. Allfrey, *J. Mol. Biol.* **196**, 379 (1987).

[24] L. K. Durrin, R. K. Mann, P. S. Kayne, and M. Grunstein, *Cell* **65**, 1023 (1991).

[25] E. Trivier, D. De Cesare, S. Jacquot, S. Pannetier, E. Zackai, I. Young, J. L. Mandel, P. Sassone-Corsi, and A. Hanauer, *Nature* **384**, 567 (1996).

[26] D. De Cesare, S. Jacquot, A. Hanauer, and P. Sassone-Corsi, *Proc. Natl. Acad. Sci. USA* **95**, 12202 (1998).

[27] P. Sassone-Corsi, C. A. Mizzen, P. Cheung, C. Crosio, L. Monaco, S. Jacquot, A. Hanauer, and C. D. Allis, *Science* **285**, 886 (1999).

[28] P. Cheung, K. G. Tanner, W. L. Cheung, P. Sassone-Corsi, J. M. Denu, and C. D. Allis, *Mol. Cell* **5**, 905 (2000).

[29] K. Merienne, S. Pannetier, A. Harel-Bellan, and P. Sassone-Corsi, *Mol. Cell. Biol.* **21**, 7089 (2001).

[30] S. Thomson, A. L. Clayton, C. A. Hazzalin, S. Rose, M. J. Barratt, and L. C. Mahadevan, *EMBO J.* **18**, 4779 (1999).

it appears that after UV irradiation Ser28 also becomes phosphorylated, possibly also by MSK1, although this event has not been fully character-ized.[31] These studies indicate that various mitogenic or hormonal stimula-tions lead to Ser10 phosphorylation, likely involving a larger number of signaling cascades and effector kinases. This notion is confirmed by the finding that the IKK-α kinase phosphorylates Ser10 in response to cytokine stimulation leading to activation of NF-κB-responsive genes.[32,33]

Programmed Cell Death

The complex gene expression program that accompanies apoptosis is likely to implicate profound changes in chromatin organization. To date, there is a lack of satisfactory information on the various histone modifica-tions that are likely involved. It seems that the type of modifications impli-cated is highly dependent on the cell type, and on the nature of the apoptotic agent used to trigger the cell death program. This notion further highlights the complexity and variability of the apoptotic processes. Yet, some common hallmarks appear to exist. Phosphorylation of histones H2A.X (a minor variant of histone H2A) and H2B (respectively on Ser139 and Ser32) seems to be induced in response to a wide variety of apoptotic agents.[34,35] The timing of these phosphorylation events seems to parallel the process of nucleosomal DNA fragmentation occurring at early stages of apoptosis.

The best characterized histone modification event that is uniquely asso-ciated with apoptotic chromatin, in species ranging from frogs to humans, is H2B phosphorylation at position Ser14. The kinase involved in this event was identified as Mst1 (mammalian sterile twenty), whose function is directly regulated by caspase-3.[36]

Other phosphorylation events have been reported, depending on the apoptotic agent used. These include Ser10 of histone H3, but also phosphorylation of histones H1 and H4.[11,37] However, some of the data

[31] S. Zhong, C. Jansen, Q. B. She, H. Goto, M. Inagaki, A. M. Bode, W. Y. Ma, and Z. Dong, *J. Biol. Chem.* **276**, 33213 (2001).

[32] Y. Yamamoto, U. N. Verma, S. Prajapati, Y. T. Kwak, and R. B. Gaynor, *Nature* **423**, 655 (2003).

[33] V. Anest, J. L. Hanson, P. C. Cogswell, K. A. Steinbrecher, B. D. Strahl, and A. S. Baldwin, *Nature* **423**, 659 (2003).

[34] E. P. Rogakou, W. Nieves-Neira, C. Boon, Y. Pommier, and W. M. Bonner, *J. Biol. Chem.* **275**, 9390 (2000).

[35] K. Ajiro, *J. Biol. Chem.* **275**, 439 (2000).

[36] W. L. Cheung, K. Ajiro, K. Samejima, M. Kloc, P. Cheung, C. A. Mizzen, A. Beeser, L. D. Etkin, J. Chernoff, W. C. Earnshaw, and C. D. Allis, *Cell* **113**, 507 (2003).

still need confirmation, as these modifications have been observed only following a limited range of treatments.

Response to Damage/Repair

An attractive link between histone phosphorylation and the damage/ repair mechanisms invokes the fascinating possibility that specific signaling pathways must be implicated in distinct damage/repair events. Following exposure to various agents that cause double-strand DNA breaks, histone H2A.X, is rapidly phosphorylated on Ser139 in mammalian cells.[38] This modification spans megabases of DNA around the lesion.[39] The same modification has been implicated in other cellular processes known to occur through genomic rearrangements, such as recombination in lymphocytes and in germ cells.[40,41] Recent studies have shown the possible implication of some members of the phosphatidylinositol-3-OH kinase-related kinase (PIKK) family in this process, notably using *in vitro* kinase assays.[42]

Methods

Chromatin Immunoprecipitation

This technique is widely used for experiments aiming at the isolation of specific subpopulations of nucleosomes, for example nucleosomes containing H3 phosphorylated at Ser10. It constitutes an useful approach to study the association of chromatin modifications with specific DNA sequences.[43] This technique can be applied also to investigate the coupling of different modifications (i.e., acetylation, methylation, or even phosphorylation on other residues) on a precise subset of nucleosomes.

1. Harvest cultured confluent cells from a 150-mm (approximately 2 $\times 10^7$ cells) in PBS, pellet the cells by centrifugation (5 min, 2000 rpm, 4°)

---

[37] P. Waring, T. Khan, and A. Sjaarda, *J. Biol. Chem.* **272,** 17929 (1997).

[38] E. P. Rogakou, D. R. Pilch, A. H. Orr, V. S. Ivanova, and W. M. Bonner, *J. Biol. Chem.* **273,** 5858 (1998).

[39] E. P. Rogakou, C. Boon, C. Redon, and W. M. Bonner, *J. Cell Biol.* **146,** 905 (1999).

[40] H. T. Chen, A. Bhandoola, M. J. Difilippantonio, J. Zhu, M. J. Brown, X. Tai, E. P. Rogakou, T. M. Brotz, W. M. Bonner, T. Ried, and A. Nussenzweig, *Science* **290,** 1962 (2000).

[41] S. K. Mahadevaiah, J. M. Turner, F. Baudat, E. P. Rogakou, P. de Boer, J. Blanco-Rodriguez, M. Jasin, S. Keeney, W. M. Bonner, and P. S. Burgoyne, *Nat. Genet.* **27,** 271 (2001).

[42] S. Burma, B. P. Chen, M. Murphy, A. Kurimasa, and D. J. Chen, *J. Biol. Chem.* **276,** 42462 (2001).

[43] A. L. Clayton, S. Rose, M. J. Barratt, and L. C. Mahadevan, *EMBO J.* **19,** 3714 (2000).

and wash in 500 μl of Buffer A (15 mM Tris–HCl, pH 7.4, 0.15 mM spermidine, 0.5 mM spermine, 15 mM NaCl, 60 mM KCl, 2 mM EDTA, 0.2 mM EGTA, 0.5 mM PMSF, 20 mM NaF) containing 0.5 M sucrose (same centrifugation parameters).

2. Lyse the cells in 1 ml of Buffer A (containing 0.5 M sucrose and 0.5% NP-40), incubate 10 min on ice.

3. Pellet nuclei by centrifugation (15 min, 2000 rpm, 4$°$) and wash them twice with Buffer A (containing 0.35 M sucrose) (same centrifugation parameters as in step 1).

4. Resuspend the nuclei in 500 μl of Digestion Buffer (50 mM Tris–HCl, pH 7.4, 0.32 M sucrose, 4 mM MgCl$_2$, 1 mM CaCl$_2$, 0.5 mM PMSF, 20 mM NaF), evaluate DNA concentration by measuring the absorbance at 260 nm of a 1:100 dilution in 2 M NaCl and dilute the nuclei suspension to a concentration of 100 μg DNA/200 μl.

5. Digest the chromatin by incubating the nuclei 10 min at 37$°$ with micrococcal nuclease (4 U/200 μl of nuclei suspension), stop the reaction by adding EDTA to a final concentration of 10 mM and cool 10 min on ice.

6. Centrifuge 10 min 13,000 rpm at 4$°$ and store the supernatant (which will be further referred as fraction S1) at 4$°$.

7. Resuspend the pellet in Lysis Buffer (1 mM Tris–HCl, pH 7.4, 0.2 mM EDTA, 0.5 mM PMSF, 20 mM NaF) (same volume as that used for Digestion Buffer) and dialyze for 12 h against 2 l of Lysis Buffer.

8. Centrifuge 1000 rpm, 10 min at 4$°$, keep supernatant (fraction S2) and resuspend pellet in Lysis Buffer (half volume of Digestion Buffer) (fraction P).

9. In order to determine the concentration of DNA, take 1/10 of fractions S1, S2, and P, add SDS to a final concentration of 1% and digest with proteinase K (final concentration 1 mg/ml) 30 min at 37$°$. After standard phenol/chloroform extraction and precipitation with ethanol, DNA will be analyzed on a 1.5% agarose gel.

10. Pool fractions S1 and S2 together, add NaCl and EDTA to final concentrations of 50 mM and 5 mM, respectively and incubate overnight at 4$°$ with rabbit polyclonal antiphospho-Ser10-histone H3 antibody (Upstate Biotechnology Incorporated, Lake Placid, USA) (1:1000 dilution).

11. Add 40 μl of a 50% slurry of Sepharose-protein A and incubate 3 h at room temperature.

12. Pellet the beads by centrifugation (2000 rpm, 5 min at 4$°$) and wash three times with 500 μl of Wash Buffer (50 mM Tris–HCl [pH 7.4], 10 mM EDTA, 50 mM NaCl, 0.5 mM PMSF, 20 mM NaF). In the case of experiments aiming at studying the coupling of different chromation modifications by Western blot, boil the beads 10 min in 30 μl of Laemmli buffer 2×.

13. Elute the DNA fragments by incubating the beads two times 15 min at room temperature in 1% SDS, pool the two fractions, perform phenol–chloroform extraction and ethanol precipitation.

The presence of modifications associated to phosphorylation on the purified nucleosomes can be analyzed following a standard Western blot approach, using antibodies raised against a modified histone (acetylation, methylation, phosphorylation on another residue, . . .).

If the experimenter has an interest in studying the association of these modified nucleosomes with the underlying DNA sequences, the fragments should be digested by nuclease S1 in order to remove all possible overhangs generated by the micrococcal nuclease digestion step, and subsequently cloned in linearized plasmids. Sequencing of the clones will thus provide information about the identity of the genes that are affected by these histone modifications. Once some of these genes are identified, quantitative PCR can be performed to assess the dynamics of histone phosphorylation. For example, in the case of cell stimulation giving rise to gene-specific histone phosphorylation and subsequent gene activation, a comparison between unstimulated and stimulated samples will give an idea about the modification status of the nucleosomes underlying the sequences of interest.

In Gel Histone Kinase Assay

This technique allows the elucidation of the molecular weights of the proteins bearing a kinase activity towards the substrate of interest. It constitutes a good starting point in order to get a clue about the identity of a kinase phosphorylating a specific substrate.[44]

Proteins from a cellular extract are separated on a SDS-PAGE gel, whose resolving gel has been polymerized in the presence of the substrate. After a cycle of denaturation/renaturation of the proteins in the gel, proteins are submitted to the kinase reaction by incubation with $[\gamma^{32}P]$-ATP. The gel is then dried and exposed (see Fig. 1).

To validate the physiological relevance of the results, several controls should be performed.

—Samples thought to contain an active kinase activity for the substrate (e.g., the cellular extract of hormone-stimulated cells) should always be compared with a sample supposed of "containing no activity" (e.g., cellular extract of an untreated cell).

—To exclude the possibility that the radioactive band is due to autophosphorylation of the protein, it is highly recommended to run

[44] M. J. Monteiro and T. I. Mical, *Exp. Cell Res.* **223,** 443 (1996).

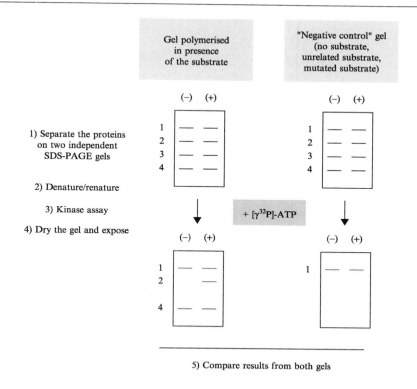

5) Compare results from both gels

Band 1 also present on the "negative control" gel: autophosphorylation of the kinase
Band 2 specific labeling of the substrate: a good candidate kinase
Band 3 no kinase activity displayed towards the substrate
Band 4 no difference with negative control

FIG. 1. Overview and interpretation of results of a representative in gel kinase assay. This technique allows the elucidation of the molecular weights of the proteins bearing a kinase activity towards the substrate of interest. Proteins from a cellular extract are separated on a SDS-PAGE gel, whose resolving gel has been polymerized in the presence of the substrate. After a cycle of denaturation/renaturation of the proteins in the gel, proteins are submitted to the kinase reaction by incubation with γ^{32}P-ATP. The gel is then dried and exposed. Examples of possible results are shown, along with their interpretation.

the same samples on a gel containing no substrate, or an unrelated substrate (e.g., BSA, ...).

—If the site of phosphorylation within the substrate has been previously identified, an effective control would be to run the samples in parallel on a gel which has been polymerized in the presence of a substrate whose phosphoacceptor residue has been mutated (serine to alanine substitution).

1. Prepare SDS-PAGE gels (8–10% acrylamide), containing in the resolving parts the substrate of interest (for histone H3 [calf thymus histone H3, Boehringer Ingelheim GmbH, Germany]: 0.1 mg/ml; BSA: 0.5 mg/ml), or even no substrate at all. For each sample, load 50 μg of protein on the gel and perform migration the same way as a standard protein electrophoresis.

2. Remove SDS from the gel by washing it twice 30 min in Washing Buffer (20% 2-propanol, 50 mM Tris–HCl, pH 8.0) at room temperature and denature the proteins by incubating the gel 20 min at 20° in Denaturing Buffer (6 M guanidine–HCl, 50 mM Tris–HCl, pH 8.0, 5 mM 2-mercaptoethanol).

3. Incubate the gel for at least 16 h at 4° in Renaturing Buffer (50 mM Tris–HCl, pH 8.0, 5 mM 2-mercaptoethanol, 0.04% Tween 40 [v/v], 100 mM NaCl, 5 mM MgCl$_2$, 1 mM DTT). Several changes of buffer will allow a better efficiency of renaturation.

4. Wash the gel three times 20 min at 20° in Kinase Buffer (40 mM HEPES, pH 7.4, 2 mM MnCl$_2$, 5 mM MgCl$_2$, 1 mM DTT, 0.2 mM EGTA).

5. Incubate the gel 4 h at room temperature in 3 ml Kinase Buffer containing 50 μM ATP and 70 μCi of [γ^{32}P]-ATP (3000 Ci/mol). In order to eliminate the unincorporated radioactivity, wash the gel four times 5 min at room temperature in 1% tetrasodium pyrophosphate-5% trichloroacetic acid.

6. Stain the gel with Coomassie blue R-250, dry the gel and expose.

As the kinase activity of a protein may undergo various regulatory steps which would not be necessarily reproduced in gel kinase assay (i.e., interaction with regulatory proteins), this approach would be useful for protein kinases whose activity is regulated by an increase in their protein amount or by a change in their modification status (e.g., phosphorylation event altering activity). Hence, the experimenter will not always be able to establish the physiological relevance of the results obtained. For example, as shown with band 3 on Fig. 1, a kinase can give no signal if it requires interaction with other proteins to be activated. On the contrary, a physiological kinase for the substrate can also give a signal in the negative control, as examplified by band 4 on Fig. 1: this situation occurs in the case of kinases activated by the release from an inhibitory protein, or in the case of kinases requiring translocation to the same cellular compartment as the substrate.

In Vitro *Kinase Assay*

This technique allows one to test the kinase activity of a given protein for a putative substrate. Histone proteins as substrate can derive from different sources.

Nucleosomes: Given the expected complexity of the "histone code," it is predictable that cells contain a wide range of different nucleosome subpopulations, all defined by unique modification states. Some prior histone modifications might affect the ability of enzymes to execute their subsequent modification. In the case of histone H3, phosphorylation at Ser10 has been shown to enhance the activity of HATs for Lys14,[28,45] but to interfere with methylation at Lys9 by Suv39h1.[46] The use of nucleosomes reconstituted *in vitro*, and thus supposedly devoided of modifications, as substrate for kinase assays will not allow the experimenter to take into account the complexity of the histone code. On the other hand, it is also possible to perform the assay on nucleosomes directly extracted from the cell, bearing in mind that possibly only a few specific subpopulations of nucleosomes remained phosphorylated. In any rate, the use of nucleosomes in the kinase assay remains the best approach that takes into account the structural requirements for the enzymatic activity.

A mix of free histones: In which all four core histones are present in equimolar amounts, but not folded as a nucleosome.

A recombinant N-terminal tail of a histone: Although reductive, the use of a recombinant protein will enable the experimenter to easily identify the modified residue, by targeted mutagenesis (e.g., Ser to Ala mutation).

The kinase to be used in the assay can either be isolated from the cell by immunoprecipitation, or generated as recombinant protein in bacteria.

1. Incubate the kinase with 15 μl 2$\times$ Kinase Buffer (40 mM HEPES, pH 7.4, 300 mM KCl, 10 mM MnCl$_2$), 5 mM NaF, 1 mM DTT, 50 μM cold ATP, 20 μM [γ^{32}P]-ATP (3000 Ci/mol) and the substrate (nucleosomes: 1–5 μg; mix of free histones: 15 μg; an N-terminal tail of the histone: 8 μg). Complete to 30 μl with distilled water, and incubate 20 min at 37°.

2. Stop the reaction by adding 30 μl of Laemmli Buffer 2$\times$, and boil 10 min.

3. Run 5 to 10 μl on a 15% acrylamide SDS-PAGE gel and rinse the gel briefly several times (8 times should be enough) with distilled water, in order to eliminate all unincorporated radioactivity.

4. Stain the gel with Coomassie blue R-250, dry it, and expose.

[45] W. S. Lo, R. C. Trievel, J. R. Rojas, L. Duggan, J. Y. Hsu, C. D. Allis, R. Marmorstein, and S. L. Berger, *Mol. Cell* **5,** 917 (2000).

[46] S. Rea, F. Eisenhaber, D. O'Carroll, B. D. Strahl, Z. W. Sun, M. Schmid, S. Opravil, K. Mechtler, C. P. Ponting, C. D. Allis, and T. Jenuwein, *Nature* **406,** 593 (2000).

GST-Pulldown Assays on Cellular Extracts

The protocol we provide below is the one we used to demonstrate the interaction between histone H3 N-terminal tail and Aurora-B, the kinase involved in mitotic H3 phosphorylation.[16] In this experiment, a cellular extract was incubated with beads previously coupled to a GST-fusion protein encompassing the first 30 aminoacids of histone H3 and the precipitate, containing histone H3 interacting partners, was analyzed for the presence of Aurora-B by Western blot. The protocol is slightly different from the standard technique due to the high insolubility of chromatin. Thus, some parameters had to be modified, especially concerning the preparation of the extract.

Although most of the standard protocols for GST-pulldown assays advise to use a physiological concentration of NaCl (between 150 and 200 mM) both during preparation of the extract and during the incubation with the GST-fusion protein, the fact that Aurora-B was tightly associated to the chromatin *in vivo*, thus displaying a high-level of insolubility, prompted us to prepare the extract using a high concentration in NaCl (600 mM) before finally lowering it to 170 mM during the incubation with the fusion protein. We therefore advise the experimenter pay special care to the NaCl concentration required to dissociate the protein from the chromatin to fully solubilize it.

Histone tails contain a high number of basic residues, which confer a high electrostatic charge and thereby high propension to nonspecific *in vitro* interactions. We advise the experimenter to use as a negative control the tail of another histone, for example H4, to discriminate between nonspecific and specific interactions. The optimal NaCl concentration for incubation and washing steps should be set up by comparing results obtained with the two tails.

1. Wash twice the cells with PBS, once with EBC buffer (50 mM Tris–HCl, pH 8.0, 50 mM NaF, 0.5 mM PMSF) containing 170 mM NaCl directly on the plate (diameter 100 mm). Lyse the cells with 1 ml of EBC buffer (containing 600 mM NaCl and 0.5% NP-40), and collect them in an Eppendorff tube.

2. Pass through a syringe several times (around 10 times), in order to solubilize the chromatin and incubate 30 min at 4° with constant stirring.

3. Centrifuge 10 min at 13,000 rpm in a microfuge, keep the supernatant (soluble fraction) and the pellet (insoluble fraction), in order to check complete solubilization of the protein.

4. Dilute the supernatant with EBC buffer (without NaCl, 0.5% NP-40) in order to bring NaCl concentration to 170 mM and measure protein concentration of the extract by Bradford assay. Dilute the extract with EBC buffer (170 mM NaCl, 0.5% NP-40) to a final concentration of 1 mg/ml.

5. For each binding reaction, preclear 1 mg of extract by incubating 1 h at $4°$ with 30 μl of beads preequilibrated in EBC buffer (170 mM NaCl, 0.5% NP-40). Centrifuge 1 min at full speed, keep the precleared extract on ice.

6. Couple the beads to the GST-fusion protein by washing 30 μl of beads/binding reaction with 500 μl BCO buffer (20 mM Tris–HCl, pH 8.0, 0.5 mM EDTA, 20% glycerol, 500 mM KCl, 1% NP-40, 1 mM DTT, 0.5 mM PMSF) four times by centrifuging 5 s at 7000 rpm in a microfuge. Resuspend the beads in 500 μl BCO buffer and 10 μg GST-fusion protein, incubate 1 h at $4°$ with constant stirring. Wash the beads four times with 500 μl BCO buffer and then twice with EBC buffer (170 mM NaCl, 0.5% NP-40).

7. Incubate the coupled beads with the precleared extract for 12 h at $4°$. Centrifuge 5 s at 7000 rpm, discard the supernatant (an aliquot of the supernatant should be kept in order to compare it with the initial extract).

8. Wash the beads three times with EBC buffer (170 mM NaCl, 0.5% NP-40) and boil the beads in 30 μl Laemmli buffer 2$\times$.

The results of the interaction can then be visualized by Western analysis, using a specific antibody to the protein of interest. Due to possible nonspecific interactions, it is also advised to include a negative control, that is a protein that is not supposed to interact with the fusion protein.

Depending on the protein to be studied, some parameters (NaCl concentration, amount of protein in the interaction assay, time of incubation, etc.) of the above protocol may be modified to suit the specific experimenter's goal.

Immunofluorescence

The visualization of the intracellular localization of a kinase with respect to its substrate is an essential step towards the validation of its natural physiological function. A number of antibodies that recognize modified histone tails have been used successfully in immunofluorescence and immunostaining analyses. The outcome depends greatly on the quality of the antibody, for some their effectiveness in other experimental approaches is not a guarantee of success for immunofluorescence studies.

Mitotic phosphorylation of histone H3 at Ser-10 results in a striking visualization of condensing and condensed chromatin. The experimental steps described below can be applied also to other antibodies. The anti-phospho-H3 is a very efficient marker of the different phases of mitosis and shows a distinct colocalization with the kinase involved in mitotic H3 phosphorylation, the passenger protein Aurora-B. Indeed, mitotic phosphorylation of histone H3 is a highly dynamic process (see Fig. 2).

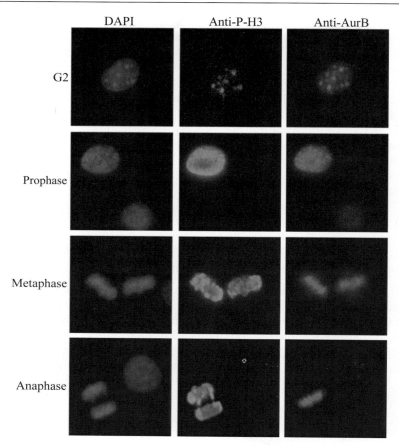

FIG. 2. The Ser-10 P-H3 signal allows an easy discrimination between the different phases of mitosis. NIH3T3 cells are fixed in paraformaldehyde 4%, then hybridized with a mouse monoclonal antibody raised against the Ser-10-phosphorylated form of histone H3 along with a rabbit polyclonal antibody raised against the Aurora-B kinase, and then stained with DAPI.

1. Grow the cells on 35 mm diameter plates.

2. Discard the medium and wash the cells twice with ice-cold PBS.

3. Fix the cells by incubating them with ice-cold PBS (4% paraform-aldehyde) 20 min on ice and wash them four times with PBS.

4. Permeabilize the cells by incubation with PBS (0.2% Triton X-100) 10 min at room temperature and wash them four times with PBS (0.05% Tween-20).

5. Incubate the cells with PBS (5% BSA) in order to avoid aspecific interactions.

6. Hybridize the cells overnight with the antibody in PBS (5% BSA) at 4°, using the appropriate dilution (rabbit polyclonal anti-phospho-Ser10-histone H3 antibody: 1:1000; mouse monoclonal anti-phospho-Ser 10-histone H3 antibody: 1:200) and wash them four times with PBS (0.05% Tween-20).

7. Incubate the cells 1 h at room temperature in the dark with the appropriate secondary antibodies conjugated to fluorophores diluted in PBS (5% BSA) and wash them twice with PBS (0.05% Tween-20).

8. Incubate the cells for 1 min in a 1:50,000 dilution of DAPI in PBS (0.05% Tween-20) and wash them with PBS.

9. Cover the cells with one drop of Vectashield and put a coverslip on top. Slides can be stored at 4° in the dark.

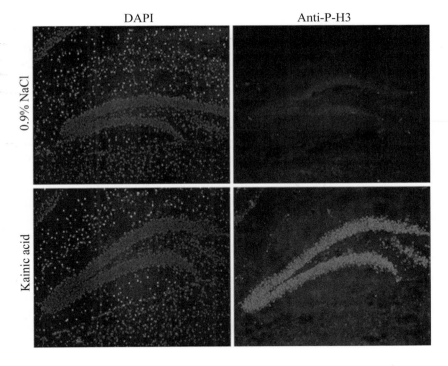

FIG. 3. Rapid induction of H3 phosphorylation in mouse hippocampal neurons in response to *in vivo* injection of pharmacological agonists to specific receptors. This is an example of immunohistochemistry with an antibody that recognizes the phosphorylated Ser-10 site on hippocampi of mice injected with 35 mg/kg of kainic acid (an agonist of glutamaergic receptors) or with a saline solution (0.9% NaCl). H3 phosphorylation is rapid and transient, following the kinetics of the immediate early response.

Analogous protocols can also be applied on tissue sections, either to identify the mitotic cells or to visualize other nonmitotic phosphorylation events. For example, a rapid and transient phosphorylation of histone H3 in hippocampal cells in response to several neurotransmitter receptor agonists can be revealed on mouse brain sections (paraffin or bouin). This event is coupled to the transcriptional induction of the immediate-early genes (e.g., *c-fos*) (see Fig. 3). As example we show phosphorylation of histone H3 on serine 10 that, differently from mitotic phosphorylation, is coupled to localized chromatin decondensation and transcriptional activation. This is elicited by injection of the mouse with kainic acid, an agonist of the glutamaergic receptors, and this event can also be easily visualized by immunocyto- or immunohistochemistry. Similar results have been obtained by studying neurons of the suprachiasmatic nucleus, a small structure in the lower hypothalamus of the mammalian brain, where the endogenous circadian clock resides. A simple pulse of light induces a rapid and transient phosphorylation at Ser10 of H3, underscoring the great plasticity of these neurons in responding to physiological stimuli that reset the endogenous clock system.[47] This event is mediated by signaling through $GABA_B$ receptors and is tightly coupled to activation of both clock and early response genes.

Concluding Remarks

Among all histone modifications, phosphorylation has the unique feature of constituting a direct link with intracellular signaling pathways. Thus, its analysis will provide further information into how epigenetic modifications occur in response to physiology, in cellular differentiation and during development. The techniques described here will allow the experimenter to study different aspects of histone phosphorylation. The in gel kinase assay and the *in vitro* kinase assay will be of great help for the identification of specific kinases phosphorylating a given residue, an aspect that could be complementary to the GST-pulldown assay. Functional information can be gathered by chromatin immunoprecipitation and immunofluorescence studies using specific antibodies directed to phosphorylated residues. These experimental approaches will shed further light on the physiological importance of the "histone code."

---

[47] C. Crosio, N. Cermakian, C. D. Allis, and P. Sassone-Corsi, *Nat. Neurosci.* **3**, 1241 (2000).

[12] Purification of Histone Methyltransferases from HeLa Cells

By JIA FANG, HENGBIN WANG, and YI ZHANG

Although discovered about 40 years ago, histone methylation has remained one of the least understood forms of posttranslation modifications until recently.[1,2] Methylation of histones occurs on both lysine and arginine residues on histone H3 and H4. The known methylation sites include lysines 4, 9, 27, 36, 79, and arginines 17, 26 on histone H3, and lysine 20, and arginine 3 on histone H4. While arginine methylation is catalyzed by members of the protein arginine methyltransferase (PRMT) family, the majority of methylation on lysine residues is catalyzed by proteins containing SET domain with the exception of Dot1.[1,3] Identification and characterization of these histone methyltransferases (HMTases) that target different lysine or arginine residues for methylation is an important step towards understanding the function of histone methylation as well as their combinations with other histone modifications, which is believed to form a "histone code."[4,5] Dissecting the "histone code" will reveal fundamental regulatory mechanisms in most, if not all, chromatin-templated processes.

Here, we describe several conventional chromatographic strategies aimed at purifying HMTases from HeLa cells. We also describe detailed protocols for preparing histone substrates and for detecting HMTase activities.

Preparation of Histone Substrates and Assay for HMTase Activities

Both core histone octamers and nucleosomes purified from micrococcal nuclease (MNase) digested HeLa cell nuclei can be used for HMTase assays. Because of the variation of concentration and quality of different batches of nuclei or micrococcal nuclease, test digestion is strongly recommended. The following protocol is adapted from a published procedure[6] and can be used for preparation of native oligonucleosomes, mononucleosomes, and histone octamers substrates from HeLa cells for HMTase assays.

[1] Y. Zhang and D. Reinberg, *Genes Dev.* **15,** 2343 (2001).
[2] T. Jenuwein and C. D. Allis, *Science* **293,** 1074 (2001).
[3] A. U. Khan and M. Hampsey, *Trends Genet.* **18,** 387 (2002).
[4] B. D. Strahl and C. D. Allis, *Nature* **403,** 41 (2000).
[5] B. M. Turner, *Bioessays* **22,** 836 (2000).
[6] J. Ausio and K. E. van Holde, *Biochemistry* **25,** 1421 (1986).

METHODS IN ENZYMOLOGY, VOL. 377

Copyright 2004, Elsevier Inc.
All rights reserved.
0076-6879/04 $35.00

Prepare and Digest Nuclei

About 20 l of cultured HeLa cells (5×10^5 cell/ml) are collected and washed once with PBS. The cell pellet is then resuspended with a type B pestle of a Dounce homogenizer in 100–150 ml buffer N1 (10 mM MES, 0.5 mM sodium metabisulfite, 0.5 mM benzamidine–HCl, 5 mM MgCl$_2$, 1 mM CaCl$_2$, 15 mM NaCl, 60 mM KCl, 0.25 M sucrose, 0.5% Triton X-100, 0.1 mM PMSF, 0.5 mM DTT, pH 6.5) to generate nuclei. After washing once with buffer N1, nuclei are collected by spinning at 4000 rpm (Sorvall RTH-750) for 10 min and resuspended in 12 ml buffer N2 (10 mM PIPES, 0.5 mM sodium metabisulfite, 0.5 mM benzamidine–HCl, 5 mM MgCl$_2$, 1 mM CaCl$_2$, 0.1 mM PMSF, 0.5 mM DTT, pH6.5). To obtain optimal MNase digestion conditions, 500 μl nuclei suspension is taken out and warmed up to 37° before adding CaCl$_2$ to 5 mM final concentration and 5 μl MNase stock (10 U/μl, Worthington Biochemical, Freehold, NJ). Every 5 min for up to 45 min, 50 μl of digested nuclei is transferred to an Eppendorf tube containing 1 μl 0.5 M EDTA to stop the reaction. Add to each tube in order 75 μl H$_2$O, 30 μl 10% SDS, 25 μl 4 M NaCl, and finally 200 μl phenol–chloroform, vortexing at each step. After centrifugation, 5 μl of the aqueous phases derived from each time point are loaded into a 1% agrose gel to examine the digestion efficiency. Appropriate digestion will yield the majority of products around 2 kb. The bulk of nuclei are then digested using the optimized conditions and stopped by adding 0.5 M EDTA to final concentration of 4 mM. Digested nuclei are centrifuged at 5000 rpm in a Sorvall SS-34 rotor for 5 min to generate the supernatant S1. Pellet is then resuspended in one pellet volume of 1 mM EDTA with 0.2 mM PMSF. The suspension is vortexed and incubated on ice for 15 min before centrifuging at 10,000 rpm (Sorvall SS-34) for 15 min min to generate supernatant S2. S1 and S2 are combined and the final salt concentration is brought to 0.75 M by adding 5 M NaCl drop by drop to remove histone H1 (omit this step if H1 is desired). The total concentration of nucleic acid can be determined by measuring A_{260} of 10 μl sample in 1 ml 1 M NaOH.

Sucrose Gradient Purification of Oligonucleosomes

Linear 5–30% sucrose gradients (35 ml) are prepared with buffer NG (10 mM Tris–Cl, 1 mM EDTA, 0.75 M NaCl, 0.3 mM PMSF, pH 7.5) using a binary gradient mixing apparatus (Hoefer SG100, Amersham) and are poured in 40 ml polyallomer tubes (Beckman). About 4–5 ml MNase digested samples are carefully layered on the top of each gradient. Gradients are centrifuged at 26,000 rpm for 16 h using a SW28 rotor (Beckman).

Fractions are then collected manually by siphoning 1 ml fractions from top to bottom of each gradient. Fifty microliters of sample from each fraction is precipitated with 100 μl 25% TCA and the purity of histone proteins is examined by running a 18% SDS-PAGE and viewed by Coomassie staining. The size of nucleosomes in each fraction is examined using the same method as that used in MNase digestion test described earlier. Fractions with good purity of histone and majority of DNA size from 1 to 2.5 kb (7–15 nucleosomes) are pooled and dialyzed against histone storage buffer (10 mM HEPES–KOH, 1 mM EDTA, 10 mM KCl, 10% glycerol, 0.2 mM PMSF, pH 7.5). Purified HeLa nucleosomes are stable for up to 6 months at 4° and can be stored for longer at $-70°$.

Purification of Mononucleosomes

Although top gradient fractions contain some mononucleosomes, we do not recommend them to be used for HMTase activity assay due to impurity. To obtain relatively pure mononucleosomes, fractions containing good purity of histones and majority of DNA size more than 2.5 kb are pooled and dialyzed against Tris–Cl (pH 8.0) buffer with 1 mM PMSF to remove EDTA and sucrose. Materials are concentrated by ultrafiltration (Ultrafree-15, Millipore) and then completely digested with MNase before subjected to sucrose gradient again as discussed earlier.

Purification of Histone Octamers

To purify histone octamers, all other histone containing fractions derived from sucrose gradient are combined and dialyzed to TE buffer with 1 mM PMSF. An appropriate volume hydroxyapatite (RioRad) column is prepared according to the total amount of protein, and equilibrated with buffer NP (40 mM Na$_2$HPO$_4$, 1 mM DTT, 0.2 mM PMSF, pH 6.8) containing 0.3 M NaCl (NP300). The dialyzed samples are loaded onto the column, and the column is washed with 6–10 column volume (cv) buffer NP500. Histone proteins are then eluted with buffer NP2500 and collected at 2 ml per fraction. The purity and the ratio of different core histone proteins are examined by running a 18% SDS-PAGE and viewed by Coomassie staining before fractions are pooled, dialyzed to histone storage buffer and aliquoted.

In addition to cultured HeLa cells, chicken blood is also a good source for preparing substrates for HMTase assay.[7] However, we noticed that the same HMTase has different activity towards histone substrates isolated

[7] C. A. Mizzen, J. E. Brownell, R. G. Cook, and C. D. Allis, *Methods Enzymol.* **304,** 675 (1999).

from different organisms. In some situations, both chicken and HeLa derived nucleosomes or histones are recommended to be used as substrates for HMTase assay. Besides native histones and nucleosomes, synthetic peptides, recombinant histones and reconstituted octamers or nucleosomes have also been successfully used as substrates for some HMTases although some HMTases, such as Dot1, prefer native nucleosome substrate.[8,9]

Histone Methyltransferase Assays

HMTase assays are performed essentially as described.[10] Briefly, protein fractions are incubated with appropriate amount of different substrates (visible in Coomassie) in total of 20–50 μl reactions containing 1/5 volume $5\times$ HMT buffer (100 mM Tris–HCl, 20 mM EDTA, 5 mM PMSF, 2.5 mM DTT, pH 8.0) and 1 μl S-Adenosyl-L-[$methyl$-^{3}H]methionine (15 Ci/mM, NEN Life Science Products) for 1 h at 30°. Reactions are stopped by addition 1/5 volume of $5\times$ SDS loading buffer (0.25 M Tris–Cl, 0.5 M DTT, 10% SDS, 0.25% bromphenol blue, 50% glycerol, pH 6.8) and histones are separated on an 18% SDS-PAGE. After Coomassie staining and destaining, gels are treated with EN^{3}HANCE or ENTENSIFY (NEN Life Science Products) for 30–45 min, dried and exposed to X-ray films (BioMax, Kodak) for an appropriate time. For quantification, the gel slices can be excised and counted with liquid scintillation.

Purification of Histone Methyltransferases from Cultured HeLa Cells

Most of HMTases discussed below were purified from HeLa nuclei. The nuclear proteins of cultured HeLa cells can be divided into nuclear extract and nuclear pellet fractions based on whether they can be easily extracted away from bulk of chromatin. Both nuclear extract and nuclear pellet fractions can be further fractionated on ion-exchange phosphocellulose P11 (Sigma) or DEAE-52 (Whatman) columns, respectively. Chromatography on these two columns is performed conventionally. Chromatography resins were pretreated following the instruction of the manufacturers before packing the columns of appropriate size using empty columns from BioRad.

[8] J. Min, Q. Feng, Z. Li, Y. Zhang, and R. M. Xu, *Cell* **112,** 711 (2003).
[9] H. H. Ng, Q. Feng, H. Wang, H. Erdjument-Bromage, P. Tempst, Y. Zhang, and K. Struhl, *Genes Dev.* **16,** 1518 (2002).
[10] H. Wang, R. Cao, L. Xia, H. Erdjument-Bromage, C. Borchers, P. Tempst, and Y. Zhang, *Mol. Cell* **8,** 1207 (2001).

Preparation of Starting Materials

HeLa nuclear proteins are separated into nuclear extract (NE) and nuclear pellet (NP) fractions essentially as previously described.[11] All the steps described below are performed in cold room on ice. Briefly, after collection and washing with PBS, cultured HeLa cells are allowed to swell in 5-pellet-volume of buffer A (10 mM Tris–Cl, 1.5 mM MgCl$_2$, 10 mM KCl, 0.2 mM PMSF, 0.5 mM DTT, pH 7.9) for 10 min. Cells are then collected by centrifugation and homogenized using a type B pestle in 2 pellet volumes of buffer A. Crude nuclei are pelleted by centrifuging at 2500 rpm (Sorvall RTH-750) for 10 min. Nuclei are resuspended in 3 ml buffer C (20 mM Tris–Cl, 0.42 M NaCl, 1.5 mM MgCl$_2$, 0.2 mM EDTA, 0.1 mM PMSF, 0.5 mM DTT, 25% glycerol, pH 7.9) per 10^9 cells and homogenized again using a type B pestle. The suspension is then stirred gently using a magnetic stirring bar for 30 min and cleared by centrifuging at 15,000 rpm in a Sorvall SS-34 rotor for 30 min. The supernatant is dialyzed against buffer D (50 mM Tris–Cl, 0.1 mM EDTA, 0.2 mM PMSF, 2 mM DTT, 25% glycerol, pH 7.9) containing 0.1 M KCl and saved as the NE fraction. Nuclear pellet (NP) needs to be solubilized as described[12] before further purification. Briefly, nuclear pellet is resuspended and homogenized in appropriate volume of buffer E (50 mM Tris–Cl, 5 mM MgCl$_2$, 0.5 mM EDTA, 0.2 mM PMSF, 5 mM DTT, 25% glycerol, pH 7.9). After adding 1/10 volume 3 M (NH$_4$)$_2$SO$_4$, the suspension is mixed immediately and DNA should be sheared by sonication until the solution is not viscous anymore. Suspension is then centrifuged at 35,000 rpm in a KAL-40.100 rotor (KOMPSPIN) for 70 min. The supernatant is diluted by addition of 2 volumes of buffer E and cleared of debris again by centrifuging at 40,000 rpm for 1 h. Proteins are precipitated by addition of 0.42 g (NH$_4$)$_2$SO$_4$ per ml supernatant and resuspended in appropriate volume buffer D.

Distribution of HMTase Activities in NE and NP Fractions from HeLa Cells

Proteins in NE fractions (6 g) are further fractionated on a lab-made 700 ml phosphocellulose P11 column equilibrated with buffer D containing 0.1 mM KCl. Proteins bound to the column are step eluted with buffer D containing 0.3, 0.6, and 1.0 M KCl, respectively. Proteins derived from NP fractions (4 g), after ammonium sulfate concentration is adjusted to 20 mM, are loaded onto a lab-made 500 ml DEAE-52 column equilibrated with buffer D containing 20 mM (NH$_4$)$_2$SO$_4$. The bound proteins are step eluted with

[11] J. D. Dignam, R. M. Lebovitz, and R. G. Roeder, *Nucleic Acids Res.* **11,** 1475 (1983).
[12] G. LeRoy, G. Orphanides, W. S. Lane, and D. Reinberg, *Science* **282,** 1900 (1998).

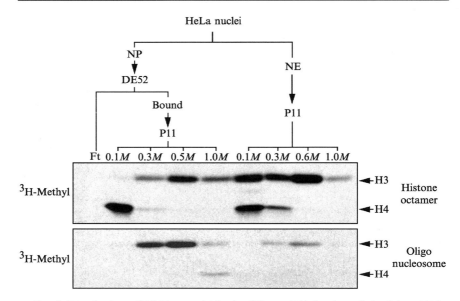

FIG. 1. Distribution of HMTase activities in different P11 fractions derived from HeLa nuclear proteins. Substrates used for HMTase assays are indicated.

0.35 and 0.5 M ammonium sulfate in buffer D. The 0.35 M elution is then dialyzed against buffer D containing 0.1 M KCl and loaded onto a 400 ml phosphocellulose P11 column. Bound proteins are step eluted with buffer D containing 0.3, 0.6, and 1.0 M KCl, respectively. Small aliquots of samples from every P11 fraction are dialyzed against buffer B (40 mM HEPES-KOH, 0.2 mM EDTA, 0.2 mM PMSF, 1 mM DTT, 1 μg/ml each of leupeptin, aprotinin, and pepstatin A, 10% glycerol, pH 7.9) containing 50 mM KCl (BC50) and then assayed HMTase activity using both histone octamers and oligonucleosomes as substrates. Results shown in Fig. 1 revealed multiple HMTase activities specific for histones H3 and H4. The fact that different patterns of activities are detected using core histones and nucleosomal histone substrates indicate that HMTases have substrate preferences.

Using the P11 fractions as starting material, by following the HMTase activity, we have purified several HMTases or HMTase complexes. In protocols described below, most chromatographic steps are performed on a ÄKTA fast protein liquid chromatography (FPLC) system (Amersham Pharmacia Biotech). In each chromatographic step, every fourth fraction (or portion of a fraction) is dialyzed to BC50 before HMTase activity assay. The same fractions are also analyzed by silver staining of a 6.5–15% linear gradient SDS-PAGE to evaluate the purity of the enzymes.

Purification of Histone H4-specific HMTases

Histone H4 tail can be methylated at arginine 3 (R3) and lysine 20 (K20) by PRMT1[13,14] and SET8/PR-SET7,[15,16] respectively. We describe how the two HMTases were purified from HeLa nuclear pellet fractions below.

Purification of Histone H4-R3 Methyltransferase PRMT1

To identify the protein responsible for histone H4 methylation, we focused on the 0.1 M P11 fraction derived from nuclear pellet because of its robust activity towards core histones (see Fig. 1). After dialyzing against buffer D containing 40 mM ammonium sulfate (BD40), the protein samples are loaded onto a high-performance liquid chromatography (HPLC) DEAE-5PW column (TosoHaas, 45 ml) which has been equilibrated with BD40. The bound proteins were eluted with 10 cv linear gradient from BD40 to BD400. Every fourth fraction was dialyzed into BC50 and analyzed for HMTase activity using core histones as substrates. The fractions containing the enzymatic activity (BD120–BD225) were combined and its salt concentration was adjusted to final of 0.5 M ammonium sulfate by addition of saturated $(NH_4)_2SO_4$ drop by drop. Samples were then loaded to a 22 ml FPLC Phenyl Sepharose column (Amersham Pharmacia Biotech) and bound proteins were eluted with a 20 cv linear gradient from BD500 to BD0. Fractions between 0.1 and 0.25 M ammonium sulfate were pooled according to the HMTase activity and protein purity analysis. A portion of the pooled material was dialyzed to buffer P (5 mM HEPES-KOH, 40 mM KCl, 0.01% Triton X-100, 0.01 mM $CaCl_2$, 0.5 mM PMSF, 1 mM DTT, 2 μg/ml each of leupeptin, aprotinin, and pepstatin A, 10% glycerol, pH 7.5) containing 10 mM K_2HPO_4–KH_2PO_4 (BP10), and loaded onto a 1 ml BP10 equilibrated hydroxyapatite (BioRad) column. Bound proteins were eluted with a 20 cv linear gradient from BP10 to BP600. HMTase activity assay coupled with silver staining of an SDS-PAGE containing the column fractions revealed a single polypeptide of 42 KDa correlates with the HMTase activity (see Fig. 2). Mass spectrometric analysis of the 42 kDa protein identified it to be the human protein arginine

[13] H. Wang, Z. Q. Huang, L. Xia, Q. Feng, H. Erdjument-Bromage, B. D. Strahl, S. D. Briggs, C. D. Allis, J. Wong, P. Tempst, and Y. Zhang, *Science* **293,** 853 (2001).

[14] B. D. Strahl, S. D. Briggs, C. J. Brame, J. A. Caldwell, S. S. Koh, H. Ma, R. G. Cook, J. Shabanowitz, D. F. Hunt, M. R. Stallcup, and C. D. Allis, *Curr. Biol.* **11,** 996 (2001).

[15] J. Fang, Q. Feng, C. S. Ketel, H. Wang, R. Cao, L. Xia, H. Erdjument-Bromage, P. Tempst, J. A. Simon, and Y. Zhang, *Curr. Biol.* **12,** 1086 (2002).

[16] K. Nishioka, J. C. Rice, K. Sarma, H. Erdjument-Bromage, J. Werner, Y. Wang, S. Chuikov, P. Valenzuela, P. Tempst, R. Steward, J. T. Lis, C. D. Allis, and D. Reinberg, *Mol. Cell* **9,** 1201 (2002).

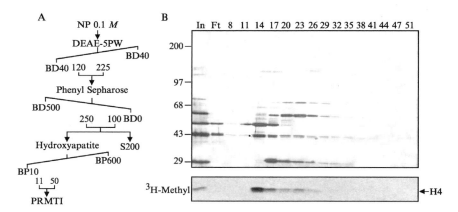

FIG. 2. Purification of the H4-R3-specific HMTase PRMT1. (A) Schematic representation of the purification steps. Numbers represent the salt concentrations (mM) at which the HMTase activity eluted from the column. (B) Analysis of fractions derived from the hydroxyapatite column by silver staining (top panel) and HMTase activity assay (bottom panel). The elution profile of the dotted band correlates with the enzymatic activity. The protein size marker on SDS-PAGE is indicated to the left of the panel.

N-methyltransferase 1 (PRMT1). Western blot analysis confirmed that PRMT1 correlated with this H4 HMTase activity in previous columns.

Purification of Histone H4-K20 Methyltransferase SET8

By following the H4-specific HMTase activity, we unexpectedly revealed that H4-R3 can be methylated both *in vitro* and *in vivo*.[13] However, our initial goal of purifying an H4-K20 HMTase was not achieved. Since all the P11 fractions capable of methylating free core histone H4 contain PRMT1, which is unable to methylate nucleosomal histone, we used nucleosome substrate to avoid contaminating activity from PRMT1. This analysis revealed 1.0 M P11 fraction from nuclear pellet (see Fig. 1) to be the only fraction capable of methylating nucleosomal H4. Thus, this fraction likely contains the H4-K20 specific methyltransferase(s).

To purify the candidate H4-K20 HMTase(s), the 1.0 M P11 nuclear pellet fraction was fractionated sequentially through four columns (see Fig. 3). The purification was monitored by HMTase assay using oligonucleosome substrates and the purity of proteins was examined by silver staining of 6.5–15% linear gradient SDS-PAGE containing column fractions. The sample was first dialyzed to BD20 and loaded onto a preequilibrated 45 ml HPLC DEAE-5PW column. The bound proteins were eluted with a 12 cv linear gradient from BD20 to BD600. Fractions containing the

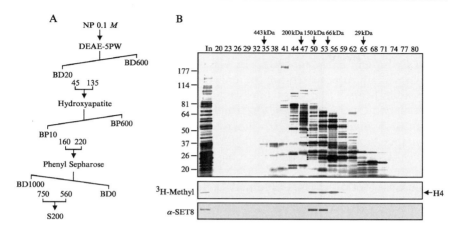

FIG. 3. Purification of the H4-K20-specific HMTase SET8. (A) Schematic representation of the purification steps. Numbers represent the salt concentrations (mM) at which the HMTase activity eluted from the column. (B) Analysis of fractions derived from the Superose 200 gel filtration column by silver staining (top panel), HMTase activity assay (middle panel), and Western blotting (bottom panel). The dotted polypeptides coelutes with the HMTase activity and SET8. The elution profile of the protein markers is indicated on the top of the panel. The protein size marker on SDS-PAGE is indicated to the left of the panel.

H4-specific HMTase activity were pooled and dialyzed against BP10 before loading to a 5 ml hydroxyapatite (BioRad) column. The bound proteins are eluted with a 12 cv linear gradient from BP10 to BP600. Fractions containing the H4-specific activity were combined and its conductivity was adjusted to BD1000 by adding saturated ammonium sulfate drop by drop with stirring. After cleared of debris by centrifugation, samples were loaded onto a BD1000 preequilibrated 1 ml FPLC Phenyl Sepharose (Amersham Pharmacia Biotech) column. Bound proteins were fractioned with a 20 cv linear gradient from BD1000 to BD0 and the H4 HMTase activity was eluted between 0.75 and 0.56 M of ammonium sulfate. After dialyzing into BC50, the enzymatic peak fractions were pooled and concentrated by binding to 0.2 ml BC50 preequilibrated P11 resin in a 1.5 ml microtube at 4° and bound proteins were eluted with 0.2 ml BC1000. Proteins were then fractionated on a 22 ml Superose 200 (Amersham Pharmacia Biotech) gel-filtration column. Analysis of the fractions derived from the last purification step indicated that the activity eluted with a native size of 50–150 kDa between fractions 50 to 56 (see Fig. 3B). Silver staining of a SDS-PAGE containing the column fractions revealed multiple proteins present in the enzymatically active fractions (see Fig. 3B, top panel). The limited amount of the sample prevented us from further purification. Since the activity begins with fraction 50 and ends with fraction 56, the candidate protein bands should be present in fraction 50 but absent in fractions 47 and 59. Using these criteria, we identified six protein bands (see Fig. 3B, top panel, indicated by dots and star) that are likely responsible for the enzymatic activity. Mass spectrometry analysis indicated that most masses obtained from the 45 kDa protein, marked by a star (see Fig. 3B, top panel), matched a SET domain-containing protein named PR/SET07 in GenBank (AAF97812). Since all the lysine-specific HMTases identified so far contain a SET domain, the 45 kDa protein is likely responsible for the detected H4-specific HMTase activity.

Because not all masses obtained from the 45 kDa protein can be accounted for by conceptual triptic digestion of PR/SET07 protein, we obtained and sequenced several human EST clones that are similar or identical to that of the PR/SET07. Conceptual translation of the cDNA from one of the EST clones (BE867579) generated an ORF (open reading frame) of 352 amino acids which contains all the peptides derived from the 45 kDa protein. Thus, we conclude that the 45 kDa protein, named SET8,[15] is distinct from but highly similar to PR/SET07 (84% identical). In an independent study, the PR/SET07 was purified and demonstrated to be an H4-K20-specific HMTase.[16]

Purification of Histone H3 HMTases

Comparing with histone H4, H3 harbors more sites that can be methylated. The documented sites including lysines 4, 9, 27, 36, 79, and arginines 17 and 26.[1,17] At least one protein responsible for methylation at each of these sites has been identified. Below, we describe procedures used in our lab that lead to the discovery of three different HMTases that target lysine 4, 9, and 27 for methylation, respectively.[10,18]

Purification of the Histone H3-K4 Methyltransferase SET7/9

While pursuing the histone H4-specific methyltransferases described earlier, we also monitored the purification of HMTase activity specific for histone H3. About 7 g of proteins derived from nuclear pellet were dialyzed against buffer D containing 20 mM ammonium sulfate and loaded onto a 900 ml BD20 equlibrated DEAE52 column. Bound proteins were step eluted with 0.1 and 0.6 M ammonium sulfate in buffer D. The salt

[17] M. Lachner and T. Jenuwein, *Curr. Opin. Cell Biol.* **14,** 286 (2002).
[18] R. Cao, L. Wang, H. Wang, L. Xia, H. Erdjument-Bromage, P. Tempst, R. S. Jones, and Y. Zhang, *Science* **298,** 1039 (2002).

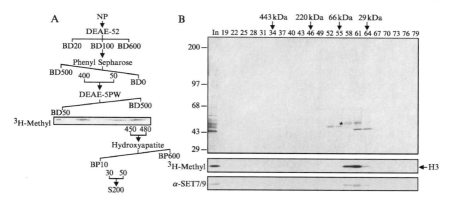

FIG. 4. Purification of the H3-K4-specific HMTase SET7. (A) Schematic representation of the purification steps. Numbers represent the salt concentrations (mM) at which the HMTase activity eluted from the column. Three HMTase activity peaks on the DEAE-5PW column are shown in the middle. (B) Analysis of fractions derived from the Superose 200 gel filtration column by silver staining (top panel), HMTase activity assay (middle panel), and Western blotting (bottom panel). The elution profile of the protein markers is indicated on the top of the panel and the protein size marker on SDS-PAGE is indicated to the left of the panel.

concentration of the 0.1 M elution was adjusted to 0.5 M using saturated ammonium sulfate, and loaded onto a 22 ml FPLC Phenyl Sepharose column. Bound proteins were eluted with a 15 cv linear gradient from BD500 to BD0. Fractions containing HMTase activity were combined and dialyzed to BD50 before loading onto a 45 ml DEAE-5PW column. Bound proteins were eluted with a 10 cv linear gradient from BD50 to BD500. Analysis of the fractions derived from this column identified three peaks of HMTase activity towards H3. The third peak was further purified on a 1 ml BP10 equilibrated hydroxyapatite column with a 20 cv linear gradient from BP10 to BP600. The active fractions, eluted between BP30 and BP50, were pooled and concentrated with 0.2 ml DEAE-52 resin, and fractionated on a Sepharose 200 gel-filtration column. HMTase assay coupled with silver staining of a SDS-PAGE containing the column fractions identified that a polypeptide of 45 kDa correlates with the enzymatic activity (see Fig. 4). Mass spectrometry analysis revealed that the protein matched a SET-domain containing protein (KIAA1717) of unknown function in GenBank. Given that this protein does not share any sequence homology with any of the six SET-domain containing proteins in *S. cerevisiae* outside the SET domain, we have named the protein SET7.[10] This same protein has also been independently purified with similar approach and has been named SET9.[19]

[19] K. Nishioka, S. Chuikov, K. Sarma, H. Erdjument-Bromage, C. D. Allis, P. Tempts, and D. Reinberg, *Genes Dev.* **16,** 479 (2002).

Purification of the Histone H3-K9 Methyltransferase SETDB1 Complex

Thus far, at least four proteins have been reported to be capable of methylate histone H3 at lysine 9. These proteins include the Suv39h1,[20] G9a,[21] ESET and its human homolog SETDB1,[22,23] and Eu-HMTase,[24] a protein similar to G9a. With the exception of Eu-HMTase, information on the native protein complexes of these proteins are currently not available. We describe the purification of the SETDB1 complex from HeLa cells below.[25]

To purify the native SETDB1 complex, we monitored the purification process with both Western blot and HMTase assays. Western blot analysis revealed that the majority of SETDB1 is in the 0.1 M P11 fraction derived from nuclear extract, which contains strong HMTase activity towards core histone octamer substrates (see Fig. 1). Therefore, the proteins in this fraction were loaded onto a 300 ml DEAE-52 column and step eluted with BC150, BC200, BC350, and BC500, respectively. The SETDB1-containing BC350 fraction was loaded onto a 22 ml Phenyl Sepharose column after its salt concentration was adjusted to 0.5 M with saturated ammonium sulfate. The bound proteins were eluted with a 20 cv linear gradient from BD500 to BD0. SETDB1 containing fractions were combined, dialyzed to BD50, and loaded onto a 45 ml DEAE-5PW column equilibrated with BD50. Bound proteins were eluted with a 5 cv linear gradient from BD50 to BD500. The SETDB1 containing fractions were pooled and concentrated with 200 μl DEAE52 resin and fractionated on a 22 ml Superose 6 (Amersham Pharmacia Biotech) gel-filtration column. Western blot and HMTase assays of the column fractions revealed that SETDB1 coelutes with the enzymatic activity as a complex of 670–2000 kDa. Further purification of the active fractions on a 1 ml BC150 equilibrated Mono Q (Amersham Pharmacia Biotech) column revealed that two protein bands coelute with the HMTase activity (see Fig. 5). Mass spectrometry analysis identified the proteins as SETDB1 and the human homolog of mouse protein mAm.[26]

Purification of the Histone H3-K27 Methyltransferase
EED-EZH2 Complex

As described earlier, the HMTase activity in the 0.5 M P11 nuclear pellet fraction split into three peaks on DEAE-5PW column when core histones were used as substrates. However, only two peaks were detected

[20] S. Rea, F. Eisenhaber, D. O'Carroll, B. D. Strahl, Z. W. Sun, M. Schmid, S. Opravil, K. Mechtler, C. P. Ponting, C. D. Allis, and T. Jenuwein, *Nature* **406,** 593 (2000).
[21] M. Tachibana, K. Sugimoto, T. Fukushima, and Y. Shinkai, *J. Biol. Chem.* **20,** 20 (2001).
[22] L. Yang, L. Xia, D. Y. Wu, H. Wang, H. A. Chansky, W. H. Schubach, D. D. Hickstein, and Y. Zhang, *Oncogene* **21,** 148 (2002).
[23] D. Schultz, K. Ayyanathan, D. Negorev, G. Maul, and F. Rauscher, *Genes Dev.* **16,** 919 (2002).
[24] H. Ogawa, K. Ishiguro, S. Gaubatz, D. Livingston, and Y. Nakatani, *Science* **296,** 1132 (2002).

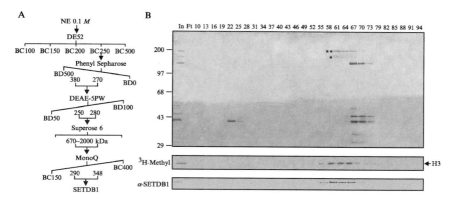

FIG. 5. Purification of the H3-K9-specific HMTase SETDB1 complex. (A) Schematic representation of the purification steps. Numbers represent the salt concentrations (mM) at which the HMTase activity eluted from the column. (B) Analysis of fractions derived from the Mono Q column by silver staining (top panel), HMTase activity assay (middle panel), and Western blotting (bottom panel). SETDB1 is indicated by a "*" and its associated protein is indicated by a "**."

when nucleosomal histone was used as substrates. The second peak fractions, eluted between 0.22 and 0.25 mM ammonium sulfate, were combined and salt concentration adjusted to BD700 using saturated ammonium sulfate before they were loaded onto a 22 ml Phenyl Sepharose column. The bound proteins were eluted with a 10 cv linear gradient from BD700 to BD0. Active fractions were pooled and dialyzed to BP10 before loading to a 1 ml hydroxyapatite column and eluted with 20 cv linear gradient from BP10 to BP600. The enzymatic peak fractions that were eluted between BP80 and BP130 were pooled and concentrated by ammonium sulfate precipitation, and further fractionated on a 22 ml Superose-6 gel-filtration column. HMTase assay coupled with silver staining of an SDS-PAGE containing the column fraction revealed that six proteins coelute with the HMTase activity between 440 and 670 kDa (see Fig. 6B). Coimmunoprecipitation confirmed that five of the six proteins exist as a protein complex (see Fig. 6C). Mass spectrometry identified these proteins as EZH2, SUZ12, AEBP2, EED, and RbAp48.[18] Using a slightly different purification scheme, a similar complex containing EZH2, SUZ12, EED, and RbAp48 was also purified from HeLa cells.[27] In addition, equivalent protein complexes were also purified from *Drosophila* embryo extracts.[28,29]

[25] H. Wang, W. An, R. Cao, L. Xia, H. Erdjument-Bromage, B. Chatton, P. Tempst, R. G. Roeder, and Y. Zhang, *Mol. Cell* **12**, 475 (2003).

[26] F. De Graeve, A. Bahr, B. Chatton, and C. Kedinger, *Oncogene* **19**, 1807 (2000).

[27] A. Kuzmichev, K. Nishioka, H. Erdjument-Bromage, P. Tempst, and D. Reinberg, *Genes Dev.* **16**, 2893 (2002).

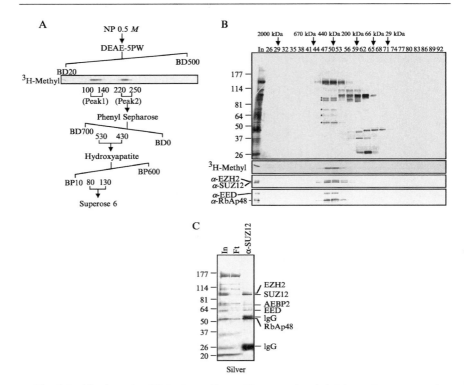

FIG. 6. Purification of an H3-K27-specific HMTase complex. (A) Schematic representation of the purification steps. Numbers represent the salt concentrations (mM) at which the HMTase activity eluted from the column. Two HMTase activity peaks on the DEAE-5PW column are shown in the middle. (B) Analysis of fractions derived from the Superose 6 gel-filtration column by silver staining (top panel). HMTase activity assay (second panel), and Western blotting (bottom two panels). Each of the five copurifying proteins is indicated by a "*". The elution profile of the protein markers is indicated on the top of the panel and the protein size marker on SDS-PAGE is indicated to the left of the panel. (C) Silver stained SDS-polyacrylamide gel demonstrating coimmunoprecipitation of the five components using anti-SUZ12 antibody. "In and Ft" represent input and flow-through, respectively. The protein size marker on SDS-PAGE is indicated to the left of the panel.

Acknowledgments

We thank Erin Henry for critical reading of the manuscript. Y. Z. is a Kimmel Scholar and is supported by NIH (GM63067–01) and ACS (RSG-00–351–01-GMC).

[28] B. Czermin, R. Melfi, D. McCabe, V. Seitz, A. Imhof, and V. Pirrotta, *Cell* **111,** 185 (2002).

[29] J. Muller, C. M. Hart, N. J. Francis, M. L. Vargas, A. Sengupta, B. Wild, E. L. Miller, M. B. O'Connor, R. E. Kingston, and J. A. Simon, *Cell* **111,** 197 (2002).

[13] Global Proteomic Analysis of *S. cerevisiae* (GPS) to Identify Proteins Required for Histone Modifications

By JESSICA SCHNEIDER, JIM DOVER, MARK JOHNSTON, and ALI SHILATIFARD

During the process of organismal development and differentiation, cells become committed to different fates through heritable, quasi-stable changes in gene expression that is regulated by RNA polymerase II. The 2-m long DNA of mammalian cells must remain functional when packaged into the cell nucleus. This process of DNA packaging is not well-understood. The first stage in this packaging process is the formation of the nucleosome core particle. Nucleosomes were first observed in electron micrographs of lysed nuclei[1] as a series of "beads on a string," the beads being the individual nucleosomes; the "string" is the linker DNA. Nucleosomes consist of 146 bp of DNA tightly wrapped around an octamer of histone proteins (H3, H4, H2A, and H2B).[2] The linker histone H1 associates with DNA between the nucleosomes to stabilize the formation of higher-order, more compacted structures.[3,4] The compaction of DNA via the histone proteins affects the binding of nonhistone proteins, such as transcription factors, by restricting access to the binding sites within the DNA.

Over the past decade there has been an explosion of information regarding the role of nucleosomes in the regulation of gene expression.[4] It is increasingly becoming clear that nucleosomes can transmit epigenetic information from one cell generation to the next.[5,6] The amino termini of histone tails protrude from the nucleosome and are available for interactions with the DNA or with other proteins,[2] and are thus the sites of many covalent modifications that alter nucleosome structure. Multiple modifications decorate each histone tail, and some amino acids in the histone tails can be modified in several different ways. Covalent modifications of histone tails include acetylation, phosphorylation, ubiquitination, and methylation. Although some of these modifications were discovered long ago, only recently have functional roles for these modifications begun to surface.[4] Each histone can undergo numerous modifications, and the

[1] R. D. Kornberg, *Science* **184,** 868 (1974).

[2] K. Luger *et al.*, *Nature* **389,** 251 (1997).

[3] G. Arents and E. N. Moudrianakis, *Proc. Nat. Acad. Sci. USA* **92,** 11170 (1995).

[4] J. L. Workman and R. E. Kingston, *Ann. Rev. Biochem.* **67,** 545 (1998).

[5] T. Jenuwein and C. D. Allis, *Science* **293,** 1074 (2001).

[6] B. M. Turner, *Cell* **111** 285 (2002).

Copyright 2004, Elsevier Inc.
All rights reserved.
0076-6879/04 $35.00

METHODS IN ENZYMOLOGY, VOL. 377

combinatorial effect of these serves to elicit a multitude of different responses. This combinatorial modification of histone tails, which has been referred to as the "histone code,"[5,6] has been proposed to play a pivotal role in the regulation of gene expression.

Most recently, covalent modification of histones by methylation has been demonstrated to be required for the regulation of gene expression by several different histone methyltransferases.[7–19] Lysine four of histone H3 is methylated by the macromolecular complex COMPASS, which contains Set1. To better define the pathways of posttranslational modification of histones in yeast, we devised a proteomic approach called Global Proteomic analysis in *S. cerevisiae* (GPS). In GPS, we test by Western blotting extracts of each of the ∼4800 nonessential yeast gene deletion mutants[20] for defects in modifications of histone H3, using as probes antibodies specific to various modified histones. Employing antibody specific to histone H3 methylated on its fourth lysine, GPS revealed that ubiquitination of lysine 123 of histone H2B by Rad6 (the E2 conjugating enzyme)[21] is required for histone methylation by COMPASS.[20] Allis and colleagues[22] have also demonstrated a requirement for Rad6 in this process. Since Rad6 is involved in ubiquitination of many other substrates, we set out to identify its E3 ligase specific for its role in transcription. Employing GPS we identified Brel and Lgel (the E3 ligase complex) and also the components of the RNA Polymerase II elongation factor Paf1 complex, which are required for this modification.[23,24]

[7] S. Rea *et al.*, *Nature* **406**, 593 (2000).
[8] H. H. Ng *et al.*, *Genes Dev.* **16**, 1518 (2002).
[9] F. van Leeuwen, P. R. Gafken, and D. E. Gottschling, *Cell* **109**, 745 (2002).
[10] B. D. Strahl and C. D. Allis, *Nature* **403**, 41 (2000).
[11] H. H. Ng *et al.*, *J. Biol. Chem.* **277**, 34655 (2002).
[12] T. Miller *et al.*, Erratum appears in *Proc. Natl. Acad. Sci. USA* **98**, 15393 (2001).
[13] A. Roguev *et al.*, *EMBO J.* **20**, 7137 (2001).
[14] P. L. Nagy *et al.*, *Proc. Nat. Acad. Sci. USA* **99**, 90 (2002).
[15] N. J. Krogan *et al.*, *J. Biol. Chem.* **277**, 10753 (2002).
[16] T. Jenuwein *et al.*, *Cell Mol. Life Sci.* **54**, 80 (1998).
[17] K. Nishioka *et al.*, *Genes Dev.* **16**, 479 (2002).
[18] M. Bryk *et al.*, *Curr. Biol.* **12**, 165 (2002).
[19] H. Santos-Rosa *et al.*, *Nature* **419**, 407 (2002).
[20] J. Dover *et al.*, *J. Biol. Chem.* **277**, 28368 (2002).
[21] K. Robzyk, J. Recht, and M. A. Osley, *Science* **287**, 501 (2000).
[22] Z. W. Sun and C. D. Allis, *Nature* **418**, 104 (2002).
[23] A. Wood *et al.*, *Mol. Cell* **11**, 267 (2003).
[24] N. J. Krogan *et al.*, *Mol. Cell* **11**, 721 (2003).

Methods of Assay

Reagents

Growing the Nonessential Gene Deletion Mutants

1. Plastic, disposable 96 well pinning devices were purchased from Genetix (catalog # X5054, UK).
2. YPD for growing yeast consisted of 1% yeast extract, 2% proteo-peptone, 2% dextrose, and 1.5% agar (Fisher Scientific). The yeast extract and proteo-peptone are autoclaved together, when the solution cools dextrose is added from a sterile 20% stock solution. When agar plates were used, they were Nunc brand, omnitray, single well containers.
3. The cells were first inoculated from frozen stocks in 96 well plates by pinning the thawed cells into liquid YPD in 2 ml 96 well plates (Costar brand) and grown for 48 h at $30°$ without shaking.

Preparation of the Extracts

1. Cells were spun down by centrifugation and resuspended in 100 μl of Lysis Buffer (20 mM Tris, pH 7.5, 1 mM dithiothreitol, 50 mM KCl, 1 mM EDTA, 0.1% NP-40, and 1 mg/ml Zymolyase 100T [US Biological].
2. The extracts were placed in Fisher brand, flexible, 96 well PCR plates, which stand in Costar brand, 96 well cell culture cluster plates.
3. 4× Laemmli loading buffer (1 ml water, 2 ml 1 M Tris, pH 6.8, 3.2 ml 100% glycerol, 0.64 g 10% SDS (Fisher Scientific), 1.6 ml 2-mercaptoethanol (Sigma), and 1 mg bromophenol blue) was added to the extracts, then heated to $95°$ in an MJ thermocycler for 5 min.

Testing the Extracts for Histone Methylation on the Fourth Lysine of Histone H3

1. The extracts were loaded onto 16% SDS gels and separated by electrophoresis in a 1:10 dilution of electrophoresis buffer (0.25 M Tris, 2 M Glycine, and 10% SDS).
2. The proteins were resolved by SDS-PAGE and blotted to Osmosis brand nitrocellulose membranes in Western transfer buffer (30 mM Tris, 0.15 M Glycine, and 20% Methanol).
3. Protein blots were probed with affinity-purified polyclonal anti-serum (1:1000 diluted) specific for methylated K4 of histone H3 (purchased from Upstate) in TBS pH 7.4 (0.1 M Tris, 1.5 M sodium chloride, 7 mM calcium chloride, 4 mM magnesium chloride) and washed three times in TBS.

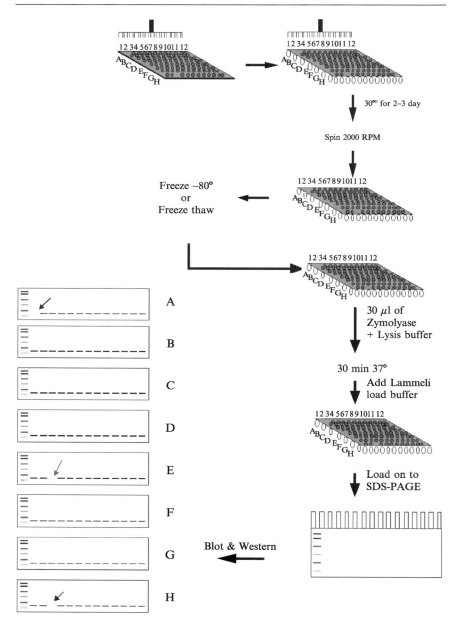

FIG. 1. Global Proteomic analysis of *S. cerevisiae* (GPS). Schematic representation of GPS employing polyclonal antibody specific to lysine 4 methylated histone H3. As described under the detailed procedure section, employing a 96-well pinning device, the entire collection of

4. The blots were visualized by exposure to IsoMax film, size 14 × 17 inches using chemilluminescence.

Detailed Procedure

1. Using a 96-well pinning device, the entire collection[12] of about 4800 yeast nonessential gene deletion mutants were inoculated from −80° stocks onto agar plates containing YPD + 200 μg/ml GENETICIN (GIBCO), allowed to grow 48 h, and used to inoculate 96-tube PCR-plates filled with 1.5 ml of YPD. After 48 h of growth at 30° the plates were centrifuged at 2000g for 10 min. Medium was removed by wrist-snap inversion and draining into absorbent towels. Plates were then covered and frozen at −80° for up to 1 week.

2. Frozen cells were thawed at room temperature, resuspended in 100 μl lysis buffer (20 mM Tris pH 7.5, 50 mM KCl, 1 mM EDTA, 1 mM DTT, 0.1% NP-40, 1 mg/ml Zymolyase), and incubated at 37° for 30 min. Fifty μl of 4× Laemmli loading buffer was added and the samples were vortexed briefly before heating at 95° for 5 min.

3. About 30 μl of each extract were applied to a 16% SDS-PAGE and the proteins were resolved at 40 mA for 1 h. The gels were then blotted to nitrocellulose paper at 500 mA for 45 min in Western transfer buffer (40 mM Tris, 10% methanol, and 10 mM glycine).

4. The presence of the methylated histone H3 was detected by the incubation of each blot in antimethyl K4 histone H3 antibodies (purchased from Upstate). The schematic diagram describing this procedure is shown in Fig. 1.

GPS Analysis in Identification of the Components of COMPASS and Rad6 Complex Involvement in Histone Methylation

To determine whether we were capable of detecting mutants defective for methylation of lysine 4 of histone H3, the plate containing the mutant missing the Cps50/Yar003 component of COMPASS, which is known to be defective in histone H3 K4 methylation, was tested. All of the other 95 mutants in this microtiter plate possess this modification (see Fig. 2), indicating that GPS can be reliably employed to identify genes required for methylation of histone H3 on K4. Screening of all other microtiter plates

nonessential gene deletion mutants in yeast were inoculated into liquid culture and allowed to grow. Cells were collected by centrifugation and resuspended in lysis buffer containing Zymolyase. Following lysis, each extract was applied to a 16% SDS-PAGE. The gels were then blotted to nitrocellulose paper and the presence of the methylated histone H3 were detected by the incubation of each blot in antimethyl K4 histone H3 antibodies. (See color insert.)

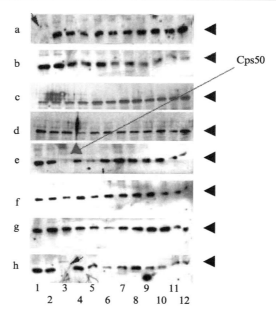

FIG. 2. Identification of the requirement of the subunits of COMPASS for the methylation of histone H3. Whole cell extracts from a microtiter plate containing 96 different yeast strains, each missing a different (nonessential) gene, one of which (row 3, lane 3) is the Cps50/Yar003w mutant, missing a subunit of COMPASS that is essential for histone H3 Lys4 methylation, were analyzed by GPS for this histone modification as described earlier. Arrows at position a1 and h3 indicate empty wells as plate markers.

revealed several other genes required for this histone modification, including Rad6 (see Fig. 3).

Since Rad6 is involved in diverse biological processes, such as N-end role pathway, DNA repair and recombination,[25-31] different ubiquitin-protein isopeptide ligases (E3 enzymes) are presumably responsible for specifying its different activities. Thus, the E3 enzyme that directs Rad6 to ubiquitinate histone H2B is expected to be a key determinant of transcriptional regulation. Employing GPS, we identified *BRE1* as one of the genes required for the ubiquitination of histone H2B (see Fig. 4A). Our analysis also demonstrated that Lge1, which is known to interact with

[25] S. Jentsch *et al.*, *Nature* **329,** 131 (1987).
[26] M. Koken *et al.*, *Proc. Natl. Acad. Sci. USA* **88,** 3832 (1991).
[27] V. Bailly *et al.*, *Genes Dev.* **8,** 811 (1994).
[28] T. Hishida *et al.*, *EMBO J.* **21,** 2019 (2002).
[29] M. Kupiec *et al.*, *Mol. Gen. Genet.* **203,** 538 (1986).
[30] X. L. Kang *et al.*, *Genetics* **130,** 285 (1992).
[31] H. Huang *et al.*, *Mol. Cell. Biol.* **17,** 6693 (1997).

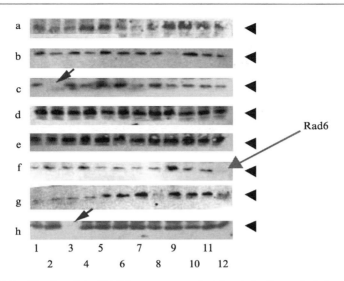

FIG. 3. Identification of the RAD6 as an enzyme that signals for the methylation of histone H3 by the ubiquitination of histone H2B. Extracts of *S. cerevisiae* mutants missing one of the approximately 4800 nonessential genes were tested by GPS for the presence of Lys4-methylated histone H3. One of the mutants lacking this histone modification is Rad6 (row F, lane 12). It has been demonstrated that Rad6 signals for the methylation of histone H3 by COMPASS by ubiquitinating lysine 123 of histone H2B.[20–22]

Bre1,[32] is partially required for methylation of K4 of histone H3 (see Fig. 4B). GPS also identified Paf1 and Rtf1, which are involved in transcriptional elongation, are required for methylation of histone H3 (on both K4 and K79). Since antibodies are available against other histone modifications, such as methylation of H3 on K36 and K79, phosphorylation of H3 on S10, and acetylation of histone H4 on K12, we have been able to use GPS to identify proteins required for these posttranslational modifications.

Potential of GPS for Defining Pathways of Posttranslational Modifications

GPS should enable identification of proteins required for any protein modification that can be detected, either by an antibody specific for the modified protein, or by the electrophoretic mobility change of the protein caused by its modification. One disadvantage of GPS is the requirement for an antiserum sensitive enough to detect the relatively small amounts of protein made available by our method. To surmount this problem, we are

[32] Y. Ho *et al.*, *Nature* **415**, 180 (2003).

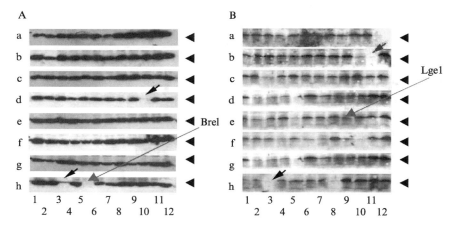

FIG. 4. Identification of the Bre1 and Lge1 as Rad6 E3 ligase complex that signals for the methylation of histone H3. Extracts of *S. cerevisiae* mutants missing one of the approximately 4800 nonessential genes were tested for the presence of Lys4-methylated histone H3 by GPS. One of the mutants lacking this histone modification is Bre1 (row H, lane 5). Arrows at position d10 and h3 indicate empty wells as plate markers.

now developing methods that do not require antibody specific for the given posttranslational modification.

Acknowledgments

We are grateful to Mark Gerber for critical reading of this manuscript. This work was supported in part by Grants from the American Cancer Society (RP69921801), National Institute of Health (1R01CA089455) and a Mallinckrodt Foundation Award to AS's laboratory. Work in MJ's laboratory was supported by the James S. McDonnell Foundation. A.S. is a Scholar of the Leukemia & Lymphoma Society.

[14] Purification of Sir2 Proteins from Yeast

By SUNIL GANGADHARAN, SONJA GHIDELLI, and ROHINTON T. KAMAKAKA

Silencing is characterized as the packaging of entire domains of chromatin into a structure that is transcriptionally repressed in a gene nonspecific manner and this state is stably propagated from a mother to the daughter cell. There are four silenced loci in the yeast *Saccharomyces cerevisiae:* the cryptic mating type loci *HML* and *HMR*, telomeres and the rDNA repeats (reviewed in Dhillon and Kamakaka[1]).

Silencing of the mating type genes at *HML* and *HMR* requires inactivation centers termed silencers that flank the *MATα* and *MAT*a genes, respectively. A silencer element is analogous to an enhancer element in that it functions in either orientation, is distance-independent and can act to repress unrelated genes. The *HM* silencers contain binding sites for ORC, Rap1p, and Abf1p.

In addition to the proteins that directly bind the silencer elements, there are other proteins that are also necessary for silencing including histones and the products of four genes *SIR1, SIR2, SIR3*, and *SIR4*.[2] It is believed that the role of the silencer and the silencer-binding components is to efficiently recruit the Sir proteins to the silenced loci. It is noteworthy that neither histones nor the Sir proteins bind DNA in a sequence specific manner. An important advance in understanding the molecular mechanism of silencing was made by the observation that Sir3p and Sir4p bind the N-terminal domains of histones H3 and H4 *in vitro*.[3,4] Also, these same Sir proteins are associated with the entire length of the silenced domain *in vivo*[5,6] reinforcing the view that these proteins are structural components of the silenced chromatin. The prevailing model therefore is that the silencer bound proteins recruit Sir1p and together these further recruit Sir3p and Sir4p. Sir2p on the other hand has been shown to interact with Sir4p[7–9] and is most likely recruited to the silenced loci via Sir4p.

Silencing at telomeres shares many similarities with silencing at *HML* and *HMR*. Telomeric silencing requires Rap1p and the Sir proteins Sir2p, Sir3p, and Sir4p but not Sir1p.[10] The multiple Rap1p binding sites at the telomere act as a silencer and Rap1p probably recruits Sir3p and Sir4p along with Sir2p resulting in a repressed domain. Since all three Sir proteins (Sir2p, Sir3p, and Sir4p) are required for silencing at *HML, HMR*, and telomeres, it is thought that silencing at these three loci is mediated by similar if not the same protein complexes.[11]

[1] N. Dhillon and R. T. Kamakaka, *Curr. Opin. Genet. Dev.* **12**, 188 (2002).

[2] J. Rine and I. Herskowitz, *Genetics* **116**, 9 (1987).

[3] A. Hecht, T. Laroche, S. Strahl-Bolsinger, S. M. Gasser, and M. Grunstein, *Cell* **80**, 583 (1995).

[4] A. A. Carmen, L. Milne, and M. Grunstein, *J. Biol. Chem.* **277**, 4778 (2002).

[5] A. Hecht, S. Strahl-Bolsinger, and M. Grunstein, *Nature* **383**, 92 (1996).

[6] S. Strahl-Bolsinger, A. Hecht, K. Luo, and M. Grunstein, *Genes. Dev.* **11**, 83 (1997).

[7] D. Moazed, A. Kistler, A. Axelrod, J. Rine, and A. D. Johnson, *Proc. Natl. Acad. Sci. USA* **94**, 2186 (1997).

[8] D. Moazed and A. D. Johnson, *Cell* **86**, 667 (1996).

[9] S. Ghidelli, D. Donze, N. Dhillon, and R. T. Kamakaka, *EMBO J.* **20**, 4522 (2001).

[10] O. M. Aparicio, B. L. Billington, and D. E. Gottschling, *Cell* **66**, 1279 (1991).

[11] G. Cuperus, R. Shafaatian, and D. Shore, *EMBO J.* **19**, 2641 (2000).

The 1–2 Mb rDNA locus is the fourth and largest repressed locus in *S. cerevisiae*.[12,13] It consists of approximately 200 copies of the 9.1 kbp rDNA repeat. *In vivo* analysis suggests that approximately half of the genes in the repeats are transcriptionally repressed at any given time. Several proteins including Sir2p affect repression at rDNA although it is not yet known exactly how Sir2p is recruited to this locus. Sir2p is part of a large molecular weight complex called the RENT-complex (Regulator of nucleolar silencing and telophase exit)[14,15] that localizes to the nucleolus and which consists of Sir2p, Cdc14p, Net1p, and Nan1p.

Sir2p is the only Sir protein that is necessary for repression at all the silenced loci and is also the only Sir protein with a known enzymatic activity. Sir2p possesses an NAD^+-dependent deacetylase activity[16,17] that can catalyze an NAD-nicotinamide exchange reaction in the presence of acetylated lysines such as those found in the N-termini of histones[18] to deacetylate histones *in vitro*. Sir2p is also the only Sir proteins to be conserved through evolution.[19]

Histones in chromatin of eukaryotic cells are reversibly acetylated and lysine residues K9 and K14 of histone H3 and K5, K8, and K16 of histone H4 are acetylated in active chromatin and deacetylated in silenced chromatin. Mutational studies indicate that K16 of histone H4 and K9, K14, and K18 of histone H3 are critically important in silencing.[3,20–22] Experiments suggest that Sir2p can specifically deacetylate K9 and K14 of histone H3 and K16 of histone H4.[16,23,24] Additionally it has been shown that over-expression of Sir2p promotes global deacetylation of histones *in vivo*.[25]

[12] J. S. Smith and J. D. Boeke, *Genes Dev.* **11,** 241 (1997).

[13] M. Bryk, M. Banerjee, M. Murphy, K. E. Knudsen, D. J. Garfinkel, and M. J. Curcio, *Genes Dev.* **11,** 255 (1997).

[14] W. Shou, J. H. Seol, A. Shevchenko, C. Baskerville, D. Moazed, Z. W. Chen, J. Jang, H. Charbonneau, and R. J. Deshaies, *Cell* **97,** 233 (1999).

[15] A. F. Straight, W. Shou, G. J. Dowd, C. W. Turck, R. J. Deshaies, A. D. Johnson, and D. Moazed, *Cell* **97,** 245 (1999).

[16] S. Imai, C. M. Armstrong, M. Kaeberlein, and L. Guarente, *Nature* **403,** 795 (2000).

[17] J. Landry, A. Sutton, S. T. Tafrov, R. C. Heller, J. Stebbins, L. Pillus, and R. Sternglanz, *Proc. Natl. Acad. Sci. USA* **97,** 5807 (2000).

[18] K. G. Tanner, J. Landry, R. Sternglanz, and J. M. Denu, *Proc. Natl. Acad. Sci. USA* **97,** 14178 (2000).

[19] C. B. Brachmann, J. M. Sherman, S. E. Devine, E. E. Cameron, L. Pillus, and J. D. Boeke, *Genes Dev.* **9,** 2888 (1995).

[20] J. S. Thompson, X. Ling, and M. Grunstein, *Nature* **369,** 245 (1994).

[21] M. Braunstein, R. E. Sobel, C. D. Allis, B. M. Turner, and J. R. Broach, *Mol. Cell. Biol.* **16,** 4349 (1996).

[22] N. Suka, Y. Suka, A. A. Carmen, J. Wu, and M. Grunstein, *Mol. Cell* **8,** 473 (2001).

[23] A. Kimura, T. Umehara, and M. Horikoshi, *Nat. Genet.* **32,** 370 (2002).

TABLE I
PURIFICATION OF Sir2p CONTAINING COMPLEXES FROM YEAST WHOLE CELL EXTRACTS

Column	Volume (ml)	Protein Conc. (mg/ml)	Total Protein (mg)
Yeast extract	200	2.5	500
SP-Sepharose	120	1.58	189
Cobalt affinity	44	0.4	17.6
Q-Sepharose			
Peak I	11	0.18	1.98
Peak II	10	0.38	3.8
Calmodulin affinity			
Peak I	7.5	0.04	0.3
Peak II	7	0.05	0.35

Purification and characterization of Sir2p containing-complexes is an important step for identifying the proteins that mediate silencing and is the first step in recapitulating the silenced domain *in vitro*. The purified Sir2p complexes can be used to carry out direct biochemical analysis such as their interactions with nucleosomal templates. The enzymatic activity of Sir2p in the complex and in particular, the activity of mutant complexes can also be examined using a broad variety of substrates.

Two nucleosome-binding complexes with distinct deacetylase activities containing Sir2p have been identified in our laboratory.[9] One of the complexes is a large multiprotein complex with an approximate molecular weight of 800 kDa, which contains Sir2p, Sir4p, and other unidentified subunits. This complex does not contain either Sir3p or Net1p. The presence of Sir4p in this complex suggested its involvement in silencing at *HML, HMR*, and the telomeres. Since Net1p is present in the second Sir2p-containing complex, this most likely represents the RENT complex.

Tagging Sir2p for Affinity Purification

We used Sir2p with an affinity tag to purify Sir2p-containing complexes from yeast strains. Strains of yeast derived from W303 that carried a His_6-HA_3 epitope tag or a His_6-HA-Calmodulin binding peptide tag (TAP tag) fused to the N-terminus of Sir2p were used. The His_6-HA-Calmodulin binding peptide tag is similar to the CHH (Calmodulin binding peptide-HA-His_6) tag reported previously for the purification of a cyclin-CDK

[24] N. Suka, K. Luo, and M. Grunstein, *Nat. Genet.* **32,** 378 (2002).
[25] M. Braunstein, A. B. Rose, S. G. Holmes, C. D. Allis, and J. R. Broach, *Genes Dev.* **7,** 592 (1993).

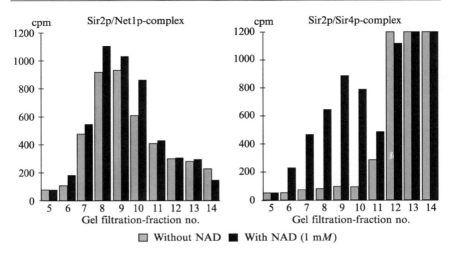

FIG. 1. Analysis of purified Sir2p containing complexes. Indicated column fraction from a Superose 6B gel filtration column were assayed for histone deacetylase activity in the presence and absence of NAD.

complex[26] and is based on the tandem affinity purification approach using the original TAP tag.[27] The tagged *SIR2* gene was expressed under the control of its own promoter at the endogenous locus. Strong over expression of the protein using inducible promoters such as the *GAL1* gene promoter is not recommended because we have found that under these conditions multiple complexes are formed presumably by nonspecific and/or unnatural interactions with host proteins. We also observe that maximal expression of proteins is obtained in strains where the gene is integrated in the genome rather than on a plasmid since a significant proportion of cells in a culture can survive without plasmid for significant lengths of time reducing yields and generating anomalous subcomplexes.

We used standard DNA procedures to introduce the N-terminal tag inframe with the coding region of *SIR2* in a pRS406 vector.[28] An appropriate restriction fragment was generated from this plasmid and used to transform a strain (JRY4571) that had a *URA3* gene downstream of the *SIR2*

[26] S. Honey, B. L. Schneider, D. M. Schieltz, J. R. Yates, and B. Futcher, *Nucleic Acids Res.* **29**, E24 (2001).

[27] G. Rigaut, A. Shevchenko, B. Rutz, M. Wilm, M. Mann, and B. Seraphin, *Nat. Biotechnol.* **17**, 1030 (1999).

[28] R. Rothstein, *Methods Enzymol.* **194**, 281 (1991).

promoter. Colonies in which the *URA3* gene was replaced with the tagged *SIR2* gene fragment were selected on a 5-FOA counter selection plate. Correct integration of the tagged gene at the appropriate locus was confirmed by PCR and Southern blot analysis. Protein blots with antibodies against Sir2p were used to check for expression of the protein.

Protein Blot Analysis of Yeast Extracts

Solutions

   Extrusion Buffer

     50 mM HEPES pH 7.0
     150 mM NaCl
     0.2 mM PMSF
     0.5 mM Benzamidine.

   Extraction Buffer

     50 mM HEPES pH 7.0
     150 mM NaCl
     0.1% NP-40
     20 μM Zinc acetate
     5 mM 2-mercaptoethanol.

   TBST buffer

     0.8% NaCl
     0.02% KCl
     0.3% Tris–Cl pH 7.5
     0.1% Tween 20.

Method

   Colonies to be screened for expression of the tagged protein are grown as 5 ml liquid YPD cultures overnight at 30° to an $A_{600nm} \sim 2$.

   Cells are centrifuged at 2500 rpm for 5 min on a bench top centrifuge. The cell pellets are first washed with 1 ml extrusion buffer, transferred to a 1.5 ml microfuge tube. The pellets are resuspended in 200 μl extraction buffer, topped with glass beads and vortexed in 6–8 bursts of 30 s each.

   The bottom of the tube was perforated with a needle and the tube was placed in another microfuge tube and spun at 2500 rpm for 1 min. To the cell suspension add 50 μl sample buffer and boil for 5 min. Spin the sample for 1 min at 13,000 rpm and remove the supernatant to a fresh tube. Load 20 μl of supernatant on to a SDS Polyacrylamide gel. Following electrophoresis, blot the gel onto Immobilon membranes using standard protein blotting techniques.

If this method fails to detect proteins expressed at low levels, a small-scale cobalt-bead affinity-binding step can be introduced since the protein is fused to a poly-Histidine tag. For this step, the lysate obtained after vortexing with glass beads is given a high-speed spin and the supernatant added to 50 μl of cobalt-beads (BD Talon-Clonetech) that has been previously equilibrated with the binding buffer (50 mM HEPES pH 7.2, 150 mM NaCl, 0.1% NP-40, 10% Glycerol). The lysate and beads are incubated for 1 h on a rotator at 4°. The beads are spun down at 2000 rpm for 10 s in a microfuge and the supernatant removed. The beads are washed three times with 1 ml of binding buffer. The supernatant is aspirated out and 40 μl of sample buffer are added to the beads and boiled for 3 min. The beads are spun down and the supernatant is loaded on a SDS polyacrylamide gel prior to protein blotting.

Genotypes of Yeast Strains

> ROY 1515: *MATα ade2-1 can1-100 his3-11 leu2-3, 112 trp1-1 ura3-1 GAL 6xHis-3xHA-SIR2 pep4Δ:: TRP1 9xMyc-NET1::LEU2*
> ROY2511: *MATα ade2-1 can1-100 his3-11 leu2-3, 112 trp1-1 ura3-1 GAL 6xHis-HA-CBP-SIR2*
> JRY3009: *MATα ade2-1 can1-100 his3-11 leu2-3, 112 trp1-1 ura3-1 GAL*
> JRY 4571: *MATα ade2-1 can1-100 his3-11 leu2-3, 112 trp1-1 ura3-1 GAL sir2Δ:: URA3.*

Assays to Test the Functionality of the Tagged Protein

Mating Assays

Patches of the yeast strain with the tagged Sir2p (ROY 1515 or ROY 2511), wild type strain (JRY 3009) and *sir2Δ* strain (JRY 4571) are grown on a YPD plate overnight at 30°. A lawn of either *MAT*a *his4* cells or *MATα his4* cells are spread on YM plates with 300 μl YPD to select for diploids that arise from mating events. The plates are allowed to dry and the yeast patches are replica plated onto the lawns and incubated overnight at 30°. Cells carrying a functional Sir2p are capable of mating and therefore grow as diploids on the selective plates. In this assay the growth of strains that had Sir2p tagged at the N-terminus with both His_6-HA_3 epitope tag and CBP-HA-His_6 tag was comparable to that of the wild type strain.

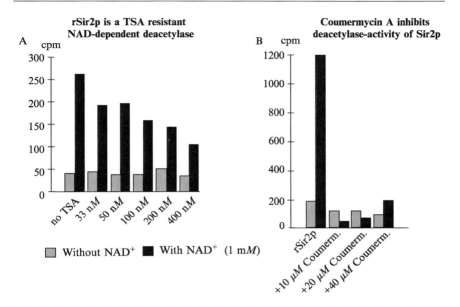

Fig. 2. Histone deacetylase activity of recombinant yeast Sir2p. (A) Effect of trichostatinA on NAD-dependent histone deacetylase activity of *E. coli* expressed recombinant yeast Sir2p. (B) Effect of CoumermycinA1 on NAD-dependent histone deacetylase activity of *E. coli* expressed recombinant yeast Sir2p.

Deacetylase Assays

Chicken erythrocyte histones were acetylated with recombinant yeast Hat1p acetylase and the acetylated histones were purified as described.[29]

Histone deacetylase activity was determined using 1 μg of acetylated histones with varying amounts of enzyme complex in a final volume of 200 μl. The reaction contained 50 mM Tris–HCl pH 9.0, 4 mM MgCl$_2$, 0.2 mM DTT with or without 2 mM NAD. The acetylated histones were added last to the reaction mix. The reaction was incubated for 30 min at 30° and the reaction was stopped by the addition of 50 μl of 0.16 M Acetic acid, 0.1 M HCl. The samples were then extracted by the addition of 600 μl of ethyl acetate, vortexed and left to stand on the bench for 5 min. Five hundred μl of the upper organic phase was removed and diluted with 5 ml scintillation fluid and counted for free acetate.

Various inhibitors can also be added into this reaction. Sir2p mediated deacetylation is resistant to 33 nM TrichostatinA but is sensitive to 10 μM CoumermycinA1.

[29] P. A. Wade, P. L. Jones, D. Vermaak, and A. P. Wolffe, *Methods Enzymol.* **304,** 715 (1999).

General Considerations While Handling Proteins

All steps are carried out at $4°$. Milli-Q water is used to make all solutions. All solutions are filtered using 0.45 μm filters. Utmost care has to be taken to reduce keratin contamination and therefore gloves are worn while performing these purification experiments. The purified fractions are stored at $-80°$ after flash freezing in liquid nitrogen. Care should also be taken to reduce foaming of the protein solution.

If the complex is to be analyzed for mass spectrometry the buffering ion should be Tris–HCl and glycerol is not recommended in the buffer. Also the concentration of detergent should be as low as possible (0.02%NP40).

$1000\times$ Protease Inhibitor Stocks

0.5 M Benzamidine
0.2 M PMSF
100 mg/ml Bacitracin
0.5 M Sodium metabisulphite
5 mg/ml Pepstatin
5 mg/ml TPCK
5 mg/ml TLCK
5 mg/ml Aprotinin
10 mg/ml Leupeptin
These are diluted thousand times to get the final concentrations for solutions.

Preparation of Yeast Whole Cell Extracts

Solutions

Extrusion Buffer

50 mM HEPES pH 7.0
150 mM NaCl
All protease inhibitors.

Extraction Buffer

50 mM HEPES pH 7.0
150 mM NaCl
0.1% NP-40
20 μM Zinc acetate
5 mM 2-mercaptoethanol
All protease inhibitors.

Materials

Coffee grinder
10 ml Plastic pipettes
Dewar flask
Dry ice.

Method

Inoculate 50 ml of YPD (yeast extract 1%, peptone 2%, dextrose 2%) medium with the yeast strain (ROY1515 or ROY2511) and grow overnight at 30°. with vigorous shaking. This overnight culture is used to inoculate 10 L of YPD and the yeast cell culture is grown to an $A_{600 \, nm} = 1{\sim}1.5$. The medium has to be supplemented with adenine if the strain used is defective in adenine biosynthesis. We use 10 two-liter baffle flasks for the growth of the culture though a fermenter can also be used. We typically obtain about 10 g (wet weight) of yeast cells per liter.

The cells are pelleted by centrifugation at 5000 rpm for 5 min at 4° using a Sorvall GS-3 rotor. They are washed with 2 L of 50 mM HEPES pH 7.0 and finally the cells are washed with 700 ml extrusion buffer. This is followed by a second wash with 200 ml extrusion buffer. Finally 5 ml of the extrusion buffer is added to the pellet and the cells resuspended as thick slurry with the aid of a glass rod. The thick cell slurry is then added drop-wise into liquid nitrogen in a Dewar flask using a pipette. The frozen cells are removed from the liquid nitrogen and may be stored at −80°.

The frozen cells are lysed in a coffee grinder, Braun-Model KSM2-household type by the method described.[30] Dry ice is added so as to just cover the blades. Approximately 20 g of the frozen cells are ground with constant shaking for 5 min in the grinder. The finely ground frozen powder is transferred to a plastic container but can also be stored at −80° at this stage.

To prepare the extract the frozen powder should be thawed on ice in extraction buffer. The choice of extraction buffer depends on the solubility of the protein. We arrived at the extraction conditions for Sir2p by a systematic analysis where isolated yeast nuclei were extracted with varying amounts of salt from 0.05 to 2.0 M NaCl and with different detergents such as NP40, Tween 20, Tween 40, and Triton X-100. The best conditions for extraction were 0.15–0.5 M NaCl plus 0.5% Tween 20. Higher salt

---

[30] M. C. Schultz, D. J. Hockman, T. A. Harkness, W. I. Garinther, and B. A. Altheim, *Proc. Natl. Acad. Sci. USA* **94,** 9034 (1997).

concentrations with detergent resulted in some of the protein being precipitated out. Varying pH of the buffers can also be tested although we used physiological conditions for pH as this would most likely preserve the complex in its native state. Zinc acetate was included in the buffers as analysis of the primary sequence of Sir2p reveals a potential zinc binding domain which was shown to be required for Sir2p function and which may be required for complex stability.

The frozen cells are placed in a beaker and thawed by the addition of 200 ml of extraction buffer containing all of the protease inhibitors. Once the cells have thawed in the extraction buffer, the lysate is mixed and passed through a Yamato homogenizer 1000 rpm for at least 5 times to ensure complete lysis of cells. The homogenate is then cleared from the cell debris by centrifugation at 10,000 rpm for 15 min at $4°$ in a GS-3 rotor. We consistently get $\sim$350 ml of extract with a protein concentration of 2–3 mg/ml.

SP-Sepharose FF Ion Exchange Chromatography

The yeast extract is subjected to an initial separation on an ion exchange column. Ion exchange chromatography is capable of separating molecules that have only small differences in charge. This technique is most suited for intermediate steps in purification to collect fractions in a concentrated and semipurified form. SP-sepharose Fast Flow is a strong cation exchanger with excellent flow properties and a high capacity for proteins of almost all pI values. The ion exchange group is a sulphopropyl group that remains charged and maintains consistently high capacities for binding over a pH range of 4–13.

Solutions

Buffer A

50 mM HEPES pH 7.2
12.5 mM MgCl$_2$
10% Glycerol
0.1% NP-40
All protease inhibitors.

Buffer B

50 mM HEPES pH 7.2
12.5 mM MgCl$_2$
10% Glycerol
0.1% NP-40

1 M KCl
All protease inhibitors.

Wash Buffer

50 mM HEPES pH 7.4
12.5 mM MgCl$_2$
1 M KCl
All protease inhibitors.

Materials

SP-Sepharose FF (Amersham Pharmacia)
XK-26 column (Amersham Pharmacia).

Method

It is important to empirically determine the binding conditions of the protein to the column such as ionic strength, temperature, pH, and requirement for divalent cations. We chose a pH within the physiological range so that the Sir2p complexes are not disrupted. Glycerol reduces the dielectric constant of the medium and helps to stabilize weak protein-protein interactions. A detergent like NP-40 minimizes nonspecific adsorption of proteins to plastic and glass. The ionic strength was maintained with KCl although NaCl could also be used. Chelating agents such as EGTA and EDTA are left out as they might affect complex stability.

Seventy-five milliliters of SP-Sepharose FF beads stored in 20% ethanol, 0.2% sodium acetate were gently washed using a sintered glass funnel in wash buffer (5–7 volumes). The resin was finally resuspended as a 50% (v/v)-slurry in the buffer. One hundred and fifty milliliters of the slurry is enough to pack an XK-26 column of dimensions, 2.6 × 14.5 cm.^{2} The column is packed by FPLC using five column volumes of buffer B at a flow rate of 6 ml/min. The column is finally equilibrated with buffer A containing 150 mM KCl. All protein purification was performed with pre-chilled buffers in the cold room at 4–6°.

The yeast extract which usually has an ionic strength of ~350 mM, was adjusted to 150 mM KCl by adding an equal volume of Buffer A prior to loading on the column. Proteins that do not bind the column during loading are collected. The FPLC is programmed to increase the concentration of KCl in steps of 0.25, 0.40 and 0.6 M. Five column volumes of each buffer concentration are used. The fraction size is approximately one fifth of the volume of the column, which in this case is 15 ml. One hundred microliters of each fraction is collected for Western blot analysis with an anti-HA monoclonal antibody HA.11 (Babco) and protein estimation.

The Sir2p-containing fractions appear in the 0.4 M KCl elution and peak fractions are pooled and used for the subsequent purification step.

Save 100 μl of the pooled fractions for protein estimation and Western analysis to determine fold purification.

Immobilized Metal Affinity Chromatography (IMAC)

IMAC is a group specific-affinity technique based on the reversible interaction between various amino acid side chains such as histidine, and cysteine with immobilized metal ions such as Ni(II), Co(II) or Zn(II). BD Talon™ resin is a cobalt based IMAC resin. The metal ion Co(II) is held by a chelator covalently attached to a sepharose CL-6B support. We prefer to use this resin over nickel based IMAC resins as there is significantly less adsorption of unwanted proteins with exposed histidine residues. Another advantage with using this resin is that the bound proteins can be eluted under milder conditions of pH and lower concentration of imidazole.

Solutions

Binding Buffer A

50 mM HEPES pH 7.2
150 mM NaCl
0.1% NP-40
10% Glycerol
All protease inhibitors.

Wash Buffer B

50 mM HEPES pH 7.2
150 mM NaCl
0.1% NP-40
10% Glycerol
10 mM Imidazole
All protease inhibitors.

Elution Buffer C

50 mM HEPES pH 7.2
150 mM NaCl
0.1% NP-40
10% Glycerol
60 mM Imidazole
All protease inhibitors.

Materials

20 ml of BD Talon™ resin (Clonetech)
Econo-pac disposable chromatography column (Biorad).

Method

Equilibrate 20 ml of the BD Talon™ resin with buffer A in a 50 ml falcon tube by resuspending in 25 ml of Buffer A, spinning briefly at 1900 rpm and discarding the supernatant. This procedure is repeated four times.

A Biorad Econopac (20 ml) disposable column is attached to a fraction collector adjusted to collect 200 drops per fraction (~4 ml each).

The pooled fractions from the SP-sepharose column containing Sir2p (0.4 M KCl eluate ~60 ml) are diluted with an equal volume of binding buffer A lacking NaCl. Following this dilution, the Sir2p containing sample is added to the equilibrated Talon resin in 50 ml falcon tubes. The tubes are rotated for 1 h at 4°, spun briefly at 1900 rpm to collect the supernatant. The supernatant is the flow through for this column and should be saved for analysis.

To the resin add an equal volume of wash buffer and pour this slurry into a disposable Biorad Econopac column. Connect a reservoir of buffer B to the column and wash the column with 200 ml of Wash Buffer B (10 column volumes); collecting all fractions. Switch the reservoir from the wash buffer to the elution buffer and continue to collect fractions. Analyze 20 μl of all collected fractions by Western blot. Sir2p is typically found in fractions 4–14 of the eluate and these are pooled and used for further chromatographic separation on a Q-sepharose column.

Save 100 μl of the pooled fractions for protein estimation and Western analysis to determine fold purification.

Q-Sepharose Anion Exchange Chromatography

Q-sepharose Fast Flow is an anion exchanger and the exchange group is a quaternary amine that remains charged over a wide range of pH.

Solutions

Buffer A

50 mM HEPES pH 7.4
12.5 mM $MgCl_2$
0.1% NP-40
10% Glycerol
All protease inhibitors.

Buffer B

1 M KCl
50 mM HEPES pH 7.4
12.5 mM MgCl$_2$
0.1% NP-40
10% Glycerol
All protease inhibitors.

Materials

Q-sepharose FF (Amersham Pharmacia)
HR-16 column (Amersham Pharmacia).

Method

An 8-ml HR16 Q-sepharose column packed on the FPLC is used for the next step of purification of Sir2p. The column is equilibrated with 40 ml of buffer with 0.08 M KCl with a flow rate not to exceed 300 cm/h.

The pooled samples from the cobalt column affinity step are diluted with buffer A so that the ionic strength is equivalent to 80 mM KCl. This sample is loaded on to an 8-ml Q-sepharose column and samples are collected as outlined above. The FPLC is programmed to increase the concentration of KCl in steps of 0.08, 0.15, 0.30, and 0.6 M and five column volumes of each buffer concentration are used to collect 2 ml fractions. Fifty microliters of each fraction are removed for protein blot analysis with the anti-HA monoclonal antibody HA.11 and protein estimation. The Sir2p reactive species appears in the 0.15 M KCl fraction and the 0.3 M KCl fractions. The peak fractions from each concentration are pooled and used for the subsequent purification step.

Save 50 μl of the pooled fractions for protein estimation and Western analysis to determine fold purification.

Heparin Sepharose Affinity Chromatography

Heparin sepharose resins are excellent for the purification of DNA binding proteins because they possess properties similar to both ion exchange and affinity resins.

Solutions

Buffer A

50 mM HEPES pH 7.0
12.5 MgCl$_2$

10% Glycerol
0.1% NP-40
All protease inhibitors.

Materials

Heparin-sepharose resin (column volume 2 ml)
HR 5/5 column 5 × 50 mm.[2]

Method

The heparin sepharose resin is packed into a HR 5/5 column and equilibrated in Buffer A with 150 mM KCl. The pooled Sir2p fractions from the Q sepharose column are adjusted to an ionic strength of 150 mM KCl and then loaded onto the equilibrated heparin-sepharose column. The column is washed with three columns of buffer A with 150 mM KCl and a linear gradient of 0.15 to 1 M KCl (10 column volumes) is applied to the column and 0.5 ml fractions collected. The proteins in each fraction are assayed by protein blotting as described above

Save 25 μl of the pooled fractions for protein estimation and additional Western analysis.

Calmodulin Affinity Chromatography

Calmodulin binds with high affinity to a calmodulin-binding peptide in the presence of calcium and removal of calcium from the medium abrogates this interaction. This interaction has been widely and successfully used in the purification of protein complexes where one subunit of the complex contains the calmodulin-binding peptide fused to the coding sequence of the polypeptide.[27,31,32] Commercially available calmodulin sepharose beads can be used for the affinity purification of any protein fused to a calmodulin binding peptide tag. Yeast strains containing Sir2p fused to the calmodulin binding peptide were further purified using calmodulin sepharose as a second affinity step and chelation of Ca(II) with EGTA results in the elution of the complex. Care should be taken during this step to determine any change in the size of the protein complex that may occur following removal of divalent cations by EGTA.

[31] W. W. Pijnappel, D. Schaft, A. Roguev, A. Shevchenko, H. Tekotte, M. Wilm, G. Rigaut, B. Seraphin, R. Aasland, and A. F. Stewart, *Genes Dev.* **15,** 2991 (2001).
[32] A. Roguev, D. Schaft, A. Shevchenko, W. W. Pijnappel, M. Wilm, R. Aasland, and A. F. Stewart, *EMBO J.* **20,** 7137 (2001).

Solutions

Binding Buffer A

50 mM HEPES pH 7.4
150 mM NaCl
10 mM 2-mercaptoethanol
1 mM Magnesium acetate
1 mM Imidazole
2 mM CaCl$_2$
0.1% NP-40
10% Glycerol
All protease inhibitors.

Elution Buffer B

50 mM HEPES pH 7.4
150 mM NaCl
10 mM 2-mercaptoethanol
1 mM MgOAc
1 mM Imidazole
2 mM EGTA
0.1% NP-40
10% Glycerol
All protease inhibitors.

Materials

Calmodulin Sepharose
Econo-pac disposable chromatography column (Biorad).

Method

Equilibrate 1 ml of calmodulin-sepharose in Buffer A as described for the Talon resin. The pooled Sir2p containing fractions are adjusted for the various buffer components of Buffer A by the addition of components that are missing from the previous step, that is, 10 mM 2-mercaptoethanol, 2 mM CaCl$_2$, and 1 mM MgOAc. This is done by adding an equal volume of Buffer A containing the missing components at twice the concentrations rather than adding the concentrated stocks directly to the Sir2p fractions.

The Sir2p-containing fractions are then added to the equilibrated calmodulin sepharose beads and the binding of Sir2p to the beads is allowed to proceed at 4° for 1 h with constant rotation. The beads are then packed into an econo-pac disposable plastic column and washed with 50 ml Binding Buffer A. Elution of the calmodulin bound fraction is carried out using

Buffer B and 1 ml fractions are collected. Remove 50 μl for protein estimation and protein blotting.

Gel Filtration Chromatography

Gel filtration chromatography, which is commonly referred to as size exclusion chromatography, is a method for the separation of molecules on the basis of their size and shape. We chose Superose 6 resin because it can withstand higher back-pressure and has low ionic or hydrophobic interactions therefore reducing loss of protein during the purification. The gel filtration chromatography also allows us to change buffers in the sample.

We initially checked the apparent molecular weight of Sir2p-containing complexes by applying yeast whole cell lysates, prepared from logarithmically growing cells directly on a Superose 6B column. Column fractions were analyzed by SDS-PAGE followed by immunoblotting using antiserum specific to Sir2p. This analysis indicated that Sir2p eluted from the column in two distinct peaks of apparent molecular weight of $\sim$900 and $\sim$70 kDa. The 70 kDa is most likely free Sir2p, which has a predicted molecular weight of 63 kDa, while the 900 kDa peak represents the major Sir2p-containing complexes. This analysis was also done initially to determine the stability of the Sir2p complexes under various conditions. A partially purified fraction of the Sir2p-containing complexes was treated with DNAse I, RNAse A, or high concentrations of KCl (500 mM) for 30 min at room temperature and was then subjected to gel filtration analysis. Neither nuclease treatment nor high salt concentration affected the overall size of the complex.

Solutions

Buffer A

50 mM HEPES 7.4
150 mM KCl
12.5 mM MgCl$_2$
0.1% NP-40
10% Glycerol
All the protease inhibitors.

Materials

Prepacked Superose 6 column (30 ml) HR 10/50 (Amersham Pharmacia).

Method

A Superose 6 column (30 ml size) was equilibrated with 150 ml of buffer A at a flow rate that did not exceed 50 cm/h. Use caution while connecting the column to avoid introducing any air bubbles. Wash the column with 3 volumes of buffer A to equilibrate the resin. Load $\sim$2 ml of the Sir2p containing fractions onto the Superose 6 column and collect 1 ml fractions after loading is complete. Remove 100 μl of each fraction for protein blotting and Western analysis. These conditions were maintained for all our analysis for ease of comparison across experiments. The Sir2p complex is usually observed in fractions 7–9.

It is important to calibrate the gel filtration column by running a set of standard molecular weight marker proteins through the column. Ensure the same volume as sample is also loaded in this case.

Red Sepharose Chromatography

Sir2p is an NAD dependant histone deacetylase. We took advantage of this to further purify Sir2p containing complexes using a Red Sepharose CL 6B matrix that has a ligand mimic of NAD, Procion Red HE-3B coupled to it. We used this procedure to further purify and concentrate the sample.

Solutions

Buffer A

50 mM HEPES pH 7.0
12.5 $MgCl_2$
10% Glycerol
0.1% NP-40
All protease inhibitors.

Buffer B

50 mM HEPES pH 7.0
12.5 $MgCl_2$
10% Glycerol
0.1% NP-40
1 M KCl
All protease inhibitors.

Materials

HR 16 column (Amersham Pharmacia).

Method

An 8-ml HR 16 column is packed with the Red Sepharose matrix equilibrated as described for the SP-Sepharose resin. The column is equilibrated in Buffer A with 100 mM KCl. The Sir2p containing fractions are loaded on to this column. The column is then washed with 5 column volumes of 300 mM KCl in buffer A. The bound protein is eluted with 0.1% SDS, 2 ml fractions are collected and analysed by Western blots. We have not been successful in eluting the Sir2p containing complex from this column in its native form.

We have also observed that Sir2p binds a Phenyl-Sepharose column with very high affinity in buffer containing 1.0 M KCl but we have once again only been able to elute these complexes from the Phenyl-Sepharose column following denaturation with 0.1% SDS.

Purification of Recombinant Sir2p from *E. coli*

Solutions

Lysis Buffer

50 mM HEPES pH 7.0
150 mM NaCl
0.1% Tween 20
20 μM Zinc acetate
All the protease inhibitors.

Wash Buffer

50 mM HEPES pH 7.0
150 mM NaCl
10% Glycerol
10 mM Imidazole HCl
All protease inhibitors.

Elution Buffer

50 mM HEPES pH 7.0
150 mM NaCl
10% Glycerol
150 mM Imidazole HCl
All protease inhibitors.

Dialysis Buffer

25 mM HEPES, K$^+$, pH 7.6

12.5 mM MgCl$_2$
0.1 mM EDTA
10% (v/v) Glycerol
0.1 M KCl
0.1% Tween 20.

Materials

20 ml of BD Talon™ resin (Clonetech)
Econo-pac disposable chromatography column (Biorad).

Method

Transform *E. coli* DE81 pLysS cells with plasmid pRO 655. The Sir2 coding region is cloned in frame with the T7 6× His tags in a pET28 vector. Select transformants on kanamycin plates.

Grow 1 L of cells to an A$_{600}$ = 0.6 in LB + Kan.

Add IPTG (1 mM final) to induce expression of the recombinant protein and grow the cells for a further 2.5 h.

The cells are pelleted by centrifugation at 5000 rpm for 5 min at 4° using a Sorvall GS-3 rotor.

The cells are washed in 1 L 20 mM Tris–HCl pH 7.8 and the cell pellets are frozen at −80°.

The cell pellets are resuspended in 25 ml of lysis buffer and the A$_{600}$ of the suspension is measured (10 μl in 1 ml water).

Sonicate the cell slurry for 2.5 min (30 s × 5 pulses).

Remeasure A$_{600}$ (10 μl in 1 ml) and the absorbance of the suspension should be approximately 25% of the original.

Spin the lysed cell extracts in a Sorval SS34 rotor at 12,000 rpm for 20 min at 4°. The supernatant should contain most of the recombinant Sir2p.

The Sir2p containing extract is added to 1.5 ml equilibrated settled BD Talon™ resin in a 50 ml falcon tube (equilibrated in 50 mM HEPES pH 7.0, 150 mM NaCl, 10% glycerol, 0.5% Tween 20).

Rotate the extract for 60 min at 4° on a rotator.

Pour this slurry into a disposable Biorad Econopac column attached to a fraction collector. Connect a reservoir of wash buffer to the column and wash the column with 15 ml of Wash Buffer (10 column volumes) collecting fractions. Switch the reservoir from the wash buffer to the elution buffer and continue to collect fractions. Analyze 20 μl of all collected fractions by SDS polyacrylamide gels.

Dialyze appropriate fractions against dialysis buffer for 3 h at 4° and freeze in aliquots at −80°.

[15]   Purification and Biochemical Properties of the
        Drosophila TAC1 Complex

By Svetlana Petruk, Yurii Sedkov, Sheryl T. Smith,
Wladyslaw Krajewski, Tatsuya Nakamura, Eli Canaani,
Carlo M. Croce, and Alexander Mazo

The *Drosophila trithorax (trx)* gene belongs to the Trithorax group (trxG) of positive regulators of homeotic genes (for review, Simon and Tamkun[1]). These genes work in conjunction with the Polycomb group (PcG) of repressors to establish a spatially restricted pattern of homeotic gene expression (for review, Brock and van Lohuizen[2]). TrxG and PcG proteins are constituents of multiprotein complexes, which are involved in alterations of chromatin structure. The mechanism of action of trxG and PcG proteins has begun to emerge in recent years due to the advances in purification of these protein complexes. TrxG proteins BRM, SNR1, MOR, and OSA have been found to be members of the *Drosophila* BRM complex, which is homologous to both the yeast and the mammalian SWI/SNF chromatin remodeling complexes.[3–6] The trxG protein GAGA factor has been shown to be required for the activity of another *Drosophila* chromatin remodeling complex, NURF.[7,8] Two protein complexes containing *Drosophila* Trx and its human homologue, ALL-1, have been purified.[9,10] Both complexes were found to possess acetyltransferase (HAT) activity, as well as histone H3 lysine 4 methyltransferase activity (HMT)[9,10] (S. Petruk unpublished, see later). The trxG protein ASH1 is

[1] J. A. Simon and J. W. Tamkun, *Curr. Opin. Genet. Dev.* **12**, 210 (2002).
[2] H. W. Brock and M. van Lohuizen, *Curr. Opin. Genet. Dev.* **11**, 175 (2001).
[3] R. T. Collins and J. E. Treisman, *Genes Dev.* **14**, 3140 (2000).
[4] M. A. Crosby, C. Miller, T. Alon, K. L. Watson, C. P. Verrijzer, R. Goldman-Levi, and N. B. Zak, *Mol. Cell. Biol.* **19**, 1159 (1999).
[5] A. J. Kal, T. Mahmoudi, N. B. Zak, and C. P. Verrijzer, *Genes Dev.* **14**, 1058 (2000).
[6] O. Papoulas, S. J. Beek, S. L. Moseley, C. M. McCallum, M. Sarte, A. Shearn, and J. W. Tamkun, *Development* **125**, 3955 (1998).
[7] T. Tsukiyama and C. Wu, *Cell* **83**, 1011 (1995).
[8] C. Wu, T. Tsukiyama, D. Gdula, P. Georgel, M. Martinez-Balbas, G. Mizuguchi, V. Ossipow, R. Sandaltzopoulos, and H. M. Wang, *Cold Spring Harb. Symp. Quant. Biol.* **63**, 525 (1998).
[9] S. Petruk, Y. Sedkov, S. Smith, S. Tillib, V. Kraevski, T. Nakamura, E. Canaani, C. M. Croce, and A. Mazo, *Science* **294**, 1331 (2001).
[10] T. Nakamura, T. Mori, S. Tada, W. Krajewski, T. Rozovskaia, R. Wassell, G. Dubois, A. Mazo, C. M. Croce, and E. Canaani, *Mol. Cell* **10**, 1119 (2002).

Copyright 2004, Elsevier Inc.
All rights reserved.
0076-6879/04 $35.00

also a component of large protein complex.[6] Although this complex has not yet been purified, it has been demonstrated that ASH1 possesses multicatalytic HMTase activities with respect to several histone residues.[11] Two PcG protein complexes have also been purified. The *Drosophila* ESC-E(Z)[12,13] and the homologous human EED-EZH2/PRC2[14,15] complexes were both found to methylate lysine 9 and lysine 27 of histone H3. Another PcG complex, PRC1, containing four known PcG proteins, has been shown to counteract the chromatin remodeling activity of the SWI/SNF complex.[16] Thus, current data suggest that trxG proteins are components of either chromatin modification or chromatin remodeling complexes.

This chapter describes the purification of trxG TAC1, a Trithorax (Trx)-containing protein complex, from *Drosophila* embryos. TAC1 is 1-MDa protein complex, which contains three proteins: Trx, and the *Drosophila* homologues of the CREB-binding protein, dCBP, and the SET-binding protein, Sbf1. Purified TAC1 possesses two types of enzymatic activities: histone acetylation and histone methylation (HMT). It can acetylate all four core histones by virtue of the HAT activity of dCBP, and it can methylate lysine 4 of histone H3 through the SET domain of Trx. We describe our TAC1 purification scheme and methods that are used to assay enzymatic activities of the TAC1 components.

TAC1 Complex Purification

The TAC1 complex is purified by conventional column chromatography from *Drosophila* embryonic nuclear extracts. The three components of TAC1 differ in their abundance in nuclear extracts. Sbf1 and especially dCBP are relatively abundant proteins. Trx, however, is an extremely low abundance protein. Therefore, in order to perform multistage column chromatography, our purification is usually initiated by using nuclear extract prepared from at least 200 g of embryos. The complete purification scheme consists of six steps. Alternatively, the material eluted after the

[11] C. Beisel, A. Imhof, J. Greene, E. Kremmer, F. Sauer, *Nature* **419,** 857 (2002).

[12] J. Muller, C. M. Hart, N. J. Francis, M. L. Vargas, A. Sengupta, B. Wild, E. L. Miller, M. B. O'Connor, R. E. Kingston, and J. A. Simon, *Cell* **111,** 197 (2002).

[13] B. Czermin, R. Melfi, D. McCabe, V. Seitz, A. Imhof, and V. Pirrotta, *Cell* **111,** 185 (2002).

[14] R. Cao, L. Wang, H. Wang, L. Xia, H. Erdjument-Bromage, P. Tempst, R. S. Jones, and Y. Zhang, *Science* **298,** 1039 (2002).

[15] A. Kuzmichev, K. Nishioka, H. Erdjument-Bromage, P. Tempst, and D. Reinberg, *Genes Dev.* **16,** 2893 (2002).

[16] Z. Shao, F. Raible, R. Mollaaghababa, J. R. Guyon, C. T. Wu, W. Bender, and R. E. Kingston, *Cell* **98,** 37 (1999).

third column can be further purified by glycerol gradient sedimentation and immunoaffinity purification. This procedure yields the highly purified TAC1 complex, although some contaminating proteins are still detected.

Collection of Staged Embryos

Drosophila melanogaster (Oregon R strain) is grown in mass culture using 2–3 large $55 \times 40 \times 30$ cm population cages. Using these cages, it is possible to collect 10–15 g of embryos per day. Embryos are collected by allowing flies to lay eggs on agar-molasses plates for a 0–18 h period. Embryos are washed off these plates and transferred into cutoff 50 ml tubes which were covered at one end with the 110 μkm mesh Nitex filter, and rinsed extensively with water. Chorions are removed by immersing and swirling gently the collected embryos in 50% Chlorox bleach for 2 min. To remove traces of Chlorox, embryos are washed with 0.7% NaCl and after that they are extensively rinsed with water. Excess of liquid is drained using paper towels. Embryos can then be used immediately for isolation of nuclei, or can be frozen for storage for several months at $-80°$.

Preparation of Nuclear Extract

The following steps are carried out on ice or at $4°$. Approximately 3 g of embryos are transferred into a 15 ml glass Dounce tissue homogenizer. Embryos are resuspended in 15 ml of ice-cold buffer A and are homogenized using a type B pestle until the pestle runs smoothly. Usually, homogenization is repeated 10 times using 3 g of embryos to obtain a total of 30 g of homogenized embryos. Homogenates are combined and filtered through a nylon mesh (100 μkm Nitex) to remove debris. To prevent clogging, the homogenate is stirred with the top end of a Pasteur pipet. To eliminate cellular debris, the homogenate is cleared by low-speed centrifugation in four 50 ml polypropylene tubes for 1 min at $500g$. The supernatant is subsequently transferred into four 50 ml tubes and nuclei are collected by centrifugation for 10 min at $2300g$. The supernatant is discarded immediately and the nuclear pellet is resuspended in 12.5 ml (per tube) of buffer A (50 ml total). Twenty-five milliliters of the supernatant is layered very slowly into two 50 ml centrifuge tubes which already contain 25 ml of buffer A-1. The material is centrifuged for 20 min at $2300g$ to precipitate nuclei. Avoiding dispersion of the hard yellow pellet, the loose pellet is gently but quickly washed with 25 ml per tube of buffer A-2, by inverting the tube several times. The resuspended pellet is immediately transferred into a new tube and nuclei are collected by centrifugation for 5 min at $2300g$. The supernatant is quickly discarded (as much as possible), leaving approximately 5 ml of pellet in each tube. Twenty-five milliliters (per tube, i.e., 5 ml per 1 ml of nuclei) of buffer E is added yielding a final salt concentration of 0.35 M KCl, and nuclear proteins are extracted by rotation for 40 min at 4°. The material is transferred into six 10 ml SW41 tubes and centrifuged for 1 h at 35,000 rpm at 2° in a Beckman SW41Ti rotor. The lipid layer is removed from the top of the tubes with Pasteur pipet. The final supernatant can be frozen and stored at −80°.

Solutions

Buffer A: 0.35 M sucrose, 15 mM HEPES, pH 7.6, 10 mM KCl, 5 mM MgCl$_2$, 0.1 mM EDTA, 0.5 mM EGTA, 1 mM DTT, 1 mM PMSF, 1 μg/ml of aprotenin, leupeptin, and pepstatin each.

Buffer A-1: Buffer A-1 is the same as buffer A except for 0.8 M sucrose.

Buffer A-2: 10 mM HEPES, pH 7.9, 1.5 mM MgCl$_2$, 10 mM KCl, 0.5 mM DTT, 0.6 mM PMSF, 1 μg/ml of aprotenin, leupeptin, and pepstatin each.

Buffer E: 20 mM HEPES, pH 7.9, 420 mM KCl, 1.5 mM MgCl$_2$, 0.2 mM EDTA, 25% glycerol, 0.5 mM DTT, 0.5 mM PMSF, 1 μg/ml of aprotenin, leupeptin, and pepstatin each.

TAC1 Fractionation Procedure by Conventional Chromatography

To monitor fractionation of the Trx complex, we used rabbit anti-Trx antibodies, N1,[17] dCBP C2, and Sbf1 S1.[9] To purify TAC1, we used six column purification steps (see Fig. 1). All procedures are performed at 4° using the FRAC 200 Fraction Collector and the Peristaltic Pump-1 (Pharmacia). All columns are prepared one day in advance. The resin is equilibrated in starting buffers C or D, packed in column (Sigma) and washed with at least 10 bed volumes of the starting buffer. After applying nuclear proteins, the column is usually washed with 10 bed volumes of the starting buffer and every subsequent elution step is performed with 10 bed volumes of the appropriate elution buffer. The amount of protein in each eluted fraction is tested by the Bradford assay, and peak fractions of each elution step are combined. Aliquots from the combined fractions of each elution step are tested by Western blot using Trx, dCBP and Sbf1 antibodies to detect the presence of these proteins in the eluted material. At later chromatographic steps, proteins from the TAC1-containing fractions are visualized by silver staining of 6 or 12% SDS gels (Silver stain kit, Sigma), to check the level of purification in the eluted fractions. A major Trx-containing fraction is selected after each purification step and applied next,

---

[17] B. Kuzin, S. Tillib, Y. Sedkov, L. Mizrokhi, and A. Mazo, *Genes Dev.* **8,** 2478 (1994).

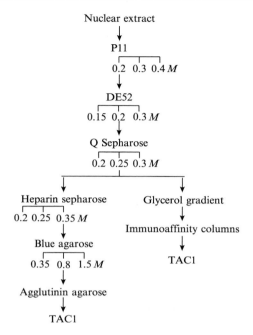

FIG. 1. The TAC1 purification scheme.

to the sequential column. Further, selected Trx-containing fractions are dialyzed to low salt concentrations prior to each column loading. This process involves two changes (4 h and overnight at $4°$) in the ice cold $100\times$ volume of the starting buffer using the Spectrapor cellulose dialysis tubing cutoff of 6–8 kD. After dialysis, the sample is clarified by centrifugation at $4500g$ for 10 min and is then applied onto the next chromatography column, which has been preequilibrated with starting buffer consisting of the same salt concentration as the dialysis buffer. The following chromatographic procedures are used:

1. Cellulose phosphate P11 (Whatman) column is prepared according to manufacturer's protocol. Combined nuclear protein extracts from 200 g of embryos are dialyzed twice against chromatographic buffer C to reduce the salt concentration to 0.1 M KCl. In the P11 step, we used buffer C with KCl, since we found in pilot experiments that most of the Trx protein does not bind efficiently to P11 equilibrated with 0.1 M NaCl-containing buffer. Dialyzed nuclear extract is applied onto a 250 ml P11 column at 0.5 ml/min. The column is washed with 10 column volumes of buffer C.

Bound proteins are eluted stepwise with buffer C (0.1 M increments, from 0.2 to 0.4 M KCl) at 2 ml/min using 10 column volumes per step. Ten milliliters fractions are collected. Trx protein is eluted at 0.2 M KCl from a P11 column. The same fraction also contains significant amounts of dCBP, and almost all nuclear Sbf1 protein.

2. Combined 80–100 ml of 0.2 M KCl P11 fractions are dialyzed against chromatographic buffer C and are applied onto a 25 ml DE-52 Cellulose (Whatman) column preequilibrated in buffer C. The column is then washed with 10 bed volumes of buffer D containing 0.1 M NaCl. Buffer C is exchanged to buffer D (KCl to NaCl), so that aliquots of the material can be directly used for Western blotting, to avoid precipitation of KCl in the presence of SDS sample buffer. Proteins bound to the DE-52 column are eluted stepwise with buffer D (0.05 M increments, from 0.15 to 0.3 M NaCl) at 1 ml/min and 2 ml fractions are collected. Trx and Sbf1 and significant part of dCBP are eluted at 0.2 M NaCl.

3. After dialysis against buffer D, the 0.2 M NaCl DE-52 fraction is applied onto a 5 ml Q Sepharose (Pharmacia) column, preequilibrated in buffer D. The column is washed with 10 bed volumes of buffer D, and bound proteins are eluted stepwise (0.05 M increments, from 0.15 to 0.4 M NaCl) with the same buffer at 0.4 ml/min. One milliliter fractions are collected. Trx, Sbf1, and dCBP are eluted in the 0.25 M NaCl fraction. Most other proteins are eluted at 0.4 M NaCl, leading to significant purification of TAC1 on this column.

4. Heparin Sepharose (Pharmacia) resin is prepared by suspending 1 g of dried powder in distilled water, washing for 15 min in 200 ml of distilled water on a glass filter. The resin is packed onto a 5 ml column and washed with 50 ml of buffer D with 0.2 M NaCl. The 0.25 M Q Sepharose fraction is dialyzed against buffer D containing 0.2 M NaCl and applied on the heparin Sepharose column. Following washing, bound proteins are eluted stepwise (0.05 M increments, from 0.25 to 0.4 M NaCl) with buffer D at 0.3 ml/min. TAC1 components are found in the 0.35 M NaCl fraction.

5. A 0.5 ml column of Cibarcon Blue 3GA Agarose (Sigma) is equilibrated with buffer D containing 0.35 M NaCl. The 0.35 M NaCl fraction from heparin Sepharose is directly loaded onto this column. Proteins are eluted with the same buffer containing from 0.5 to 1.5 M NaCl in 0.15 M NaCl increments at a rate of 0.05 ml/min, 0.25 ml fractions are collected. TAC1 is eluted at 0.8 M NaCl.

6. The 0.8 M NaCl Blue Agarose fraction is incubated for 2 h with agitation at 4° with 0.2 ml of the wheat germ agglutinin agarose (Vector) suspension, which is preequilibrated with buffer D containing 0.8 M NaCl. The suspension is then poured through the Bio-Spin chromatography column (Bio-Rad). Material which is passed through this column contains

the purified TAC1 complex. This step is performed to eliminate any residual, contaminating glycoproteins.

Solutions

Buffer C: 20 mM HEPES, pH 7.6, 0.1 M KCl, 0.2 mM EDTA, 20% glycerol, 0.5 mM DTT, 1 mM PMSF.
Buffer D: 20 mM HEPES, pH 7.6, 0.1 M NaCl, 0.2 mM EDTA, 20% glycerol, 0.5 mM DTT, 1 mM PMSF.

TAC1 Purification by Immunoaffinity Chromatography

An alternative procedure to purify the TAC1 complex includes size fractionation of the 0.25 M Q Sepharose fraction using glycerol gradient, followed by immunoaffinity purification. This procedure yields highly purified TAC1, although some residual, contaminating proteins are also detected in this material.

Size Fractionation of TAC1

In the first step of this procedure, 10 ml of 25–45% glycerol gradient is prepared in SW40 tubes, using the Hoefer SG Gradient maker by mixing 5 ml of buffer D containing 0.25 M NaCl and 25% glycerol with 5 ml of same buffer containing 45% glycerol. One milliliter of the 0.25 M NaCl Q Sepharose fraction containing 20% glycerol is carefully layered on top of 10 ml of a 25–45% linear glycerol gradient and centrifuged for 72 h at 35,000 rpm at 4° in an SW40 rotor (Beckman Instrument). Following centrifugation, 0.5 ml fractions are collected manually from the bottom of each tube, and aliquots from these fractions are tested for the presence of TAC1 components by Western blotting. The molecular weight of TAC1 is determined by comparing the size of fractions containing TAC1 with that of fractions from parallel 20–45% glycerol gradients containing the following markers:

1. 1 ml of HMW native protein marker (Amersham) in buffer D containing 0.25 M NaCl and with 20% glycerol is layered over 10 ml of 20–45% glycerol gradient (see above). The same number of 0.5 ml fractions are collected. Aliquots of these fractions are separated by SDS-PAGE and stained with Coomassie to visualize proteins and to identify which fractions contain standard protein of certain molecular weight.

2. 1 ml of nuclear extract in buffer D is layered over 10 ml of 20–45% glycerol gradient. Aliquots from 0.5 ml fractions are tested by Western blotting with antibodies against Enhancer of Zeste (E[Z]), a component of

the 0.6 MD complex) and SNR1 (a component of the 2 MD BRM complex).

TAC1 is found to be larger than both the 0.6 MD E(Z) complex and the 0.67 kDa protein marker, but it is smaller than the 2 MD SWI/SNF (BRM) complex. We, therefore, estimate the size of the TAC1 complex at 1 MD.

Immunoaffinity Purification of TAC1

As a second step of the alternative purification procedure, we employed immunoaffinity column chromatography using N1 Trx and C2 dCBP antibodies. For these experiments, we used fractions from the glycerol gradient described earlier. This procedure yields a very significant purification of TAC1, although some remaining contaminants are also detected. Therefore, with some restrictions, this procedure can be used as an alternative scheme for TAC1 purification.

Protein A-antibody affinity column is prepared using a standard protocol[18] with slight modifications:

1. 50–100 μg of affinity purified Trx or dCBP antibodies are bound to protein A Sepharose 4 Fast Flow (Amersham). Antibody is mixed in 0.5 ml of buffer F containing 0.1 ml of the protein A Sepharose resin, preequilibrated in buffer F, and incubated for 2 h at 4° with rotation in a 1.5 ml tube. The beads are washed twice with 1.5 ml of 0.1 M sodium borate, pH 9.0 (Sigma), collected by low-speed cenrifugation and resuspended in 1.5 ml of 0.1 M sodium borate, pH 9.0. Dimethylpimelimidate (Sigma) in 0.1 M sodium borate, pH 9.0, is added to a final concentration of 20 mM to cross-link antibody to the protein A Sepharose resin. Beads are incubated for 45 min at room temperature. Fresh dimethylpimelimidate is added to the beads, and they are incubated for an additional 45 min. The beads are washed twice with 1.5 ml of 0.2 M ethanolamine, pH 8.0, and is incubated for 2 h at room temperature with 0.2 M ethanolamine, pH 8.0 to stop the reaction. After incubation, the mix is poured into a Bio-Spin chromatography column (Bio-Rad). The column is equilibrated with buffer F. To decrease noncovalent binding of antibody to protein A, the column is also washed with 5 bed volumes of the elution buffer, followed by washing with 10 bed volumes of buffer F. This step is repeated two more times before the column is ready for use.

2. Aliquots from the peak TAC1-containing glycerol gradient fractions are applied to either N1 or C2 affinity columns and incubated overnight at 4° in binding buffer. Unbound proteins are removed by extensive washing

[18] E. Harlow and D. Lane, "Antibodies: A Laboratory Manual." Cold Spring Harbor Laboratory Press, 1988.

with binding buffer with increasing concentrations of NaCl: 0.25, 0.5, and 0.7 M NaCl. Bound proteins are eluted with the elution buffer, concentrated, resolved by 6% or 12% SDS-PAGE, and visualized by silver staining or Western blotting.

Solutions

Buffer D: 20 mM HEPES, pH 7.6, 0.25 M NaCl, 0.2 mM EDTA, 0.5 mM DTT, 1 mM PMSF.
Buffer F: 25 mM HEPES, pH 7.3, 0.15 M NaCl.
Elution buffer: 0.1 M glycine, pH 2.8.
Binding buffer: 50 mM HEPES, pH 7.3, 5% glycerol, 1.5 mM MgCl$_2$, 1 mM EDTA, pH 8.0, 0.1% NP-40, 0.25 M NaCl.

Acetylation of Histones by TAC1

Trx is an extremely low abundance protein and thus the yield of the highly purified TAC1 is low. The level of histone acetylation activity of TAC1 is also relatively low. Therefore, in most acetylation experiments, we used either the 0.25 M NaCl Q Sepharose fraction, or peak fractions of the glycerol gradient containing peak fractions of TAC1. To demonstrate that histone acetylation activity detected in these fractions is specific to TAC1, we performed either depletion experiments with antibodies against Trx and dCBP (see later and described in Ref. 9), or we tested the acetylation activity of the affinity purified TAC1 (see Fig. 2). Histone acetylation tests were performed using the four core histones, and oligo- and mono-nucleosomes (for preparation, see below). TAC1 is shown to acetylate all four free core histones and histones in mono and oligonucleosomes (see Fig. 2 and Ref. 9).

The histone acetylation assay is performed as described previously with some modifications.[19] As a substrate we used either 10 μg per reaction of core histones (Roche), or 2 μg of the N-terminal peptide of histone H4 (Upstate Biotechnology), or 2 μg of either mono-, or oligonucleosomes, isolated as described later. Fifty nanograms of 0.25 M NaCl Q Sepharose fraction (measured by silver staining), or the material from peak fractions of the glycerol gradient, or the affinity resin with attached TAC1 fractions (see immunoaffinity purification, discussed previously) are incubated in 30 μl of HAT buffer for 30 min at 30° with 0.1 μCi of ^{14}C-acetyl CoA (52 mCi/mmol, Amersham). As a negative control, the reaction is

[19] V. V. Ogryzko, R. L. Schiltz, V. Russanova, B. H. Howard, and Y. Nakatani, *Cell* **87,** 953 (1996).

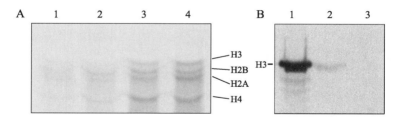

FIG. 2. Analyses of the acetylation and methylation activities of TAC1. (A) Acetylation of histones by TAC1 attached to the N1 anti-Trx (lane 3) and to the C2 anti-dCBP (lane 4) affinity resins. No acetylation of histones is seen when TAC1 was eluted from the N1 (lane 1) and C2 affinity resins (lane 2). (B) Lane 1, methylation of histone H3 by 0.1 μg of the SET domain of Su(Var)3–9. Lane 2, methylation of histone H3 by the 0.25 M Q Sepharose material. Signal of similar intensity was detected when 0.1 μg of the bacterially expressed GST-fused Trx SET was used (not shown). Lane 3, control, no protein added.

performed without adding proteins. The reaction is stopped by boiling in the SDS-sample buffer and the proteins are resolved by 18% SDS-PAGE. After electrophoresis the gel is stained with Coomassie, destained in 10% acetic acid for 2 h at room temperature with gentle agitation with several changes of buffer. The gel is then dried on a piece of Whatman paper using the Fisher Gel Dryer Vacuum system, and exposed for 7–14 days. Alternatively, 15 μl of the reaction mixture is spotted onto Whatman P81 cellulose paper disc. The disc is washed in the 0.05 M sodium bicarbonate buffer, pH 9.2, for 30 min with several changes of buffer, air dried, and radioactivity is measured in a liquid scintillation counter.

Depletion Experiments

To test whether HAT activity of the 0.25 M NaCl Q Sepharose fraction is due to TAC1, we performed depletion experiments using N1 Trx and C2 dCBP antibodies and compared the acetylation of core histones in control and experimental reactions. As a control we used protein A Sepharose alone and protein A Sepharose with bound mock (β-galactosidase) antibody. In these experiments, 100 ng of the 0.25 M NaCl Q Sepharose fraction is incubated with 0.5–5 μg of the N1, or C2 antibodies, or anti-β-galactosidase antibody, or without antibody for 4 h at 4°. During this time, 20 μl of protein A Sepharose Fast Flow (Amersham) are preblocked with 10% BSA in buffer D (see previously) containing 0.25 NaCl for 1 h at 4°. After incubation, samples are added to the preblocked protein A Sepharose and are incubated with rotation for 1 h at 4°. Protein A Sepharose is discarded by low-speed centrifugation, and supernatants are used for the HAT assay.

Isolation of H1 Stripped Oligonucleosomes and Nucleosome Core Particles from Chicken Erythrocytes

All of the following procedures are performed on ice or at $4°$:

1. 1 ml of the erythrocyte pellet is resuspended in 20 ml of lysis buffer I, containing 1 tablet of the "Complete protease inhibitor" cocktail (Boehringer Mannheim), and mixed well by vortexing for 2 min. Nuclei are collected by centrifugation at $1500g$ for 10 min at $4°$.

2. Nuclei are resuspended in 20 ml of lysis buffer II and collected by centrifugation at $1500g$ for 10 min at $4°$. This step is repeated two more times.

3. Nuclei are resuspended in the micrococcal nuclease buffer to an $OD_{600} = 10$ (approximately 3–5 ml). 1–10 U/ml of Micrococcal nuclease (Sigma) are added (see Note), and nuclei are incubated with shaking for 10 min at room temperature. EGTA is added to a final concentration of 2 mM.

4. Nuclei are collected by centrifugation at $2500g$ for 10 min, resuspended in 1 ml of 0.25 mM EDTA and incubated on ice for 1 h with periodic shaking. Nuclear debris is discarded after 10 min centrifugation in an Eppendorf centrifuge.

5. NaCl is added to the supernatant to a final concentration of 80 mM and mixed well. Sixty milligrams of dry Sephadex CM25 (Sigma) is added, and suspension is incubated with shaking for 2 h at $4°$.

6. Solution is cleared by a 10 min centrifugation in the Eppendorf centrifuge. Glycerol is added to the supernatant to a final concentration of 80%. The final material is stored at $-20°$.

Note: Digestion of nuclei with Micrococcal nuclease results in oligonucleosome fragments of different lengths. To obtain single nucleosomes or oligonucleosomes of the desired length, it is necessary to try different concentrations of nuclease in pilot experiments. Note that units of Micrococcal nuclease obtained from different suppliers may vary.

Solutions

Stock solution of histones: dissolve core histones (Roche) in water to a final concentration of 10 mg/ml. HAT buffer: 50 mM Tris–HCl, pH 8.0, 40 mM NaCl, 1 mM EDTA, 10 mM sodium butyrate (Upstate), 1 mM PMSF (added fresh just before use).

Buffer F: 25 mM HEPES, pH 7.3, 0.15 M NaCl.

Lysis buffer I: 50 mM Tris–HCl, pH 8.0, 25 mM KCl, 25 mM NaCl, 4 mM MgCl$_2$, 10 mM Na-butyrate, 5 mM β-mercaptoethanol,

0.2 mM PMSF, 0.5 M sucrose, 0.5% NP-40 (v/v) (Sigma), 10 ng/ml of Trichostatin A.

Lysis buffer II: 50 mM Tris–HCl, pH 8.0, 50 mM NaCl, 4 mM MgCl$_2$, 10 mM Na-butyrate, 5 mM β-mercaptoethanol, 0.2 mM PMSF, 0.34 M sucrose, 0.5% NP-40 (v/v).

Micrococcal nuclease buffer: 50 mM Tris–HCl, pH 8.0, 50 mM NaCl, 4 mM MgCl$_2$, 1 mM CaCl$_2$, 5 mM β-mercaptoethanol, 0.2 mM PMSF, 10% glycerol, 0.05% NP-40 (v/v).

Methylation of Histones by TAC1

Methyltransferase activity of the Trx SET domain and the subpurified TAC1 is very low compared to that of the bacterially expressed SET domain of Su(var)3–9 (see Fig. 2). Trx SET domain can methylate histone H3 (see Fig. 2). Specifically, Trx SET methylates lysine 4 of histone H3, as demonstrated by Edman degradation analysis of methylated histone H3 (S. Smith, submitted). The histone methyltransferase assays with TAC1 fractions and Trx SET domain are performed essentially as described previously with slight modifications.[20]

Ten micrograms of core histones are incubated in methylase activity buffer with 300 nCi of S-adenosyl-L-[methyl-^{14}C]methionine (25 μCi/ml) (Amersham) and with the 0.25 M Q Sepharose fraction for 1 h at 30°. Reaction is stopped by adding SDS sample buffer, and proteins are resolved by 15% SDS-PAGE. The gel is stained with Coomassie, washed in 10% acetic acid for 2 h at room temperature with gentle agitation, and with several changes of buffer. The destained gel is dried and exposed to film for 14–18 days.

Solutions

Methylase activity buffer: 50 mM Tris–HCl, pH 8.8, 10 mM MgCl$_2$, 20 mM KCl, 10 mM β-mercaptoethanol, 250 mM sucrose, 1 mM PMSF.

Stock solution of histones: dissolve core histone (Roche) in water to final concentration 10 mg/ml, store at $-20°$.

Acknowledgments

We thank Elizabeth Cho for critically reading the manuscript. This work was supported by a grant (CA50507) from the National Cancer Institute to Alexander Mazo.

[20] S. Rea, F. Eisenhaber, D. O'Carroll, B. D. Strahl, Z. W. Sun, M. Schmid, S. Opravil, K. Mechtler, C. P. Ponting, C. D. Allis, and T. Jenuwein, *Nature* **406,** 593 (2000).

[16] Isolation and Characterization of CHRASCH, a Polycomb-Containing Silencing Complex

By DER-HWA HUANG and YUH-LONG CHANG

Polycomb group (Pc-G) consists of more than a dozen genes that have been genetically identified for their abilities to confer transcriptional silencing on Drosophila homeotic genes.[1-3] Studies of transgenic constructs have shown that this silencing effect is not only exerted on the adjacent promoter, but also spreads onto distant promoters.[4] More surprisingly, this effect becomes stronger when homologous chromosomes are paired[5] and can sustain through cell division.[6] The molecular mechanism(s) by which Pc-G achieves these silencing effects remains poorly understood. To gain insights into the underlying molecular mechanisms, efforts have been made recently to purify these proteins and characterize their biochemical functions.

Previous studies suggested that Pc-G might act as multiprotein complexes.[7] This notion has been substantiated by the identification and isolation of two physically distinct protein complexes.[8-11] Although the exact composition of these complexes are yet to be resolved, it is clear that one complex contains Pc, Polyhomeotic (Ph), and Posterior sex combs (Psc),[8,11] while the other contains Enhancer of zeste (E[Z]) and Extra sex combs (ESC).[9,10] The observations that several Pc-Gs are not included in these complexes suggest that additional complexes may exist. Despite the existence of multiple protein complexes, Pc-G as a group apparently acts on defined regulatory sequences, since its silencing effect can be recapitulated by

[1] J. A. Kennison, *Annu. Rev. Genet.* **29,** 289 (1995).
[2] V. Pirrotta, *Cell* **93,** 333 (1998).
[3] J. A. Simon and J. W. Tamkun, *Curr. Opin. Genet. Dev.* **12,** 210 (2002).
[4] C.-S. Chan, L. Rastelli, and V. Pirrotta, *EMBO J.* **13,** 2553 (1994).
[5] J. A. Kassis, *Genetics* **136,** 1025 (1994).
[6] G. Cavalli and R. Paro, *Cell* **93,** 505 (1998).
[7] A. Fanke, M. DeCamillis, D. Zink, N. Cheng, H. W. Brock, and R. Paro, *EMBO J.* **11,** 2941 (1992).
[8] Z. Shao, F. Raible, R. Mollaaghababa, J. R. Guyon, C.-T. Wu, W. Bender, and R. E. Kingston, *Cell* **98,** 37 (1999).
[9] J. Ng, C. M. Hart, K. Morgan, and J. A. Simon, *Mol. Cell. Biol.* **20,** 3069 (2000).
[10] F. Tie, T. Furuyama, J. Prasad-Sinha, E. P. Jane, and P. J. Harte. *Development* **128,** 275 (2001).
[11] Y.-L. Chang, Y.-H. Peng, I.-J. Pan, D.-S. Sun, B. King, and D.-H. Huang, *Proc. Natl. Acad. Sci. USA* **98,** 9730 (2001).

Copyright 2004, Elsevier Inc.
All rights reserved.
0076-6879/04 $35.00

specific DNA fragments from homeotic genes in transgenic animals[4] and cultured cells,[12] and since many Pc-Gs can be visualized on these DNA fragments inserted at ectopic sites on polytene chromosomes.[4] These sequences are known as Pc-G response element[13] (PRE). Two motifs have been shown to be critical for the function of PRE: a (GA)n motif[14–16] and the binding motif of Pleiohomeotic,[17] a known Pc-G. The exact roles of these motifs begin to be elucidated.

In addition to homeotic genes, Pc-G acts on many other target genes. The role of Pc-G on these genes is not well understood. It has been reported that the association of Pc-G with some genes does not correlate with their silenced state.[18] There also appears to be significant variation in the relative abundance of Pc-G on different target genes.[18] Furthermore, studies of mutant embryos have revealed large variations in tissues and developmental stages susceptible to different members of Pc-G.[19] Thus, Pc-G may form somewhat heterogeneous complexes with distinct functions in different cellular, developmental, and gene contexts.

In this chapter, we describe an immuno-affinity method to purify a functional complex from a stably transformed *Drosophila* Schneider S2 cell line expressing an epitope-tagged Pc. The use of a cell line has an obvious advantage over embryos, since it eliminates the problems caused by the heterogeneity of cell types and developmental stages. S2 cells were chosen, because many Pc-Gs are abundantly expressed and most endogenous homeotic genes are silenced in these cells.[11,20] To avoid complications that could lead to the isolation of a subgroup of Pc-G complexes during fractionation, we preferred a single-step approach to maximize the isolation of proteins associated with Pc under moderate ionic strength. This preparation allowed us to identify factors previously unknown for homeotic gene silencing, including factors involved in chromatin modification[11] (i.e., histone deacetylase 1) and binding to the (GA)n motif[16] (i.e., Pipsqueak [Psq]). In principle, any novel factor identified by biochemical methods

[12] Y.-L. Chang, B. O. King, M. O'Connor, A. Mazo, and D.-H. Huang, *Mol. Cell. Biol.* **15,** 6601 (1995).

[13] J. Simon, A. Chiang, W. Bender, M. J. Shimell, and M. O'Connor. *Dev. Biol.* **158,** 131 (1993).

[14] B. Horard, C. Tatout, S. Poux, and V. Pirrotta, *Mol. Cell. Biol.* **20,** 3187 (2000).

[15] J. W. Hodgson, B. Argiropoulos, and H. W. Brock, *Mol. Cell. Biol.* **21,** 4528 (2001).

[16] D.-H. Huang, Y.-L. Chang, C.-C. Yang, I.-J. Pan, and B. King, *Mol. Cell. Biol.* **22,** 6261 (2002).

[17] J. L. Brown, D. Mucci, M. Whiteley, M.-L. Dirksen, and J. A. Kassis, *Mol. Cell* **1,** 1057 (1998).

[18] H. Strutt and R. Paro, *Mol. Cell. Biol.* **17,** 6773 (1997).

[19] M. C. Soto, T.-B. Chou, and W. Bender, *Genetics* **140,** 231 (1995).

[20] V. Orlando and R. Paro, *Cell* **75,** 1187 (1993).

should only be considered as a "candidate" until its functional relevance can be proved by *in vivo* studies. The available *Drosophila* genetics and cytology provided us powerful tools to show that both HDAC1 and Psq are indeed involved in one important function of Pc-G, that is, silencing of homeotic genes.[11,16] The functional entity of these activities is henceforth referred to as CHRASCH—CHRomatin-Associated Silencing Complex for Homeotics.

Isolation of CHRASCH

Epitope-Tagged Pc Construct

We used a pMt/Hy vector[21] for construction of an epitope-tagged Pc plasmid pMT/Pc-FH. This vector contains a metallothionine promoter the expression of which can be controlled by different concentrations of copper ion, ranging from 0.7 mM to less than 0.05 mM with a linear response[22] (see also D.-H. Huang, unpublished observation). It also contains a bacterial hygromycin-B phosphotransferase gene driven by Copia LTR for the selection of stably transformed cells. pMT/Pc-FH contains the entire coding sequences of Pc with the following modifications: The termination codon, TGA, was changed to TTG and the immediate downstream sequences replaced by DNA sequences encoding FLAG epitope (in bold) and hexa-histidine (underlined): ATATCAGATCTG**ATGGACTACAA-GGACGATGACGAT**AA GAACGCGTCC CACCATCACCACCA-TCACTAGGATCT. Consequently, the following peptide sequences were added to the C-terminus of wild-type Pc: LISDLMDYKDDDD-KNASHHHHHH. The plasmid was prepared by the standard CsCl double banding method.[23]

Stable Cell Lines

To establish Pc-expressing cell lines, we transfected S2 cells with pMT/Pc-FH plasmid by the CaCl$_2$-phosphate method, originally described by Wigler *et al.*[24] and subsequently adopted by Di Nocera and Dawid[25] for

[21] M. R. Koelle, W. S. Talbot, W. A. Segraves, M. T. Bender, P. Cherbas, and D. S. Hogness. *Cell* **67**, 59 (1991).

[22] T. A. Bunch, Y. Grinblat, and L. S. B. Goldstein, *Nucleic Acids Res.* **16**, 1043 (1988).

[23] J. Sambrook and D. W. Russell, "Molecular Cloning: A Laboratory Manual," 3rd Ed. Cold Spring Harbor Laboratory Press, Cold Spring Harbor, New York, 2001.

[24] M. Wigler, A. Pellicer, S. Silverstain, R. Axel, G. Urlaub, and L. Chasin, *Proc. Natl. Acad. Sci. USA* **76**, 1373 (1979).

[25] P. P. Di Nocera and I. B. Dawid, *Proc. Natl. Acad. Sci. USA* **80**, 7095 (1983).

Drosophila cells. Following the selection of stable cell lines, we took an additional step to enrich the population of Pc-expressing cells (see step 6). The standard medium we used for growing S2 cells was Schneider medium supplemented with 10% heat-inactivated fetal bovine serum and 50 unit/ml penicillin, and streptomycin with or without hygromycin B. Several other media have also been reported for growing S2 cells.[26]

Reagents

2 M $CaCl_2$.

2× HeBS: 50 mM HEPES, 280 mM NaCl, 10 mM KCl, 2.5 mM Na_2HPO_4, 12 mM dextrose, pH was adjusted to 7.1. Keep small aliquots at $-20°$.

Phosphate-buffered saline (PBS): one liter of PBS contains 8.0 g NaCl, 0.2 g KCl, 1.44 g Na_2HPO_4, and 0.24 g KH_2PO_4, pH was adjusted to 7.4.

0.1 × TE: 1 mM Tris–HCl (pH 8.0), 0.1 mM EDTA.

Schneider medium (Sigma).

All solutions were made in milli-Q water and sterilized by passing through 0.22 μm filters.

Fetal bovine serum (Gibco) was heat-inactivated for 30 min at 56°.

Hygromycin B (50 mg/ml, Roche Applied Science).

1. S2 cells were grown in the T flask at 25° without CO_2. Approximately 20 h before transfection cells were seeded at 8 × 10^5 cells/ml into a T-25 flask.

2. Two hundred and fifty microliters of DNA-$CaCl_2$ solution was prepared in a sterile microfuge tube by mixing 2 M $CaCl_2$ with a solution containing 0.1 × TE and 12.5 μg plasmid. The final $CaCl_2$ concentration was 250 mM. The DNA solution was then added drop-wise to a sterile microfuge tube containing 250 μl 2× HeBS with constant tapping. The mixture was left at room temperature for 30 min and then added drop-wise to cells.

3. Twenty-four hours after transfection, cells were diluted 1:4 with fresh medium and spilt into four new T-25 flasks. Hygromycin B was added to cells to a final concentration of 200 μg/ml after another 24 h. Large amount of cells died within the first week of drug selection.

4. After 7–10 days, cell debris was removed from each flask by spinning down and resuspending intact cells with fresh medium plus hygromycin B. The volume of medium was adjusted according to the amount of recovered cells. In cases of low recovery, cells were combined from different flasks. Alternatively, transfection was repeated.

---

[26] L. Cherbas, R. Moss, and P. Cherbas, *Methods Cell Biol.* **44,** 161 (1994).

5. Significant cell division was observed in the following week. For flasks containing actively growing cells, cells were split 1:4 again with selection medium at the end of the week. Each flask was kept independently until protein expression was examined. Cell lines were considered to be stably established when they grew in a normal 22–24 h cell cycle under drug selection. This took at least a month.

6. Expression of tagged Pc protein in these cells was examined by cell staining, as described in next section. Since these cells were not derived from a clone, variable levels of Pc expression were not uncommon. If the populations of Pc-expressing cells in these flasks were not satisfactory, cells from the best flask were then diluted and split to establish several new sublines. This step was repeated several times until satisfactory results were obtained.

7. We noted that the population of Pc-expressing cells was significantly reduced after several months of passage or when cells were amplified from a frozen vial, despite their drug resistance. In these cases, the enrichment step, described in step 6, was used to make improvements.

Cell Staining

It was crucial to monitor the expression of tagged Pc proteins during the establishment of cell lines. We also routinely checked protein expression by cell staining during long-term growth of cell lines. All these steps are conducted at room temperature unless specified.

Reagents

10% NP-40.
0.1% poly-lysine (Sigma).
0.5 M $CuSO_4$.
37% Formaldehyde.
PBSB: PBS containing 2% BSA.
AP buffer: 0.1 M Tris–HCl (pH 9.5), 0.1 M NaCl, 50 mM $MgCl_2$.
Monoclonal FLAG M2 antibody (originally from Kodak, now available from Sigma).
Staining reagents: 5% NBT, 5% BCIP in 70% dimethylformamide.

1. Approximately 2 ml of cells from each flask were seeded for at least 6 h on a poly-lysine-coated cover glass in six-well plates at $\sim 1 \times 10^6$ cells/ml. The cover glasses were prepared by soaking in 0.01% poly-lysine solution for 15 min, followed by an extensive wash with deionized H_2O and overnight baking at 180° inside a glass Petri dish.

2. $CuSO_4$ was added to each well at a final concentration of 0.7 mM with gentle swirling. The induction continued for about 16 h. After

removal of medium, cells were washed three times with PBS. To avoid detaching cells from the cover glass, the solution was added along the wall of the well.

3. Cells were fixed for 20 min with 3.7% formaldehyde in PBS, followed immediately by a wash with PBS.

4. Cells were then permeabilized with 0.5% NP-40 for 5 min, followed by two washes with PBS.

5. Cells were incubated for 30 min in PBSB.

6. FLAG antibody (1:500) was freshly diluted in PBSB and 100 μl added to each well. After 1-h incubation, cells were washed three times with PBS for 5 min. Anti-mouse antibody conjugated with alkaline phosphatase was diluted (1:200) in PBSB and added to the wells.

7. After 1-h incubation and PBS washes, cells were washed twice with AP buffer for 5 min. During the second wash, staining solution was prepared by adding 4.4 μl 5% NBT and 3.3 μl 5% BCIP to 1 ml AP buffer. 100–200 μl staining solution was then added to each well. Excessive solution may result in high background.

8. The reaction was occasionally monitored under a dissection microscope. The staining solution was removed when desired intensity was reached. The slides were then washed extensively with PBS and mounted with cover glasses in PBS containing 25% glycerol.

Protein Purification

Reagents

Buffer A: 25 mM Tris–HCl (pH 7.6), 10 mM KCl, 2 mM MgCl$_2$, 0.1 mM EDTA, 0.5 mM EGTA.

Buffer B: 250 mM Tris–HCl (pH 7.6), 1 M KCl, 50 mM MgCl$_2$, 1 mM EDTA, 5 mM EGTA.

Buffer C: Buffer A and B at 9:1 ratio.

Buffer D: 25 mM Tris–HCl (pH 7.6), 150 mM NaCl, 0.1 mM EDTA.

Buffer E: Buffer D plus 0.1 mM DTT and four freshly diluted protease inhibitors (see later).

(NH$_4$)$_2$SO$_4$: solid (NH$_4$)$_2$SO$_4$ (SigmaUltra or equivalent grade) was dissolved in 50 mM Tris–HCl (pH 7.5) with mild heating on a hot plate until a saturated solution was obtained. The solution was then passed through a 0.45 μm filter.

Protease inhibitors: 200 mM PMSF in 100% ethanol; 2 mg/ml Pepstatin A in 100% methanol; 10 mg/ml Aprotinin and 10 mg/ml Leupeptin in sterile water. These were kept at $-20°$ until use. Aprotinin and Leupeptin were kept in small aliquots and left at 4° once thawed. They were freshly diluted into the buffers as indicated to final concentrations of 1 mM for PMSF and 5 μg/ml for the others. Vigorous shaking was needed when PMSF was added into the buffer.

1. S2 cells grown in T flasks were readily adapted to spinner flasks for large-scale preparations. We kept cell density at 2×10^6 to 4×10^6 ml^{-1}. Fresh medium was added everyday to dilute cells to $\sim 2 \times 10^6$ ml^{-1}. The stirring speed was maintained at approximately 30 rpm. There was an upper limit for the volume of cells in the flask. For example, in a standard 3-L flask, optimal growth was achieved at 2-L capacity, while larger volumes (e.g., 2.5-L) slowed down cell doubling significantly.

2. When a desired quantity of cells was obtained, $CuSO_4$ was added into freshly diluted culture to a final concentration of 0.1 mM. The culture was kept for 67 h without further replenishment. The cell density usually doubled during this period.

3. Cells were harvested by centrifugation at 3000 rpm in a GS-3 rotor (Sorvall) for 10 min. The pellet was gently resuspended with 1/10 volume of cold PBS, followed by centrifugation. From now on, precooled solutions and utensils were used. The pellet was washed once with PBS before being washed again with 1/20 volume of Buffer A. The pellet was then resuspended in 1/100 volume of Buffer A supplemented with 1 mM DTT and the four protease inhibitors.

4. Cells were then broken in a Dounce homogenizer by 15 strokes with a type "A" pestle. 1/1000 volume of Buffer B was immediately added, followed by five more strokes. Nuclei were collected by centrifugation at 10,000 rpm for 10 min in a SS-34 rotor (Sorvall) and resuspended in 1/100 volume of Buffer C plus 1 mM DTT and protease inhibitors.

5. The exact volume of the protein suspension was measured and taken as the volume reference for the following step. To the suspension was now added 1/10 volume of saturated $(NH_4)_2SO_4$ solution, drop-by-drop, with constant stirring for 20 min. After centrifugation at 50,000 rpm for 1 h in a Ti60 rotor (Beckman), the nuclear extract was carefully removed to avoid disturbing the chromatin pellet. Saturated $(NH_4)_2SO_4$ solution was slowly added to the nuclear extract to a final concentration of 40% and stirred for 20 min. The protein pellet was then collected by centrifugation at 15,000 rpm (SS-34 rotor) for 15 min.

6. The pellet was resuspended in 1/2 volume of Buffer E. The Dounce homogenizer was used to help dissolve the pellet. One sixth the volume of Buffer E containing 0.4% Tween-20 was then added to the protein solution. After centrifugation at 15,000 rpm (SS34 rotor) for 5 min to remove insoluble material, the protein extract was loaded onto a FLAG antibody column (M2, Kodak) equilibrated in Buffer E containing 0.1%

Tween-20. We routinely used 0.5 ml resin packed in a $0.5 \times 10 \text{ cm}^2$ column (Econo-column, Bio-Rad) for extracts made from up to 4 L of cells. The flow rate was adjusted to about 12 drops/min. The extract was then passed over the resin at least three times, the resin was extensively washed, first with Buffer E containing 0.1% Tween-20, and then with Buffer E alone. To elute the protein, we first added 1 ml of Buffer E containing freshly diluted FLAG peptide (300 μg/ml). When most of the solution entered the bed, the flow was stopped for 20 min to allow the release of bound proteins. We then added 2 ml of peptide-containing solution, followed by a wash with peptide-free solution. The elution of protein parallels that of the peptide and was monitored by UV absorption. Fractions were collected and to each, DTT and glycerol were immediately added to final concentrations of 1 mM and 10% respectively. These fractions were frozen at $-70°$ before use. The column was regenerated immediately by washing with about 10 ml glycine solution (0.1 M, pH 3.0) and subsequently equilibrated in Buffer D. Note that glycine wash needs to be done in 20 min or less to avoid damaging the resin.

7. A control experiment was performed with regular S2 cells to determine copurified nonspecific proteins. Several common proteins around 55 kD and 45 kD were observed for regular S2 cells and tagged cells (see Fig. 1A). The 55 kD protein was probably the immunoglobulin heavy chain that dissociated from the resin.

Histone Deacetylase (HDAC) Assay

Radioactively labeled histone substrate could be prepared either from *in vivo* labeling of cultured cells[27] or from *in vitro* acetylation reactions with intact histone acetyltransferase (HAT) or the HAT domain. The latter is more cost-effective. However, it should be kept in mind that the acetylated residues are limited by the specificity of the HAT used in such reactions. We used the HAT domain of mouse p300/CBP[28] (codons 1195 to 1673) purified from bacteria, because it appears to have wide specificity toward all four core histones.

Preparation of Acetylated Histone

We carried out a 1-ml reaction to make a stock of actylated-histone substrate.

[27] M. J. Hendzel, G. P. Delcuve, and J. R. Davie, *J. Biol. Chem.* **266,** 21936 (1991).

[28] V. V. Ogryzko, R. L. Schiltz, V. Russanova, B. H. Howard, and Y. Nakatani, *Cell* **87,** 953 (1996).

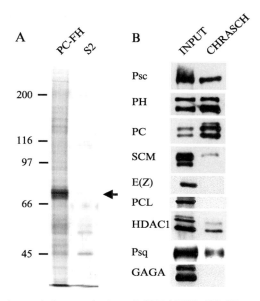

FIG. 1. Purification and characterization of CHRASCH. (A) Silver staining of proteins purified from regular S2 cells (S2) or cells containing tagged Pc construct (PC-FH). Peak fractions eluted from FLAG antibody column were shown. The arrow indicates the position of Pc proteins. Note that the staining intensity is somewhat exaggerated in this region by the co-migration of several proteins in this gel. The sizes of molecular weight standard (in kD) are indicated at left. (B) Protein blots of CHRASCH. Column input (INPUT) and peak fraction (CHRASCH) were blotted and probed with following antibodies: monoclonal antibody against Posterior sex combs (Psc, a gift of P. Adler), rabbit antibodies against Polyhomeotic (PH), Sex comb on midleg (SCM, a gift of J. Simon), Enhancer of zeste (E[Z], a gift of R. Jones), Polycomblike (PCL), HDAC1, Pipsqueak (Psq, a gift of C. Berg), and GAGA factor (a gift of C. Wu). Pc protein was detected by FLAG antibody (M2). HDAC1, PSQ and GAGA were not previously known to be involved in homeotic gene silencing. (1A and 1B were reproduced or adapted with permission from NAS and ASM, respectively.)

Reagents

$5\times$ buffer: 250 mM Tris–HCl (pH 7.8), 5 mM DTT.

5 mg/ml core histone (Sigma type IIA) in 50 mM Tris–HCl (pH 7.8).

0.1 M sodium butyrate in deionized H_2O.

These solutions are kept frozen in small aliquots.

[^{3}H] acetyl Co-A (4.9 Ci/mmol, Amersham).

50% (v/v) glycerol.

purified p300/CBP HAT domain ($\sim$0.1 mg/ml).

The reaction contained $1\times$ buffer, 10% glycerol, 10 mM sodium buty-rate, 10 μCi [^{3}H] acetyl Co-A, 400 μg histone, and 8 μg p300/CBP. After 30 min incubation at 30°, the reaction was terminated by addition of

100% TCA to a final concentration of 2%. The solution was left on ice for about 1 h, followed by centrifugation for 10 min in an Eppendorf microfuge at top speed. To the supernatant, TCA was added to a final concentration of 25%. After standing on ice for 1 h, labeled histone was again precipitated by centrifugation. The protein pellet was resuspended with 500 μl H_2O and precipitated again by 25% TCA. This step was repeated twice. The pellet was air-dried, dissolved in H_2O and then dialyzed extensively against H_2O at room temperature with two changes of H_2O. The recovery of labeled histones was based on their staining intensity on a SDS-polyacrylamide gel. The specific activity was estimated to be 25,000 dpm/μg. We found that the limiting factor for *in vitro* acetylation was the amount of [^{3}H] acetyl Co-A present in the reaction. Higher specificity could be obtained by adding more [^{3}H] acetyl Co-A.

Histone Deacetylation Assay

We used the method originally described by Hendzel *et al.*[27] with some modifications. We routinely performed 50-μl reactions with 1 μg of labeled histones. The reaction buffer contained 25 mM Tris–HCl (pH 7.5), 150 mM NaCl, 10 mM EDTA and 10 mM β-mercaptoethanol. 2–5 μl aliquots of column fractions were assayed in duplicate. The reactions were incubated at 37° for 60 min and terminated by adding acetic acid and HCl to final concentrations of 0.12 and 0.72 N, respectively. Two volumes of ethyl acetate were then added to extract free [^{3}H]-acetate. Following a brief centrifugation to separate the aqueous phase, 100 μl of supernatant was recovered and 50 μl subjected to scintillation counting. Two control reactions were conducted. A negative control was done alone with the column buffer to determine the background level of free acetate. In a separate control, trichostatin A (100 ng/ml), a HDAC specific inhibitor, was included in parallel reactions to determine the level of HDAC activity.

DNA Binding Assays

Electrophoretic mobility shift assays were employed to study the DNA binding activity of CHRASCH. We have previously shown by transfection assays that a minimal PRE resides in a ~440 bp B-151 fragment, at approximately 20 kb upstream of a homeotic gene *Ultrabithorax Ubx*.[6] The right half of this fragment, that is, bxd-b, contains high affinity binding sites of CHRASCH and was chosen for this DNA binding study. Various fragments of B-151 have been cloned into pBluescript. Taking the AflII site at the left end of B-151 as +1, these subclones contain following DNA sequences: +1 to +255 (bxd-a), +259 to +446 (bxd-b), −17 to +113 (bxd-1),

+92 to +216 (bxd-2), +197 to +369 (bxd-3), and +350 to 461 (bxd-4). They can be recovered by cleaving at flanking restriction sites (XbaI and XhoI for bxd-a and bxd-b, SacII and XbaI for bxd-1 to 4) and subsequently used for either end-labeling of the binding probe or for competition assays.[15]

Reagents

5× binding buffer: 100 mM Tris–HCl (pH 7.9), 300 mM KCl, 0.5 mM EDTA, 5 mM DTT, 50% glycerol, 0.5% Tween-20. Store as small aliquots at $-70°$.

Poly (dI-dC): 0.5 mg/ml stock was prepared in 10 mM Tris–HCl (pH 7.9), 1 mM EDTA, 100 mM NaCl. Store as small aliquots at $-70°$.

50 × TAE buffer: 2 M Tris–acetate, 25 mM EDTA.

40% acrylamide solution (37.5 : 1, Bio-Rad).

A standard 10-μl binding reaction contained 1× binding buffer, 5 μg BSA, 0.1 μg poly(dI-dC), 0.5–1 ng ^{32}P-labeled probe (approximately 10,000 cpm) and 1–3 μl of protein fractions eluted from the column. Binding reactions were carried out at 30° for 20 min. Samples were immediately loaded on a 3.5% poly-acrylamide gel and run at ∼250 V for ∼3 h, 4°. Both gel buffer and electrophoresis buffer contained 1 × TAE. Preelectrophoresis was carried out for at least half an hour at ∼250 V at 4°. Control reactions were carried out without protein fractions or with protein fractions from regular S2 cells. In addition, competition experiments were conducted by adding ∼30 ng (in 1 μl) of DNA fragments purified from bxd-1, 2, 3, or 4 to the reaction. After electrophoresis, gels were vacuum-dried and subjected to autoradiography.

In Vivo Studies

As discussed earlier, Pc-G complexes might exist in different forms with distinct compositions and regulatory function toward different targets. Thus it was important to demonstrate that the activities under study were directly related to specific functions of Pc-G *in vivo*. In the case of homeotic gene silencing, it was required to show that the lack of these activities *in vivo* could cause ectopic expression of homeotic genes. In this section, we provide a protocol used to examine mis-expression of homeotic genes in the larval tissues called imaginal discs (see Fig. 2). These morphologically distinct discs are derived from primordial cells clustered in different segmental units of the embryo. They will ultimately give rise to adult body parts. The expression patterns of homeotic genes in these discs have been well characterized.[29] These discs are highly sensitive to gene dosage effects and can be readily prepared. In addition, we provide a protocol to examine protein colocalization on giant polytene chromosomes from the salivary gland of third instar larvae. This method was originally developed by Silver

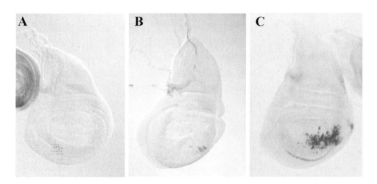

FIG. 2. Ectopic expression of Ubx in mutant discs. Ubx antibody staining of wing discs from wild-type (A), or from mutants heterozygous for Pc (B) or for both $Hdac1$ and Pc (C) are shown. $Hdac1$ mutant alone did not show significant Ubx mis-expression in wing disc. However, when combined with Pc mutation, it strongly enhanced the degree of Ubx mis-expression.

and Elgin[30] and later improved by Zink and Paro.[31] It provides a powerful tool to show that both the HDAC1 and Psq we identified are acting together with Pc-G on homeotic genes *in vivo*.

Imaginal Disc Staining

Reagents

0.7% NaCl.

37% formaldehyde.

PBTN: 1 × PBS containing 0.15% Triton X-100, 0.2% Tween-20 and 0.2% NP-40.

PBTNBS: PBTN containing 3% BSA (Sigma, fraction V) and 5% normal goat serum.

ABC kit (Vector).

5 mg/ml diaminobenzidine (DAB) in 0.1 M Tris–HCl (pH 7.6), stored in 0.5 ml aliquots at $-20°$. Handle DAB with care for its potential carcinogenicity. Diaminobenzidine solution and containers in contact with it can be detoxified by soaking in diluted laundry bleach, followed by extensive wash.

30% H_2O_2.

[29] S. M. Cohen, *in* "The Development of *Drosophila melanogaster*" (M. Bate and A. Martinez-Arias, eds.), Vol. 2, p. 747. Cold Spring Harbor Laboratory Press, Cold Spring Harbor, New York, 1993.

[30] L. M. Silver and S. C. R. Elgin, *Proc. Natl. Acad. Sci. USA* **73,** 423 (1976).

[31] B. Zink and R. Paro, *Nature* **337,** 468 (1989).

1. Several wandering larva from an uncrowded vial were picked and left in 0.7% NaCl for a few minutes to wash off the food. This and subsequent steps were carried out in a 9-well Pyrex plate at room temperature unless otherwise specified. Larvae were transferred to a clean well containing 0.7% NaCl. The mouth was tweezed apart with watch forceps under the dissection microscope. The cuticle was torn apart along the trachea from the position about 2/3 of the body. Although most internal tissues were lost during subsequent steps, imaginal discs remained attached to the cuticle.

2. Discs were washed with PBS and then fixed in PBS containing 3.7% formaldehyde for 20 min with shaking.

3. The discs were washed three times, for 5 min each, in PBS and then washed twice with PBTN for 5 min.

4. The discs were now incubated for 1 h in PBTNBS.

5. The PBTNBS solution was then replaced with freshly diluted monoclonal antibodies against Ubx[32] (1:2, culture supernatant of FP3.38) in PBTNBS and incubated for 3 h.

6. The discs were washed with PBTN for 10 min at least six times and freshly diluted (1:100) biotinylated secondary antibody in PBTNBS was incubated for 3 h.

7. The discs were washed again with PBTN for 10 min at least six times.

8. While performing the last step in this assay, an avidin-biotin complex was prepared by mixing 1% reagent A and 1% reagent B (ABC kit) in PBS plus 0.1% Tween and leaving on a rotating rack for 30 min.

9. The discs were now incubated with avidin-biotin complex for 90 min.

10. The discs were washed with PBTN for 10 min at least six times.

11. 0.1 M Tris–HCl (pH 7.6) was now used to wash the discs twice for 5 min.

12. The chromogenic reaction was carried out with 0.5 mg/ml DAB in 0.1 M Tris–HCl (pH 7.6) plus 0.03% H_2O_2.

13. The reaction was stopped by washing the discs three times with PBTN for 5 min.

14. The discs were transferred to PBS containing 50% glycerol.

15. Finally individual disc was removed from the cuticle and mounted for photography.

[32] R. A. H. White and M. Wilcox, *Cell* **39,** 163 (1984).

Polytene Staining

This protocol is adopted from Zink and Paro[33] with some modifications.

Reagents

16% formaldehyde (EM grade).
Solution 1: 0.1% Triton X-100 in PBS (pH 7.4).
Solution 2: 2.5% formaldehyde, 1% Triton X-100 in PBS (pH 7.4).
Solution 3: 2.5% formaldehyde in 50% acetic acid.
Both Solutions 2 and 3 were freshly prepared.
Solution 4: 3% BSA, 10% nonfat dry milk, 0.2% NP 40, 0.2% Tween-20 in PBS.
Solution 6: 0.4 M NaCl, 0.2% NP-40, and 0.2% Tween-20 in PBS.
Solution 7: 0.5 M NaCl, 0.2% NP-40, and 0.2% Tween-20 in PBS.

1. Wandering larvae were rinsed in Solution 1 as described earlier and a pair of salivary glands was removed under a dissection microscope.

2. A tungsten needle was then used to transfer the glands to a drop of Solution 2 on a siliconized slide. The glands were thoroughly fixed for 30 s by shaking the needle in the solution.

3. The glands were then transferred to a drop of Solution 3 on a siliconized cover glass and fixed for 90 s.

4. Steps 4 and 5 were done in 90 s. The cover glass was placed onto a poly-lysine coated slide by gently touching the cover glass with the slide.

5. The cover glass was tapped with the blunt edge of pencil and rubbed in a circular motion a few times.

6. The cover glass was scratched several times with the blunt-end of forceps to spread out the chromosome. Polytene chromosomes were then examined under a compound microscope with phase contrast. The slide containing well-spread chromosomes was turned upside down over a paper towel and pressed by thumb. The position of the cover glass was marked by pencil on the edges of slide.

7. The slides were frozen in liquid nitrogen.

8. The cover glasses were removed from the slides with a razor blade.

9. The slides were then washed twice with PBS for 15 min and kept in 100% methanol at 4° for up to a week until use.

10. The slides were rehydrated by two PBS washes for 15 min.

11. The samples were blocked for 1 h in Solution 4 and then incubated for 1 h or longer in a humid chamber with about 50 μl of freshly diluted primary antibodies in Solution 4. The example shown in Fig. 3 used

[33] B. Zink and R. Paro, *Dros. Inform. Serv.* **72,** 196 (1993).

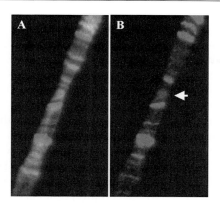

FIG. 3. Colocalization of Psc and Psq at the Bithorax complex on polytene chromosomes. Polytene chromosomes from wild-type third instar larvae were simultaneously stained with antibodies against Psq and Psc. The chromosome arm shown here contains the Bithorax complex (indicated by arrow in B) where three homeotic genes including *Ubx* reside. (A) Triple color image is shown with DNA in green, Psq in red and Psc in blue. The region where three signals merge is shown as turquoise color. (B) Double color image of the same chromosome arm. The merged signal is shown as purple color. (See color insert.)

affinity-purified rabbit antibody against Psq (1:2000 dilution) and mouse monoclonal antibody against Psc[34] (1:5 dilution of culture supernatant). The exact dilution of the antibody and the incubation time were empirically determined, since the efficacy of antibodies was batch-dependent.

12. Slides were now washed twice with PBS for 5 min, followed by three washes with Solution 6 for 5 min.

13. The slides were then washed three times with Solution 7 for 5 min. This step was optional, dependent on the degree of background staining.

14. The slides were finally washed twice with PBS for 5 min and incubated for 1 h with secondary antibodies diluted in Solution 4 containing 2% normal serum—RRX-conjugated antirabbit antibody (1:100) and Cy5-conjugated anti-mouse antibody (1:500).

15. The slides were then washed as before and incubated with PBS containing 0.1 μM Sytox in PBS for 10 min to stain the DNA.

16. Finally the slides were washed six times with PBS for 5 min and mounted in Fluoromount-G (EMS) for fluorescent microscopy.

---

[34] E. C. Martin and P. N. Adler, *Development* **117,** 641 (1993).

Concluding Remarks

The procedures described here allowed us to isolate CHRASCH, a complex that contains a subset of Pc-G and several novel activities involved in homeotic gene silencing.[11,16] For example, HDAC1 can remove the acetyl group from the lysine residue on the N-terminal tails of histones, thus leading to more compact chromatin structure and gene silencing. Pipsqueak is a DNA binding protein that can bind to the (GA)n motif in PRE.[16,35] The physiological relevance of these factors have been demonstrated by genetic and immuno-localization studies. A simple mechanism we can envision for homeotic gene silencing is that CHRASCH becomes first targeted to PRE through the binding of Psq to the (GA)n motifs. Subsequently, HDAC1 causes hypo-acetylation of local chromatin. Additional activities are likely to exist in this complex. It is possible that some might not be directly related to homeotic gene silencing, since Pc may represent a common subunit of multiple complexes. However, our approach should provide a direct means to identify relevant ones.

Acknowledgments

We thank members of our laboratory for participation in development of the procedures. We also thank Ken Den for critical reading of this manuscript. These works were funded to D.-H. H. by grants from NSC and Academia Sinica, Republic of China.

[35] M. Lehmann, T. Siegmund, K. Lintermann, and G. Korge, *J. Biol. Chem.* **273,** 28504 (1998).

[17] Biochemical Analysis of Mammalian Polycomb Group Protein Complexes and the Identification of Genetic Elements that Block Polycomb-Mediated Gene Repression

By RICHARD G. A. B. SEWALT, TED H. J. KWAKS,
KARIEN HAMER, and ARIE P. OTTE

Chromatin plays an important role in the regulation of different processes such as replication, recombination and transcription. Unravelling components of chromatin-associated protein complexes is of importance to understand the mechanism underlying these processes. In recent years many components of distinct chromatin complexes have been identified

Copyright 2004, Elsevier Inc.
All rights reserved.
0076-6879/04 $35.00

by either large scale biochemical purification or for instance by two-hybrid screens. Examples are ATP-dependent chromatin remodelling complexes and complexes that are associated with chromatin-associated gene repression. To the latter classes belong the Polycomb-group (PcG) proteins[1] and heterochromatin protein 1 (HP1)[2] containing complexes. When identified and analyzed by means of two-hybrid analyses the *in vivo* relevance of the observed interactions is a key question. One method to verify two hybrid protein-protein interactions is an *in vivo* coimmunoprecipitation experiment. Although coimmunoprecipitation experiments seem straightforward, they are not in practice. Many different protocols are available but not all of them are useful for coimmunoprecipitation of chromatin associated proteins. This is for instance because the large protein complexes are also held together by DNA. This DNA can be the cement that holds together a protein complex, without bona fide protein-protein interactions being present. This justifies the establishment of protocols that describe stringent methods to identify *in vivo* protein-protein interactions of chromatin-associated protein complexes. Here we describe a protocol for coimmunoprecipitation, based on our experience with many different Polycomb-group proteins (PcG) and associated factors. This protocol is not only useful for PcG proteins, but should be applicable for other chromatin-associated proteins, such as HP1 containing complexes, as well.

The identification of components of chromatin-associated protein complexes increases the understanding of the role of chromatin in processes such as transcription. One ongoing discussion is whether these complexes act locally or whether they can spread over large regions of the genome. Related to this question is whether chromatin-associated protein complexes are within the genome blocked by specific DNA-protein elements, such as insulators or boundary elements. In yeast for instance, distinct chromatin domains have been identified that are either transcriptional active or inactive. Manipulation of the sequences in between results in a shift of the inactive heterochromatin towards the active euchromatin, indicating that these sequences are able to block heterochromatin.[3] With such ideas in mind we have developed a screening method to identify sequences that are able to block repression mediated by the heterochromatin associated proteins Polycomb and HP1. This method has resulted in the identification of a novel class of anti-repressor elements that potently block repression mediated by PcG proteins or HP1. Besides practical applications of these

[1] D. P. E. Satijn and A. P. Otte, *Biochim. Biophys. Acta* **1447,** 1 (1999).
[2] J. C. Eissenberg and S. C. Elgin, *Curr. Opin. Genet. Dev.* **10,** 204 (2000).
[3] K. Noma, C. D. Allis, and S. I. Grewal, *Science* **293,** 1150 (2001).

elements in for instance transgenesis, the methodology can also be applied to identify genome-wide genetic elements that act as counterparts of chromatin-associated repressors.

Immunoprecipitation of Polycomb Group Proteins

One method to detect protein-protein interactions *in vivo* is to overexpress proteins of interest with an added tag (FLAG, His, T7, myc)[4] and use antibodies against the respective tag for coimmunoprecipitations. Advantages of this method are that specific antibodies against the proteins of interest are not needed and that in principle any cell line can be used to overexpress the tagged proteins. This method is useful for a first indication for determining *in vivo* protein-protein interactions. A serious drawback is, however, that high, "artificial" expression levels of two proteins can lead to detection of interactions that do not reflect the *in vivo* situation. If, for instance, the *in vivo* stoichiometry of protein A and B is 10:1, this endogenous interaction may be very difficult to detect. When both tagged proteins are then overexpressed to similar levels, the detected interaction may have nothing to do with the *in vivo* situation. Finally, it is not an automatism that a tagged protein incorporates into a multimeric chromatin-associated proteins complex. Furthermore, the *in vivo* stoichiometry of the complex may even be changed. Hence it is preferable to perform coimmunoprecipitation experiments with endogenous proteins. This requires high quality and specific antibodies, which is the major drawback of this approach. Chromatin-associated proteins are often highly conserved and hence not very immunogenic, which hampers the efficiency of raising high titre, specific antibodies.

When opting for coimmunoprecipitation of endogenous chromatin-associated proteins, the choice of cell lines in which the endogenous proteins are to be detected is very important because of the relative abundance of the respective proteins. If one of the respective proteins is not, or at a low level expressed in a particular cell line, coimmunoprecipitation experiments are useless or very hard to perform. Therefore, a first step is to analyse several cell lines by Western analysis for the relative abundance of the proteins that are being analysed. For instance, the expression levels of human PcG proteins differs considerably in distinct common cell lines. Examples are shown Fig. 1. In U-2 OS and 293 cells for instance all the analysed PcG proteins are expressed at a relative high level, whereas in HeLa cells PcG proteins could hardly be detected. This makes U-2 OS and 293 cells ideal cells for coimmunoprecipitation experiments of PcG proteins

---

[4] K. Terpe, *Appl. Microbiol. Biotechnol.* **60,** 523 (2003).

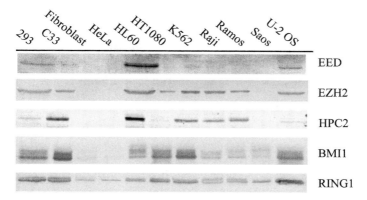

FIG. 1. Western blot analysis of a panel of human cell lines with Polyclonal antibodies directed against human Polycomb-group proteins show that the PcG proteins EED, EZH2, HPC2, BMI1, and RING1 are differentially expressed in various cell types.

and HeLa a poor choice. It has become clear that there are at least two distinct PcG complexes, the PcG I (or PRC1) complex containing HPC, HPH, RING1, and BMI1[5–9] and the PcG II complex containing EED, EZH2, and YY1.[10,11] Associated with PcG I complex we found the corepressor CtBP.[12] For immunoprecipitation of PcG I complex proteins we preferably use U-2 OS cells, since U-2 OS cells is one of the few cell lines that expresses all described PcG proteins.[5–7,10–12] In contrast, other groups have used HeLa cells as a source for the analysis of PcG proteins.[13,14] However,

[5] M. J. Gunster, D. P. E. Satijn, K. M. Hamer, J. L. den Blaauwen, D. de Bruijn, M. J. Alkema, M. van Lohuizen, R. van Driel, and A. P. Otte, *Mol. Cell. Biol.* **17,** 2326 (1997).

[6] D. P. E. Satijn, M. J. Gunster, J. van der Vlag, K. M. Hamer, W. Schul, M. J. Alkema, A. J. Saurin, P. S. Freemont, R. van Driel, and A. P. Otte, *Mol. Cell. Biol.* **17,** 4105 (1997).

[7] D. P. E. Satijn, D. J. Olson, J. van der Vlag, K. M. Hamer, C. Lambrechts, H. Masselink, M. J. Gunster, R. G. A. B. Sewalt, R. van Driel, and A. P. Otte, *Mol. Cell. Biol.* **17,** 6076 (1997).

[8] Z. Shao, F. Raible, R. Mollaaghababa, J. R. Guyon, C. T. Wu, W. Bender, and R. E. Kingston, *Cell* **98,** 37 (1999).

[9] A. J. Saurin, Z. Shao, H. Erdjument-Bromage, P. Tempst, and R. E. Kingston, *Nature* **412,** 655 (2001).

[10] R. G. A. B. Sewalt, J. van der Vlag, M. J. Gunster, K. M. Hamer, J. L. den Blaauwen, D. P. E. Satijn, T. Hendrix, R. van Driel, and A. P. Otte, *Mol. Cell. Biol.* **18,** 3586 (1998).

[11] D. P. E. Satijn, K. M. Hamer, J. den Blaauwen, and A. P. Otte, *Mol. Cell. Biol.* **21,** 1360 (2001).

[12] R. G. A. B. Sewalt, M. J. Gunster, J. van der Vlag, D. P. E. Satijn, and A. P. Otte, *Mol. Cell. Biol.* **19,** 777 (1999).

[13] S. S. Levine, A. Weiss, H. Erdjument-Bromage, Z. Shao, P. Tempst, and R. E. Kingston, *Mol. Cell. Biol.* **22,** 6070 (2002).

[14] R. Cao, L. Wang, H. Wang, L. Xia, H. Erdjument-Bromage, P. Tempst, R. S. Jones, and Y. Zhang, *Science* **298,** 1039 (2002).

HeLa cells are a poor source of PcG proteins (Fig. 1). For immunoprecipitation of PcG II complex proteins we have used 293 cells (Fig. 1). In nuclear extracts of 293 cells we have also found that histone deacetylases (HDAC) proteins are associated with the PcG II complex proteins.[15]

All coimmunoprecipitation experiments were performed according to the following protocol. It should be stressed, however, that this protocol is not only useful for coimmunoprecipitation experiments of PcG proteins and associated proteins but should be useful for other chromatin proteins as well.

Coimmunoprecipitation Experiments

1. The relative expression levels of the proteins of interest in different cell lines is first determined by Western blot analyses (see Fig. 1). Nuclei which are prepared from human 293 or U-2 OS cells, which were grown to confluence. Generally a confluent 75 cm^2 flask is sufficient to obtain enough material for at least five immunoprecipitations. Nuclei are prepared by 10 strokes with a glass douncepestle in 1.5 ml buffer containing 20 mM HEPES [pH 7.0], 1.5 mM MgCl$_2$, 10 mM KCl, 0.5 mM DTT, 0.5 mM PMSF and protease inhibitors leupeptin 2 μg/ml, benzamidine 10 μg/ml and aprotinin 2.5 μg/ml.

2. Nuclei are pelleted by centrifugation at 1000g at 4° for 10 min. The nuclei are lysed in 200 μl lysis buffer (250 mM NaCl, 0.1% NP-40, 50 mM HEPES [pH 7.0], 5 mM EDTA, 0.5 mM dithiothreitol, 1 mM PMSF and the protease inhibitors leupeptin 2 μg/ml, benzamidine 10 μg/ml and aprotinin 2.5 μg/ml (Note 1).

3. The lysate is sonicated three times with bursts of 10 s and the supernatant is centrifuged at 42,000 rpm (167,000g) for 1 h in a Kontron TST54 rotor.

4. After centrifugation, the supernatant is incubated with antibodies directed against one of the PcG proteins or another chromatin-associated protein of interest, for 2 h at 4°. The amount of material and the antibody has to be determined empirically, because they can differ considerably amongst proteins of interest and the available antibodies.

5. Goat antirabbit or mouse immunoglobulin G (lgG) antibody coupled to agarose beads (Sigma, A-6531 and A-1027) is added and the volume is increased by adding up to 300 μl lysis buffer

[15] J. van der Vlag and A. P. Otte, *Nat. Genet.* **23,** 474 (1999).

containing protease inhibitors. The slurry is incubated for 1 h at $4°$ under continuous mixing.

6. Centrifuge the tubes for 30 s at 3000 rpm at $4°$ and remove the supernatant carefully without disrupting the beads.

7. Add 900 ul lysis buffer, resuspend the beads carefully and centrifuge the tubes for 30 s at 3000 rpm at $4°$ and remove the supernatant. Repeat this step six times.

8. The immunoprecipitate is separated by SDS-polyacrylamide gelelectrophoresis (PAGE) after addition of loading buffer and heating for 5 min at $95°$. To remove the beads, spin down the beads for 30 s at 3000 rpm. Load the supernatant onto the gel. Subsequently, the proteins are transferred to a nitrocellulose membrane.

9. For detection, the blots are probed with primary antibodies directed against the different PcG proteins (RING1, BMI1, EED, EZH2, HPC2, HPH).[5–7,10–12] For antibodies directed against different HDACs and YY1 we used commercially available antibodies: HDAC1 (Santa Cruz, SC-7872), HDAC2 (Santa Cruz, SC-7899), and YY1 (Santa Cruz, SC-7341).

10. Goat antirabbit conjugated with alkaline phosphatase (AP) (Jackson ImmunoResearch Laboratories, 111–056–003) or Goat antimouse conjugated with AP (Biorad, 170–6520) are used as secondary antibody at a dilution of 1:10,000 or 1:3000, respectively. 4-Nitro blue tetrazolium chloride (NBT, Roche, 1087479)/5-bromo-4-chloro-3-indolyl-phosphate (BCIP, Roche, 1017373) is used as substrate for detection. Bands should appear in between 30 min and 2 h. Do not exceed 2 h since this might result in aspecific staining (Note 2).

Note 1: An important option is to add ethidium bromide to a final concentration of 50 μg/ml. This is thought to disrupt protein-DNA interactions. Immunoprecipitation of large chromatin structures has the intrinsic danger that many proteins are held together by the DNA strands. Ethidium bromide will help to overcome this problem.[15]

Note 2: Enhanced chemiluminescence (ECL Western blotting detection reagents, Amersham Biosciences, RPN 2209) can be used as well. However, we observed much more background staining using ECL than using AP.

Methods to Isolate Antirepressor Elements

Chromatin-associated repression plays a pivotal role in cell differentiation and maintenance. There is ample evidence that genomic elements exist that delimit the action of these repressors.[16] Potentially, these elements can be insulators or boundary elements.[17–20] However, insulators and boundary elements have only limited ability to block the action of repressors.[16] To identify novel elements that do have potent antirepressor characteristics, we developed a screening methodology to isolate these elements. This methodology is described later.

Components of the selection system are cell lines with inducible expression of LexA-repressor fusion proteins and a genomic DNA library in episomal selection vectors containing LexA operators upstream of a selection gene. LexA-repressor fusion proteins are targeted chromatin-associated repressors, such as Polycomb group proteins, HP1 and MeCP2. High expression levels of the LexA-repressor fusion proteins results in effective repression of a target gene, even 2 kbp downstream of the site of targeting.[16] If a reporter gene is replaced by a selection gene, this will also result in repression of this gene and subsequent addition of the selection agent to the culture medium will result in killing of the cell. This is, unless a DNA fragment is present between the targeted repressors and the selection gene that blocks these repressors. Now the selection gene will be expressed and cells will survive after administering the selection agent. The plasmids can be recovered, resulting in the identification of DNA elements that in principle potently block chromatin-associated repression (see Fig. 2).

Construction of a Library of Genomic DNA Fragments

1. The backbone of the selection vector is the pREP4 vector (Invitrogen, V004–50). This vector provides the Epstein Barr oriP origin of replication and EBNA-1 nuclear antigen for high-copy episomal replication in primate cell lines (Note 1). The vector also encompasses the hygromycin resistance gene with the thymidine kinase promoter for selection in mammalian cells; and the ampicillin resistance gene for maintenance in *Escherichia coli*. The selection vector further contains four consecutive LexA operator sites.[21] Embedded between the LexA operators and a *Nhe*I site a polylinker is placed that consists of the following restriction sites: *Hind*III-*Asc*I-*Bam*HI-*Asc*I-*Hind*III. Between the *Nhe*I site and a *Sal*I site the

---

[16] J. van der Vlag, J. L. den Blaauwen, R. G. A. B. Sewalt, R. van Driel, and A. P. Otte, *J. Biol. Chem.* **275,** 697 (2000).

[17] A. Udvardy, E. Maine, and P. Schedl, *J. Mol. Biol.* **185,** 341 (1985).

[18] R. Kellum and P. Schedl, *Mol. Cell. Biol.* **12,** 2424 (1992).

[19] J. H. Chung, A. C. Bell, and G. Felsenfeld, *Proc. Natl. Acad. Sci. USA* **94,** 575 (1997).

[20] H. Cai and M. Levine, *Nature* **376,** 533 (1995).

[21] C. A. Bunker and R. E. Kingston, *Mol. Cell. Biol.* **14,** 1721 (1994).

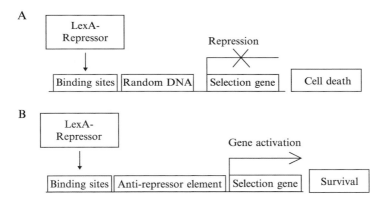

FIG. 2. Schematic representation of the antirepressor screen. LexA-repressor fusion proteins are targeted to lexA binding sites upstream of stretches of human genomic DNA. A library of such plasmids is made, consisting of human genomic DNA fragments ranging from 500 to 2000 bp. (A) Addition of Zeocin to the culture medium results in cell-death if the DNA fragment is not able to block the action of the repressor protein. (B) Addition of Zeocin to the culture medium results in cell survival if the DNA fragment is able to block the action of the repressor protein. This particular DNA is then referred to as an anti-repressor element.

zeocin resistance gene with the SV40 promoter is placed; this is the selectable marker for the antirepressor screen (see Fig. 3).

2. Random DNA libraries are constructed by partial digestion of human genomic DNA, purified from placenta (BD Biosciences, 636401). The DNA is size-fractionated, and the 0.5–2 Kb size fraction (Note 2) is ligated into BamHI-digested pSelect vector. Usage of 4-cutters MboI and Sau3AI will result in easy cloning since they have compatible overhangs with BamHI. $E.\ coli$ colonies are scraped off the LB-plates and pooled for plasmid-DNA isolation. A single 15 cm LB-plate can contain up to 5000 colonies. For a typical screen approximately 1×10^6 independent colonies (from 200 15 cm LB plates) can be recovered from $E.\ coli$.

3. Small DNA libraries from specific genomic loci can be made using the specific BAC- or PAC-clones (BAC/PAC resources). However, a significant number of the clones in these libraries will contain either vector or genomic $E.\ coli$ DNA (up to an estimated 80%). Therefore, coverage of a single 100 Kb BAC/PAC-clone requires isolation of at least 500 independent colonies (with an average insert size of 1 Kb) (Note 3).

$Note\ 1:$ The integration frequency to obtain stable cell colonies is too low (~1%) to achieve sufficient coverage in the screen. Usage of

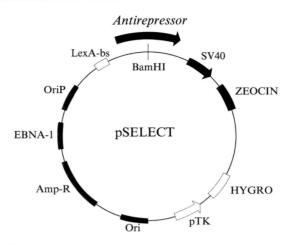

FIG. 3. The pSelect vector for screening antirepressor elements. Human genomic DNA fragments are cloned between the LexA binding sites and the SV40 driven Zeocin expression cassette. The vector contains a TK driven Hygromicin (HYGRO) cassette for selection in mammalian cells. An EBNA-1 nuclear antigen and the Epstein Barr oriP origin of replication are present for high copy episomal replication. Furthermore, the pSelect vector contains an ori and ampicillin resistance gene for propagation in bacterial cells.

episomally replicating vectors results in a much higher number of stably transfected cells. Since the episomally replicating vectors have proper chromosomal structure they are suitable for this screen.[16] Also the isolation of these vectors, containing an antirepressor element is much facilitated by usage of episomally replicating vectors. One drawback is, however, that only a limited number of cell lines can be used because these vectors replicate as an episome. Reportedly human, primate and canine cell lines are of use, whereas murine cell lines are not.[22]

Note 2: Smaller fragments will probably also deliver good antirepressor elements. Fine mapping of a number of antirepressor elements showed that the antirepressor ability of the DNA elements may reside in stretches smaller than 500 bp.[23] On the other hand, fragment sizes larger than 2 kbp will result in less efficient repression of the zeocin resistance gene by the LexA-repressor fusion proteins.[16] Incorporation of fragments larger than 2 kbp will therefore result in false positive, surviving cell colonies.

[22] J. L. Yates, N. Warren, and B. Sugden, Nature 313, 812 (1985).
[23] T. H. J. Kwaks, P. Barnett, W. Hemrika, T. Siersma, R. G. A. B. Sewalt, D. P. E. Satijn, J. F. Brons, R. van Blokland, P. Kwakman, A. Kruckeberg, A. Kelder, and A. P. Otte, Nat. Biotech. 21, 553 (2003).

Note 3:　BAC- or PAC-clone DNA libraries will contain vector as well as genomic *E. coli* DNA. In our experience, about 80% of all rescued clones contain bacterial sequences. In particular bacterial sequences containing oriP sequences are able to block PcG and HP1 repressors very effectively, which results in false positives in terms of human or mouse DNA sequences. It is therefore advisable to obtain DNA that is largely free of bacterial sequences.

Construction of Host Strains

1. In order to be able to fully employ the EBNA based, episomally replicating selection library this cell line must be of human, canine or primate origin.[22] The EBV-derived vectors obtain a bona fide chromatin structure and after transfection and selection they can be isolated from cell lines for further analysis. The screen for antirepressor elements is performed in a cell line that contains a plasmid encoding a LexA-fusion protein with a chromatin-associated repressor. High levels of LexA-repressor proteins can have overall negative effects on mammalian cells and this may result in lower growth rates and/or cell death compared to the normal situation. Therefore an inducible expression system for the LexA-repressor fusion proteins is applied.

2. The U-2 OS human osteosarcoma cell line (American Type Culture Collection HTB-96) is stably co-transfected with the pTet-Off plasmid (BD Biosciences, 631017), encoding a protein chimera consisting of the Tet-repressor DNA binding domain and the VP16 transactivation domain, and pBabe-puro[24] containing the puromycin resistance gene. The cell line is subsequently stably transfected with fusion protein genes containing the LexA DNA binding domain, and the coding region of a chromatin associated repressor. The LexA-repressor genes are under control of the Tet-Off transcriptional regulatory system.[25] Transfectants can be selected by using Geneticin (G418; Roche 1464990).

3. Cell clones are selected in which the expression of LexA-repressor proteins is induced to a high level upon removal of the tetracycline analogue doxycyclin. In the presence of high doxycyclin concentrations no detectable expression of LexA-repressor protein should be detectable on a Western blot. Detection of these fusion proteins

---

[24] J. P. Morgenstern and H. Land, *Nucleic Acids Res.* **18,** 3587 (1990).
[25] M. Gossen and H. Bujard, *Proc. Natl. Acad. Sci. USA* **89,** 5547 (1992).

can be accomplished by either specific antibodies against, for instance, HPC2, or against LexA (Santa Cruz sc-7544)

4. In our experience, particular LexA-HP1 and LexA-HPC2 are useful repressor proteins to perform a screen with. LexA-MeCP2 as a repressor is slightly less useful because in this case even cells which express high levels of LexA-MeCP2 protein appear hard to be killed. Since high levels of repressor proteins like LexA-HP1 can have a general negative effect on cell-growth and survival, the optimal doxycyclin concentration used for LexA-repressor protein induction is to be determined empirically. Usually, we find a 0.1 ng/ml doxycyclin concentration useful.

Library Screening, Transfection, and Selection Scheme

Day 1   Seed cells about 24 hours before transfection to 75 cm^2 flasks in order to obtain 10 to 25% confluency ($\sim$1 × 10^6 cells) on the day of transfection (day 2). This means a 1:10 to 1:15 dilution for the various lexA-repressor cell-lines in the case of confluent 75 cm^2 flasks. Use standard DMEM medium (Gibco 52100–039) including L-Glutamine (Gibco 25030–024), penicillin/streptomycin (Gibco 15140–148), Fetal Bovine Serum (FBS, GibcoBRL 10500–064), 100 μg/ml geneticin (G418; Roche 1464990) and 0.5 μg/ml puromycin (SIGMA, P-8833) and 10 ng/ml doxycyclin (Clontech 8634–1). Puromycin is needed as selection agents for pBabe-puro.

Day 2   The pSelect DNA libraries are transfected into the U-2 OS/Tet-Off/LexA-repressor cell line by calcium phosphate precipitation as recommended by the supplier of the transfection reagent (GibcoBRL 18306–019). From this point use commercial available standard DMEM medium (Invitrogen, 11965) to ensure constant pH values, add pen/strep, G418 and puromycin. Use 20 μg of DNA per 75 cm^2 flask. In addition to the library selection vectors transfect the negative control (pSELECT-SV40-ZEO) in one flask (Note 1). The number of transfections or flasks depends on the number of independent clones in your library. A guideline is that at a 10–25% confluency about 500,000–1,500,000 cells will attach. With a transfection efficiency of about 20% this will result in 100,000 to 300,000 transfected cells. In the case of a library of 1,000,000 independent clones about 10 independent 75 cm^2 flask transfections are required. Transfection efficiency can be determined by co-transfection of 2 μg pcDNA3-GFP reporter for one or two flasks. Transfection efficiency can be determined on day 4 by performing FACS-analysis on these specific flasks.

Day 3   After 20–24 h gently wash transfections with 10 ml PBS and add fresh DMEM medium with all additives including 100 μg/ml G418 and 0.5 μg/ml puromycin. Add 10 ng/ml doxycyclin separately.

Day 4   About 48 h after transfection wash flasks with PBS, trypsinize cells using 1 ml Trypsin-EDTA (Gibco 25300–054) and transfer all cells at different dilutions to petri dishes (range: 1:5 to 1:200). Use DMEM with all standard additives including 100 μg/ml G418, 0.5 μg/ml puromycin and 25 μg/ml hygromycin (Sigma H-0654). Add 10 ng/ml doxycyclin separately. Hygromycin is the selection agent for the pSelect plasmids.

Day 6   The day scheme starting from here is a guideline for the expected selection procedure. Inspect dishes to monitor the number of attached, growing and death cells. Wash dishes to remove death cells and refresh medium with DMEM with all standard additives including 100 μg/ml G418, 0.5 μg/ml puromycin and 25 μg/ml hygromycin. Add 10 ng/ml doxycyclin separately.

Day 7–17   Inspect dishes/cells regularly and refresh medium at least once a week. When colonies contain about 50 to 100 cells, wash dishes and replace the medium with standard DMEM medium with all additives including 100 μg/ml G418, 0.5 μg/ml puromycin and 25 μg/ml hygromycin. Add now Tet-system approved Foetal Calf Serum (BD Biosciences 8630–1) instead of the normal FCS and lower the doxycyclin concentration to 0.1 ng/ml. This starts the induction of the respective LexA-repressor fusion proteins.

Day 18   To ensure that a new steady state doxycyclin concentration in the medium is achieved, the medium has to be refreshed the day after a change of doxycyclin concentration.

Day 19–21   Inspect colonies/cells. Cells start to die approximately 4 days after the change to medium containing lower doxycyclin concentrations. The extent differs between different LexA-repressors.

Day 21   Replace the hygromycin selection medium with zeocin selection medium (Note 1). This is DMEM medium containing all additives including test-system approved Foetal Calf

Serum, 100 μg/ml G418, 0.5 μg/ml puromycin and 0.1 ng/ml doxycyclin. Prior to use add 100 μg/ml zeocin (Invitrogen 45–0430) (Note 2).

Day 24   Inspect petri dishes and monitor the cells. Increase the zeocin concentration to 250 μg/ml, keep the doxycyclin on 0.1 ng/ml. From now on refresh the medium twice a week.

Day 26–39   Inspect petri dishes/cells. The majority of the colonies should look sick. If too many cells start to form large colonies, decrease the doxycyclin concentration further to 0.01 ng/ml. Maintain the zeocin concentration at 250 μg/ml. Refresh the medium the day after a change in doxycyclin concentration. During zeocin selection large senescent cells and colonies of loosely associated cells will appear. Usually all these cells will die, either in the petri dishes or after successive transfer of the colonies. Only colonies consisting of healthy looking, compact cells as in confluently growing cultures can be considered to be good candidates that stably express the zeocin resistance gene.

Day 40   Wash petri dishes with PBS and transfer individual colonies (1–3 mm in diameter) to 24-wells plates by pipetting/scraping single colonies with 15 μl of PBS. To help cells to survive at this point the doxycyclin concentration can be raised to 10 ng/ml just during the period required for subsequent attachment (2 days). After attachment of the cells the doxycyclin concentration should be lowered again.

Day 44   Transfer cells from 24-wells plates to 6-wells plates by normal trypsinization procedures. Note that the growth rate of different colonies may vary. This means that the transfer of 24-wells to 6-wells may be out of phase.

Day 50   Transfer cells from 6-wells to two 25 cm^2 flasks by normal trypsinization procedures.

Day 55   After growth to confluency freeze the cells of one 25 cm^2 flask in two vials in liquid nitrogen. Cells from the other 25 cm^2 flask will be used for plasmid isolation. pSelect plasmid DNA can be isolated by normal lysis procedures followed by a normal miniprep (e.g., Qiagen spin mini kit) and transformed to *E. coli* (Stratagene XL10-gold 200314, bacteria should be plated on LB-zeocin plates since amp selection will result in a high background). Several independent clones originating from each mammalian clone should be characterized. Because each cell can contain up to ~6 selection vectors it cannot be excluded that different library selection plasmids are present in one positive mammalian clone. The candidate antirepressor elements in the rescued DNA are analyzed by restriction endonuclease mapping and DNA sequence analysis.[26] Also, individual pSelect plasmids are retransfected into the LexA-repressor expressing cell line in which it was originally selected. The following selection procedure is much quicker than the original screen, on average the selection takes ~28 days (Note 3). The growth rate of Zeocin resistant clones is indicative to the relative strength of the antirepressor element (see Fig. 4).

Note 1: The selection vectors and the genomic DNA libraries in the selection vectors need to be highly-purified for optimal transfection efficiencies (e.g., CsCl or Qiagen-columns).

Note 2: Zeocin is a good choice because Zeocin acts in a 1:1 stoichiometric fashion with the Zeocin resistance protein, thus preventing it from binding to DNA. This property allows efficient titration and thus gives the possibility to simply increase the concentration of zeocin in the medium if during selection the repression of the zeocin-resistance gene is not complete. In principle selection of the surviving colonies can be achieved by direct selection of Zeocin. In our experience this, however, is too harsh for a transfected library. Therefore a 'milder' step-wise approach is recommended, in which all pSelects plasmids are first selected on hygromycin. Once hygromycin-resistant colonies are forming and the LexA-repressor

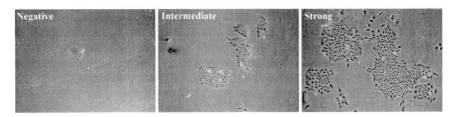

Fig. 4. Analysis of cells during selection. Left panel: Negative cells that are not able to grow due to the absence of an antirepressor element. Middle and right panel: The growth rate of zeocin resistant clones is indicative to the relative strength of the antirepressor element. The presence of a strong antirepressor results in fast growing cells (right panel), highly resistant to zeocin, whereas the presence of an intermediate antirepressor (middle panel) results in smaller and slow-growing colonies.

[26] J. Sambrook, E. F. Fritsch, and T. Maniatis, "Molecular Cloning." Cold Spring Harbor Laboratory Press, New York, 1989.

proteins have been induced, selection is changed towards Zeocin. Care should be taken, however, that the hygromycin colonies do not become too large. Large colonies are hard to kill upon subsequent addition of Zeocin in the medium. The timing of hygromyicn selection, induction of repressor proteins and the last step of Zeocin selection is therefore critical.

Note 3: The screens we performed with human genomic DNA yielded elements that displayed antirepressor activity upon retransfection of the pSelect-based clones into the host U-2 OS human osteosarcoma cell line (indicating that the antirepressor activity expressed in the initial screen is plasmid-specific and not due to artefactual changes in the host cells). The elements contain noncoding DNA sequence that matches sequence in the human genome sequence database (indicating that the clone does not contain contaminating DNA sequence, from for example, bacterial or vector sources).

Section III

ATP-Dependent Chromatin Remodeling Enzymes

[18] Purification and Functional Analysis of the Mammalian SWI/SNF-Family of Chromatin-Remodeling Complexes

By Tianhuai Chi, Zhijiang Yan, Yutong Xue, and Weidong Wang

The SWI/SNF family of complexes utilizes the energy of ATP hydrolysis to remodel chromatin structure, thereby facilitating access of transcription factors to DNA.[1] The human SWI/SNF family of complexes, including BAF and PBAF, are expressed in all tissues and cell types examined. Their crucial roles in development and differentiation are revealed by mutations in their components that cause embryonic lethality or increased risks of cancer. Purification of SWI/SNF complexes has been accomplished from a number of source tissues and cell lines, which has allowed subsequent structural and functional analysis of their many components. Here, we describe methods used in our groups for purification and analyses of these complexes.[2–6] The strategy and methods should aid those interested to isolating and studying novel remodeling complexes. In particular, the complete human genome contains at least 20 SWI2/SNF2-like ATPases, many of which likely form as yet uncharacterized complexes waiting to be discovered.

Purification of SWI/SNFs

The Strategy

The most difficult obstacle to studies of a chromatin-remodeling complex such as SWI/SNF is its isolation. The complex has to be purified in sufficient quantity and purity for subsequent analysis. The model SWI/SNF

[1] W. Wang, *Curr. Top. Microbiol. Immunol.* **274,** 143 (2003).
[2] W. Wang, J. Cote, Y. Xue, S. Zhou, P. A. Khavari, S. R. Biggar, C. Muchardt, G. V. Kalpana, S. P. Goff, M. Yaniv, J. L. Workman, and G. R. Crabtree, *EMBO J.* **15,** 5370 (1996).
[3] W. Wang, T. Chi, Y. Xue, S. Zhou, A. Kuo, and G. R. Crabtree, *Proc. Natl. Acad. Sci. USA* **95,** 492 (1998).
[4] K. Zhao, W. Wang, O. J. Rando, Y. Xue, K. Swiderek, A. Kuo, and G. R. Crabtree, *Cell* **95,** 625 (1998).
[5] Y. Xue, J. C. Canman, C. S. Lee, Z. Nie, D. Yang, G. T. Moreno, M. K. Young, E. D. Salmon, and W. Wang, *Proc. Natl. Acad. Sci. USA* **97,** 13015 (2000).
[6] T. H. Chi, M. Wan, K. Zhao, I. Taniuchi, L. Chen, D. R. Littman, and G. R. Crabtree, *Nature* **418,** 195 (2002).

is a multisubunit complex of 1–1.5 MDa, with about 10 subunits, and care must be taken to avoid its dissociation and degradation. For example, all procedures must be carried out at 4°, and all buffers should have protease inhibitors. Traditional purification by a succession of chromatographic steps is slow and also has a higher risk of degradation and dissociation of components of complexes. We have increasingly settled on the alternative of immunoaffinity-based methods, such as immunopurification or immuno-precipitation (abbreviated as IP) as the key purification approach. Advantages of IP over the conventional chromatography include: first, it is fast and efficient. A good antibody can easily achieve 1000-fold of purification in a single step, which is far superior to conventional chromatography (which usually gives 5–10-fold purification at each column step). Several remodeling complexes, including SWI/SNFs, have been isolated to good purity by this method alone. Second, it minimizes dissociation and proteolytic degradation. Third, it is highly user-friendly: people with minimal experience working with proteins can learn it in a relatively short period. Here, we describe two IP approaches to purify mammalian SWI/SNFs.

Purification with Cognate Antibodies Against Components of SWI/SNF

Pros and Cons. The first approach uses polyclonal antibodies against different components of SWI/SNFs for IP. An advantage of this approach is that polyclonal antibodies are easy to make. Once a good antibody is obtained, it can be used to isolate SWI/SNFs from any tissue or cell. A disadvantage is that because polyclonal antibodies bind the complex at multiple epitopes, highly disruptive conditions are needed to break up the multivalent antibody-antigen complexes. Unfortunately, these conditions often separate intrinsic components of the complex and inactivate its activity. To negate this potential problem, we performed many *in vitro* assays with SWI/SNFs immobilized on beads. However, certain assays may require SWI/SNF in solution. A monoclonal antibody developed against SWI/SNF components can obviate such difficulties, because the monovalent interaction between the antibody and its antigen can be disrupted under physiological conditions by a short epitope-containing peptide. Regrettably, recovering a good monoclonal antibody is time- and labor-intensive. The second approach below utilizes a trick to side step these difficulties.

Developing a Polyclonal Antibody. High quality antibody is the key for the success of this procedure. We infer from multiple instances that if good antibodies can be made, essentially any chromatin-remodeling complex can be purified. To increase the likelihood of obtaining a good antibody, we usually make two or three antibodies against different regions of a given protein. If multiple antibodies obtained are all of good quality, all can be used for IP to increase confidence in the integrity of the preparation.

Selecting and making immunogens. We have found that synthetic peptides as immunogen have a low success rate for obtaining good antibodies. Therefore, we use recombinant fusion proteins produced in *E. coli* as immunogen. These fusion proteins contain regions, usually between 50 and 200 amino acid residues in length, selected from BAF subunits. If the selected region is much longer, the fusion protein may give low yield, or become insoluble. We pick regions with a high density of charged amino acids (Lys, Arg, Asp, Glu), which are more immunogenic and more likely to be on the surface of the protein, accessible for antibody-binding.

We have used both glutathione transferase (GST) and maltose-binding protein (MBP) as the fusion partner for the protein of interest. These fusion proteins can be easily produced and isolated from *E. coli* lysate by affinity purification. Between the two, we prefer the MBP-fusion proteins (New England Biolabs), which are more soluble than GST-fusion proteins. We always use soluble fusion proteins for immunization, because the antibodies produced are more likely to recognize the native form of the protein, a prerequisite for IP use. In contrast, antibodies produced against insoluble proteins usually only react with the denatured form of the proteins and are not suitable for IP.

Immunization and serum-testing. We normally inject 2 rabbits for each immunogen, with 0.5 mg of immunogen for each injection. The rabbits are injected 5 times during a 70-day period, though titers often increase by immunizing the rabbits up to 100 days. Sera are tested by immunoblotting analysis, using the nuclear extracts from HeLa cells as the antigen source. A usable antiserum should recognize the cognate protein at 1:1000 dilution. There should preferably be no cross-reactivity to other nuclear proteins; but in cases where the antiserum has some cross-reactivity, as long as the expected protein is the major immunoreactive species and the cross-reactive polypeptides are relatively few (preferably <10), the serum may still be usable. This is because most cross-reactivity can be strongly reduced after affinity purification of the antibody.

Affinity purification of the antibody. The most common source of contamination in IP is derived from antibody cross-reactivity. Sometimes, cross-reactivity is an intrinsic property of the antibody. In most cases, however, the cross-reactivity derives from antibodies that are also present against other immunogens in serum. Thus, cross-reactivity can generally be significantly reduced by affinity-purification of the serum with the specific antigen. We have noted that most of the antisera must be purified before using them for IP.

Our antibody purification procedure follows a previous protocol.[7] Basically, the MBP-fusion protein (5–20 mg) used for immunization is directly cross-linked to cyanogen bromide (CNBr)-activated agarose beads (1–1.5 ml) by following the manufacturer's protocol (Pierce). These beads are packed in an antigen column. The serum containing the antibody (20–50 ml) is then applied to the column. Only the antibody specific to the antigen should be retained. After extensive washing to remove nonspecific antibodies and serum proteins (we use PBS buffer containing 0.5 M NaCl for washing), the antibody is eluted under acidic conditions. We normally use 0.1 M glycine (pH 2.5) as the elution buffer. The eluted antibody fractions are immediately neutralized by adding 1/10 volume of Tris–HCl (1 M, pH 8.0). The final pH should be between 7 and 8. The concentration of each antibody fraction is determined by standard Bradford assay (New England Biolabs) and tested for its specificity by Western blotting. If the affinity purification is successful, the purified antibody should have a higher titer and lower cross-reactivity than that of the crude serum.

Because the entire MBP-fusion protein is used for immunization as well as for purification of the antibody, the purified antibody fractions are directed against MBP as well as protein of interest. The anti-MBP antibodies can be removed by incubating antibody fractions with MBP-LacZ fusion protein cross-linked to CNBr-activated beads.

When making antigen-coupled beads, it is important to use buffers that lack primary amine groups (which will react with and quench the CNBr). We use phosphate buffers for purification of the antigen. An advantage of using MBP-fusion proteins as antigens is that after elution from their affinity matrix by maltose, they can be used directly for cross-linking. In contrast, elution of GST fusion proteins requires glutathione, which contains amine groups and must be removed before GST-proteins can be cross-linked to beads.

Some groups use protein A or protein G–conjugated agarose beads to purify the IgG fractions of the serum and use that for IP. This procedure, however, does not remove the large amount of the nonspecific IgG antibodies, which may result in high backgrounds during purification. In our hands, it has worked only with antisera with excellent specificity and high titer.

Use Existing Antibodies to Purify Epitope-Tagged SWI/SNF

Pros and Cons. One disadvantage of the method above is that SWI/SNFs obtained by polyclonal antibodies cannot be eluted under

[7] E. Harlow and D. Lane, "Antibodies: A Laboratory Manual." Cold Spring Harbor Laboratory, New York, 1988.

physiological conditions, which precludes certain *in vitro* analyses of function. A second approach solves these problems by using standard monoclonal antibodies that have defined epitopes. The protein of interest, such as a subunit of SWI/SNF, is linked to these epitopes (called "epitope-tagged") through recombinant DNA technology. The tagged genes are ectopically expressed in cells, and complexes containing the tagged subunit are then immunoisolated by the monoclonal antibody against the tag. The complexes can be eluted with a synthetic peptide containing the epitope sequence under physiological conditions.

The tag approach can be very powerful and efficient because it avoids the lengthy procedure of developing an antibody. Another benefit of the approach is that one can introduce mutations in the tagged gene and isolate complexes containing the mutant subunits for structural and functional studies.[3] However, this method does not always work. Its success depends on multiple factors. First, a cell line has to be established to stably express the tagged protein at a level suitable for purification. Second, the tagged protein has to incorporate into the complex. Third, the tag has to be accessible to the antibody.

Because of these limitations, it is useful to try both approaches. In our experience, for smaller proteins, such as BAF57 (57 kd), the second approach can be faster and easier, because the likelihood of obtaining a stable cell line is high. For large proteins, such as BRG1, it is best to try the first approach, because it is difficult to obtain a cell line that expresses a large exogenous gene at high levels.

Establishing Cell Lines Stably Expressing Tagged Protein. Our group has used a monoclonal antibody against the HA-epitope,[3] whereas Kingston and colleagues[8] utilize one against the Flag-epitope. Both antibodies are commercially available (Upstate and Sigma, respectively) and work well in IP. For example, we clone HA-tagged BAF57 into a mammalian expression vector, pBJ5neo, which contains a selectable marker, neomycin. Other commercially available vectors, such as pcDNA3.1 from Invitrogen, pIRESneo2 from Clonetech, pREP4 from Invitrogen, can also be used. The vector is transfected into mammalian cell lines (we use Jurkat T-cells). The cells are grown in the presence of neomycin to kill those that do not carry the plasmid. Those resistant to the drug are selected (cloned out) and grown to sufficient quantity (>1 million cells) for screening purposes. The cells are lysed with RIPA buffer (150 mM NaCl, 1% NP-40, 0.5% deoxycholate, 0.1% SDS, 50 mM Tris–HCl, pH 8.0) to yield a whole cell extract. Western blotting is done using both the antitag antibody and

[8] S. Sif, P. T. Stukenberg, M. W. Kirschner, and R. E. Kingston, *Genes Dev.* **12**, 2842 (1998).

antibody to BAF57. Clones that show a high level of expression are grown in large quantities (>3 L) to make nuclear extract for further studies.

Immunoprecipitation and Immunopurification

Immobilizing the Antibodies

We use protein A beads (Amersham-Pharmacia Biotech) to immobilize the antibody. In our hands, protein A is the only matrix that has low background binding activity for nuclear proteins. Protein G and CNBr-beads have relatively high nonspecific binding activity to nuclear proteins, particularly to human SWI/SNFs. They should therefore be avoided for co-IP studies involving human SWI/SNFs.

Some groups add antibodies to extract, and then collect the immunoprecipitate by adding protein A beads. We have rather let the antibodies bind protein A beads first, and then washed away the unbound antibodies and serum-derived proteins before IP. This extra step helps to purify the antibody further.

The IP activity of an antibody depends on several factors, including the instrinsic activity of the antibody itself (as judged by its titer using Western blotting), the level of its antigen in the extract, and the accessibility of epitopes. Therefore, the amount of antibodies used in each IP must be determined empirically. Up to 2 μg IgG can be bound to 1 μl of protein A beads. We usually use between 0.050 and 0.5 μg of antibodies for 1 μl of protein A beads. Binding can employ many buffers, such as phosphate buffered saline (PBS) buffer, for 30 min at room temperature or 2 h at 4°. Unbound antibodies can be washed away with PBS. Antibody-bound beads can be stored at 4° in the presence 0.01% NaN_3.

Covalently Linking Antibodies to Protein A Beads. We often cross-link the antibodies to protein A beads for purification of many complexes, including SWI/SNFs. The major advantage of cross-linking is that antibodies will remain associated with protein A beads after elution, and will not contaminate the purified complex. This is important for complexes such as SWI/SNFs that have several components (BAF57 and BAF53) about the size of IgG heavy chain (55 kd). Without cross-linking, these components cannot be resolved from the heavy chain on SDS-PAGE (see Fig. 1B). Another advantage of cross-linking is that the beads are reusable up to 5 times without losing appreciable activity. Some antibodies, particularly monoclonal antibodies, may lose activity after cross-linking; but polyclonal antibodies usually retain enough activity. To compensate any loss of activity from cross-linking, we usually add 3 to 10 times excess antibody when we cross-link it to protein A beads.

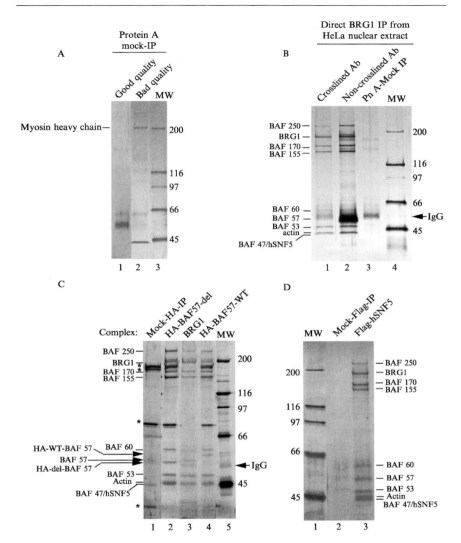

FIG. 1. Immunopurification of human SWI/SNF related BAF complex. (A) A silver-stained SDS-gel showing two examples of the polypeptides isolated by mock-immunoprecipitation with protein A beads. If the IP conditions are not optimized, various contaminating polypeptides will be isolated by protein A beads only (lane 2). The most common one is the 200 kd myosin heavy chain, which was regarded as a contamination indicator in our hands. (B) A silver-stained SDS gel showing human BAF complex immunoisolated directly from HeLa nuclear extract using a BRG1 antibody. (C) A silver-stained SDS-gel showing BAF complex immunoisolated directly from Jurkat cell nuclear extract using an HA antibody against HA-tagged BAF57 (lane 4). A mutant BAF complex containing BAF57 deleted of its

We use dimethylpyrimilidate (DMP from Pierce) as the cross-linker, and follow the published protocol.[7] Briefly, protein A-bound antibody beads (100 μl) are washed twice with 1 ml of sodium borate (0.1 M, pH 9.0), and resuspended in 1 ml of the same buffer. DMP (5.2 mg) is added to the mixture and the tube is rotated at room temperature for 30 min on a rotating wheel. The beads are collected by centrifugation, washed once with 0.2 M ethanolamine (pH 8.0), resuspended in the same buffer, and rotated overnight at 4° to block nonreactive groups. The beads are then washed four times with 0.1 M glycine (pH 2.5) to remove antibodies that are not cross-linked, and are neutralized by washing twice with PBS.

Preparation of Nuclear Extract. Preparation of nuclear extract is another important step for IP. Several complexes, including SWI/SNFs, can be immunoisolated to good purity by a single IP step from extracts of high quality (see Fig. 1B–D). We have favored standard cell lines, such as HeLa or Jurkat, to purify SWI/SNF and other complexes. Large quantities of these cells can be purchased from the National Cell Culture Center. The advantage of cultured cells is that they are homogeneous and very easy to extract. We can reproducibly obtain high quality nuclear extracts that show little or no degradation of BRG1. We basically follow a published straightforward method to prepare the nuclear extract.[9]

Several additional features of the procedure may help to obtain a good extract. First, we use a dounce homogenizer linked to a drill press to homogenize the cell pellet. A drill press can be run at very low speed, which is important to avoiding bubbles and heating. Second, after homogenizing the cells, we use a large volume (10 times the volume of cell pellet) of hypotonic buffer A (10 mM HEPES, pH 7.9, 1.5 mM MgCl$_2$, 10 mM KCl) to wash the pelleted nuclei. This allows efficient removal of cytoplasmic material, the presence of which may cause contamination during IP. Third, when we use high salt buffer C (20 mM HEPES, pH 7.0, 25% glycerol, 0.42 M NaCl, 1.5 mM MgCl$_2$, 0.2 mM EDTA) to extract nuclei, we usually do it twice consecutively. We have noted that many nuclear

[9] J. D. Dignam, P. L. Martin, B. S. Shastry, and R. G. Roeder, *Methods Enzymol.* **101**, 582 (1983).

HMG-domain was also shown (lane 2). The polypeptides marked with asterisks are present in mock-isolated experiment and therefore are contaminants. This figure is adapted from W. Wang, T. Chi, Y. Xue, S. Zhou, A. Kuo, and G. R. Crabtree, *Proc. Natl. Acad. Sci. USA* **95**, 492 (1998). (D) A silver-stained SDS-gel showing the BAF complex immunoisolated directly from HeLa nuclear extract using the flag antibody against the flag-tagged hSNF5. This figure is adapted from Y. Xue, J. C. Canman, C. S. Lee, Z. Nie, D. Yang, G. T. Moreno, M. K. Young, E. D. Salmon, and W. Wang, *Proc. Natl. Acad. Sci. USA* **97**, 13015 (2000).

proteins that bind tightly to chromatin are significantly enriched in the second extraction. Fourth, we always include freshly-prepared protease inhibitors in all buffers to avoid proteolysis (at least PMSF and leupeptin; or in some cases, the protease inhibitor cocktail from Boeringer-Mannheim). We also add 0.1 mM dithiothreitol to all solutions. The quality of the nuclear extract should be tested before its use by immunoblotting with antibody to BRG1. This large protein is a sensitive indicator of protease action.

We have also used animal tissues, such as rat liver and calf thymus, to make nuclear extracts for purification of SWI/SNFs. In this case, we use a protocol involving ultracentrifugation to isolate nuclei.[10] This protocol is more time-consuming and the extract obtained usually has certain degrees of degradation of BRG1. Thus, we recommend the use of cultured cells for purification whenever possible.

Immunopurification. Protein A-antibody beads can be packed into a column for affinity chromatography or put into a test tube for batch IP. The first approach can isolate 20–100 μg of SWI/SNFs, which is suitable for purpose such as Edman degradation sequencing. However, this approach uses a considerable amount of antibody (0.3 to 1 mg). Recently, mass spectrometry technology has been developed that can identify proteins with as low as 0.1 μg. Thus, we now routinely do IP in test tubes. The following steps are included in our IP.

1. The nuclear extract is first cleaned by ultracentrifugation (Beckman Optima TLX ultracentrifuge, 50,000–1,00,000 rpm, 10–20 min, once or twice until it is clear of debris). This helps remove contaminants derived from insoluble aggregates, which nonspecifically associate with protein A beads.

2. The nuclear extract can be mixed with custom IP buffers to improve the antibody specificity. For the isolation of human SWI/SNFs, this step is not absolutely necessary. The nuclear extract contained in buffer C can be directly used (see Fig. 1C, D). However, adding detergent to nuclear extract buffers can help reduce the nonspecific antibody interactions. For example, nuclear extract can be diluted using RIPA buffer before IP.[4] For the SWI/SNF isolation illustrated in Fig. 1B, the nuclear extract was mixed with 4 volumes of IP buffer containing 20 mM HEPES (pH 7.9), 0.3 M KCl, 0.1% NP-40, 0.25 mM EDTA, 10% glycerol.

3. The nuclear extract is then added to protein A-antibody beads and rotated on a wheel from 2 h to overnight at 4°. For Fig. 1B (lane 2), 100 μl of nuclear extract (6 mg/ml) was first mixed with IP buffer, and then

[10] K. Gorski, M. Carneiro, and U. Schibler, *Cell* **47,** 767 (1986).

incubated with 0.3 μg of BRG1 antibody-loaded protein A beads (10 μl). In Fig. 1C, 1 ml of nuclear extract prepared from cells that express HA-tagged BAF57 was directly incubated with 100 μl of cross-linked protein A-HA antibody (30 μg) beads. In Fig. 1D, 1 ml of nuclear extract from HeLa cells expressing Flag-hSNF5 was incubated with 100 μl of Flag-antibody-conjugated beads (Sigma).

4. The immunoprecipitate on the antibody beads is washed at least four times with buffers (at least 10× the volume of beads) to remove contaminating proteins. Because SWI/SNF is highly stable in salt, we sometimes use more stringent washing conditions, such as 0.5 M buffer D (20 mM HEPES, pH 7.9, 0.5 M KCl, 25 mM EDTA, 10% glycerol, 0.1% Tween-20) (for Fig. 1C, D). SWI/SNF is also an example of a "Sticky" complex that can nonspecifically associate with protein G and several column matrices. The use of detergent (Tween-20 or NP-40) can reduce such nonspecific interactions. With some highly quality SWI/SNF antibodies (such as a BRG1 antibody used in Fig. 1B), 0.3 M buffer D was used and the result is still satisfactory. But for other antibodies of lesser quality, we include a strong detergent, 0.2 M guanidine hydrochloride, in washing. We usually use 0.1 M buffer D as a last wash to lower salt levels. The SWI/SNF obtained can be used directly for remodeling assays, or eluted for SDS-PAGE and silver-staining analyses.

We have found that successful IP requires not only a good antibody and nuclear extract, but also optimization of the IP conditions. Here are some suggestions.

1. One way to judge whether the IP condition is optimal is to do a mock IP using protein A beads without loaded antibody. If the condition is not optimized or if the nuclear extract is of poor quality, there will be high background (see Fig. 1A).

2. The amount of nuclear extract used should contain more than enough antigen compared to the antibody used. Otherwise, remaining antibodies may bind other proteins nonspecifically, resulting in contamination. Whether there is enough antigen in the extract can be determined by immunoblotting the supernatant after IP.

3. It is always helpful to run a Superose 6 gelfiltration column to determine the native molecular weight of a given protein. If it is less than 500 kd, a possibility exists that the protein is not in a complex. But if it is larger than 670 kd marker, the protein is likely part of a complex. This step may also reveal any diversity of target complexes. One can also do IP using the fractions after Superose 6 fractionation to isolate each complex. Even if a protein is in a single complex, IP after fractionation can help to reveal polypeptides that not only co-IP, but also cofractionate, with the protein of

interest. These polypeptides are the most likely candidates as subunits of the complex containing the studied protein.

4. If a complex is stable in high salt, running one or two ion-exchange columns helps to improve purity. In the case of human SWI/SNFs, ion-exchange columns help to separate the two major SWI/SNF complexes, BAF and PBAF.[2,5]

5. It is useful to employ multiple antibodies against the protein of interest to do IP. One can compare the polypeptides isolated by different antibodies to identify those obtained by multiple antibodies. These common polypeptides are thus supported as components in the same complex as the protein of interest.

6. Once other components of the complex are identified, it is important to do reciprocal IP using antibodies against these components. We have shown that antibodies against many BAF subunits immunoisolate the same set of polypeptides, providing strong corroborative evidence that these polypeptides are components of the same complex.

Analysis of the Immunoisolated Complex

Silver-Staining Analysis

The immunoisolated SWI/SNF should be eluted from the antibody beads to determine its purity. The elution can be accomplished using acidic solutions at $4°$, such as glycine (pH 2.5, 0.1 M), and the eluted complex should be immediately neutralized by 1/10th volume of Tris–HCl (Tris–HCl, 1 M, pH 8.0). Alternatively, the elution can use Laemmli SDS-gel loading buffer. The eluted polypeptides are resolved on SDS-PAGE (8% Tris–glycine gel, Invitrogen) and visualized by silver-staining. If there are over 100 major proteins on the silver-stained gel, or if the expected antigen protein is a minor polypeptide, IP conditions must be improved to achieve better purity. One can try a different antibody or include conventional chromatography before IP.

In many published co-IP-Western studies, silver-stained gels are not shown. In such cases, it is impossible to judge the quality of these IP experiments. For example, if over 500 proteins are isolated by the IP, or if the antigen is a minor polypeptide among the isolated species ($<1\%$), the likelihood that an isolated protein is a contaminant rather than an interacting partner is very high. Reciprocal IP-Western experiments can help reduce the possibility of error; but if the quality of the reciprocal IP is similarly low, the results will not be reliable. Therefore, we highly recommend silver-staining analysis for each IP experiment.

Immunoblotting

Immunoblotting experiments should be performed to determine whether the antigenic protein, such as BRG1 or other BAFs, has been isolated by IP. If it is absent, the IP should be considered a failure. Even if a band with the expected size for the antigen protein is present, subsequent verification is still required before concluding that the IP is a success. For example, the immunoreactive species may be derived from a cross-reactive polypeptide. One way to exclude this possibility is to use a different antibody for immunoblotting. For IP using noncross-linked antibodies, it is helpful to include a control blot without the primary antibody, which should rule out the possibility that the immunoreactivity is not derived from the secondary antibody. Of course, the most convincing way to show that the expected antigen protein is in the IP is to identify the protein by mass spectrometry or Edman sequencing.

Immunoblotting can in fact reveal how many cross-reactive polypeptides are isolated by IP. If immunoreactive polypeptides other than the antigen protein are present, one should match these polypeptides with those on the silver-stained gel. These polypeptides could be degraded or modified forms of the antigen protein. Alternatively, they could be contaminants isolated through antibody cross-reactivity.

It is always helpful to determine the levels of the antigen in the extract before and after the IP. If the level remains unchanged, one can reduce the amount of the extract. If the antigen protein is completely depleted from the extract, one may need to use more extract. One can analyze the levels of the associated partners of the antigen in the same experiment. If these partners associate with the antigen exclusively, depletion of the antigen should deplete these partners in a comparable way.

Coomassie-Blue Staining and Mass Spectrometry Analysis

Coomassie-blue staining should be done to estimate the stoichiometry and amount of the complex. We often use Colloidal Blue (Invitrogen), which is more sensitive than similar dyes. Although Coomassie is less sensitive than silver-staining, the intensity of its staining is a better measure of the amount of proteins in the gel. Silver-staining usually cannot be used for this purpose, because it depends on the idiosyncratic number of cysteine residues in a protein.

The polypeptides that can be visualized by Coomassie-staining are usually in sufficient quantity for mass spectrometry identification. Many people only use immunoblotting for protein identification. A band on a gel recognized by an antibody on immunoblotting is frequently inferred to represent the corresponding antigenic protein. However, immunoblotting reaction can result from antibody cross-reactivity. Moreover, immunoblotting cannot determine whether the antigenic protein is the major or a minor species in the observed band. Therefore, mass spectrometry or Edman-sequencing should be done to determine the identity of the band.

Protein samples, in the form of gel slices after Coomassie-staining, can be sent to special mass spectrometry facilities for peptide mapping by MALDI-TOF or identification by LC-MS/MS. Facilities include those at Harvard University, Yale University, University of Virginia, and City of Hope in Los Angeles.

A Chromatin-Remodeling Assay

We describe a commonly-used chromatin-remodeling assay, the mono-nucleosomes disruption assay. This assay was first described for yeast SWI/SNF,[11] and was later found to work well with mammalian SWI/SNF and NURD complexes.[3,12] In this assay, a 176 bp fragment of 5 S ribosomal DNA containing a nucleosome-positional sequence was assembled into a rotationally phased mononucleosome. DNase I digestion of the nucleo-some produces a pattern of distinctive 10 bp ladders on denaturing PAGE (see Fig. 2, lane 2). In the presence of ATP, SWI/SNF disrupts these 10 bp ladders, which can be easily seen by the appearance of new bands between the ladders, as well as reduced intensity of the ladders (see Fig. 2). This pattern mimics that of naked DNA, and is an indication that the rotational phasing of the nucleosome becomes randomized after remodeling. The basic steps are as follows:

Preparation of 5S DNA fragment: The plasmid encoding the 5S DNA (pG5208–10, 20 μg) is digested with Ava I (40 units) to release the 200 bp 5S DNA fragment. This fragment and the vector are end-labeled with ap32-dCTP (200 μCi, Amersham) by Klenow, in a mixture containing dGTP and dTTP (2.5 mM each). The fragment is purified through a sepha-dex G50 spin column to remove the unlabeled nucleotides. The labeled fragment should be at least 1×10^7 cpm. The fragment is then digested with Sca I to remove p32 at one end, which produces a new fragment of 176 bp. The fragment is purified by native PAGE (6%).

Nucleosome reconstitution by octamer transfer: We follow a published protocol to prepare the H1-depleted nucleosome cores and reconstitute nucleosomes.[13] For the reconstitution, the 5S fragment ($>5 \times 10^5$ cpm) is

[11] T. Owen-Hughes, R. T. Utley, J. Cote, C. L. Peterson, and J. L. Workman, *Science* **273,** 513 (1996).

[12] Y. Xue, J. Wong, G. T. Moreno, M. K. Young, J. Cote, and W. Wang, *Mol. Cell* **2,** 851 (1998).

[13] J. Cote, R. Utley, and J. L. Workman, *Meth. Mole. Genet.* **6,** 108 (1995).

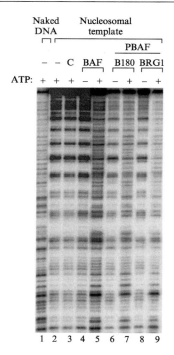

FIG. 2. Human SWI/SNF-related BAF and PBAF complexes have ATP-dependent mononucleosomes disruption activity. An autoradiograph showing the results of the mononucleosomes disruption assay. The BAF complex used in this study was immunoisolated with BAF250a antibody. PBAF was isolated by either BAF180 or BRG1 antibodies. The presence or absence of ATP is indicated. C, a control using antibody beads without complex loaded. This figure is adapted from Y. Xue, J. C. Canman, C. S. Lee, Z. Nie, D. Yang, G. T. Moreno, M. K. Young, E. D. Salmon, and W. Wang, *Proc. Natl. Acad. Sci. USA* **97,** 13015 (2000).

incubated with varying amounts of HeLa H1-depleted oligonucleosomes (2–10 μg) in the presence of 1 M NaCl. Oligonucleosomes become destabilized under such high salt condition. After decreasing concentration of the salt by serial dilution, the histone octamers will be transferred from the donor DNA to the 5S DNA. The efficiency of the transfer can be determined by native PAGE (4%) based on the differential mobility of nucleosome compared to the free probe. If the transfer is not complete, more nucleosome cores should be added.

Nucleosome disruption by SWI/SNF: The reaction mixture (10 μl) contains 1 mM ATP, 4 mM MgCl$_2$, 0.2 mg/ml BSA, 0.05 mM PMSF, 0.05% β-mercaptoethanol, 25 mM HEPES (pH 7.4), 50 mM NaCl, 5% glycerol, and reconstituted 5S nucleosomes (>5000 cpm). The mixture is added to SWI/SNF complexes on antibody beads (1–5 μl). Usually, 10–50 ng of SWI/SNFs are present in 1 μl of beads (the amounts of complex on the beads can be estimated by silver or Coomassie staining-compared to molecular weight markers of known quantity). Protein A beads can be added to maintain the total volume of beads at 5 μl. The reaction proceeds at 30° for 60 min, with frequent tapping of the tube to keep the beads in suspension. DNase I (0.15–0.2 unit, Boeringer-Mannheim) is then added to the mixture and further incubated for 1 min at room temperature. The reaction is terminated by adding 10 μl of stop buffer (20 mM Tris–HCl, pH 7.5, 50 mM EDTA, 2% SDS, 0.25 mg/ml yeast tRNA, 0.2 mg/ml proteinase K) and incubated at 50° for 60 min. The mixture is then extracted with phenol and chloroform. The DNA is precipitated by ethanol, dissolved in sequencing gel loading buffer, heat-denatured, and resolved by denaturing PAGE (6%). Signals are visualized by autoradiography.

For this assay to work well, it is important to have efficient transfer of nucleosomes to the 5 S probe, because the free probe will produce background bands that obscure the signals from disruption. In addition, one should also avoid adding too much donor nucleosome cores, which can function as competitors for SWI/SNFs and reduce the sensitivity of the assay. If the assay does not work, one can also test whether the complex is active in a direct ATPase assay.

Detecting Physiological Binding Targets of Mammalian SWI/SNFs Using Chromatin Immunoprecipitation (ChIP) Assay

ChIP is a powerful method to detect the physiological targets of DNA binding proteins *in vivo*. In this assay, DNA-binding proteins are cross-linked to DNA with formaldehyde. The chromatin is then sheared into small fragments by sonication. The fragments that associate with the protein interest are separated from the bulk of chromatin by immunoprecipitation. The cross-links in the DNA fragments obtained are then reversed and subsequently characterized on microarrays spotted with known DNA sequences,[14] or by PCR which we used to detect association of the CD4 silencer to SWI/SNFs in thymocytes.[6]

Chromatin Cross-Linking, Shearing, and Purification

1. To cross-link protein to DNA, 1.4 ml of 37% formaldehyde (Aldrich) is added directly to 50 ml media containing 100 million thymocytes, and the mixture is incubated at 37° for 30 min. Cells are

[14] A. S. Weinmann and P. J. Farnham, *Methods* **26,** 37 (2002).

collected by centrifugation at 3000 rpm for 5 min at $4°$, washed once with 15 ml PBS, and resuspended in 4 ml of TE buffer (10 mM Tris–HCl, pH 8.0, EDTA 1 mM) containing 0.5 mM PMSF in a 15 ml conical tube.

2. To fragment chromatin, the cells are sonicated for 30 s in an ice bath at a power setting of 6 (the upper limit for micro-probe) with a Branson 450 sonicator. The mixture is then chilled on ice for 30 s, and the process repeated 10 times to shear chromatin to fragments about 3 kb. The length of sheared DNA can be measured by agarose gel electrophoresis, after reversing the cross-links (see later).

3. Chromatin is solubilized by adding 10% Sarkosyl (N-Lauroylsarcosine, Sigma) to the mixture to a final concentration of 0.5%. The tube containing the mixture is rotated at room temperature for 30 min. Insoluble materials, including cell debris, are removed by centrifugation at 13,000 rpm for 5 min at room temperature.

4. The DNA:protein complexes in the supernatant are purified in a CsCl gradient. Specifically, the supernatant from step 2 is transferred to a 15 ml conical tube. CsCl (2.83 g) and TE buffer containing 0.5% Sarkosyl are added into a final volume of 5 ml. The solution is centrifuged at 40,000 rpm in a Beckman SW50 rotor for 72 h at $20°$. The cross-linked chromatin is visible as a thin gray layer. Fractions (0.5 ml each) are collected and their optical density ($OD_{260\ nm}$) measured. Peak fractions (about 1.5 ml total) are pooled and dialyzed against TE buffer containing 0.5 mM PMSF at $4°$ for 3 h. The chromatin sample is stored at $-20°$.

Immunoprecipitation and DNA Purification

1. To obtain "precleared" chromatin, 100 μl protein A beads (bed volume) are incubated with 100 μl preimmune serum and 100 μl PBS at RT for 15 min with rotation. The beads are washed twice with 1 ml RIPA buffer (1% Triton-X, 0.1% NaDOC, 0.1% SDS, 10 mM Tris pH 7.4, 1 mM EDTA, 0.5 mM PMSF, 1 mM DTT) containing 0.3 M NaCl. 100 μl of chromatin sample obtained as above is mixed with 100 μl TE and 20 μl 10× RIPA buffer containing 3 M NaCl. This mixture is added to the preimmune serum-loaded protein A beads and rotated at $4°$ for 1 h. The supernatant is then collected by centrifugation.

2. Immunoprecipitation is performed by mixing precleared chromatin with salmon sperm DNA (2 mg/ml final concentration), bovine serum albumin (1 mg/ml), and affinity-purified J1 antibody (0.1–0.3 μg). The mixture is incubated overnight at $4°$, and the immunoprecipitate collected by adding 15 μl protein A beads. The protein A-bound chromatin is washed four times with RIPA buffer containing 0.3 M NaCl, once with

RIPA buffer with no NaCl, and once with TE. All washes are performed at 4° for 10 min.

3. To purify DNA, the beads are resuspended in 100 μl TE containing Proteinase K (0.5 μg/μl) and SDS (0.25%). The mixture is incubated at 50° for 6 h to allow digestion of proteins. Cross-links are reversed by incubating the mixture at 65° for 6 h. The mixture is then extracted with phenol–chloroform, and DNA precipitated with ethanol in the presence of 5 μg glycogen. The sample is resuspended in 50 μl water. For PCR analysis 1–5 μl of the sample is used.

Because formaldehyde can only cross-link specific functional groups that are in direct contact, the efficiency of cross-linking depends on the nature and stability of the DNA:protein contacts. As the SWI/SNF–like complexes may interact with different target DNAs in different ways, the enrichment of DNA sequences in the ChIP assays may vary from target to target. It is important to keep in mind that formaldehyde can also cross-link proteins, and thus the binding sites detected by the ChIP assay may be indirect targets of particular immunoprecipitated proteins. Furthermore, the resolution of the assay is limited by the length of sheared chromatin. Finally, control IP using protein A beads alone, preimmune serum or other irrelevant antibodies should always been included in this experiment. An example of the ChIP assay is shown in Fig. 3, demonstrating specific interaction of BRG1-containing complexes with the CD4 silencer.

In summary, mammalian cells contain at least 20 SWI2/SNF2-like ATPases, many of which likely constitute as yet uncharacterized chromatin-remodeling complexes. The methods described here for SWI/SNF studies should help the purification and analysis of other complexes.

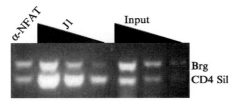

FIG. 3. BAF complexes bound the CD4 silencer in thymocytes. Shown are multiplex PCR measuring the abundance of CD4 silencer in cross-linked chromatin before (input) or after (IP) immunoprecipitation with J1 anti-BRG antibody. A region in Brg genomic sequence (Brg) was coamplified to control for DNA yields. The amplification was in the linear range of PCR as shown by a 3-fold titration of IP input. This figure is adapted from T. H. Chi, M. Wan, K. Zhao, I. Taniuchi, L. Chen, D. R. Littman, and G. R. Crabtree, *Nature* **418,** 195 (2002).

Acknowledgments

We thank Dr. D. Schlessinger for critical reading of this manuscript. We also thank Dr. G. Crabtree for his support for the project. W. W. has received Grants from the Ellison Medical Foundation and Rett Syndrome Research Foundation.

[19] Isolation and Assay of the RSC Chromatin-Remodeling Complex from *Saccharomyces cerevisiae*

By YAHLI LORCH and ROGER D. KORNBERG

RSC is the more abundant of two SWI/SNF-related, chromatin-remodeling complexes found in yeast, and conserved in humans and other eukaryotes.[1] Comprising 15 subunits, many encoded by genes essential for yeast cell growth,[2,3] RSC has been implicated in both gene activation and repression.[4] Purified RSC exhibits both a DNA-dependent ATPase activity and the capacity to perturb the structure of nucleosomes,[1] which is conveniently monitored by gel electrophoretic mobility shift analysis,[5] by enhanced accessibility to nuclease digestion,[1,5] or by histone octamer transfer from nucleosomes to naked DNA.[6]

Preparation of RSC

RSC can be purified from yeast extracts by conventional chromatography[1] or by tagging and affinity chromatography. In as much as RSC exists in two distinct forms,[7] one containing only Rsc1 and the other only Rsc2, isolation on the basis of a tag on one of these subunits is expected to yield a homogeneous preparation. Isolation of the Rsc2-containing form is advantageous, since Rsc2 is more abundant than Rsc1 in yeast. A procedure

[1] B. R. Cairns, Y. Lorch, Y. Li, L. Lacomis, H. Erdjument-Bromage, P. Tempst, B. Laurent, and R. D. Kornberg, *Cell* **87**, 1249 (1996).

[2] B. C. Laurent, X. Yang, and M. Carlson, *Mol. Cell. Biol.* **12**, 1893 (1992).

[3] Y. Cao, B. R. Cairns, R. D. Kornberg, and B. C. Laurent, *Mol. Cell. Biol.* **17**, 3323 (1997).

[4] M. Damelin, I. Simon, T. I. Moy, B. Wilson, S. Komili, P. Tempst, F. P. Roth, R. A. Young, B. R. Cairns, and P. A. Silver, *Mol. Cell* **9**, 563 (2002).

[5] Y. Lorch, B. R. Cairns, M. Zhang, and R. D. Kornberg, *Cell* **94**, 29 (1998).

[6] Y. Lorch, M. Zhang, and R. D. Kornberg, *Cell* **96**, 389 (1999).

[7] B. R. Cairns, A. Schlichter, H. Erdjument-Bromage, P. Tempst, R. D. Kornberg, and F. Winston, *Mol. Cell* **4**, 715 (1999).

for the isolation of RSC with the use of a TAP tag is presented here. Only aspects unique to RSC preparation are described, as general protocols for TAP-tagging and for the isolation of TAP-tagged proteins are available and in common use.[8]

RSC was TAP-tagged at the C-terminus of the second largest subunit of the complex, Rsc2.

TAP-tagging of Rsc2 was accomplished by integration of a PCR fragment encoding the TAP tag and a selectable marker in the yeast genome.[9] PCR was performed with pBS1479, containing the TAP tag and TRP1 marker of *Kluyveromyces lactis,* and the synthetic oligonucleotides 5′ACGGCGCACAGACTCTCTATGCTGCGGCCTCCTTCGTCGTC-TTCATCCATGGAAAAGAGAAG3′ with 45 residues at the 5′-end encoding the C-terminal 15 amino acids of Rsc2, and 5′AGGGTA-ATGCGCAATGGGAAGATATTATGCTGCCATTGCTTTTACTAC-GACTCACTATAGGG3′ with 45 residues at the 5′-end representing the reverse complement of the sequence from 12 to 67 residues downstream of the *RSC2* open reading frame (not including the termination codon).

The PCR product was transformed into the protease-deficient yeast strain CB010 (MAT a pep4::HIS3 prbl::LEU2 prc::HISG can1 ade2 trpl ura3 his3 leu2–3, 112 cir^o GAL$^+$ RAF$^+$ SUC$^+$) by the lithium acetate method. Successful integration and expression of the TAP-tagged protein were confirmed by PCR analysis of genomic DNA and immunoblot detection of a polypeptide of appropriate size (102 kDa).

A typical protein preparation begins with 200 g of cells, harvested from 7.5 L of culture, grown overnight at 30° in YPD medium. Cells are suspended in 100 ml of 0.45 M HEPES, pH 7.6, 0.15 M potassium acetate, 3 mM EDTA, 3 mM DTT, 30% glycerol, 3× protease inhibitors, and broken in a bead beater as described.[10] Debris is removed by centrifugation at low speed (e.g., 7500 rpm in a Beckman JA-10 rotor for 30 min) and a clarified supernatant obtained by further centrifugation at 42,000 rpm in a Beckman Ti45 rotor for 30 min. The supernatant (9 g total protein, by Bradford determination) is applied to a 300 ml Bio-Rex 70 column equilibrated with of 0.02 M HEPES, pH 7.6, 0.1 M potassium acetate, 1 mM EDTA, 1 mM DTT, 20% glycerol, 1× protease inhibitors, washed with the same buffer containing 200 mM potassium acetate, and eluted with the same buffer containing 600 mM potassium acetate. The

[8] O. Puig, F. Caspary, G. Rigaut, B. Rutz, E. Bouveret, E. Bragado-Nilsson, M. Wilm, and B. Seraphin, *Methods* **24,** 218 (2001).
[9] G. Rigaut, A. Shevchenko, B. Rutz, M. Wilm, M. Mann, and B. Seraphin, *Nat. Biotechnol.* **17,** 1030 (1999).
[10] W. J. Feaver, O. Gileadi, and R. D. Kornberg, *J. Biol. Chem.* **266,** 19000 (1991).

eluate (protein-containing fractions, 220 ml) was diluted with the same buffer containing no potassium acetate to a conductivity of 100 μS at a dilution of 1:200 and purified by IgG and calmodulin affinity chromatography exactly as described[8] with the use of 300 μl of affinity beads.

Assay of RSC Activity

Electrophoretic Mobility Shift. RSC binds naked DNA and nucleosomes with comparable affinities and the resulting complexes can be revealed by gel electrophoresis[5] A system developed for the electrophoresis of nucleosomes, employing gels of high porosity and low ionic strength,[11] has proved effective. While gel shifted complexes of RSC with DNA and nucleosomes have comparable mobilities, only the RSC-nucleosome complex is further shifted by treatment with ATP (Protocol 1, Fig. 1).

DNase I Digestion. The first evidence for the perturbation of nucleosome structure by SWI/SNF complex came from an effect on the DNase I digestion pattern of the nucleosome. An alternating pattern of cleavage and protection, characteristic of the nucleosome, was replaced upon treatment with SWI/SNF complex and ATP by a more uniform pattern of digestion, similar to that of naked DNA. Treatment with RSC and ATP produces the same effect.[1] For assay by this approach, the nucleosomal DNA must be end-labeled and following digestion it must be analyzed by

PROTOCOL 1
Gel electrophoretic mobility shift assay

Sample number	1	2	3
^{32}P-labeled nucleosomes, 3.5 ng DNA/μl	1.8	1.8	1.8
1 mg/ml BSA	1.5	1.5	1.5
250 mM HEPES, pH 7.5, 50 mM MgCl$_2$	1.5	1.5	1.5
200 mM potassium acetate	3	2.7	2.7
10 mM ATP	–	1	–
20 mM ATPγS	–	–	0.5
RSC, 1 mg/ml in 375 mM potassium acetate	0.15	0.15	0.15
H$_2$O	12.1	11.4	11.9

Procedure: Nucleosomes are prepared as described (Lorch *et al.*[5] #2368). All volumes are in microliters. Incubate 20 min at 30°. Add 1.6 μl of 50% glycerol and subject to electrophoresis in a 3.2% polyacrylamide gel in 10 mM Tris, pH 7.5, 1 mM EDTA.

[11] A. J. Varshavsky, V. V. Bakayev, and G. P. Georgiev, *Nucleic Acids Res.* **3**, 477 (1976).

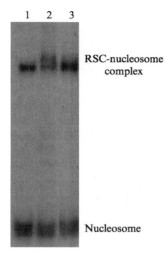

FIG. 1. Gel electrophoretic mobility shift assay (Protocol 1).

PROTOCOL 2
DNase I digestion assay

Sample number	1	2	3	4	5	6
^{32}P-end-labeled nucleosomes, 3.5 ng/μl	–	–	6	6	6	6
^{32}P-end-labeled DNA	5.3	5.3	–	–	–	–
17-residue oligonucleotide, 3.3 mg/ml	0.25	0.25	0.25	0.25	0.25	0.25
1 mg/ml BSA	2.5	2.5	2.5	2.5	2.5	2.5
250 mM HEPES, pH 7.5, 50 mM MgCl$_2$	2	2	2	2	2	2
200 mM potassium acetate	5.6	1.85	5.6	1.85	4.65	3.7
10 mM ATP	–	1	–	–	1	1
RSC, 1.4 mg/ml in 375 mM potassium acetate	–	2	–	2	0.5	1
H$_2$O	4.35	5.1	3.65	5.4	3.1	3.55
Incubate 20 min at 30°C						
DNase buffer	4	4	2.2	2.2	2.2	2.2
DNase I, 13 μg/ml in DNase buffer	1	1	2.8	2.8	2.8	2.8
Incubate rm. temp. for sec indicated	6	6	40	40	40	40

Procedure: Nucleosomes and DNA fragment were prepared as described. (Cairns *et al.*, 1996) The 17-residue oligonucleotide can be of any base sequence. DNase buffer is 25 mM HEPES, pH 7.5, 50 mM KCl, 5 mM MgCl$_2$, 3 mM CaCl$_2$, 0.2 mg/ml BSA, 20% glycerol. All volumes are in μl. DNase I digestion is stopped by the addition of 2.5 μl 100 mM EDTA. DNA is extracted and analyzed in a sequencing gel.

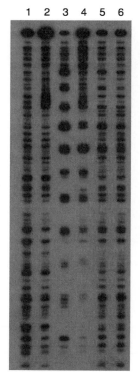

FIG. 2. DNase I digestion assay (Protocol 2).

PROTOCOL 3
Restriction enzyme digestion assay

Sample number	1	2	3	4
^{32}P-labeled nucleosomes, 3.6 ng/μl	1.7	1.7	1.7	1.7
1 mg/ml BSA	1.5	1.5	1.5	1.5
250 mM HEPES, pH 7.5, 50 mM MgCl$_2$	1.2	1.2	1.2	1.2
200 mM potassium acetate	2	1.7	1.7	1.7
10 mM ATP	–	1	1	1
RSC, 1 mg/ml in 375 mM potassium acetate	–	0.15	0.15	0.15
H$_2$O	13.6	12.8	12.8	12.8
Incubate 15 min at 30°				
Dra I, 40 units/μl	1.5	1.5	1.5	1.5
Incubate 30° for minutes indicated	60	20	40	60

Procedure: Nucleosomes are prepared as described (Lorch *et al.*[5] #2368). All volumes are in microliters. DNA is extracted and analyzed in a 7% polyacrylamide gel in TBE buffer.

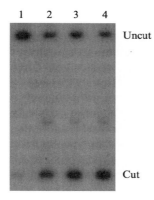

FIG. 3. Restriction enzyme digestion assay (Protocol 3).

PROTOCOL 4
Histone octamer transfer assay

Sample number	1	2	3	4	5
^{32}P-labeled DNA, 3.7 ng/μl	2	2	2	2	2
Nucleosome core particles, 0.19 mg DNA/ml	1	1	1	1	1
1 mg/ml BSA	1.1	1.1	1.1	1.1	1.1
250 mM HEPES, pH 7.5, 50 mM MgCl$_2$	0.9	0.9	0.9	0.9	0.9
10 mM ATP	0.75	0.75	0.75	0.75	0.75
RSC, 0.29 mg/ml in 500 mM potassium acetate	0.1	0.1	0.1	0.1	0.1
H$_2$O	7.9	7.9	7.9	7.9	7.9
Incubate 5 min at 30°					
Incubate at 30° for minutes indicated	5	10	20	40	60
Bacterial plasmid DNA, 3 mg/ml	0.4	0.4	0.4	0.4	0.4

Procedure: ^{32}P-labeled DNA, a 154-bp fragment, was prepared as described.[6] Nucleosome core particles were prepared as described.[12] All volumes are in microliters. Following reaction and incubation with unlabeled plasmid DNA, glycerol was added and electrophoresis was performed as in Protocol 1.

gel electrophoresis in single stranded form. A 17-residue oligonucleotide may be used as "carrier" in the DNase I digestion, since it is not expected to interfere with RSC-nucleosome interaction, as judged from the failure of such an oligonucleotide to stimulate RSC ATPase activity[1] (Protocol 2, Fig. 2).

Restriction Endonuclease Digestion. Perhaps the most convenient assay of RSC activity is based on restriction endonuclease digestion. Restriction sites normally protected in the nucleosome are exposed by treatment with

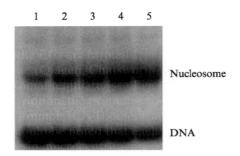

FIG. 4. Histone octamer transfer assay (Protocol 4).

RSC and ATP.[5] The restriction enzyme must be present simultaneously with the RSC and ATP, as exposure of sites by RSC is only transient. Following digestion, nucleosomal DNA is extracted and analyzed by non-denaturing gel electrophoresis (Protocol 3, Fig. 3).

Histone Octamer Transfer. RSC catalyzes the transfer of a histone octamer from a nucleosome core particle to a naked DNA fragment, in an ATP-dependent manner.[6] The DNA fragment is radiolabeled, and following the reaction, RSC is removed by the addition of excess unlabeled DNA, so the free, labeled nucleosome that is generated can be detected by gel electrophoresis in the same manner as for the electrophoretic mobility shift analysis described earlier (Protocol 4, Fig. 4).

---

[12] R. D. Kornberg, J. W. LaPointe, and Y. Lorch, *Methods Enzymol.* **170,** 3 (1989).

[20] DNA Translocation and Nucleosome Remodeling Assays by the RSC Chromatin Remodeling Complex

By JACQUI WITTMEYER, ANJANABHA SAHA, and BRAD CAIRNS

The packaging of DNA into nucleosomes compacts and organizes the genome and also restricts access to enzymes involved in DNA and RNA metabolism. Chromatin has a highly dynamic structure; regions of nucleosomal DNA must be accessible at certain times and inaccessible at others. This selected and controlled access to chromatin is tightly regulated by ATP-dependent chromatin remodeling complexes. Chromatin remodelers

METHODS IN ENZYMOLOGY, VOL. 377

regulate the access of factors, such as the transcription machinery, to nucleosomal DNA by altering the structure and positioning of nucleosomes. Though remodelers are multi-subunit complexes, they contain a single catalytic ATPase subunit that utilizes the energy of ATP hydrolysis to disrupt histone–DNA contacts within the nucleosome, which mobilizes the DNA relative to the surface of the histone octamer. The additional subunits appear to perform roles in the regulation and targeting of the complex.[1]

The majority of the highly conserved ATP-dependent chromatin remodeling complexes belong to either the SWI/SNF or the ISWI class; both types have been extensively characterized genetically and biochemically. RSC (remodels the structure of chromatin) is one of two complexes in *Saccharomyces cerevisiae* that belong to the highly conserved SWI/SNF class of ATP-dependent remodelers; the other is the founding member of the family, ySWI/SNF. The RSC complex is more abundant than SWI/SNF and, unlike SWI/SNF, the majority of its subunits are essential for cell viability. Both complexes behave nearly identically in biochemical assays.[1]

This chapter describes assays used to monitor the biochemical activity of the RSC remodeling complex. These assays include those for DNA-stimulated ATPase activity, DNA translocation (triple-helix displacement), nucleosome binding, and endonuclease access of nucleosomal DNA. Protocols for the TAP-purification of RSC and the reconstitution of nucleosomes using recombinant yeast histones and Nap1 (nucleosome assembly protein) are also described. Also, refer to the two other chapters (Volume 375, Part A, Chapters 2 and 6) in this series by Karolin Luger and Toshio Tsukiyama and their colleagues that include additional protocols for the reconstitution of nucleosomes.

Purification of TAP-RSC

TAP (tandem affinity) purifications utilize two affinity resins.[2] The TAP tag consists of a calmodulin-binding peptide, followed by a TEV protease cleavage site, proceeded by the IgG-binding portion of Protein A. The complex containing the TAP-tagged protein is purified from a whole cell lysate initially by binding to IgG-Sepharose, followed by cleavage with the highly specific TEV protease (which leaves the bulky Protein A portion of the tag behind), then bound to calmodulin beads in the presence of calcium, and eluted in the presence of EGTA.

[1] R. E. Kingston and G. J. Narlikar, *Genes Dev.* **13,** 2339 (1999).
[2] O. Puig, F. Caspary, G. Rigaut, B. Rutz, E. Bouveret, E. Bragado-Nilsson, M. Wilm, and B. Seraphin, *Methods* **24,** 218 (2001).

RSC is purified from *S. cerevisiae* cells expressing a carboxy-terminally TAP-tagged version of Rsc2 at the endogenous locus. Typically, 10 liters of cells are grown in $2\times$ YPD at $30°$ until the $OD_{600} = 6$, collected by centrifugation, then washed in a buffer containing $0.5\times$ PBS and 5% glycerol (v/v). The pellet is suspended in an equal volume of $2\times$ lysis buffer, where $1\times$ lysis buffer consists of 50 mM HEPES, pH 7.5; 250 mM potassium acetate, pH 7.5; 20% glycerol (v/v); 10 mM Na-EDTA; 0.5 mM DTT; and protease inhibitors. A $100\times$ protease inhibitor cocktail consisting of 0.03 mg/ml leupeptin, 0.14 mg/ml pepstatin, 0.02 mg/ml chymostatin, 8.5 mg/ml phenylmethanesulfonyl fluoride, 33 mg/ml benzamidine is solubolized in ethanol, then added to buffers at a final concentration of $1\times$. The cell suspension can be frozen in liquid nitrogen and stored at $-80°$.

Cells are lysed using 0.5 mm glass beads (Biospec #11079105) in a 330 ml bead beater (Biospec #1107900) using the following relative amounts of components: 100–200 ml of cell suspension (1:1) in lysis buffer, 120 ml of beads, then filling the chamber with $1\times$ lysis buffer. Once sealed, the chamber is surrounded in an ice/salt water bath. Cells are mechanically sheared by beating for 30 s pulses with 90 s rests in between for 60 cycles (2 h total), changing the ice/salt water as required.

Centrifuge the lysate at $15,000g$ for 10 min at $4°$, then increase the salt concentration of the supernatant to 400 mM potassium acetate (using a 3 M stock), assuming that the salt concentration of the lysate is 250 mM. Precipitate the nucleic acids by drop-wise addition of 10% polyethylenimine, pH 7.2 (v/v) to a final concentration of 0.1% while continuously stirring the lysate at $4°$. Centrifuge the sample at $200,000g$ in a Ti 45 rotor for 45 min at $4°$.

IgG-Sepharose Affinity Chromatography

All of the following steps can be done in batch format. Incubate the cleared lysate with IgG-Sepharose beads (Pharmacia; 1–2 ml beads/g of total protein, that have been prewashed in lysis buffer) for 3–4 h at $4°$ with rotation. Wash the beads 4 times with 10 mM Tris–HCl, pH 8.0; 400 mM potassium acetate; 0.1% NP-40 (v/v); 10% glycerol (v/v); 0.5 mM Na-EDTA; 0.5 mM DTT; and protease inhibitors; then 4 times with TEV cleavage buffer containing 10 mM Tris–HCl, pH 8.0; 200 mM potassium acetate; 0.1% NP-40; 10% glycerol; 0.5 mM Na-EDTA; and 0.5 mM DTT. **It is crucial to eliminate protease inhibitors at this stage, as they will inhibit the TEV protease that cleaves the TAP-tagged protein from the resin.** Mix the beads with an equal volume of TEV cleavage buffer, add TEV protease (Invitrogen, 100–150 U/ml of beads) and incubate at $4°$ for 1 h, then at $15°$ for 1 h, with rotation. Collect the TEV eluate, and then rinse the beads with an equal volume of TEV cleavage buffer, pooling the supernatants.

Calmodulin Affinity Chromatography

To the TEV eluate, add 1 M $CaCl_2$ to achieve 3 mM final, then add 3 volumes of CBP buffer containing 25 mM Tris–acetate, pH 7.5; 250 mM potassium acetate; 1 mM magnesium acetate; 1 mM imidazole; 2 mM $CaCl_2$; 10% glycerol (v/v); 0.1% NP-40 (v/v); 0.5 mM DTT; and protease inhibitors. Incubate the sample with CBP beads (Stratagene; 0.7 ml beads/g of total protein in the original cleared lysate, that have been pre-equilibrated in CBP buffer) for 3 h at 4°, with rotation. Wash the beads 4 times in 25 mM Tris–acetate, pH 7.5; 400 mM potassium acetate; 1 mM magnesium acetate; 1 mM imidazole; 2 mM $CaCl_2$; 10% glycerol (v/v); 0.1% NP-40; 0.5 mM DTT; and protease inhibitors. The initial washes can be done in batch format, but the final wash and elution steps should be done in column format to achieve the highest purity and concentration.

Wash the column with several column volumes of wash buffer as listed above, then with 25 mM Tris–HCl, pH 7.5; 150 mM NaCl; 2 mM $CaCl_2$; 10% glycerol; 0.1% NP-40; 0.5 mM DTT; and protease inhibitors. Elute the bound proteins with 25 mM Tris–HCl, pH 7.5; 150 mM NaCl; 2 mM EGTA; 10% glycerol; 0.1% NP-40; 0.5 mM DTT; and protease inhibitors, collecting fractions equivalent to ~0.3 column volumes.

If the concentration of the complex is dilute, supplement with BSA (Sigma, protease-free) to 1 mg/ml. Aliquot into convenient quantities, freeze in liquid nitrogen, and store at $-70°$. RSC can undergo several freeze-thaw cycles (~3–5), without significant loss in activity, but repeated freeze-thaw cycles should be avoided. Typical yields are ~0.1–0.2 mg of RSC per liter of cells.

Remodelers Couple DNA-Stimulated ATPase Activity to DNA Translocation

All chromatin remodeling complexes contain an ATPase subunit whose activity is crucial for the biochemical alterations in nucleosome structure. These subunits are DNA-stimulated ATPases and share additional sequence identity with DEAD box helicases[3–5]; but the remodelers lack helicase activity, as assayed by displacement of a radiolabeled oligonucleotide

[3] B. C. Laurent, X. Yang, and M. Carlson, *Mol. Cell. Biol.* **12,** 1893 (1992).

[4] J. A. Eisen, K. S. Sweder, and P. C. Hanawalt, *Nucleic Acids Res.* **23,** 2715 (1995).

[5] T. M. Lohman and K. P. Bjornson, *Annu. Rev. Biochem.* **65,** 169 (1996).

within a DNA duplex.[6] However, DNA helicases are translocases that couple the energy of ATP hydrolysis to translocate, or track, along the DNA.[7–10] Translocation of remodelers on nucleosomal DNA could provide the mechanical force required to disrupt the histone–DNA contacts observed in remodeling.[11,12] The following assays monitor the DNA-stimulated ATPase and DNA translocation activities of the RSC complex and its catalytic subunit, Sth1.

ATPase Assay

This assay provides a fast, sensitive, colorimetric assay for determining many of the parameters of the DNA-stimulated ATPase activity. ATP hydrolysis releases ADP and a free phosphate ion. The phosphate ion interacts with an ammonium molybdate/malachite green complex and changes its absorbance properties, providing a quantitative colorimetric assay for free phosphate concentration.[13] All of the samples tested must be phosphate-free, because the assay is sensitive in the nanomolar range.

Reactions (25–50 μl) are performed in a buffer containing 10 mM HEPES, pH 7.3; 20 mM potassium acetate; 5 mM $MgCl_2$; 0.5 mM DTT; 0.1 mg/ml BSA; 5% glycerol (v/v); 1 mM ATP; and protease inhibitors. DNA is added at a concentration of ~30 μM nucleotide substrate (10 ng/μl of pBluescript, Stratagene) to elicit V_{max}. Note that all DNA concentrations are given in micromolar nucleotide to normalize for length and compositional differences. Reactions are incubated at 30° for 30 min.

All of the following steps are done at room temperature (RT). To measure the phosphate release, add 0.8 ml MGAM solution (see below) and incubate 1 min, then add 0.1 ml of 34% sodium citrate (tribasic dihydrate), mix and incubate for 10 min, then take readings at 650 nm. A standard curve is prepared with a sodium phosphate solution of known concentration. Reaction parameters such as enzyme concentration and time should be adjusted so that readings fall within the linear range (2–15 nM of

[6] J. Cote, J. Quinn, J. L. Workman, and C. L. Peterson, *Science* **265,** 53 (1994).

[7] M. S. Dillingham, D. B. Wigley, and M. R. Webb, *Biochemistry* **39,** 205 (2000).

[8] M. S. Dillingham, D. B. Wigley, and M. R. Webb, *Biochemistry* **41,** 643 (2002).

[9] P. Soultanas, M. S. Dillingham, P. Wiley, M. R. Webb, and D. B. Wigley, *EMBO J.* **19,** 3799 (2000).

[10] S. S. Velankar, P. Soultanas, M. S. Dillingham, H. S. Subramanya, and D. B. Wigley, *Cell* **97,** 75 (1999).

[11] A. Saha, J. Wittmeyer, and B. R. Cairns, *Genes Dev.* **16,** 2120 (2002).

[12] I. Whitehouse, C. Stockdale, A. Flaus, M. D. Szczelkun, and T. Owen-Hughes, *Mol. Cell. Biol.* **23,** 1935 (2003).

[13] B. R. Cairns, Y. J. Kim, M. H. Sayre, B. C. Laurent, and R. D. Kornberg, *Proc. Natl. Acad. Sci. USA* **91,** 1950 (1994).

phosphate). Note that after the absorbance initially plateaus at an accurate value, it then increases slowly with time, so the addition of reagents and subsequent readings should be coordinated and staggered appropriately.

Preparation of MGAM solution: make a solution of 0.045% malachite green-HCl in 0.1 N HCl and stir overnight (stable indefinitely at RT), and a separate solution of 4.2% ammonium molybdate in 4 N HCl. Mix 3 volumes of the malachite green solution with 1 volume of the ammonium molybdate solution and stir for 20 min at RT, filter through a 0.2 μm polycarbonate phosphate-free membrane (Nucleopore Inc., #111106). The solution should be clear amber and is stable for 2 weeks stored in the dark at 4°.

The ATPase activity of RSC is stimulated equally well by either single-stranded or double-stranded DNA.[11] There is a triphasic nature to the length dependence of the DNA-stimulated ATPase activity of RSC: (1) DNA fragments less than 15–20 bp do not stimulate the ATPase, representing the minimal fragment length of DNA required for binding by the remodeler, (2) fragments ranging from ∼20 to 80 bp stimulate the ATPase proportionate to their length, and (3) fragments 80 bp or longer stimulate the ATPase maximally, representing the processive length of DNA that the remodeler can translocate along before it dissociates (see Fig. 1). The isolated ATPase subunit, Sth1, behaves similarly to the RSC complex.[11]

Triple-Helix Displacement Assay

Triple-helix strand displacement has been used to demonstrate translocation for both the endonuclease *Eco*AI and the SV40 T-antigen.[14,15] The assay relies on the propensity for a homopurine–homopyrimidine repeat to form a three-strand triple-helix with a complementary homopyrimidine oligonucleotide at pH 5.5.[16] When shifted to neutral pH, the triple-helix remains stably bound. However, if the third strand is displaced, it will not reform a triple-helix at neutral pH. In the triple-helix, the homopyrimidine oligonucleotide occupies the major groove of the mirror repeat, with each base of the oligonucleotide forming a Hoogsteen base pair with the Watson–Crick base pairs.[17] The basis for the translocation assay is the displacement of the homopyrimidine oligonucleotide, which is bound to

[14] V. Kopel, A. Pozner, N. Baran, and H. Manor, *Nucleic Acids Res.* **24,** 330 (1996).

[15] K. Firman and M. D. Szczelkun, *EMBO J.* **19,** 2094 (2000).

[16] H. Htun and J. E. Dahlberg, *Science* **243,** 1571 (1989).

[17] M. J. van Dongen, J. F. Doreleijers, G. A. van der Marel, J. H. van Boom, C. W. Hilbers, and S. S. Wijmenga, *Nat. Struct. Biol.* **6,** 854 (1999).

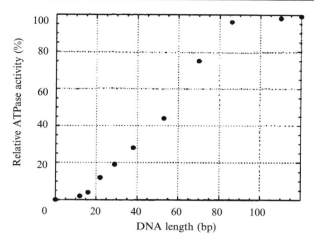

FIG. 1. Length dependence of the DNA-stimulated ATPase activity of RSC. Maximal velocity of RSC ATPase activity with varying length of double-stranded DNA at 30° for 30 min. Each DNA was tested at a uniform nucleotide concentration of 30 μM. All values are reported relative to the V_{max} elicited by plasmid DNA (pBluescript, 3 kb).

the duplex by weaker Hoogsteen base pairs, as the translocase proceeds through the triple-helix.

A triple-helix of 40 bp, centrally located on a 190 bp double-stranded DNA, was formed using a 40 base homopyrimidine oligonucleotide. This triple-helix is highly stable when shifted to pH 7–8, and is released efficiently when heated briefly at 90° (see Fig. 2, lanes 1 and 2).

Formation of the Triple-Helix Substrate

A DNA fragment containing a $d(GA)_{20} \cdot d(TC)_{20}$ tract was cloned into pBluescript, at the *BamHI/HindIII* sites. Duplex DNA (190 bp) is prepared by PCR amplification using this modified plasmid, containing the centrally located 40-bp $d(GA) \cdot d(TC)$ tract as template, and T3 and T7 primers. The DNA products are organic extracted [using an equal volume of phenol:chloroform:isoamyl alcohol (25:24:1), then an equal volume of chloroform], ethanol precipitated and gel purified. The third strand, consisting of a 40 nucleotide single-stranded homopyrimidine repeat $(dTC)_{20}$, is end-labeled with $[\gamma^{-32}P]$-ATP and T4 kinase.[18] Triple-helices are formed by incubating the radiolabeled $(dTC)_{20}$ with a twofold molar excess of the

---

[18] F. M. B. Ausubel, R. E. Kingston, D. D. Moore, J. G. Seidman, J. A. Smith, and K. Struhl, eds., "Current Protocols in Molecular Biology." Wiley, Boston, 1995.

Triple helix displacement

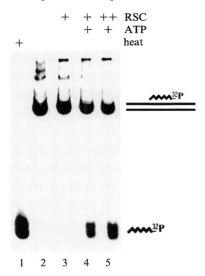

Fig. 2. Triple-helix displacement assay. Substrates (30 ng) consisting of a 40-base triple helical region centrally located on a 190 bp duplex DNA were prepared and treated with 8 nM RSC and 1 mM ATP, as indicated, for 30 min at 30°. The products were separated on a 15% native acrylamide gel and autoradiographed.

190 bp dsDNA in a buffer containing 33 mM Tris–acetate, pH 5.5; 66 mM potassium acetate; 100 mM NaCl; 10 mM MgCl$_2$; and 0.4 mM spermine.[14]

The double-stranded DNA for the blunted triple-helix (40 bp) is prepared by hybridizing complimentary oligonucleotides, containing the (TC)$_{20}$ repeats and a 14 bp GC clamp at one end (which is required for proper positioning) with an AluI restriction site in between. Following hybridization, the GC clamp is cleaved off by digestion with AluI, and the (TC)$_{20}$ repeat double-stranded DNA is purified using a native acrylamide gel.[19] The third strand is then annealed as described for the extended triple-helix (above).

[19] J. Sambrook, E. F. Frítsch, and T. Maniatis, eds., "Molecular Cloning: A Laboratory Manual." Cold Spring Harbor Laboratory Press, Cold Spring Harbor, New York, 1989.

Triple-Helix Displacement Assay

Triple-helix displacement reactions (25 μl) contain 30 ng triple-helix DNA and 200 ng RSC (8 nM) in a buffer containing 36 mM Tris–acetate, pH 7.3; 20 mM potassium acetate; 8 mM MgCl$_2$, 5% glycerol (v/v), 0.5 mM DTT, and 1 mM ATP where indicated, for 30–60 min at 30°. Reactions are quenched with 0.5% SDS, 15 mM Na-EDTA, 3% glycerol (v/v), 2 mM Tris–HCl, pH 8.0; and 25 ng unlabeled (dTC)$_{20}$ (to prevent reannealing of the displaced oligonucleotide during subsequent electrophoresis). The products are resolved in a 15% native acrylamide gel (29:1 acryl:bis) in 40 mM Tris–acetate, pH 5.5; 5 mM sodium acetate; and 1 mM MgCl$_2$ at 200 V for 12 h at 4° (until the bromphenol blue migrates ~14 cm). The gels are dried and analyzed by phosphoimaging (Molecular Dynamics, Image Quant).

Association of the triple-helix-forming oligonucleotide with the DNA duplex retards its migration in a native acrylamide gel (see Fig. 2, lanes 1 and 2). RSC displaces the triple-helix in proportion to its concentration, and requires the presence of a hydrolyzable form of ATP (see Fig. 2, lanes 4 and 5).[11] The triple-helix is not displaced if a blunted triple-helix, one that has no flanking DNA duplex on either side, is used as substrate, suggesting that the displacement of the triple-helix is not simply a result of the remodeler binding to the DNA triplex.[11] The remodeler and ATP can displace the triple-helix when it is located either centrally as shown in Fig. 2, or at either end of the DNA duplex.[11] These experiments suggest that the remodeler translocates along one strand of the DNA duplex to physically displace the triple-helix. A detailed model for how DNA translocation is coupled to the movement of DNA on the surface of nucleosomes is described elsewhere,[11,12] and is discussed briefly in the concluding remarks.

Reconstitution of Yeast Nucleosomes Using Nap1

This section describes the reconstitution of nucleosomes using recombinant yeast histones, yeast Nap1, and purified 5S DNA (Sea Urchin). This protocol relies on information in the large-scale nucleosome reconstitution protocol of Luger and Richmond,[20] modifications by Marnie Gelbart and Toshio Tsukiyama, as well as those of our own, to arrive at a protocol that provides high purity and yield at a scale that is appropriate for biochemical studies.

In brief, the yeast histones are individually overexpressed in a bacterial strain that contains tRNA codons for arginine, isoleucine, and leucine that

---

[20] K. Luger, T. J. Rechsteiner, and T. J. Richmond, *Methods Enzymol.* **304,** 3 (1999).

are rare in prokaryotes, but common in eukaryotic ORFs. This eliminates the cumbersome task of modifying the codon usage for each ORF to optimize for expression in bacteria. The subunits are individually purified from inclusion bodies using a tandem ion exchange chromatography step while eliminating the gel filtration step used in the original protocol.[20] Octamers are reconstituted by simultaneously refolding all 4 subunits in equimolar ratios, similar to the method of Luger and Richmond,[20] but the nucleosomes are reconstituted using yNap1, instead of salt-dialysis, and purified using gradient sedimentation. This method of nucleosome reconstitution works well for small quantities (10–100 μg), and eliminates the requirement to reposition the nucleosome particle post-assembly.

Bacterial Overexpression of Yeast Histones

S. cerevisiae histones H3, H4, H2A, and H2B were PCR amplified and cloned into the *Nde*I-*Bam*HI sites of pET11a (Stratagene) using restriction enzyme sites that were engineered into the PCR primers. The integrity of the ORFs were confirmed by DNA sequencing.

Expression constructs are transformed into BL21-Codonplus RIL cells (Stratagene) that carry extra copies of the tRNA codons for arg-, ile-, and leu-. These are maintained in the BL21 cells by selection in chloramphenicol (25 μg/ml), and once transformed, the expression constructs are kept under selection in ampicillin (100 μg/ml). For the best results, freshly transform the expression constructs prior to each large-scale growth and induction.

Fresh transformants, or a single colony from the streak of a glycerol stock, are used to inoculate a 10 ml culture of LB containing ampicillin and chloramphenicol and grown overnight at 37°. This culture is used to inoculate 1 liter of culture, which is grown to an OD_{600} of 0.4–0.5 and induced with 1 mM IPTG for 3–4 h. Cells are harvested at 5000g for 5 min, and then suspend the pellets in 25 ml/liter culture of TW buffer (50 mM Tris–HCl, pH 7.5; 100 mM NaCl; 1 mM Na-EDTA; 5 mM 2-mercaptoethanol; protease inhibitors, and 1% Triton X-100 [w/v]). Freeze the suspension in liquid nitrogen and store at $-70°$.

The yield of histone H4 (8–12 mg/liter) is generally several-fold lower than that of histones H3, H2A, and H2B (30–60 mg/liter). The histones are insoluble and found in inclusion bodies.

Inclusion Body Purification

Thaw the cell suspensions and sonicate them in a beaker on ice for 30 s on 50% duty cycle for a total of 6 times, pausing at least 1 min in between each round of sonication (Misonix probe, setting = 7–8). Pellet the inclusion bodies by centrifuging at 20,000g for 20 min at 4°. Rinse the pellets with 25 ml of TW buffer by pipeting up and down while digging at the pellets with a 25 ml pipet and pipet aid (the material in the pellets does not fully go into solution, but try to get them as homogeneous as possible). Repeat the centrifugation step, discarding the supernatant. Repeat the suspension of the pellets and centrifugation steps twice more with wash buffer, TW minus the triton X-100 (25 ml/liter culture). Freeze the drained pellets and store at −70°.

Purification of the Histone Subunits

The gel filtration step described in the original protocol[20] is not necessary. For all of the subsequent steps, buffers containing guanidinium hydrochloride, urea, or reductants should be made fresh and used within a day. Freshly made urea stocks are deionized with 25 g/liter of AG 501-X8 resin (BioRad) while stirring for 30 min at RT, then filtered. It is not necessary to add protease inhibitors during the subsequent purification steps.

Thaw the inclusion body pellets and soak them in 0.2 ml dimethyl sulfoxide (per 1 liter culture). Mince the pellet with a spatula, then add 6.5 ml unfolding buffer (20 mM Tris–HCl, pH 7.5; 7 M guanidinium hydrochloride, 10 mM DTT), and rotate for 1 h at RT. Remove the cell debris by centrifuging at 20,000g for 20 min. Transfer the supernatant to a new tube, and rinse the pellet with 1.5 ml unfolding buffer, then centrifuge, and combine the two supernatants. Dialyze the solubolized histones against 1–2 liter of urea dialysis buffer (10 mM Tris–HCl, pH 8; 7 M urea; 1 mM Na-EDTA; 5 mM 2-mercaptoethanol containing 100 mM NaCl for histones H2A and H2B and 200 mM NaCl for histones H3 and H4) using 6000–8000 MWCO tubing, for several hours with 1–2 buffer changes at RT, or at 4° if a step is done overnight.

While the sample is dialyzing, equilibrate a Q- and SP-Sepharose column (5 ml HiTrap ion exchange columns, Pharmacia) arranged in tandem, with the Q-column first in line. Column buffers contain 10 mM Tris–HCl, pH 8; 7 M urea; 1 mM Na-EDTA; 1 mM DTT (buffer A), or with the addition of 1 M NaCl (buffer B). All chromatography steps are done using a Pharmacia LC (Gradi-Frac) or FPLC system at 4°. The columns are finally equilibrated in 10% buffer A for histones H2A and H2B or 20% buffer A for histones H3 and H4 at a flow rate of 0.6 ml/min.

Remove the dialysate and check the conductivity. If the conductance of the sample is higher than that of the starting chromatography buffer, either dialyze it longer, or dilute the sample with chromatography buffer A to equal the conductance of the starting chromatography buffer. Filter the sample using a 0.45 μm cellulose acetate membrane. If it is too viscous to filter, sonicate it to shear the residual DNA, then filter it. The sample is loaded at a flow rate of 0.5 ml/min, washed with several column volumes of starting buffer (10% B for H2A/H2B, 20% B for H3/H4) until the UV trace returns to baseline.

Before eluting the protein that is bound to the SP-column, remove the Q-column from the system. Elute the SP-bound proteins with a salt gradient at 0.5–1.0 ml/min. The salt gradients are as follows: for H2A/H2B, load and wash at 10% B, then elute with a gradient of 10–40% B over 20 column volumes; and for H3/H4, load and wash at 20% B, then elute with a gradient of 20–50% B over 15 column volumes. Save the flow through and collect 1.5–2.0 ml fractions. Analyze the peak protein fractions (UV trace) by 15% SDS-PAGE and CB staining. Pool the peak fractions from the purification of the individual subunits and dialyze it against 2 liters of 10 mM Tris–HCl, pH 8 and 5 mM 2-mercaptoethanol for several hours to overnight, with two additional buffer changes.

Save an aliquot of the dialysate for quantitation. Lyophilize the remaining histones in conveniently sized aliquots (1–4 mg each) and store them at $-20°$. To quantitate the histones, take a reading at A_{280}, using the final dialysis buffer as a blank. The calculated extinction co-efficients (mg/ml) for the *S. cerevisiae* histones are as follows: H3 = 0.17, H2B = 0.45, H2A = 0.28, and H4 = 0.46, according to Gill and von Hippel[21] and based on the calculated molecular weights of: H3 = 15,225, H2B = 14,121, H2A = 13,858, and H4 = 11,237. The concentrations are then verified by SDS-PAGE followed by CB staining, comparing the band intensities to those of known standards.

Reconstitution of the Histone Octamer

Suspend the histones (2–4 mg of each) in unfolding buffer to 2 mg/ml and rotate for 2–3 h at RT. Mix the 4 histones in equimolar ratios, using a slight excess (10–15%) of histones H2A/H2B relative to H3/H4, and adjust the volume to 1 mg/ml with unfolding buffer. (Since the octamer does not separate thoroughly from the tetramer in the subsequent gel filtration step, the excess dimer subunits are added to drive as much tetramer into octamer as possible.)

Dialyze the sample against 2 liters of refolding buffer (20 mM Tris–HCl, pH 7.5; 2 M NaCl; 1 mM Na-EDTA; 5 mM 2-mercaptoethanol) using 3500 MWCO tubing (very important) at $4°$ for a total of 4 buffer changes, allowing the 2nd and 4th steps to proceed overnight. Concentrate the sample to 0.5 ml using a 10,000 MWCO Centricon-plus 20 (Amicon) at

---

[21] S. C. Gill and P. H. von Hippel, *Anal. Biochem.* **182,** 319 (1989).

$4°$. Filter the sample using a 0.45 μm cellulose acetate membrane, then load it at 0.2 ml/min onto a Superdex 200 HR (10/30) column (Pharmacia) that has been preequilibrated with refolding buffer, collect 0.3 ml fractions. Assay the peak fractions (A_{280}) (see Fig. 3A) by 10–20% SDS-PAGE and CB staining.

Pool the fractions containing the octamer (equal CB stain intensity of all 4 histones, Fig. 3B, left panel). Precoat a 10,000 MWCO Centricon concentrator (Amicon) with refolding buffer containing 0.5 mg/ml protease-free BSA (Sigma), rinse with refolding buffer, then use these to concentrate the octamers to ∼0.2–0.3 ml. Collect the octamers and adjust the buffer to 50% glycerol while keeping the NaCl at 2 M by adding an appropriate volume (1.85× sample volume) of ice-cold 76% glycerol (w/v) in 2 M NaCl and 20 mM Tris pH 7.5, mix by gently pipetting up and down, store at $-20°$ for extended periods (months to years). Alternatively, the octamers can be left in the refolding buffer (without the adjustment to 50% glycerol), aliquoted, frozen in liquid nitrogen, and stored at $-70°$

Purification of Nap1

Histidine-tagged Nap1 is overexpressed from pET28-yNap1 (from Toshio Tsukiyama) in BL21-Codonplus-RIL cells (1 liter) induced with 1 mM IPTG at an OD_{600} = 0.4 for 3–4 h at $37°$. Cells are lysed by sonication (20 s, 50% duty-cycle, 7–8 times, with 1 min rests in between) in a beaker on ice in 10 ml lysis buffer (25 mM HEPES, pH 7.6; 0.3 M KCl; 2 mM MgCl$_2$; 1 mM 2-mercaptoethanol; 0.02% NP-40 [w/v]; 10% glycerol [w/v]; and protease inhibitors). Spin the lysate at 20,000g for 20 min at $4°$.

All of the following chromatography steps are done at $4°$. Adjust the lysate to 20 mM imidazole, using a 1 M stock, and incubate with 5 ml of Ni^{2+}-agarose (Qiagen) that has been preequilibrated 3 times with Ni^{2+} buffer (50 mM sodium phosphate, pH 8; 100 mM KCl; 2 mM MgCl$_2$; 10% glycerol [w/v]; 0.1% NP-40; 1 mM 2-mercaptoethanol; and protease inhibitors), and rotate gently for 2 h at $4°$. Collect the resin by centrifugation, and wash it twice in batch with Ni^{2+} buffer plus 20 mM imidazole and 0.3 M KCl, then pour a column and continue washing with this buffer until no protein elutes (monitored using Bradford reagent). Wash with Ni^{2+} buffer containing 50 mM imidazole (some Nap1 protein elutes here, but the vast majority remains bound) until at or near baseline for protein elution, then elute the remaining Nap1 with Ni^{2+} buffer plus 250 mM imidazole. Assay fractions by SDS-PAGE and CB staining. Pool the Nap1 fractions from the 250 mM imidazole elution.

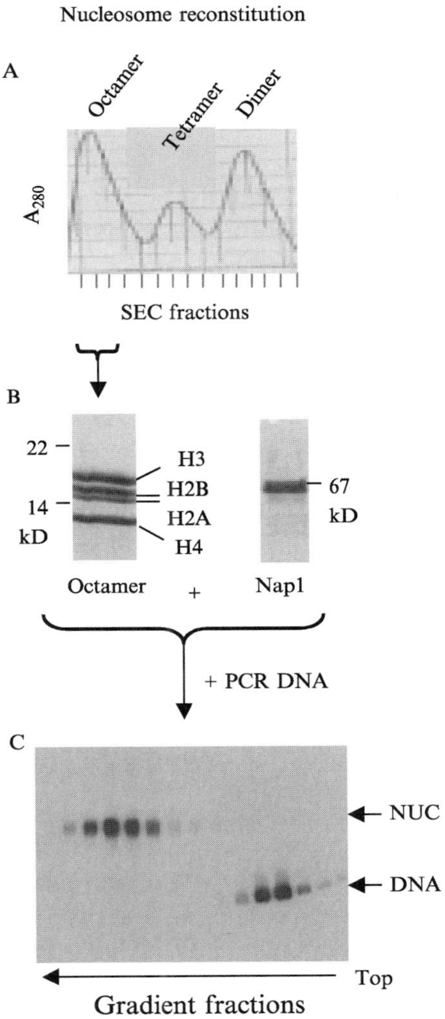

Nucleosome reconstitution

FIG. 3. Reconstitution of yeast nucleosomes using Nap1. (A) UV trace from the gel filtration purification of the octamer, (B) left panel, peak octamer fractions in a CB stained 10–20% SDS-PAGE and, right panel, purified Nap1 in a CB stained 7.5% SDS-PAGE, and (C) sucrose gradient fractions (10 μl each) run on a 4% native acrylamide gel and autoradiographed.

Preequilibrate a 5 ml HiTrap Q-Sepharose column with buffer (20 mM Tris, pH 8; 0.1 M KCl; 0.5 mM Na-EDTA; 1 mM DTT; 10% glycerol [w/v] and protease inhibitors) at 0.5 ml/min. Load the sample, then wash with the same buffer for several column volumes, until the UV trace is at baseline. Run a gradient from 0.1 to 0.6 M KCl over 60 ml, Nap1 elutes at 0.4–0.5 M KCl. Assay fractions by SDS-PAGE and CB staining. Pool the Nap1 fractions (see Fig. 3B, right panel) and dialyze into 25 mM HEPES, pH 7.6; 50 mM KCl; 2 mM MgCl$_2$; 10% glycerol (w/v); 1 mM 2-mercaptoethanol and protease inhibitors. Quantitate by A$_{280}$ ($\varepsilon_{1\ mg/ml} = 0.63$)[21] and by comparing CB stained bands in SDS-PAGE to known standards. Yields are generally 20–30 mg/liter. Aliquot and freeze in liquid nitrogen and store at $-70°$. Can be frozen/thawed multiple times without loss of activity.

Reconstitution of Nucleosomes Using Nap1

The DNA is prepared by PCR amplification using a plasmid containing 1 copy of the 202 bp Sea Urchin 5S sequence as template, and various primers containing additional sequence to achieve a given length. Alternatively, the 5S sequence can be restriction enzyme digested out of a plasmid (pIC) containing 5 copies of this sequence. Products are gel purified using 2.5% Metaphor agarose (ISC) in TBE and extracted using QIA-quick columns (Qiagen). The ends of the DNA can either be 5'-end labeled with [γ^{32}P]-ATP and T4 kinase or cut with a restriction enzyme site near the end and filled in with an appropriate [α^{32}P]-dNTP and Klenow, or left unlabeled. If the DNA is enzymatically manipulated, organic extract and ethanol precipitate the DNA. Final suspension is in 10 mM Tris–HCl, pH 8. Quantitate at A$_{260}$.

Mix together, in the following order: (1) 4× assembly buffer to achieve 1× (10 mM HEPES, pH 7.6, 50 mM KCl, 5 mM MgCl$_2$, 5–10% glycerol [w/v], 2 mg/ml BSA [protease-free, Sigma], and protease inhibitors), (2) Nap1, (3) histone octamer, and (4) DNA, in the amounts described below. This reconstitution works well for quantities between 10 and 120 μg, and reaction volumes between 50 and 400 μl. (Note that the octamer is in 2 M NaCl and 50% glycerol, so that if the volume of octamer in the reaction is a substantial fraction of the total, the reaction should be diluted with assembly buffer to bring the total salt below 0.4 M and the total glycerol to 10%.)

Use an amount of histone octamer equivalent to 1- to 1.5-fold molar excess relative to the DNA, and an amount Nap1 (48 kD) in 1 to 1.5-fold molar excess relative to the octamer (110 kD). Note that any unincorporated dimers/tetramers and Nap1 will be removed in subsequent purification steps. Incubate the assembly at 30° for 4 h.

Small-scale pilot assemblies ($\sim$1 μg in 5–10 μl) can initially be done to determine the optimal ratio of DNA : octamer : Nap1 for the large-scale reactions. These can be analyzed directly post-assembly in a 4% native gel (see below) and visualized by ethidium bromide staining. The larger assemblies can then be modified accordingly.

Removal of the Nap1

The removal of the Nap1 post-assembly is crucial for the stability of the nucleosomes in subsequent reactions. Removal of histidine-tagged Nap1 is achieved via three successive depletions using Ni^{2+}-agarose resin. First, pre-equilibrate a batch Ni^{2+}-agarose resin ($\sim$75–150 μl/assembly) by washing it 3 times with buffer containing 10 mM HEPES, pH 7.6; 300 mM NaCl; 5 mM $MgCl_2$, 0.5 mg/ml BSA; 0.1% NP-40 (w/v), and protease inhibitors. Adjust the assembly reactions to 20 mM imidazole, using a 1 M stock, and 0.1% NP-40 (w/v), using a 10% stock, then add this to one third of the resin (use an excess of resin relative to the amount of Nap1 in the reaction, generally 20–40 μl resin/reaction) that has been pre-equilibrated and drained, rotate gently 1–1.5 h at RT. Pellet the resin, remove the supernatant, rinse the beads with 20 μl of Ni^{2+} wash buffer containing 20 mM imidazole and 0.1% NP-40 for 10 min, pellet, and combine the supernatants. Repeat these steps twice more with the remaining Ni^{2+}-agarose, combining the supernatants from each reaction, making certain that all of the resin is completely removed in the final sample.

Alternatively, Nap1 can be efficiently pre-coupled to cyanogen bromide-activated agarose (Sigma) according to the manufacturer's instructions, and used as a suspension for nucleosome assemblies (with rotation during the 4 h incubation at 30°). The efficiency of Nap1-coupled agarose in assembling nucleosomes decreases slightly (within 2-fold) relative to Nap1 in solution, but more Nap1-agarose can be used per reaction, then simply centrifuged out post-assembly. Store the Nap1-agarose in a suitable buffer containing 0.02% sodium azide at 4°.

Purification of the Nucleosomes

Purify the nucleosomes from unincorporated input by gradient sedimentation. Make a 12 ml 10–30% sucrose gradient containing 25 mM HEPES, pH 7.6; 0.5 mM Na-EDTA, 0.5 mg/ml BSA (protease-free), and protease inhibitors using a gradient maker, peristaltic pump, and 0.1 ml micropipets (VWR #53432–921) to attach to the tubing and deliver the solution (1–2 ml/min) to the bottom of a centrifuge tube (Beckman #331372, polyallomer). Carefully load the Nap1-cleared sample ($<$0.45 ml) onto the top of the prechilled gradient by slowly pipetting it along the side of the tube. The total volume of the solution in the tube should be 2–3 mm from the top of the tube.

Centrifuge in an SW 41 rotor at 41,000 rpm at 4° for 21–24 h. Collect fractions (~5 drops; 350–400 μl) from the gradient by puncturing the bottom of the tube with a fraction collector set-up with a needle and tube holder support. Load 10 μl of the peak fractions (generally #8–28; nucleosomes run ~50–75% into the gradient) in a 4% native acrylamide gel (15:1 acryl:bis) containing 0.5× TBE,[19] 5% glycerol (w/v), and 2 mM MgCl$_2$ while running the gel at 60 V in 0.5× TBE, then increase the voltage to 180 V for 1.5 h, or until the bromphenol blue has migrated ~10 cm.

Either stain the gel with ethidium bromide (sensitive to 10–20 ng/band), or dry and autoradiograph it, to visualize the fractions containing nucleosomes and free DNA. Generally, 65–95% of the DNA is assembled into nucleosomes (see Fig. 3C).

Pool the fractions containing nucleosomes. If necessary, they can be concentrated up to 10-fold or more using a Centricon (10,000 MWCO) concentrator that has been precoated/concentrated with 20% sucrose gradient buffer. Quantitate the nucleosomes by counting an aliquot in a scintillation counter if the DNA is radiolabeled. If the DNA is unlabeled, run several quantities of the sample in a 4% native acrylamide gel (as described above) and compare the ethidium bromide stain intensity of the bands to known concentrations of radiolabeled nucleosomes and/or DNA (note that ethidium bromide staining of nucleosomal DNA is slightly reduced relative to that of free DNA, but is within 2-fold). Freeze in liquid nitrogen and store at −70° for extended lengths of time. Nucleosomes are stable with repeated freeze-thaw cycles, so there is no need to aliquot them. Always thaw and maintain them on ice, freeze them in liquid nitrogen and store at −70°.

Nucleosome Binding Assay

In the absence of ATP, RSC binds equally well to either DNA or nucleosomes. In the presence of ATP, the affinity of RSC for the nucleosome increases (see Fig. 4, lanes 2 and 3), whereas its affinity for DNA does not change.[22] The increased affinity of the remodeler for the nucleosome requires the presence of a hydrolyzable form of ATP, suggesting a conformation change in either the nucleosome or the remodeler, or both.

[22] Y. Lorch, B. R. Cairns, M. Zhang, and R. D. Kornberg, *Cell* **94**, 29 (1998).

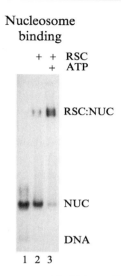

Nucleosome
binding

FIG. 4. RSC binds to nucleosomes with higher affinity in the presence of ATP. Nucleosomes (1 nM) were incubated with 10 nM RSC and 1 mM ATP, where indicated, in a 5 μl reaction for 20 min at 30°, then run in a 3.2% native acrylamide gel and autoradiographed.

Binding reactions (5 μl) are performed in a buffer containing 20 mM Tris–acetate, pH 7.9; 6% glycerol (w/v); 10 mM magnesium acetate; 50 mM potassium acetate; 1 mM DTT; and 0.1 mg/ml BSA, 1 nM nucleosomes, with or without 1 mM ATP and 10 nM RSC. Reactions were incubated at 30° for 20 min.

Polymerize a 3.2% native acrylamide gel (38.9:1.1 acryl:bis) in 10 mM Tris–HCl, pH 8.0 and 1 mM Na-EDTA at RT, then wrap and store it overnight at 4°. Polymerizing the gels overnight increases the resolution of the samples in the gel. Run the gel at 4° in 10 mM Tris–HCl, pH 8.0 and 1 mM Na-EDTA while circulating the buffer. Load the samples while running the gel at 60 V, then increase the voltage to 150 V for $\sim$4 h, or until the bromphenol blue has migrated $\sim$10 cm. The gel is dried and autoradiographed.

Endonuclease Access Assays

ATP-dependent remodelers, like RSC, increase factor access to nucleosomal DNA.[6,22–26] This increased access can be monitored by measuring either increased factor binding[6,23] or changes in the access of endonucleases to the nucleosomal DNA.[22,24–26]

Here we describe two commonly used assays to monitor changes in factor access to nucleosomal DNA. The first one, restriction enzyme access assay, monitors changes within one region of the nucleosome, whereas the second assay, DNase I sensitivity, monitors alterations in accessibility throughout the entire length of the nucleosomal DNA. The data that are shown represents assays performed with RSC. The isolated ATPase, Sth1, can also alter the accessibility of both types of endonucleases to nucleosomal DNA in the presence of ATP[11] (data not shown), at a several-fold reduced specific activity relative to that of the RSC complex.

Restriction Enzyme Access Assay

A restriction enzyme site within the nucleosomal DNA is relatively inaccessible to endonucleases (see Fig. 5, lane 2), because of the contact of the DNA with the octamer surface. In the presence of RSC and ATP, the site becomes fully accessible (see Fig. 5, lane 4), demonstrating that the DNA has been displaced from the surface of the octamer.

Remodeling reactions (20–30 μl) containing 20 ng of either DNA (5 nM) or nucleosomes (186 bp DNA with a centrally located Sea Urchin 5 S sequence) in a buffer containing 20 mM Tris–acetate, pH 7.9; 50 mM potassium acetate; 10 mM magnesium acetate; 1 mM DTT; and 0.1 mg/ml BSA; with or without 1 mM ATP, 2–10 nM RSC, and 20 U of *Dra*I (NEB), are allowed to proceed at 30° for 60 min. Reactions are quenched by adding of 0.4 ml of 10 mM Na-EDTA, followed by organic extraction, addition of 2 μg of carrier DNA, and ethanol precipitation of the DNA. The dried pellets are suspended in 20 μl of 5% glycerol (w/v), then run in an 8% native acrylamide gel (29:1 acryl:bis) in TBE[19] at 180 V until the bromphenol blue reaches the bottom of a 10 cm long gel (~1.3 h). The gel is dried and autoradiographed.

DNAse I Sensitivity Assay

Another commonly used endonuclease to monitor alterations in nucleosomal DNA accessibility is DNase I. When treated with DNase I, nucleosomal DNA is cut once every ~10 bp, since DNase I cuts in the minor groove of the DNA which is exposed once every ~10 bp on the surface of the nucleosome (see Fig. 6 left panel and lane 3 of right panel). In

[23] T. Owen-Hughes and J. L. Workman, *Crit. Rev. Eukaryote Gene. Expr.* **4**, 403 (1994).

[24] B. R. Cairns, Y. Lorch, Y. Li, M. Zhang, L. Lacomis, H. Erdjument-Bromage, P. Tempst, J. Du, B. Laurent, and R. D. Kornberg, *Cell* **87**, 1249 (1996).

[25] C. Logie and C. L. Peterson, *EMBO J.* **16**, 6772 (1997).

[26] G. Schnitzler, S. Sif, and R. E. Kingston, *Cell* **94**, 17 (1998).

Restriction enzyme access

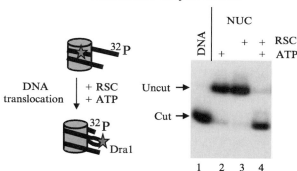

FIG. 5. Restriction enzyme access assay. Twenty nanograms of either DNA (5 nM) or nucleosomes were incubated with 20 units of *Dra1*, with the addition of 1 mM ATP and 10 nM RSC, as indicated, and allowed to proceed at 30° for 60 min. Following organic extraction and ethanol precipitation, the DNA is resolved in an 8% native acrylamide gel and autoradiographed.

the presence of ATP and increasing amounts of RSC, the DNase I digestion pattern of the nucleosomal DNA changes to a random digestion pattern resembling that of free DNA (see Fig. 6, compare lanes 5–7 to lanes 1 and 2), suggesting that the DNA has been liberated from the surface of the histone octamer and reflecting remodeling of the nucleosome.

Set up remodeling reactions (20 μl) containing 20 ng of either free or nucleosomal DNA (radiolabeled at one end) in a buffer containing 20 mM Tris–acetate, pH 7.9; 50 mM potassium acetate, 10 mM magnesium acetate, 1 mM DTT, and 0.1 mg/ml BSA, with or without 1 mM ATP, 5–20 nM RSC, and incubate them at 30° for 20 min.

Prior to initiating the remodeling reactions, make dilutions of DNase I in 5× DNase I buffer (250 mM Tris–HCl, pH 7.5; 50 mM MgCl$_2$, 2.5 mg/ml BSA protease-free, Sigma) from a storage stock of DNAse I (Worthington #06333, suspended in 10 mM Tris pH 8 and 1 mM CaCl$_2$, then adjusted to 50% glycerol [w/v], to a final concentration of 6 U/μl, and stored at −20° for extended periods). The amount of DNase I to be used for both the DNA and nucleosomes needs to be titrated for each stock of DNase I, but once titrated, the activity of the DNase I is linear in a concentration and time dependent manner. A good starting range is a 1:200 dilution of the DNase I stock (6 U/μl) in 5× DNase I buffer for a 10 s incubation with DNA (20 ng), and a 1:40 dilution of DNAse I stock (6 U/μl) for a 30 s incubation with nucleosomes at 37° (titrating the amount of DNase I within 1.5 to 2-fold of the recommended dilutions).

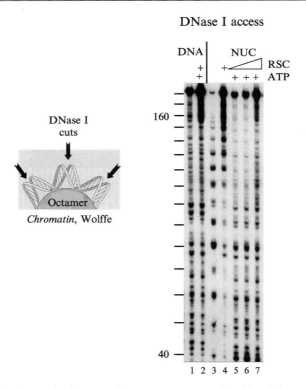

Fig. 6. DNase I sensitivity assay. Twenty nanograms of end-labeled DNA (8 nM) or nucleosomes were incubated with 1 mM ATP and RSC (40 nM, lanes 4 and 7; 20 nM, lane 6; or 10 nM, lane 5) as indicated in 20 μl for 20 min at 30°, then treated with 5 μl of DNase I solution at a 1:200 dilution (0.03 U) for 10 s for DNA (lanes 1 and 2), or at a 1:40 dilution (0.15 U) for 30 s for nucleosomes (lanes 3–7) at 37°. Following organic extraction and ethanol precipitation, the DNA is resolved in a 5% sequencing gel and autoradiographed.

After the remodeling reactions have incubated for 20 min, add 5 μl of the diluted DNase I solutions to the remodeling reactions, using staggered time points, and incubate at 37° for either 10 s (DNA) or 30 s (NUCs). Quench the reactions by adding 0.4 ml of 15 mM Na-EDTA, organic extract the DNA, add 2 μg of carrier DNA, then ethanol precipitate the DNA. Suspend the pellets in 7 μl of formamide sample buffer[18] and heat at 80° for 5 min. Prerun a 5% acrylamide sequencing gel (Long Ranger gel solution, BioWhittaker Molecular Applications) in 7 M Urea and 1× TBE at 60 W for 30 min. A good standard is the 10 bp DNA ladder (GIBCO/BRL) that has been end-labeled with [γ-^{32}P]-ATP and T4 kinase. Load the hot samples into every other lane and run the gel at 45 W for 80 min, or until the bromphenol blue runs to the bottom of a 40 cm gel. The gel is dried and autoradiographed.

Concluding Remarks

ATP-dependent chromatin remodeling complexes, like RSC, alter the structure of nucleosomes by disrupting histone-DNA contacts within the nucleosome. The ATPase subunit of the RSC complex, Sth1, couples the energy of ATP hydrolysis to translocation along one of the strands of the DNA duplex. If the remodeler remains tethered to the nucleosome, while the ATPase domain translocates along the DNA, the linker DNA would effectively be pulled into the nucleosome. Because of the helical nature of DNA, translocation of a nucleosome-bound remodeler, assuming the remodeler remains in a relatively fixed position on the nucleosome, would result in the generation of superhelical torsion in the DNA. These changes in superhelical torsion could contribute to the disruption of the histone-DNA contacts within the nucleosome. Changes in the path of the DNA duplex relative to the surface of the octamer can be monitored by altered access of endonucleases to the nucleosomal DNA.

Assays to monitor the ATPase activity and DNA translocation of the remodelers on a DNA substrate, TAP-purification of the RSC complex, the reconstitution of yeast nucleosomes on a scale suitable for biochemical analysis, and the use of these nucleosomes in remodeling assays to monitor changes in nucleosome structure by detecting changes in the sensitivity of endonucleases to nucleosomal DNA in the presence of a remodeler and ATP are described here.

Acknowledgments

We thank Marnie Gelbart and Toshio Tsukiyama for useful suggestions on modifying the pioneering protocol on nucleosome reconstitution by Luger and Richmond, and for supplying the yNap1 expression construct, Roger Kornberg for the TAP-tagged RSC2 strain, and Craig Kaplan for helpful comments on the manuscript.

[21] A Nucleosome Sliding Assay for Chromatin Remodeling Factors

By ANTON EBERHARTER, GERNOT LÄNGST, and PETER B. BECKER

ATP-dependent nucleosome remodeling emerges as a major principle endowing chromatin with dynamic properties. A family of energy-dependent nucleosome remodeling complexes powered by dedicated ATPases of the SWI2/SNF2 superfamily are involved in all nuclear processes with chromatin substrates in eukaryotes.[1,2] These enzymes change the interactions between nucleosomal DNA and histones to render DNA accessible to potential interacters. Although a variety of different assays have been employed to visualize nucleosome "remodeling" and a corresponding diversity of phenomena has been described, one shared feature of all remodelers so far is their ability to induce the relocation of nucleosomes on DNA.[3] Since this relocation apparently occurs in the absence of overall disruption of the histone octamer, it has been called "nucleosome sliding."[4,5] Nucleosome movements can occur in nucleosomal arrays[5,6] but visualizing them is complicated and measurements are not quantitative. By contrast, monitoring the mobility of a single histone octamer on defined DNA fragments can be assayed much more easily and quantitatively.[4,5] The assay serves to characterize the potential of known and candidate nucleosome remodeling machines and allows a quantitative evaluation of nucleosome sliding capacity under defined *in vitro* conditions. The assay makes use of the fact that mononucleosomes that occupy distinct positions on a longer DNA fragment will have different mobilities when resolved on native polyacrylamide gels ("bandshift gels"). The closer a histone octamer resides to the end of a DNA fragment, the faster it will run.[7,8] The "mobility assay" itself therefore essentially consists in a gel electrophoresis to determine nucleosome positions before and after the action of a remodeling ATPase. However, in order to reach this state some preparations are necessary, most of which we will describe here. First, a mononucleosome substrate

[1] C. L. Peterson, *EMBO Rep.* **3,** 319 (2002).
[2] R. E. Kingston and G. J. Narlikar, *Genes Dev.* **13,** 2339 (1999).
[3] P. B. Becker, *EMBO J.* **21,** 4749 (2002).
[4] A. Hamiche *et al., Cell* **97,** 833 (1999).
[5] G. Längst *et al., Cell* **97,** 843 (1999).
[6] P. D. Varga-Weisz, T. A. Blank, and P. Becker, *EMBO J.* **14,** 2209 (1995).
[7] W. Linxweiler and W. Horz, *Nucleic Acids Res.* **12,** 9395 (1984).
[8] S. Pennings, G. Meersseman, and E. M. Bradbury, *J. Mol. Biol.* **220,** 101 (1991).

needs to be prepared. We describe the generation and purification of the body-labeled DNA fragment, and refer to other publications that describe sources of histones. We detail the generation of nucleosome substrates following one convenient nucleosome reconstitution strategy. Critical to some applications may be the purification of substrates with defined histone octamer positions. We also describe the expression of two remodeling enzymes with well characterized nucleosome sliding potential, which may serve as controls or reference enzymes. Finally, we explain how a nucleosome sliding reaction is performed. Similar reactions have been established in other laboratories[4,9] and variant procedures may be found in these original papers.

Collecting the Ingredients

Generating the DNA Fragment

In general, for reconstituting the sliding substrates, different types and lengths of DNA can be used. To observe clear nucleosomal mobility a DNA fragment should be minimally 180 bp long[10] but longer fragments allow better resolution. The longer the fragment, the more different positions will be generated, and hence the complexity of the banding pattern may increase, but fragments below 400 bp have been used successfully.

In our laboratory we most frequently use a 248 bp DNA fragment spanning sequences between -232 and $+16$ relative to the transcription start site of the mouse rDNA, which contains a prominent bend. Nucleosomes form on this DNA at two predominant positions, that can be separated by native gels: in addition to abutting one DNA end[11] they also cluster around a central position. We generate the 248 bp fragment by PCR and include $(\alpha\text{-}^{32}P)dCTP$ in the PCR reaction to achieve uniform labeling. For large amounts of the DNA we usually perform a 1 ml PCR reaction containing 100 ng template; 500 pmol of each primer; 100 nmol each of dATP, dGTP, and dTTP; 20 nmol dCTP and 16.7 pmol $(\alpha\text{-}^{32}P)dCTP$. The desired DNA fragment is then purified as follows:

1. Precipitate DNA from the PCR reaction, wash twice with 70% ethanol and finally dissolve in 100 μl of buffer EX50 (10 mM Tris–HCl, pH 7.6; 50 mM KCl; 1.5 mM MgCl$_2$; 0.5 mM EDTA; 10% glycerol).

[9] I. Whitehouse *et al.*, *Nature* **400,** 784 (1999).
[10] S. R. Kassabov *et al.*, *Mol. Cell. Biol.* **22,** 7524 (2002).
[11] G. Längst and P. B. Becker, *Mol. Cell* **8,** 1085 (2001).

2. Resolve DNA by electrophoresis on a 4.5% polyacrylamide gel in $0.5\times$ TBE. After electrophoresis expose the wet gel to film for 5 min, cut out the desired fragment from the gel and place into an Eppendorf tube containing 600 μl of DNA-elution buffer (0.5 M ammonium acetate; 10 mM MgCl$_2$; 0.1% SDS; 1 mM EDTA).
3. Elute DNA by diffusion at 37° under constant shaking: after 3 h remove supernatant and store; then add further 600 μl of DNA-elution buffer for a second, overnight, elution.
4. To the combined eluted DNA add 1/4 volume 5 M NaCl; 1/9 volume 10% SDS, vortex for 1 min, add 1 volume chloroform, vortex again and immediately put everything on ice.
5. Centrifuge (13,000 rpm; 4°; 5 min) and collect the aqueous phase carefully avoiding the white interphase.
6. Precipitate DNA with 3 volume ethanol (1 h, -20°), wash the DNA twice with 70% ethanol and then dissolve it in 100 μl of TE. This DNA is usually at a concentration of about 0.4–0.5 μg/μl and is subsequently used in nucleosome assembly reactions.

Histone Sources

Bulk histones can be obtained from any convenient source. The isolation procedure follows a protocol described by Simon and Felsenfeld[12] and has to be adapted to the source. We purify bulk histones from nuclei of 12–24 h old *Drosophila* embryos. For the isolation of histones it is very important to keep all buffers, rotors and reagents at 4° and perform all steps in the cold with protease inhibitors added freshly. The advantage of the procedure is that intact octamers can be eluted from chromatin in 2 M KCl, and the four histones are, therefore, already bound to their partners in the appropriate stoichiometry. They are concentrated by centrifugation through Filtron devices with a molecular weight cut-off of 10 kDa for several hours and it is therefore advisable to change the filtration devices. After adding one volume of 100% glycerol the histone octamers are in 1 M KCl and can be stored at -20°.

Bulk histones have the disadvantage that they are heterogeneous, containing unknown mixtures of histone variants and isoforms. For many applications this is entirely appropriate. If, however, a homogenous source of histones is required, they can be best obtained as recombinant proteins after expression in *E. coli* following the procedure detailed by Richmond and co-workers.[13] A further advantage of recombinant histones is that

---

[12] R. H. Simon and G. Felsenfeld, *Nucleic Acids Res.* **6**, 689 (1979).
[13] K. Luger, T. J. Rechsteiner, and T. J. Richmond, *Methods Enzymol.* **304**, 3 (1999).

mutated histones can be created at will, which is an asset for the mechanistic analysis of nucleosome sliding.[14,15] A disadvantage of the procedure is that the four histones are separately expressed and octamers have to be reconstituted by careful titration of all four components. Following the excellent protocol by Luger et al.,[13] we emphasize the importance of accurate determination of the histone concentrations by measurement of the optical density at 276 nm. It is very important that the four histones are mixed in equimolar ratios. This will subsequently lead to high yield of reconstituted octamers and low amounts of undesired aggregates, hexamers or nonreconstituted H2A/H2B dimers on the Superdex S-200 size exclusion chromatography. Reconstituted octamers are stable for months in 1 M KCl 50% glycerol at $-20°$.

Reference Remodelers: Expression of ISWI and ACF via Baculovirus Vectors

In order to obtain homogenous remodeling factors we express the ATPase ISWI[16] alone or in association with its partner Acf1[17,18] after infection of Sf9 cells with appropriate baculovirus vectors. The factors are purified via a unique FLAG-tag on ISWI or Acf1. For proper levels of ISWI and ACF one has to evaluate the amount of virus DNA used for protein expression. This is especially critical for the stoichiometric coexpression of ISWI (with a FLAG-tag) and Acf1 (without FLAG-tag) to generate a functional ACF complex. Sufficient amounts of purified proteins can be obtained after infection of 2.4×10^8 Sf9 cells:

1. Collect cells by centrifugation at 1000g in a 15 ml polypropylene tube and wash once with cold PBS.
2. Resuspend cell pellet in 5 ml HEMG500 (25 mM Hepes, pH 7.6; 500 mM KCl; 0.5 mM EDTA; 12.5 mM MgCl$_2$; 10% glycerol; 0.05% NP-40; 1 mM DTT; 0.2 mM PMSF and protease inhibitors).
3. Freeze/thaw cells twice in liquid nitrogen and sonify (amplitude 40%) twice for 15 s on ice. After sonification spin down nonsoluble material at 10,000g for 20 min.
4. To the supernatant add 100 μl of anti-FLAG agarose (M2, SIGMA A-2220), equilibrated in HEMG500. Incubate under constant mixing for at least 1 h.

[14] C. R. Clapier et al., Mol. Cell. Biol. **21**, 875 (2001).
[15] C. R. Clapier, K. P. Nightingale, and P. B. Becker, Nucleic Acids Res. **30**, 649 (2002).
[16] D. F. V. Corona et al., Mol. Cell **3**, 239 (1999).
[17] T. Ito et al., Genes Dev. **13**, 1529 (1999).
[18] A. Eberharter et al., EMBO J. **20**, 3781 (2001).

5. After binding of the FLAG-tagged proteins to the affinity resin a series of washes are performed, each for 10 min: 3× with 10 ml of HEMG500, 3× with 10 ml HEMG 1000, and finally 3× with 10 ml HEMG100.
6. Transfer the beads into a 1.5 ml reaction tube, add 100 μl HEMG100, and 5 μl FLAG peptide (stock solution: 5 μg/μl). Sequential elutions are performed twice for at least 3 h under constant mixing. The concentrations of ISWI and ACF are usually at approx. 100–200 ng/μl.

Nucleosome Assembly and Purification

Mononucleosome reconstitution by salt gradient dialysis or salt/urea gradient dialysis protocols have already been described in several previous protocols in this series.[13,19,20] They can all be used to assemble the desired sliding substrate. However, the relative yield of certain nucleosome positions may vary depending on the time allowed for reconstitution. Also, it may sometimes be advisable to obtain nonequilibrium positions by rapid procedures. Here, we decribe in detail a rapid procedure that is routinely used in our laboratory using polyglutamic acid as a histone carrier.

Polyanions can be used to facilitate the assembly of nucleosomes from histones at low ionic strengths. Stein[21] introduced and optimized polyglutamic acid (PGA) as a vehicle and we and others have adapted the method successfully.[5,18] Histones are first bound to PGA before the mix (HP-mix) is added to DNA. Upon incubation, nucleosomes are formed. In a first series of small scale test assemblies one has to determine an appropriate ratio between the purified DNA and the HP-mix. Exact titrations in these assemblies are important, since these ratios are scaled up 100-fold subsequently in order to obtain large quantities of mononucleosomes. The method has the disadvantage that the resulting nucleosomes need to be purified before use since contaminating PGA may affect nucleosome sliding reactions. However, once established, the procedure is a reliable source of good quality nucleosomes.

Prepare HP mix: Mix well 60 μg (about 30 μl) of histone octamers, 18 μl PGA (SIGMA P4886, dissolved in water at 10 mg/ml) and 1150 μl

[19] B. Neubauer and W. Horz, *Methods Enzymol.* **170,** 630 (1989).
[20] P. D. Varga-Weisz, E. J. Bonte, and P. B. Becker, *Methods Enzymol.* **304,** 742 (1999).
[21] A. Stein, *Methods Enzymol.* **170,** 585 (1989).

TE (10 mM Tris–HCl, 0.1 mM EDTA, pH 8.0). Incubate this mix at room temperature (RT) for 1 h. Eliminate precipitates by centrifugation at 10,000 rpm for 5 min (RT). This is especially important when using recombinant (slightly unbalanced) histones. Store supernatant (HP-mix) in aliquots at $-20°$.

Test assembly:

1. Dilute HP-mix 1:5 in EX50 + BSA (SIGMA A7906); EX50 buffer: 10 mM Tris–HCl, pH 7.6; 50 mM KCl; 1.5 mM MgCl$_2$; 0.5 mM EDTA; 10% glycerol; EX50 + BSA: EX50 + 0.2 μg/ml BSA, sterile filtered.
2. Prepare DNA premix containing: 3 μl of purified, labeled DNA, 1 μl BSA (20 mg/ml), 96 μl KE buffer (100 mM KCl, 0.2 mM EDTA, pH 7.6).
3. Pipette 4.5, 4, 3.5, 3, 2.5, 2, 1.5, 1, 0.5 μl of KE into a series of 1.5 ml reaction tubes and add to each tube 10 μl of DNA premix.
4. Add 1.5, 2, 2.5, 3, 3.5, 4, 4.5, 5, 5.5 μl of diluted HP-mix into the reaction mix.
5. Incubate for 2 h at 30°.
6. Add to each reaction 1.5 μl of 50% glycerol and mix gently.
7. Resolve 6 μl on a native 4.5% polyacrylamide gel in 0.4× TBE at 10 mA/cm gel at RT. As a reference for the length of separation, the loading dye orange G can be run to the bottom of the gel in a separate lane.
8. Dry and expose the gel for 1 h at $-80°$ using an intensifying screen. Develop autoradiography and determine the optimal ratio between DNA and HP-mix

Figure 1 presents a typical result for a series of test assemblies. The asterisk marks the ratio, which is subsequently used for the preparative assembly.

Preparative assembly: For preparative assembly a test reaction should be scaled up at least 100-fold (133-fold in our example below), but the histone input should be reduced by about 15% (relative to the smale scale test). Electrophoresis and gel elution of the positioned nucleosomes have to be carried out at 4°. It is very important to keep all buffers cold and use siliconized tubes at all steps. In addition, all tubes should be coated with BSA (2 mg/ml) before use.

1. Combine 40 μl DNA, 34 μl HP-mix (not diluted), 5 μl BSA, and 71 μl KE and mix well by pipetting up/down. Incubate reaction at 30° for 2 h.

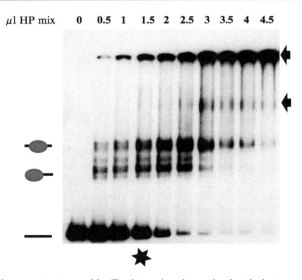

FIG. 1. Nucleosome test assembly. To determine the optimal ratio between DNA versus HP-mix small scale test assemblies are performed as described. The test assemblies are separated on a native polyacrylamide gel, an autoradiography of which is shown. The free DNA and the nucleosomal positions are indicated to the left. Aggregates are depicted by arrows. The appropriate ratio between DNA and HP-mix is marked by an asterisk. These conditions are then used for the large-scale assembly to obtain preparative amounts of nucleosome substrate.

2. Add 15 μl 50% glycerol and mix gently. Load the entire reaction on a 15 cm long, native polyacrylamide gel (4.5% polyacrylamide, 0.2× TBE, 5% glycerol; gel is 1 mm thick; slots are 25 × 20 mm^2). Run gel at 50 mA for 15 h in the cold room. Cover wet gel in Saran wrap and expose for 15 min.

3. Meanwhile, prepare 500 μl reaction tubes: coat with BSA, puncture a tiny hole in the bottom and cut off the lids. Place them in siliconized and BSA-coated 1.5 ml reaction tubes. Add 50 μl of EX50 + BSA to the small tube and keep the set-up on ice.

4. Guided by the autoradiography, cut out the bands corresponding to free DNA and the positioned nucleosomes using a sharp scalpel and carefully place the gel pieces into the small Eppendorf tube containing EX50 + BSA. Keep everything on ice.

5. Crush the isolated gel slices by centrifugation at 4° for 30 s into the 1.5 ml Eppendorf tube and put on ice.

6. Immediately add 500 μl of cold EX50 + BSA. Shake tubes for 90 min at $4°$ in an Eppendorf shaker.
7. Spin down at 10,000g for 5 min at $4°$ and collect the supernatant into a fresh tube, carefully avoiding small gel pieces. Keep on ice.
8. Elute a second time by adding another 500 μl of EX50 + BSA. Proceed as described in 6 and 7.

The isolated positioned nucleosomes are kept at $4°$ where they are stable for weeks. One microliter of isolated material usually contains about 60 fmol of nucleosome assembled on a 248 bp DNA fragment. These isolated positioned nucleosomes are of high quality. They can be used in various assays such as pull down experiments, band shift reactions and sliding assays.

Nucleosome Sliding Assay

Nucleosome sliding assays can be set up with mixtures of mononucleosomes occupying different positions[4,5,22] or with substrates with homogenous positions derived from gel purification as described. In the former case the analysis is more complex and it is not easy to keep track of various histone octamers. We prefer to work with mononucleosomes with homogenous positioning since we discovered that different remodeling enzymes may selectively mobilize octamers from some positions to others, with apparent directionality.[18] This is illustrated in Fig. 2, which presents a typical result of a mobility assay. Figure 2A shows the ability of ISWI to mobilize a centrally positioned nucleosome to a fragment end. Under those conditions it is unable to move the octamer back to more internal positions. The association of Acf1 with ISWI to form ACF makes all the difference: ACF is able to move the nucleosome from end to internal positions, but not back.[18]

1. 10 μl sliding reactions contain 0.5–1 μl of positioned nucleosome (approx. 30–60 fmol); 0.1 μl BSA (0.2 μg/ml); 0.1 μl ATP (1 mM); and EX50 buffer up to 10 μl. It is straightforward to make a master mix for $n + 1$ reactions, in siliconized tubes.
2. Make several dilutions of your remodeling factor in an appropriate buffer and add 1 μl of the diluted factor to each reaction. Incubate at $26°$ for 30–90 min.
3. Stop the reaction by adding 200 ng (ideally in 0.5–1 μl) of cold plasmid competitor DNA to each reaction and incubate for further 5 min.

[22] T. Tsukiyama, *Genes Dev.* **13**, 686 (1999).

4. Load 10 μl of each reaction on a native 4.5% polyacrylamide gel in 0.4× TBE and run gel at 10 mA/cm (after 3–4 h the orange G dye is at the bottom of the gel).
5. Dry and expose gel overnight using an intensifying screen.

Concluding Remarks

The number of nucleosome remodeling ATPases, known or suspected, and novel complexes they reside in, is steadily increasing. We presented protocols and summarized methods that will help to set up a fast and reliable assay to analyse their ability to mobilize nucleosomes. So far, all remodeling ATPases have been active in this assay. The availability of a standardized assay facilitates the direct comparison between different enzymes, both in semiquantitative as well as in qualitative terms. Understanding the reason for different directionality of nucleosome mobilization may help to explain in detail how different remodeling machineries approach and manipulate their substrate.

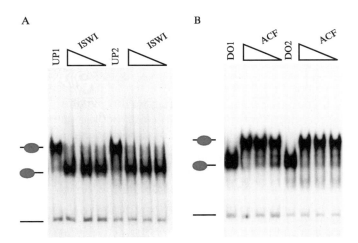

FIG. 2. Nucleosome sliding assay. (A) 60 fmol of isolated center-positioned nucleosomes (= the slower migrating "UP" nucleosome) are incubated with 10, 5, and 2.5 fmol of affinity purified ISWI. UP1 and UP2 represent the two subsequent elutions of the center nucleosome after the preparative gel. (B) For ACF sliding, 60 fmol of isolated end-positioned nucleosomes (= the faster migrating "DO" nucleosome) are incubated with 10, 5, and 2.5 fmol of affinity purified ACF. The positions of the two isolated nucleosomal particles and of the free DNA are shown on the side. The reactions are separated on a native 4.5% polyacrylamide gel, an autoradiography of which is shown.

Acknowledgments

Work on nucleosome remodeling in the authors' laboratory was supported by Deutsche Forschungsgemeinschaft through SFB/TR5 and SFB594 and the Fonds der Chemischen Industrie to P.B.B. We thank Drs. D. Fyodorov and J. Kadonaga for an Acf1-expressing baculovirus.

[22] Methods for Analysis of Nucleosome Sliding by *Drosophila* NURF

By ALI HAMICHE and HUA XIAO

Drosophila NURF is a member of the ISWI class of ATPase-containing chromatin remodeling complexes that uses the energy of ATP hydrolysis to alter chromatin structure and mobilize nucleosomes.[1] Previous studies of NURF demonstrated a complex of four distinct subunits,[2] the 301 kD largest subunit NURF301,[3] the ISWI ATPase,[4] NURF55, a WD-40 repeat protein also found in other protein complexes involved in histone metabolism,[5] and NURF38, an inorganic pyrophosphatase.[6] These studies also revealed a number of distinct biochemical properties of NURF. The ATPase activity of NURF is stimulated only by nucleosomes, not by free DNA or free histones (see Fig. 1). NURF complex requires the flexible N-terminal histone H4 tail, more specifically amino acids 16–19 of the H4 tail, for stimulating ATPase activity and for inducing nucleosome mobility.[7,8] NURF complex mediates nucleosome "sliding," the movement of histone octamers in *cis* without permanent diplacement from DNA.[9,10] These

[1] M. Vignali, A. H. Hassan, K. E. Neely, and J. L. Workman, *Mol. Cell. Biol.* **20**, 1899 (2000).
[2] T. Tsukiyama and C. Wu, *Cell* **83**, 1011 (1995).
[3] H. Xiao, R. Sandaltzopoulos, H. M. Wang, A. Hamiche, R. Ranallo, K. Y. Lee, D. Fu, and C. Wu, *Mol. Cell* **8**, 531 (2001).
[4] T. Tsukiyama, C. Daniel, J. Tamkun, and C. Wu, *Cell* **83**, 1021 (1995).
[5] M. A. Martinez-Balbas, T. Tsukiyama, D. Gdula, and C. Wu, *Proc. Natl. Acad. Sci. USA* **95**, 132 (1998).
[6] D. A. Gdula, R. Sandaltzopoulos, T. Tsukiyama, V. Ossipow, and C. Wu, *Genes Dev.* **12**, 3206 (1998).
[7] A. Hamiche, J.-G. Kang, C. Dennis, H. Xiao, and C. Wu, *Proc. Natl. Acad. Sci. USA* **98**, 14316 (2001).
[8] C. R. Clapier, K. P. Nightingale, and P. B. Becker, *Nucl. Acids. Res.* **30**, 649 (2002).
[9] A. Hamiche, R. Sandaltzopoulos, D. A. Gdula, and C. Wu, *Cell* **97**, 833 (1999).
[10] G. Langst, E. J. Bonte, D. F. Corona, and P. B. Becker, *Cell* **97**, 843 (1999).

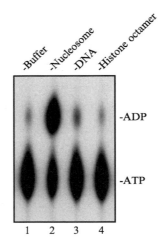

FIG. 1. Nucleosome-stimulated ATPase activity of *Drosophila* NURF. ATPase reactions with purified recombinant NURF were as described in ATPase assays section. All reactions contained the same amount of purified recombinant NURF (15 ng ISWI equivalent). The other components were as indicated: lane 1, buffer alone without DNA or core histones; lane 2, assembled mononucleosomes on the 359 bp hsp70 heat shock promoter DNA (50 ng of DNA equivalent); lane 3, the 359 bp DNA alone, and lane 4, histone octamer alone.

properties distinguish NURF from other ATPase-containing chromatin remodeling complexes. For instance, the ATPase activity of SWI/SNF is stimulated by either free or nucleosomal DNA.[4,11–14] Furthermore, the SWI/SNF or RSC complexes not only mobilize nucleosomes but can also transfer a histone octamer from one DNA fragment to another.[15] These differences indicate that there are diverse mechanisms by which nucleosome structure can be altered to increase nucleosome dynamics and DNA accessibility. However, the underlying principles of nucleosome movements by various chromatin remodeling proteins and protein complexes are yet to be fully understood.

[11] J. Cote, C. L. Peterson, and J. L. Workman, *Proc. Natl. Acad. Sci. USA* **95,** 4947 (1998).
[12] D. F. Corona, G. Langst, C. R. Clapier, E. J. Bonte, S. Ferrari, J. W. Tamkun, and P. B. Becker, *Mol. Cell* **3,** 239 (1999).
[13] A. Brehm, G. Langst, J. Kehle, C. R. Clapier, A. Imhof, A. Eberharter, J. Muller, and P. B. Becker, *EMBO J.* **19,** 4332 (2000).
[14] D. Guschin, P. A. Wade, N. Kikyo, and A. P. Wolffe, *Biochemistry* **39,** 5238 (2000).
[15] Y. Lorch, M. Zhang, and R. D. Kornberg, *Cell* **96,** 389 (1999).

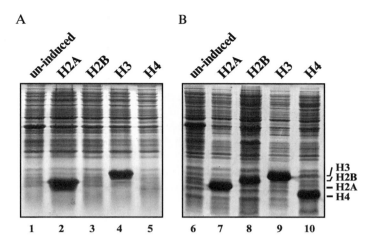

FIG. 2. Expression of recombinant core histones. *Drosophila* core histones were expressed in BL21-CodonPlus-RIL cells and run in a 12% SDS-polyacrylamide gel. Wild type core histones H2B and H4 showed little expression in *E. coli* (Fig. 2A, lanes 3 and 5). Insertion of the ATA codon (Ile) immediately after the ATG dramatically increases their production to the same levels as wild type histone H2A and H3 (see Fig. 2B, lanes 8 and 10).

Here we provide methods that we have been using in our analysis for nucleosome mobilization by *Drosophila* NURF. These methods are straightforward and easy to follow, though in some cases extensive practice may be needed to obtain the best results. Thus, they may provide a good starting assay system for analysis of nucleosome movements by other chromatin remodeling proteins and protein complexes.

Preparation of Substrates for Chromatin Remodeling

Expression and Purification of Recombinant Core Histones

Expression plasmids for the *Drosophila* core histones and their N-terminally truncated versions based on the T7-expression system have been described previously.[7] Wild type histone H2B and H4 could not be expressed in *E. coli*. Insertion of the ATA codon (Ile) immediately after the ATG dramatically increases their production to the same levels as wild type histone H2A and H3 (see Fig. 2).

BL21-CodonPlus-RIL (Stratagene) cells harboring expression plasmids are routinely grown at 37° on 2× TY medium containing 100 μg/ml ampicillin, 10 μg/ml chloramphenicol, and 0.1% glucose. Expression is induced when the culture reaches an A_{600} of 0.6 by addition of IPTG to a final concentration of 1 mM and the culture is incubated for 2 h at 37°. One hundred and fifty μg/ml of rifampicin is then added and the culture is incubated for another 3 h. Cells are pelleted at 3500g for 15 min. Individual histones are purified from inclusion bodies and assembled into octamers as described by Luger et $al.$[16] Briefly, equal amounts of purified individual histones are mixed and folded into octamers by dialysis from 7 M guanidinium-HCl to 2 M NaCl. Assembled octamers are purified from high molecular mass aggregates and other partially assembled subcomplexes by gel filtration. Aliquots of histone preparations (1 mg/ml) are stored at −80° until use.

DNA Preparation

DNA fragments of 250–350 bp give the best results for mononucleosome assembly. DNA fragments of interest are first cloned into a plasmid vector. After restriction enzyme digestion of purified plasmid DNA, the fragment insert is separated from the plasmid by electrophoresis at room temperature on a 4% acrylamide gel (acrylamide/bisacrylamide, 19:1, w/w) in 20 mM sodium acetate, 2 mM EDTA, 40 mM Tris–acetate (pH 7.8). The band corresponding to the DNA fragment is excised from the gel and crushed with a plastic pestle in an Eppendorf tube. The slurry is suspended in 400 μl of 0.5 M ammonium acetate, 10 mM MgCl$_2$, 0.1% SDS, 1 mM EDTA. The sample is vortexed and incubated for 2 h at 37°. After centrifugation, the supernatant containing DNA is adjusted to 1% SDS, 1 M NaCl and extracted with 1 volume of chloroform/isoamylalcohol (24:1, v/v). The DNA is ethanol precipitated, vacuum dried and re-suspended in 10 mM Tris–HCl (7.65), 1 mM EDTA.

Other DNA purification methods have also been successfully used. DNA fragments recovered from agarose gels using QIAEX II gel extraction kit (QIAGEN) or the Bio-Rad Prep Cell are also suitable for mononucleosome assembly and sliding assays. Purified DNA fragments can be end-labeled with ^{32}P by T4 polynucleotide kinase (after dephosphorylation) or by filling in the sticky ends of the restriction enzyme digestion with Klenow polymerase using standard procedures.[17]

[16] K. Luger, T. J. Rechsteiner, A. J. Flaus, M. M. Waye, and T. J. Richmond, *J. Mol. Biol.* **272**, 301 (1997).
[17] J. Sambrook, E. F. Fritsch, and T. Maniatis, "Molecular Cloning: A Laboratory Manual." Cold Spring Harbor Laboratory, Cold Spring Harbor, New York, 1989.

More recently, we use polymerase chain reaction (PCR) to generate large quantities of DNA fragments. This method is efficient and hundreds of micrograms of any defined fragment can be easily produced using the Titanium polymerase (Clontech/BioSicences). An advantage of using PCR is that the DNA fragment will contain only the sequence of interest, while linker DNA for a restriction enzyme site may need to be added to both ends of a fragment for its cloning into a plasmid vector. DNA fragments generated by PCR can be further purified by gel electrophoresis as described earlier.

Nucleosome Reconstitution

Mononucleosomes are assembled on linear DNA according to the "salt jump" method.[18] The following is a brief protocol:

1. Set up the assembly reaction (10 μl) in a 1.5 ml Eppendorf tube
   0.2 μl of 1 M Tris–HCl, pH 7.65 (10 mM final concentration)
   4 μl of 5 M NaCl (2 M final concentration)
   1 μl of 1 mg/ml BSA (100 μg/ml final concentration)
   1–2 μg of carrier DNA (plasmid DNA containing the fragment insert)
   0.5 μl of ^{32}P labeled DNA fragment (20,000 to 100,000 cpm)
   0.6–1.2 μg of histone octamer for a histone/DNA ratio rw = 0.6 (This ratio gives optimal reconstitution of mononucleosomes, with little formation of dinucleosomes.)
   Adjust the volume to 10 μl with water.
2. Incubate the reaction at 37° for 10 min.
3. Add 30 μl of 10 mM Tris–HCl, pH 7.65, 1 mM EDTA, 100 μg/ml BSA. This brings to a final concentration of 25 ng/μl DNA and 0.5 M NaCl.
4. Incubate the reaction at the same temperature for 30 min.
5. Dialyze the reaction mixture at 4° against 1× TE buffer (10 mM Tris–HCl, 1 mM EDTA, pH 7.65) for at least 1.5 h. Dialysis is performed in the original 1.5 ml Eppendorf or Treff tubes secured with a dialysis membrane or mini-dialysis units (Pierce). For experiments involving measurements of the histone/DNA ratio (two-dimensional gel electrophoresis, see later) or ATPase activity (see later), purified nonradioactive DNA fragment should be used in place of the carrier plasmid DNA.

---

[18] A. Stein, *J. Mol. Biol.* **130,** 103 (1979).

Nonradioactive Detection of Nucleosomal DNA

We find that SYBR Green I (Bio Whittaker Molecular Applications) and other fluorescent stains are good alternatives to radioactive labeling for in-gel detection of nucleosomal DNA after native gel electrophoresis of mononucleosomes. In this case purified DNA fragments need to be used instead of tracer amount of ^{32}P-labeled fragment plus plasmid carrier for mononucleosome assembly. This has recently become the detection method of choice in our laboratory since large quantities of DNA fragments can be produced by PCR amplification and efficiently recovered after gel electrophoresis.

Native Gel Electrophoresis

Electrophoresis of mononucleosomes is performed at room temperature in 4 to 6% polyacrylamide (acrylamide/ bisacrylamide, 29:1, w/w) slab gels (0.15 cm $\times$ 17 cm $\times$ 18 cm) in TE buffer (10 mM Tris–HCl, 1 mM EDTA, pH 7.5). Ten microliters samples are mixed with 2.5 μl of loading buffer (10 mM Tris–HCl, 1 mM EDTA, pH 7.65, 60% sucrose, 0.01% xylene cyanol). Samples are loaded onto the gel after a one-hour preelectrophoresis at 250 V and the gel is electrophoresed at the same voltage for 3 h with extensive buffer circulation to prevent salt exhaustion due to the low ionic strength of 1$\times$ TE buffer. Gels are dried and auto-radiographed at $-80°$. For preparative purposes, gels are wrapped with a saran wrap and auto-radiographed wet at 4°. TE gels give the most satisfactory results, but extensive practice may be needed to achieve optimal results. Special attention should be paid to the way acrylamide is mixed and polymerized. All the solutions should be equilibrated to room temperature before use and gently mixed to avoid bubbles. The glass plates should be free of any trace of detergent. They are usually cleaned with 7$\times$ detergent (ICN), rinsed thoroughly with tap water, then with distilled water and finally wiped dry with paper towel. The user should avoid finger contact with the inside surface of the plate.

Alternatively, for a quick analysis, mononucleosomes are electrophoresed in 4 to 6% polyacrylamide (acrylamide/bisacrylamide, 29:1, w/w) mini-gels (0.1 cm $\times$ 10 cm $\times$ 8 cm) in TGE buffer (25 mM Tris base, 190 mM glycine, 1 mM EDTA). For mini-gels, preelectrophoresis is 15 min at 120 V, and the gel is then electrophoresed at the same voltage for 1.5 h without buffer circulation.

For histone-DNA quantification, the mixed population of nucleosomes is first fractionated on a native gel as above. Nucleosome bands are excised and loaded directly onto the second dimension 15% SDS-polyacrylamide gel and embedded in place with a melted agarose solution (in 25 mM Tris base, 190 mM glycine, 1 mM EDTA). DNA and histones are visualized by silver staining.[19]

Mononucleosome Purification and Mapping

Sucrose Gradient Purification

Mononucleosomes are separated from free DNA on sucrose gradient. Samples of reconstituted mononucleosomes are layered on top of linear 5 to 20% (w/v) sucrose gradients containing 10 mM Tris–HCl (pH 7.65), 1 mM EDTA, 50 μg/ml BSA and centrifuged in the Beckman SW50.1 rotor at 32,000 rpm/min for 15 h at 5°. Fractions (150 μl) are collected from the bottom and directly analyzed in the ^{3}H channel of a liquid scintillation counter. Mononucleosome fractions are verified by gel electrophoresis and pooled.

Gel Purification of Positioned Nucleosomes

Positioned mononucleosomes can be resolved by native gel electrophoresis according to their positions on DNA. Gel-fractionated mononucleosome species are excised from the gel and the gel slices are crushed with a plastic pestle in 1.5 ml microcentrifuge tubes (Eppendorf). The slurry is suspended and incubated for 30 min on ice in 100–150 μl of 10 mM Tris–Cl (pH 7.6), 0.2 mM EDTA, 1 mM MgCl$_2$, 5% glycerol, 0.02% bovine serum albumin, containing 50 μg/ml of sheared sperm salmon DNA or core particles purified from native chromatin. The eluate is recovered from the supernatant after centrifugation for 10 min at 4° (14,000 rpm).

Mapping Nucleosome Positions

The bands corresponding to single nucleosome positions are excised and nucleosomes are eluted as described above. The supernatant containing eluted nucleosomes is adjusted to 10 mM Tris–HCl (pH 7.6), 50 mM NaCl, 3 mM MgCl$_2$, 1 mM β-mercaptoethanol, and digested with 400 U/ml of Exo III for 1 to 4 min at 37°. The reaction is immediately stopped with 10 mM EDTA and DNA is extracted with phenol according

[19] W. Wray, T. Boulikas, V. P. Wray, and R. Hancock, *Anal. Biochem.* **118,** 197 (1981).

to standard procedures. The samples are Cerenkov counted and resuspended in 90% (v/v) deionized formamide, supplemented with bromophenol blue and xylene cyanol, incubated at 96° for 2 min, and quenched in ice. The samples are run in a 6% sequencing gel with a size marker.

Expression and Purification of Recombinant NURF

To reconstitute *Drosophila* NURF, cDNAs encoding each of the four subunits, NURF-301, ISWI, NURF-55, and NURF-38 were cloned into a Baculovirus vector. In our earlier reconstitution experiments, individual recombinant subunits were expressed in either Hi-5 or SF9 cells and purified to apparent homogeneity. Purified recombinant NURF301 was mixed with a 3-fold molar excess of the other NURF subunits, and the mixture was incubated at 4° for at least 2 h. The mixture was loaded onto a 17–35% glycerol gradient and centrifuged at 50,000 rpm for 10–12 h to separate the fully assembled NURF complex from partial reconstitutes and individual subunits. 200 μl fractions were collected and assayed for ATPase and nucleosome sliding activities. Active fractions were further analyzed by protein blotting to confirm the presence of each of the four NURF subunits. Parallel gradient centrifugation with native NURF showed that recombinant NURF complex has an apparent molecular size similar to that of the native NURF complex.

So far the most quantitative way to reconstitute NURF is by coexpression of all four NURF subunits in SF9 cells (Fig. 3). We found that, of the four NURF subunits, the least expressed and unstable subunit in SF9 cells is NURF-301, and that coexpression of NURF-301 with the other subunits greatly stabilized NURF-301 from proteolysis. Based on this observation, we have established a one-step protocol for affinity-purification of recombinant NURF complex from coinfected SF9 cell extracts through affinity pull-down of a FLAG-tagged NURF-301. Coexpression dramatically enhanced the production of recombinant complex (more than 100-fold), likely as a result of stabilization of the largest subunit NURF-301.

Our standard purification protocol uses one liter of SF9 cells at 1.0×10^6 to 1.5×10^6 cells/ml and a multiplicity of infection (MOI) of 2–3 of each of the four recombinant viruses. Infected cells are grown at 27° and harvested 48–60 h after infection. Cells are resuspended in 20 ml of ice-cold HEGN buffer (25 mM HEPES, pH 7.6, 1 mM EDTA, 10% glycerol, 0.02% NP-40) containing 400 mM KCl, 1 mM β-mercaptoethanol, and 1× complete proteinase inhibitors (Roche). The cell suspension is transferred to a 40 ml tissue grind tube (Kontes Glass) on ice and homogenized

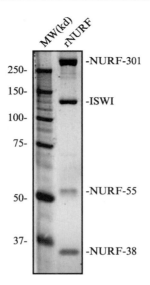

FIG. 3. An SDS gel (8%) showing *Drosophila* NURF reconstituted by coexpression of all four subunits in SF9 cells and purified to apparent homogeneity.

with pestle B, 15 strokes. All further manipulations are conducted at $4°$. The homogenate is centrifuged in a Beckman JA20 rotor at 15,000 rpm for 30 min. The supernatant is transferred to a 50 ml polypropylene conical tube and diluted 1:1 with HEGN buffer without KCl so the final concentration KCl is 200 mM. Depending on the level of expression of the recombinant complex, 1–2 ml of anti-FLAG M2 agarose resin (Sigma) is added to the extracts. The extracts are then incubated with anti-FLAG resin for 2 h with gentle mixing. FLAG beads are spun down in Beckman JA5 rotor at 1000 rpm for 1–2 min. The beads are washed 3×10 min with 30 ml of HEGN containing 400 mM KCl, 1 mM DTT, $0.5\times$ complete proteinase inhibitors (Roche), and 1×10 min with 30 ml of HEGN containing 100 mM KCl, 3 mM MgCl$_2$, 1 mM DTT, and $0.5\times$ complete proteinase inhibitors. The beads are then transferred to 1.5 ml Eppendorf tubes, and bound protein complex is eluted with 1 ml of HEGN containing 100 mM KCl, 3 mM MgCl$_2$ and 500 μg/ml $3\times$ FLAG peptide (Sigma). Protein concentrations are obtained by scanning SYPRO Orange-stained gels with a known protein marker as a reference. The yield of recombinant NURF complex from one liter of SF9 cells ranges from 300 to 500 μg.

Analysis of ATP-Dependent Nucleosome Sliding by NURF

ATPase Assays

ATPase assays were performed using α-P^{32}-labeled ATP as described by Hamiche *et al.*[9]

1. Set up the reactions (5 μl)

   0.1 μl of 500 mM HEPES-KOH, pH 7.6 (final 10 mM)
   0.1 μl of 2.5 M KCl (final 50 mM)
   0.1 μl of 5 mM EDTA (final 0.1 mM)
   0.1 μl of 100 mM MgCl$_2$ (final 2 mM)
   0.1 μl of 25 mM DTT (final 0.5 mM)
   0.5 μl of 75% glycerol (final 7.5%)
   0.1 μl of 0.5% NP-40 (final 0.01%)
   0.1 μl of 1.5 mM ATP (final 30 μM)
   50 ng nucleosomes (DNA weight equivalent)
   5 μCi [α-P^{32}] ATP (300 Ci/mmole, Amersham)
   1–5 ng of purified NURF
   Add H$_2$O to 5 μl.
2. Incubate the reactions at 37° for 30 min.
3. Separate ADP product of ATP hydrolysis by TLC (polyethylene-imine cellulose on polyester, Sigma) using 0.75 M KH$_2$PO$_4$ (spot 1 μl reaction mixture on the plate).

We found that both nucleosomal core particles purified from native sources and mononucleosomes assembled with purified DNA fragments and core histones are good stimulators of NURF ATPase activity.

Nucleosome Sliding Assay

Nucleosome mobilization was performed at 37° for 30 min in Nucleosome Sliding Buffer (10 mM Tris–HCl, pH 7.6, 50 mM NaCl, 3 mM MgCl$_2$, 1 mM β-mercaptoethanol, 1 mM ATP) with 100–200 ng DNA equivalent of assembled nucleosomes and 0.5–2 μl of purified NURF fractions. The standard (10 μl) remodeling reaction contains:

   4 μl of mononucleosomes (see Nucleosome Reconstitution section)
   1 μl of 0.1% BSA
   1 μl of 10× sliding buffer
   1 μl of 10 mM ATP
   0.5 to 2 μl of purified NURF fractions H$_2$O to 10 μl.

The reaction is incubated at 37° for 10 to 30 min and then stopped by adding 2.5 μl of loading buffer (10 mM Tris–HCl, 1 mM EDTA, pH 7.65, 60% sucrose, 0.01% xylene cyanol). Native gel electrophoresis is

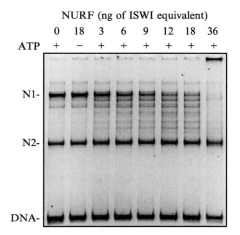

Fig. 4. ATP-dependent nucleosome sliding by *Drosophila* NURF. For mononucleosome assembly, the 282 bp DNA fragment of Clone 601 described by Lowary and Widom[20] was PCR-amplified and assembled with purified recombinant *Drosophila* core histones as described in Nucleosome Reconstitution section. The nucleosome sliding reactions contained 80 ng DNA equivalent of assembled mononucleosomes and went for 30 min at 30° with increasing amounts of purified recombinant NURF as indicated (see also Nucleosome Sliding Assay section for the assay protocol). Mobilized mononucleosomes were then analyzed on a native 6% polyacrylamide gel in 0.5× TG buffer (12.5 mM Tris, 96 mM glycine, pH 8.3). After electrophoresis, the gel was stained with SYBR Green I nucleic acid stain (Bio Whittaker Molecular Applications) and the fluorescent image was captured with a Fujifilm CCD camera (Fuji Medical Systems).

performed as described above. Figure 4 shows an example of native gel electrophoresis of NURF-mobilized mononucleosomes.

Acknowledgments

We would like to thank Dr. Carl Wu for his support and the methods described in this chapter have been largely established or refined over the past several years in his laboratory at the National Cancer Institute, National Institutes of Health, USA.

---

[20] P. T. Lowary and J. Widom, *J. Mol. Biol.* **276,** 19 (1998).

[23] Methods for Preparation and Assays for *Xenopus* ISWI Complexes

By MATTHEW B. PALMER, STUART ELGAR, and PAUL A. WADE

The African clawed toad *Xenopus laevis* has provided a classic system for the study of nuclear events ranging from transcription and replication to assembly of nuclei. *Xenopus* oocytes and eggs serve as stockpiles of material for replicating nuclei, in preparation for early development in which the fertilized egg undergoes a rapid series of cleavages to produce thousands of nuclei within hours of fertilization. Extracts from these specialized cells have been utilized to investigate the dynamic nature of chromatin. Plasmid DNA is packaged into chromatin with regularly-spaced nucleosomes in an ATP-dependent fashion in oocyte extracts.[1] Biochemical fractionation of these chromatin assembly extracts from *Xenopus* oocytes led to the identification of an ATP-dependent nucleosome spacing activity.[2] *Drosophila* embryo extracts also possess ATP-dependent activities competent for assembling regular arrays of chromatin[3,4] as well as reorganizing local chromatin architecture.[5,6] Detailed biochemical analysis of these extracts resulted in the isolation of three ATP-dependent chromatin remodeling complexes, CHRAC,[7] ACF,[8] and NURF.[9] The SNF2 superfamily member ISWI is the catalytic, ATP-hydrolyzing subunit of these and other chromatin remodeling complexes. The cloning and initial characterization of a *Xenopus* ISWI homolog led to the discovery that, as with *Drosophila*, ISWI activity is necessary for ATP-dependent chromatin assembly in oocyte extracts.[10] Further biochemical characterization of *Xenopus* ISWI by fractionation of egg extract revealed at least three distinct protein complexes, each with functional counterparts in other species: ACF,[10] WICH,[11]

---

[1] G. Glikin, I. Ruberti, and A. Worcel, *Cell* **37,** 33 (1984).

[2] D. J. Tremethick and M. Frommer, *J. Biol. Chem.* **267,** 15041 (1992).

[3] P. Becker and C. Wu, *Mol. Cell. Biol.* **12,** 2241 (1992).

[4] R. Kamakaka, M. Bulger, and J. T. Kadonaga, *Genes Dev.* **7,** 1779 (1993).

[5] T. Tsukiyama, P. B. Becker, and C. Wu, *Nature* **367,** 525 (1994).

[6] P. Varga-Weisz, T. A. Blank, and P. B. Becker, *EMBO J.* **14,** 2209 (1995).

[7] P. Varga-Weisz, M. Wilm, E. Bonte, K. Dumas, M. Mann, and P. B. Becker, *Nature* **388,** 598 (1997).

[8] T. Ito, M. Bulger, M. J. Pazin, R. Kobayashi, and J. T. Kadonaga, *Cell* **90,** 145 (1997).

[9] T. Tsukiyama and C. Wu, *Cell* **83,** 1011 (1995).

[10] D. Guschin, T. M. Geiman, N. Kikyo, D. J. Tremethick, A. P. Wolffe, and P. A. Wade, *J. Biol. Chem.* **275,** 35248 (2000).

[11] L. Bozhenok, P. A. Wade, and P. Varga-Weisz, *EMBO J.* **21,** 2231 (2002).

METHODS IN ENZYMOLOGY, VOL. 377

and CHRAC. In each complex, ISWI associates with a large protein of the BAZ family: ACF1 in the case of ACF[10] and CHRAC, WSTF in WICH.[11] The BAZ family proteins are characterized by a stereotypical domain architecture, including WAKZ, PHD, and Bromodomains. We present below a strategy for purification of the ACF, WICH, and CHRAC complexes from *Xenopus* egg extract as well as protocols for functional activity assays.

Purification of *Xenopus* ISWI Complexes

Upon fertilization, *Xenopus laevis* embryos undergo a series of synchronous divisions over a period of 4–5 h to generate several thousand nuclei at the midblastula transition and the commencement of zygotic transcription. As a consequence of this mode of development, *Xenopus* eggs contain an enormously concentrated store of maternal factors involved in nuclear replication. This makes *Xenopus* eggs an unusually convenient and abundant source of the biochemically diverse subset of ISWI complexes (described earlier) in a concentrated form. Further, while the egg contains sufficient material for several thousand nuclei, it contains only one cell equivalent of nuclear DNA. Thus, one can study chromatin constituents without the necessity of extracting these factors from chromosomes.

Biochemical purification of ISWI complexes is achieved by initial extraction of soluble egg protein from lipids, yolk platelets, and other insoluble material by ultracentrifugation. Following this step, cation and anion exchange chromatography are utilized to resolve the ISWI complexes. Individual ISWI complexes are subsequently purified to near homogeneity using high-resolution cation exchange and heparin chromatography. During purification, ISWI is most conveniently detected by immunoblot.[10] Beginning with 200 ml egg extract, one should obtain 100–250 μg of each ISWI complex following purification, sufficient material for a series of analytical experiments.

Egg Extract Preparation

1. Egg extract preparation follows a previously described protocol.[12] Briefly, mature *Xenopus laevis* females are primed using 50 units HCG (Human chorionic gonadotropin) (Sigma) into the dorsal lymph sac and then induced after 7–10 days by injection of a further 500 units HCG.

2. Eggs are collected into 0.1 M NaCl.

---

[12] P. A. Wade, P. L. Jones, D. Vermaak, and A. P. Wolffe, Vol. 304, p. 715, this series.

3. All subsequent steps are conducted at $4°$. The eggs are dejellied using 2% cysteine (pH 7.8–8.0) and rinsed with chilled buffer A(100) (20 mM HEPES pH 7.5, 5 mM KCl, 1.5 mM MgCl$_2$, 1 mM EDTA, 10% glycerol, 10 mM β-glycerophosphate, 0.5 mM dithiothreitol, 1 mM PMSF, 2 μg/ml pepstatin A and 1 μg/ml leupeptin) (figures in parentheses refer to the NaCl concentration of the buffer).

4. To prepare a high speed egg extract, SW-41 tubes are half filled with eggs, rinsed 3 times using buffer A(100) before filling the tubes with buffer A and centrifugation in an SW-41 rotor at 38,000 rpm for 2 h at $4°$.

5. The clear, straw-colored supernatant is then extracted through the side of the tube using a 21-gauge needle. The volume of supernatant yielded by 50 ml of packed eggs should be expected to be approximately 75–80 ml. Total protein concentration should be determined using the BioRad protein assay using BSA as a standard. Typical protein concentrations are in the 10–15 mg/ml range.

6. The high speed supernatant and all subsequent fractions can be stored at $-70°$ between steps. Under these conditions ISWI complexes remain intact for years.

Extract Fractionation

BioRex 70

1. A BioRex 70 (50–100 mesh, sodium form—BioRad) column is preequilibrated using buffer A(100) and 10 mg of the high speed egg extract is loaded per 1 ml bed volume.

2. The column is washed with 5 column volumes buffer A(100) to remove unbound protein. Bound proteins are eluted using 5 column volumes buffer A(500).

3. Approximately 80% of the total protein eluted in the buffer A(500) step is pooled and dialyzed against buffer A(0) until the NaCl concentration is between 100 and 50 mM.

MonoQ

1. The dialyzed and filtered (0.2 μm) buffer A(500) fraction from the BioRex 70 column is loaded onto a MonoQ 10/10 anion exchange matrix (Amersham Pharmacia) that has been preequilibrated with buffer A(100).

2. The matrix is then washed with 5 column volumes buffer A(100) followed by a 20 column volume linear gradient from 100 to 500 mM NaCl in buffer A.

3. 0.5 column volume fractions are collected, analyzed by immunoblot, and fractions containing ISWI are pooled. ISWI-B, C, and D complexes elute at 290, 330, and 390 mM NaCl, respectively.
4. Prior to further purification, ISWI-B, C, and D fractions should be dialyzed versus buffer A(0) until the NaCl concentration is 50–100 mM and then filtered at 0.2 μm.
5. The ISWI-A peak contains the least abundant ISWI species and demonstrates no detectable ISWI-dependent chromatin remodeling activity. It fails to coelute with any known BAZ family member and is most likely to be composed of free ISWI monomer. ISWI-A fractions can be purified further using sucrose gradient sedimentation.[10]

MonoS

1. The ISWI-B, C, and D complexes (corresponding to WICH, ACF, and CHRAC, respectively) are further purified by chromatography on a MonoS 5/5 cation exchange column (Amersham Pharmacia) that has been preequilibrated with buffer A(100).
2. The loaded matrix is washed with 5 column volumes buffer A(100) followed by a 20 column volume linear gradient from 100 to 500 mM NaCl in buffer A and finally 5 column volumes buffer A(500).
3. Five hundred microliter fractions are collected with the ISWI-B (WICH), C (ACF), and D (CHRAC) complexes eluting at 360, 350, and 330 mM NaCl, respectively.

Heparin

1. Final purification is achieved using a 1-ml heparin-Sepharose hi-trap column (Amersham Pharmacia). The ISWI-B, C, and D fractions from MonoS are dialyzed against buffer A(0) containing 0.1% Triton X-100 until the NaCl concentration is 50–100 mM, filtered, and then loaded onto a column that has been preequilibrated with buffer A(100).
2. The column is then washed with 5 column volumes buffer A(100), and eluted with a 20 column volume linear gradient from 100 mM to 1 M NaCl in buffer A and finally a 5 column volume wash with buffer A(1000) (all heparin-Sepaharose column buffers are supplemented with 0.1% Triton X-100).
3. Five hundred microliter fractions are collected with the ISWI-B, C and D complexes eluting from the heparin-Sepharose column at 540, 570, and 540 mM NaCl respectively (see Fig. 1).
4. Following purification, the ISWI complexes are stored at -80° where they remain active for multiple freeze thaw cycles.

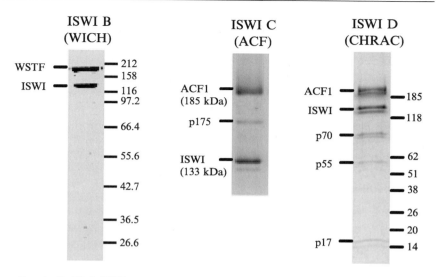

Fig. 1. Purified ISWI complexes from *Xenopus* egg extract. ISWI complexes were purified to near homogeneity and separated using 8% (ISWI B, C) and 4–20% (ISWI D) SDS-PAGE. Proteins were visualized by silver staining. Known ISWI complex subunits are identified. Protein molecular weight marker sizes are given in kDa.

Assays for *Xenopus* ISWI Complexes

Three defining and related activities of the xISWI protein are its ability to hydrolyze ATP, to alter the position of a histone octamer relative to nucleosomal DNA, and to assemble regular, physiologically-spaced arrays of nucleosomes. Following is a description of each of these assays.

ATPase Assay

ISWI and its complexes, like other nucleosome remodeling factors, catalyze the hydrolysis of ATP specifically in the context of nucleosomes.[9–10,13] Neither DNA nor histones alone can stimulate ATPase activity, implying that these complexes recognize a unique structural feature of the nucleosome that is not present in DNA or histones alone. The nucleosome specificity of this ATP hydrolysis can be assessed *in vitro* in an ATPase assay consisting of defined components. In the presence of γ-^{32}P-ATP, the purified ISWI complex of interest is incubated with chromatin

---

[13] D. Corona, G. Langst, C. Clapier, E. J. Bonte, S. Ferrari, J. W. Tamkun, and P. B. Becker, *Mol. Cell* **3**, 239 (1999).

(either mononucleosomes or long chromatin), DNA, or histones. The hydrolyzed phosphate is then resolved from ATP by thin layer chromatography and quantified. Substrates in this assay include salmon sperm DNA, chicken erythrocyte mononucleosomes, or purified chicken erythrocyte core histones. Long chromatin fragments lacking linker histones are prepared from chicken erythrocytes as described.[14] Core histones are prepared from chicken erythrocytes.[15]

Procedure

1. A reaction cocktail is assembled containing 60 ng DNA or molar equivalent of chromatin or histones. Buffer and ATP concentrations are as follows: 6.6 mM HEPES pH 7.6, 0.66 mM EDTA, 0.66 mM β-mercaptoethanol, 0.033% NP-40, 1.1 mM MgCl$_2$, 33 μM ATP, 5 μCi γ-^{32}P-ATP. Final reaction volume is 10 μl.

2. Add purified ISWI complex or recombinant ISWI (1 μl).

3. Incubate at room temperature for 30 min.

4. Spot reactions (5 μl) onto PEI-cellulose TLC plates (Sigma).

5. Develop plates in 1 M formic acid, 0.5 M LiCl in a closed glass container.

6. Expose to film (15 min exposure is generally sufficient). ATP hydrolysis can be quantitated either through scintillation counting of regions of the TLC plate cut with a razor blade or using a PhosPhor Imager.

Nucleosome Mobility Assay

A defining feature of ISWI and its complexes is the ability to catalyze the movement of histone octamers relative to the associated nucleosomal DNA. This process can be monitored *in vitro* using purified components in an assay that relies on differential migration of DNA-protein complexes in native polyacrylamide gels. Mononucleosomes assembled on a specific fragment of DNA possess identical mass and charge, but will migrate differentially according to the position of the histone octamer on the DNA strand.[16] When nucleosomes are deposited by salt dialysis on a 250-bp fragment from the *Xenopus* thyroid hormone receptor βA gene (TR βA), four distinct bands can be resolved on a gel representing four preferred locations for nucleosome positioning. These positions have been mapped by micrococcal nuclease digest.[17] Individually positioned nucleosomes can be eluted from the gel, incubated with ISWI or any of its complexes in

---

[14] R. Kornberg, J. W. LaPointe, and Y. Lorch, Vol. 170, p. 3, this series.

[15] R. Simon and G. Felsenfeld, *Nucl. Acids Res.* **6**, 689 (1979).

[16] G. Meersseman, S. Pennings, and E. M. Bradbury, *EMBO J.* **11**, 2951 (1992).

[17] D. Guschin, P. A. Wade, N. Kikyo, and A. P. Wolffe, *Biochemistry* **39**, 5238 (2000).

the presence or absence of ATP, and again resolved by electrophoresis to reveal repositioning of the remodeled nucleosomes (see Fig. 2). Under these conditions, all three *Xenopus* ISWI complexes move nucleosomes towards the center of this particular DNA fragment.

Reagents

10× Klenow buffer: 500 mM Tris pH 8, 50 mM MgCl$_2$, 10 mM DTT
Deoxynucleotide triphosphate mix
α-^{32}P-ATP (10 mCi/ml; 3000 Ci/mmol)
Nucleosome Mobilization Buffer: 10 mM HEPES 7.5, 0.2 mM EDTA, 5% glycerol, 2 mM MgCl$_2$, 1 mM ATP (Mg salt) or analog
Gel Elution Buffer: 10 mM Tris-Cl (7.6), 0.2 mM EDTA, 1 mM MgCl$_2$, 5% glycerol, 0.02% BSA, 50 μg/ml of sheared salmon sperm DNA
Purified chicken erythrocyte histone octamers
250 bp Ase1-Sty1 fragment of *Xenopus* TR βA gene[17]

Procedure

1. Isolate the Asel-Sty1 fragment (250 bp) from the *Xenopus* TR βA gene. Radiolabel the recessed 3′ ends using Klenow polymerase and standard protocol:

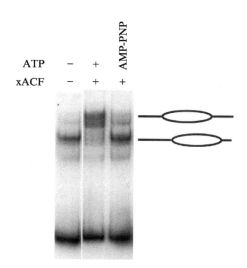

FIG. 2. In a nucleosome mobility assay, the *Xenopus* ACF complex alters the position of a histone octamer relative to DNA in an ATP-dependent manner. A uniquely positioned nucleosome is extracted from a gel, incubated with purified ISWI-containing complex and ATP, and resolved by native polyacrylamide gel electrophoresis to reveal the repositioned nucleosome.

Incubate DNA, Klenow (1 unit), 5 μl of α-^{32}P-ATP (10 mCi/ml; 3000 Ci/mmol), cold dNTPs (0.25 mM final), and diluted 10× Klenow buffer for 15 min at 30°. Stop the reaction at 70° for 10 min.

2. Labeled DNA is reconstituted with histone octamers by salt dialysis as follows. Equimolar ratios of purified chicken histone octamers (the concentration of purified histones can be determined by absorbance at 230 nm, with A_{230} of 4.3 for 1 mg/ml octamer solution) and labeled DNA fragment are mixed and NaCl added to a final concentration of 2 M. This is dialyzed against stepwise reductions in NaCl concentration from 2 M through 1.5, 1, 0.75, and finally 0 M in the presence of the protease inhibitor PMSF (0.2 mM).

3. Reconstituted nucleosomes are resolved in nondenaturing polyacrylamide gel electrophoresis, 5% polyacrylamide in 0.5× TBE buffer.

4. A selected positional isomer is isolated by excising a distinct band from the gel and eluting by crushing with a plastic pestle in a microcentrifuge tube. This mixture is suspended in 100 μl of 10 mM Tris 7.6, 0.2 mM EDTA, 1 mM MgCl$_2$, 5% glycerol, 0.02% BSA, 50 μg/ml sheared salmon sperm DNA and incubated on ice for 30 min. The eluted nucleosome is recovered by centrifugation.

5. The mononucleosome is incubated with a purified ISWI complex or recombinant ISWI for 1 h at 30° in 10 mM Tris-HCl (pH 7.6), 50 mM NaCl, 3 mM MgCl$_2$, 1 mM β-mercaptoethanol, 1 mM ATP.

6. Remodeled nucleosomes are again resolved in 5% polyacrylamide gel, 0.5× TBE. Gels are dried and exposed to film.

Nucleosome Spacing Assay

Related to the ability of ISWI complexes to reposition a single nucleosome on a short fragment of DNA is the ability to reposition an array of nucleosomes on a longer DNA template, such that the nucleosomes are positioned equidistant from one another.[7,8,10] This organization of regular nucleosomal arrays is the basis for a third functional assay for ISWI complexes. When exogenous DNA is incubated in *Xenopus* oocyte extract supplemented with ATP, it is assembled into chromatin with regular nucleosomal spacing.[1] The spacing activity is abolished by the addition of an antibody against ISWI to the extract, but recovered when purified xACF is restored to the reaction[10] (see Fig. 3). Spacing regularity is manifest by a precisely repeating banding pattern on gel electrophoresis of chromatin partially digested by micrococcal nuclease. This enzyme makes double-stranded cuts in DNA between nucleosomes, while DNA that is tightly associated with histones is protected from cleavage. Limiting the digest by

FIG. 3. Nucleosome spacing assays. Periodicity of nucleosomal arrays assembled in *Xenopus* oocyte extract revealed by micrococcal nuclease digest. Regular spacing is lost when oocyte extract is treated with anti-ISWI antibody, but restored upon addition of purified xACF complex.

time generates a series of DNA fragment lengths which, in the absence of spacing activity, appears as a smear on gel electrophoresis. When spacing between nucleosomes is equal, these fragments pool together in populations of uniform lengths corresponding to mononucleosomes, dinucleosomes, trinucleosomes, etc., revealing a ladder upon gel electrophoresis. Nucleosome spacing activity by ISWI complexes can be demonstrated in two contexts: with chromatin assembled by oocyte extract, or with salt-dialyzed chromatin from purified components.

Nucleosome Spacing in Xenopus Oocyte Extract

The oocyte extract is prepared according to the method described in an earlier volume.[18] This approach involves surgically isolating the ovaries and treating with collagenase to release the oocytes. After several washes, the oocytes are lysed by high-speed centrifugation in hypotonic buffer. The resulting extract is a rich source of nuclear proteins that, when ATP

---

[18] D. J. Tremethick, Vol. 304, p. 50. this series.

is added, assembles exogenous DNA into regularly spaced chromatin. Inclusion of ISWI antibody in the reaction abolishes this spacing regularity.

Buffers

OR2 buffer: 2.5 mM KCl, 82.5 mM NaCl, 1 mM CaCl$_2$, 1 mM MgCl$_2$, 5 mM HEPES, 1 mM Na$_2$HPO$_4$, pH 7.6

Extraction buffer: 1 mM EGTA, 5 mM KCl, 1.5 mM MgCl$_2$, 10% (v/v) glycerol, 20 mM HEPES, 10 mM β-glycerophosphate, 0.5 mM DTT, 10 μg/ml PMSF, 2 μg/ml leupeptin, 2 μg/ml pepstatin, pH 7.5

ATP buffer: 10 mM MgCl$_2$, 30 mM ATP, 400 mM creatine phosphate, 10 mM β-glycerophosphate, 0.5 mM DTT, 1 mM EGTA, 5 mM KCl, 10% glycerol, 20 mM HEPES, 10 μg/ml PMSF, 2 μg/ml leupeptin, 2 μg/ml pepstatin, pH 7.5

Protocol

1. Prepare the oocyte extract
   A. Collect the ovaries from seven adult female frogs. Anesthetize the frogs in tricaine and make an anterior/posterior incision on each side of the ventral abdomen. Pull out the ovaries through each incision with forceps and place them into 200 ml of OR2 buffer. After swirling, place the ovaries into another 200 ml of OR2 buffer. Repeat this washing 4 times.
   B. Place the ovaries into a flask with 200 ml of OR2 buffer plus 0.15 g collagenase (>400 U/mg protein). Shake the flask in a 25° water bath for 2 h.
   C. Wash the oocytes to remove collagenase and immature oocytes 5 times in 200 ml OR2 buffer. In each wash, disperse the oocytes by swirling. When the large, mature oocytes begin to settle, decant the buffer above containing the small oocytes and tissue debris.
   D. Transfer the oocytes into 50 cc Falcon tubes until half-full, and fill the remainder with extraction buffer. Wash by inverting the tubes twice and removing the buffer above after settling. Repeat three washes in extraction buffer.
   E. Using a transfer pipet, put the oocytes into 6 SW-41 ultracentrifuge tubes at 4° until 1 cm below the top. Remove the buffer and refill with fresh extraction buffer.
   F. Centrifuge at 36 K for 2 h at 4°.
   G. Recover the extract by inserting a 19G needle (on a 5-ml syringe) just above the pellet. Avoid the pellet and the lipid layer at the top. Add protease inhibitors to the extract and aliquot. The extract can be stored at −70° for 3 years.

2. Add increasing amounts (50–500 ng) of circular relaxed plasmid DNA to an appropriate number of Eppendorf tubes in a total volume of 10 μl, on ice, using the extraction buffer as a dilution buffer.

3. Add 5 μl creatine kinase (1 ng/μl) followed by 5 μl of ATP buffer.

4. Add ISWI antibody to 30 μl of oocyte extract. Slowly introduce this extract into the reaction mix (final volume 50 μl) and incubate for 5 h at 27°.

5. Perform limited micrococcal nuclease digest. Samples are digested with 0.05 units of MNase per microgram of DNA in the presence of 1 mM CaCl$_2$. Take aliquots at 2 and 5 min and quench the reactions by adjusting to 1/5 volume of 5% SDS, 25% glycerol, 10 mM Na$_2$EDTA, 0.3% bromophenol blue, 0.25 μg/μl proteinase K. Incubate for 30 min at 37° to deproteinize.

6. Electrophorese in 1% agarose gel, followed by incubation in ethidium and UV visualization.

Salt Dialysis

Plasmid DNA can be reconstituted as chromatin in the presence of purified core histones by gradually decreasing salt concentration. The histones assemble onto DNA in a stepwise fashion as the salt concentration falls from 2 to 0 M,[19] first as H3/H4 tetramers followed by H2A/H2B dimers. The procedure described below recapitulates these events by stepwise dialysis against decreasing salt concentration. When such dialyzed chromatin is visualized by gel electrophoresis after micrococcal nuclease digest, a closely packed spacing of nucleosomes is observed. ISWI complexes have the effect of reorganizing this spacing to a more extended, regular array (see Fig. 4). Distinct bands of nucleosome-protected DNA fragments are apparent, representing oligonucleosomes with uniform spacing.

Procedure

1. To reconstitute plasmid DNA with histone octamers by salt dialysis, molar amounts of histones and of nucleosomal DNA should be equal. The concentration of purified histones can be determined by absorbance at 230 nm, with A$_{230}$ of 4.3 for 1 mg/ml octamer solution.[20] Since each octamer associates with $\sim$200 bp of nucleosomal DNA, the molarity of nucleosomal DNA equivalents in the plasmid can be calculated by assuming 200 bp to be 1 molar unit. Equimolar ratios of DNA fragment

[19] J. Hansen, K. E. van Holde, and D. Lohr, *J. Biol. Chem.* **266,** 4276 (1991).
[20] A. Stein, *J. Mol. Biol.* **130,** 103 (1979).

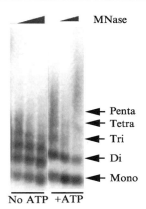

FIG. 4. Nucleosomal arrays assembled by salt dialysis. When arrays are incubated with xACF but without ATP, nucleosomes assume a close-packed configuration. Upon addition of ATP, nucleosomes are spaced in an extended, regular array. The positions of DNA fragments resulting from mono-, di-, tri-, tetra-, and penta-nucleosomal particles are indicated.

and purified chicken histone octamers are mixed and NaCl added to a final concentration of 2 M. This is dialyzed against stepwise reductions in NaCl concentration from 2 M through 1.5, 1, 0.75, and finally 0 M in the presence of the protease inhibitor PMSF (0.2 mM).

2. The reconstituted chromatin is incubated with purified ISWI complex for 1 h at 30° in 10 mM Tris-HCl (pH 7.6), 50 mM NaCl, 3 mM MgCl$_2$, 1 mM β-mercaptoethanol, 1 mM ATP.

3. Remodeled chromatin is analyzed by partial micrococcal nuclease digest. Samples are digested with 0.05 units of micrococcal nuclease per microgram of DNA in the presence of 1 mM CaCl$_2$. Aliquots are taken at 2 and 5 min and quenched by adding to 1/5 volume of 5% SDS, 25% glycerol, 10 mM Na$_2$EDTA, 0.3% bromophenol blue, 0.25 μg/μl proteinase K. Samples are incubated for 30 min at 37° to deproteinize.

4. Electrophorese in 1% agarose gel, followed by incubation in ethidium and UV visualization.

[24] Functional Analysis of ISWI Complexes in Mammalian Cells

By LUDMILA BOZHENOK, RAYMOND POOT, NADINE COLLINS, and PATRICK VARGA-WEISZ

*I*mitation *Swi*tch (ISWI) is an abundant nucleosome remodeling ATPase conserved from budding yeast to human (reviewed in Längst and Becker[1]). ISWI does not mediate the disruption of nucleosomes but causes the sliding of nucleosomes along the DNA template.[2,3]

ISWI forms several distinct chromatin remodeling complexes interacting with various other proteins. In *Drosophila* ISWI was found to be the "engine" of three distinct nucleosome remodeling factors: *Nu*cleosome *Re*modeling *Factor*[4] (NURF), *Ch*romatin *A*ccessibility *C*omplex[5] (CHRAC) and *A*TP-utilizing *C*hromatin remodeling and assembly *Factor*[6] (ACF). In mammalian cells two highly related isoforms of ISWI were identified, SNF2L (SNF2-"like") and SNF2H (SNF2-"homologue"). To date, only complexes containing SNF2H have been described. The analysis of extracts from the human HeLa cell line revealed the existence of human ACF[7,8] and CHRAC[9] that both contain the ACF1 subunit, and the WSTF-*I*SWI *Ch*romatin remodeling (WICH-) complex[10] that contains *Wi*lliams-*Syn*drome *Tr*anscription *Fa*ctor (WSTF) that is related in domain structure to ACF1. In addition, HeLa cells contain *R*emodeling and *S*pacing *Factor* (RSF), a complex of SNF2H and a protein of unknown sequence, p325, that was shown to have activity in nucleosome assembly *in vitro*.[11,12] *N*ucleolar *R*emodeling *C*omplex (NoRC), purified from mouse cells, contains

[1] G. Längst and P. B. Becker, *J. Cell. Sci.* **114,** 2561 (2001).
[2] G. Längst, E. J. Bonte, D. F. Corona, and P. B. Becker, *Cell* **97,** 843 (1999).
[3] A. Hamiche, R. Sandaltzopoulos, D. A. Gdula, and C. Wu, *Cell* **97,** 833 (1999).
[4] T. Tsukiyama, C. Daniel, J. Tamkun, and C. Wu, *Cell* **83,** 1021 (1995).
[5] P. D. Varga-Weisz, M. Wilm, E. Bonte, K. Dumas, M. Mann, and P. B. Becker, *Nature* **388,** 598 (1997).
[6] T. Ito, M. Bulger, M. J. Pazin, R. Kobayashi, and J. T. Kadonaga, *Cell* **90,** 145 (1997).
[7] D. A. Bochar, J. Savard, W. Wang, D. W. Lafleur, P. Moore, J. Cote, and R. Shiekhattar, *Proc. Natl. Acad. Sci. USA* **97,** 1038 (2000).
[8] G. LeRoy, A. Loyola, W. S. Lane, and D. Reinberg, *J. Biol. Chem.* **275,** 14787 (2000).
[9] R. A. Poot, G. Dellaire, B. B. Hulsmann, M. A. Grimaldi, D. F. Corona, P. B. Becker, W. A. Bickmore, and P. D. Varga-Weisz, *EMBO J.* **19,** 3377 (2000).
[10] L. Bozhenok, P. A. Wade, and P. Varga-Weisz, *EMBO J.* **21,** 2231 (2002).
[11] G. LeRoy, G. Orphanides, W. S. Lane, and D. Reinberg, *Science* **282,** 1900 (1998).
[12] A. Loyola, G. LeRoy, Y. H. Wang, and D. Reinberg, *Genes Dev.* **15,** 2837 (2001).

SNF2H and TIP5 (TTF-1-interacting protein 5), a protein that targets nucleolar sites and is distantly related to ACF1.[13] Another complex containing SNF2H, subunits of the NuRD complex and cohesin subunits was recently shown to be involved in loading of cohesin onto human chromosomes.[14]

Whereas the *in vitro* analysis of ISWI-complex activities has provided important insights into the mechanisms of ATP-dependent nucleosome remodeling and potential activities of these complexes, little is known about their *in vivo* activities, especially in mammalian cells. The use of dominant negative forms of ISWI and ACF1 has provided insights into the *in vivo* function of these proteins.[14,15] Also, over-expression of ISWI-interacting TIP5 has been used to study its *in vivo* function.[16,17] Potentially the most promising approach to gain functional insights *in vivo* is the use of small interfering RNA (siRNA) oligonucleotides that target the mRNA of the protein of interest via the RNA interference (RNAi) pathway resulting in the depletion of the protein.[18,19] Within cells, the RNA-induced silencing complex uses siRNAs to guide the sequence specific degradation of mRNAs.[20]

RNA interference with siRNA allows the rapid, transient depletion of a protein in cells. Protein amounts can be significantly reduced by RNAi, but the proteins may not get eliminated ("knockdown" rather than "knock out"). This is important to keep in mind when studying an enzyme such as ISWI that may perform its biological function even at very reduced levels. However, this may be an advantage when studying an essential protein as ISWI in *Drosophila*[21] and the analysis will reveal what happens when the protein becomes limiting. siRNA may be expressed in the cell from expression plasmids or viral vectors, but we found transfection of

[13] R. Strohner, A. Nemeth, P. Jansa, U. Hofmann-Rohrer, R. Santoro, G. Langst, and I. Grummt, *EMBO J.* **20,** 4892 (2001).

[14] M. A. Hakimi, D. A. Bochar, J. A. Schmiesing, Y. Dong, O. G. Barak, D. W. Speicher, K. Yokomori, and R. Shiekhattar, *Nature* **418,** 994 (2002).

[15] N. Collins, R. A. Poot, I. Kukimoto, C. Garcia-Jimenez, G. Dellaire, and P. D. Varga-Weisz, *Nat. Genet.* **32,** 627 (2002).

[16] Y. Zhou, R. Santoro, and I. Grummt, *EMBO J.* **21,** 4632 (2002).

[17] R. Santoro, J. Li, and I. Grummt, *Nat. Genet.* **32,** 393 (2002).

[18] S. M. Elbashir, J. Harborth, W. Lendeckel, A. Yalcin, K. Weber, and T. Tuschl, *Nature* **411,** 494 (2001).

[19] S. M. Elbashir, J. Harborth, K. Weber, and T. Tuschl, *Methods* **26,** 199 (2002).

[20] J. Martinez, A. Patkaniowska, H. Urlaub, R. Luhrmann, and T. Tuschl, *Cell* **110,** 563 (2002).

[21] R. Deuring, L. Fanti, J. A. Armstrong, M. Sarte, O. Papoulas, M. Prestel, G. Daubresse, M. Verardo, S. L. Moseley, M. Berloco, T. Tsukiyama, C. Wu, S. Pimpinelli, and J. W. Tamkun, *Mol. Cell* **5,** 355 (2000).

chemically synthesized[21] nucleotide RNA duplexes a rapid and easy way to deplete SNF2H and ACF1 from mammalian cells.[15]

In this chapter, we describe the use of RNA interference to obtain functional insights about ISWI and ISWI-interacting protein ACF1 in mammalian cells. Even though, many aspects of this technique are generic, each protein requires specific conditions for depletion. We explain in detail some of the analysis of the effects of depletion of these proteins on DNA replication. In addition, we describe techniques for the localization analysis of ISWI (SNF2H) to replication foci.

Depletion of SNF2H and hACF1 by RNAi

Selection and Preparation of siRNA Oligonucleotides

We obtain chemically synthesized siRNA oligonucleotides 2'-deprotected and desalted by the manufacturer (Dharmacon Research, Lafayette, CO). The sequences are selected as recommended by the manufacturer (see: www.dharmacon.com). In brief, the sequence of the siRNA were chosen at least 75 bases downstream from the start codon and start with a AA dimer followed by 19 nucleotides and are ~50% in G/C content. Selected sequences were subjected to a BLAST-search analysis (NCBI database) against EST libraries to rule out the targeting of another gene. The siRNA duplex contains a 19-nucleotide duplex region and dTdT overhangs at each 3'-terminus. We initially select at least two candidate siRNA oligonucleotides for each protein. We found that approximately half of the selected siRNA duplexes turned out to be efficient in depleting the target mRNA. Sense sequence of oligonucleotides used: ACF1, aacacugugaaccacaagaug (nucleotides 689–709 from open reading frame); SNF2H, aagaggaggaugaagagcuau (nucleotides 443–463). We used as a control siRNA duplexes that matched but do not deplete hACF1 (aaguccucccugcaucacauc, nucleotides 413–433) or that have the central 4 nucleotides inversed for SNF2H (aagagga*uaggg*aagagcuau). The hACF1 siRNA duplex was found to be also effective in depleting ACF1 from mouse (3T3) cells despite of one mismatch to the target sequence, albeit with somewhat lesser efficiency.

Complementary single-stranded siRNA oligonucleotides are annealed according to the manufacturer's recommendations: to obtain 20 μM of siRNA duplex we mixed together both strands in RNase-free 100 mM potassium acetate, 20 mM HEPES-KOH pH 7.4, 2 mM magnesium acetate and incubated 1 min at 90°, briefly spun and further incubated for 1 h at 37°. Annealed siRNA duplexes are used at room temperature and can be kept on the bench during the transfection procedure. They are stored at $-20°$ and subjected to maximal 5 thaw/freeze cycles.

Transfection of siRNA Duplexes into Cells

Different cell types and cell lines show different susceptibility to siRNA mediated protein depletion. HeLa cells are very well suited for this approach, possibly because of their low content in RNAses and easy transfectability.[20] However, we also depleted ACF1 from 3T3 cells.

We grow cells as monolayer in Dulbecco's MEM with GLUTAMAX-I and pyruvate (Invitrogen) and 10% fetal bovine serum. Omission of antibiotics may increase transfection efficiency, but we have successfully performed siRNA mediated depletions in the presence of antibiotics as well. We routinely deplete an estimated 80–95% of ACF1 from HeLa cells 2 days after transfection, as judged by Western blot analysis of whole cell extracts (see Fig. 1A). Depletion of SNF2H is less efficient (see Fig. 1B), but sufficient enough to show a clear effect in DNA replication. An analysis of the depletion efficiency using immunofluorescence microscopy of cells indicates that the depletion does not affect cells uniformly, but there are patches of cells that are less affected by the siRNA (see Fig. 1B for depletion of SNF2H).

Procedure

1. Seed cells at ~30% confluence (~3 × 10^5 cells) in a 3.5 cm diameter dish (here we use 6 well-dishes) in medium without antibiotics. Experiments can also be performed in 12 well, 2 cm diameter dishes, but then use half the volumes.
2. Let the cells attach (2–3 h in CO_2 incubator).
3. For each well to be transfected, dilute 10 μl RNA duplex from the 20 μM stock in 175 μl of OPTI-MEM-1 medium (with GLUTA-MAX-I, Invitrogen) and incubate at room temperature for 5–10 min in an 1.5 ml Eppendorf tube.
4. Dilute 4 μl OLIGOFECTAMINE (Invitrogen) in 11 μl of OPTI-MEM-1 and incubate at room temperature for 5–10 min in an 1.5 ml Eppendorf tube.
5. Add diluted duplex (step 3) to diluted OLIGOFECTAMINE (step 4), gently mix and let the mixture sit at room temperature for another 20–25 min in the Eppendorf tube.
6. Take cells from the incubator, pipet off medium and add 800 μl of fresh medium to each well.
7. Spread the 200 μl of transfection mixture drop by drop onto the cells, mix gently and put cells back to the incubator and leave overnight (16–20 h).
8. After 16–20 h transfection, wash cells with fresh growth medium twice, add 1.5 ml medium to the well and leave them for another

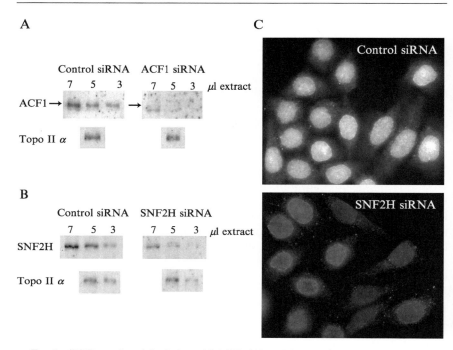

Fig. 1. siRNA mediated depletion of hACF1 (A) and SNF2H (B and C) in HeLa cells. Indicated amounts of whole cell extract prepared as described in the text were analyzed by Western blot with antibody against hACF1 (A), SNF2H (B), and Topoisomerase II alpha as loading control. The running position of ACF1 is indicated by an arrow. A protein of slightly higher mobility cross reacts with the crude anti-ACF1 antiserum used here. (This band is not detected with the affinity-purified antibody.) (C) Analysis of siRNA mediated depletion of SNF2H in Hela cells with immunofluorescence microscopy. Cells were treated with control siRNA (upper panel) or SNF2H siRNA (lower panel) and fixed as described in the text. SNF2H was visualized with an antibody against SNF2H.

26–52 h (2–3 days post transfection) before further analysis. It is important to remove the transfection reagent with at least two washes of medium, because it interferes with the normal cell cycle progression.

9. Confirm depletion of protein by Western blot analysis of whole cell extract as outlined here.

Preparation of Whole Cell Extracts for Western Analysis

1. Collect 5–10 μl of cell pellet (after trypsinization) in an Eppendorf tube (usually all cells from one well).
2. Wash the cells with PBS, remove any excess liquid.

3. Take up the cells in 50–100 μl of 2× SDS loading buffer (100 mM Tris–HCL, pH 6.8, 20% glycerol, 4% SDS, 200 mM DTT, 0.2% Bromophenol Blue) and heat up to 95°, 5 min.
4. Before loading on a gel sonicate the samples to fragment DNA in a sonicator water bath for 15 s (e.g., Pulsatron, Kerry).

Analysis of the Effect of ACF1 Depletion on S Phase Progression by Flow Cytometry

We observed that depletion of ACF1 and SNF2H by RNAi causes defects in cell cycle progression.[15] Depletion of SNF2H slows progression through S phase in general. Depletion of ACF1 by RNAi delays cell cycle progression at the late S stage.[15] This is best monitored by flow cytometry of synchronized cells. Flow cytometry allows the determination of the number of cells containing a specific DNA content, representing cells in G1 ($2n$), S ($2n$–$4n$), and G2 ($4n$). The experiment shown in Fig. 2 was performed on day 3 (72 h) posttransfection of the siRNA. Aphidicolin is a DNA polymerase inhibitor and blocks cell cycle progression of the majority of cells at the G1/S border. We found that a single block by aphidicolin is sufficient to synchronize more than 80% of the HeLa cells at the G1/S border and to observe an effect of ACF1 depletion in HeLa cells.

Cell Arrest at the G1/S Border and Synchronization

1. Prepare a stock solution of aphidicolin (Sigma) at 3 mg/ml in methanol. This can be stored at 4° for 2 months.
2. Add medium containing 3 μg/ml aphidicolin to exponentially growing HeLa cells. Exposure of the cells to the drug for 15–18 h results in accumulation of more then 80% of cells at the G1/S border.
3. Release aphidicolin-arrested cells from the G1/S block by washing out the aphidicolin by three changes of fresh, warm medium. Released cells synchronously start S-phase.

Fixation of Cells for Flow Cytometry

1. Rinse cells in dish with PBS (phosphate buffered saline, prepared from tablets, Sigma, 1–2 ml per well).
2. Add 0.5 ml of trypsin to each well (Trypsin-EDTA, Invitrogen), incubate in CO_2 incubator for 5 min.
3. Obtain single cell suspension by gently pipetting cells in the trypsin solution.

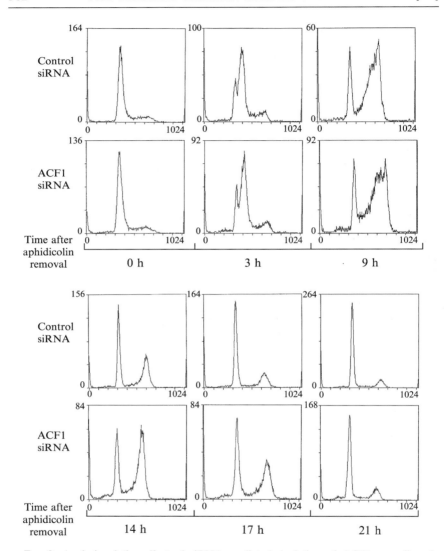

FIG. 2. Analysis of the effect of siRNA mediated depletion of ACF1 on cell cycle progression by flow cytometry. HeLa cells transfected with ACF1 siRNA or control siRNA were subjected to analysis at indicated times after release from an aphidicolin block. S-phase progression is essentially identical between the ACF1 depleted cells and the control cells at the earlier time points (3 and 9 h after wash out of aphidicolin). However, there is an accumulation of the ACF1 depleted cells at late S/G2 compared to the control cells (14 and 17 h). The x-axis is for DNA content, y-axis for cell number.

4. Collect cells in 30 or 50 ml "universal" or "Falcon" tubes (plastic tubes with conical bottoms) with medium containing serum to inactivate trypsin, and pellet cells at 1000 rpm for 5 min at $4°$.
5. Resuspend cells in 100 μl of PBS. Add 900 μl of cold $(-20°)$ methanol drop-wise and mix well (vortex). Immediately put the sample at $4°$ for 0.5 h. Samples can be stored in methanol at $4°$ for a few weeks.
6. Spin down fixed cell at 1000–2000 rpm for 5 min at $4°$.
7. Resuspend cells in 100 μl of PBS. Add 900 μl of DNA stain (Beckman Coulter), mix well and leave for 0.5 h at RT. The samples are now ready for analysis by flow cytometry. We analyze 10,000 cells per sample with the EPICS XL flow cytometer (Beckman Coulter) using standard settings.

Analysis of Subnuclear Localization of SNF2H in Combination with PCNA or BrdU Staining

The analysis of localization of SNF2H and SNF2H interacting proteins ACF1 and WSTF suggested a function of these proteins during replication of pericentromeric heterochromatin.[10,15] We monitor DNA replication using the nucleotide analogue Bromo-deoxyuridine (BrdU) or Proliferating Cell Nuclear Antigen (PCNA).

We studied subnuclear localization of ISWI (SNF2H) and ISWI-interacting proteins using antipeptide antibodies against those proteins and fusion proteins with the Green Fluorescent Protein (GFP) expressed from transfected vectors.[10,15] The antipeptide antibodies were raised in rabbit against Keyhole Limpet Hemocyanine coupled synthetic peptides of 15–20 amino acids matching either C- or N-terminus of the proteins. We affinity purify the antibodies usually for immunocytochemistry using the immunogenic peptides coupled to Sepharose beads in a standard protocol.

Although SNF2H is an abundant protein in NIH 3T3 cells, immunolocalization studies for SNF2H, especially its localization in heterochromatin, are prone to potential fixation artefacts. Studies using an SNF2H-GFP fusion protein (GEP at the C-terminus) showed, for example, that fixation with formaldehyde abolishes the localization to pericentromeric heterochromatin while fixation with acidic ethanol resulted in a localization to the nucleoli.

The correct fixation method for SNF2H, as judged by analyzing live cells expressing SNF2H-GFP, is outlined here and uses fixation with cold methanol. It can be used for cells expressing SNF2H-GFP, or to immunostain cells with an anti-SNF2H antibody. In this protocol the anti-SNF2H immuno-staining or direct visualization of SNF2H-GFP fluorescence is

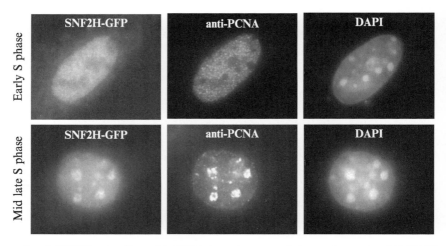

FIG. 3. SNF2H colocalizes with PCNA in pericentromeric heterochromatin of NIH 3T3 cells at mid late S phase. The localization of SNF2H-GFP fusion protein was analyzed using the GFP auto-fluorescence, an antibody against PCNA and DAPI stain.

combined with staining against PCNA, as illustrated in Fig. 3, or against incorporated BrdU.

Paraformaldehyde fixation (1%) was essential to visualize ACF1 localization in the pericentromeric heterochromatin of 3T3 mouse cells with an anti-ACF1 antibody.[15] ACF1-GFP can be visualized in the pericentromeric heterochromatin after methanol or formaldehyde fixation.

In the protocol described here for colocalization analysis of PCNA with SNF2H, we do not subject the cells to a Triton X-100 extraction step that is often used to eliminate nonreplication associated PCNA staining.[22] We found that this extraction step was not necessary to visualize replication foci associated PCNA in the pericentromeric heterochromatin of the 3T3 cells.

Visualization of BrdU incorporation requires a DNA denaturation step that compromises SNF2H-GFP auto fluorescence. The method below is especially adapted to preserve the GFP fluorescence during the staining process.

Procedure

1. Seed NIH 3T3 cells on ethanol washed 16 mm diameter glass cover slips (Thickness No. 1, Borosilicate glass, BDH) in a 12-well plate

---

[22] R. Bravo and H. Macdonald-Bravo, *J. Cell Biol.* **105,** 1549 (1987).

(NUNC) in medium as described earlier at approximately 30% confluence.

2. After the cells have attached (2 h), they can be transfected with 1 μg of pEGFP-N3-SNF2H plasmid DNA. pEGFP-N3-SNF2H is a derivative of pEGFP-N3 (Clontech) and expresses SNF2H fused at the C-terminus to Enhanced Green Fluorescent Protein (EGFP). We use FuGENE-6 (Roche) as transfection reagent with a protocol recommended by the supplier. For SNF2H immuno-staining, leave cells overnight before fixation and staining (see later).

3. Replace the transfection-medium with fresh medium next morning after one wash with 37° warm serum-free medium. Let cells grow 16 h to allow sufficient expression, but not more than 24 h after transfection, because prolonged expression of SNF2H-GFP is toxic to the cells.

4. For BrdU staining, add 1 ml of 37° warm medium containing 10 μM BrdU (Sigma, the stock solution, 10 mM in PBS, can be stored frozen for 6 months and freeze-thawed many times; after thawing take care to dissolve BrdU precipitates completely by incubation at 37°). Incubate cells immediately in 37° CO_2-incubator for at least 10 min.

5. Wash cells once with 1 ml PBS and fix them by addition of 1 ml -20° cold methanol. Add the methanol slowly and with minimal disturbance of the cells (the cells detach at this stage if the methanol is added too violently), leave 10–20 s, and then move the plate carefully to 4° and leave for at least 20 min.

6. Wash out the methanol with 3 rinses of 1 ml PBS, each.

7. For staining with anti-SNF2H and/or anti-PCNA, add 1 ml of PBS with 10% FBS (blocking solution) and slowly rock for at least 30 min at room temperature, for SNF2H-GFP + anti-BrdU staining go to step 16.

8. Remove PBS with 10% FBS completely by suction with a fine pipet tip, place cover slip in the middle of the well.

9. Add 50–70 μl of anti-SNF2H antibody diluted 1:100 in PBS with 10% FBS. Anti-PCNA antibody (dilution 1:100, PC10 mouse monoclonal cell line, Santa Cruz Biotechnology, Santa Cruz, CA) can be mixed with the anti-SNF2H antibody. For PCNA staining with this antibody, methanol fixation is obligatory. The antibody solution needs to cover the coverslip as a large drop without spreading through the whole well. If this happens remove all solution with a fine pipet tip and add antibody again.

10. Incubate for 30 min at room temperature.

11. Rinse with PBS, then incubate twice with 1 ml PBS for 5 min on a rocking platform.

12. Rinse coverslip in 1 ml PBS with 10% FBS.
13. Add 50–70 μl antirabbit antibody coupled to fluorescein isothiocyanate (FITC, Vector Laboratories, Burlingame, CA) diluted 1:100 in PBS with 10% FBS, together with antimouse antibody coupled to Texas Red (Vector Laboratories, Burlingame, CA) 1:100, to detect the PCNA staining. Incubate 30 min at room temperature in the dark.
14. Wash as in step 11.
15. For SNF2H/PCNA stain alone, the cover slips can be mounted onto slides: add a small drop of DAPI-containing mounting medium (Vector Laboratories, Burlingame, CA) on a slide, drop the cover slip, face-down, into the mounting medium, remove excess fluid with absorbent paper, seal with nail-polish.
16. For BrdU detection: add 1 ml 4% paraformaldehyde in PBS (made fresh by dissolving paraformaldehyde powder in PBS at 55°, this can be stored at $-20°$), 10 min at room temperature in fume hood (paraformaldehyde is volatile and toxic).
17. Rinse twice with 1 ml PBS.
18. Permeabilize cells with an incubation in PBS with 0.1% Triton X-100, 5 min at room temperature.
19. Rinse twice with 1 ml PBS.
20. Incubate for 30 min in 2 M HCl (stock solution should be kept airtight).
21. Rinse 3× with 1 ml PBS.
22. Block with PBS containing 10% FBS, 30 min at room temperature.
23. Add 50–70 μl mouse monoclonal anti-BrdU antibody (PharMingen/Becton Dickenson, San Diego, CA) 1:100 in PBS with 10% FBS, incubate 30 min at room temperature. Continue staining as outlined above (point 11) using antimouse antibody coupled to Texas Red (1:100, Vector Laboratories, Burlingame, CA).
24. SNF2H-GFP auto fluorescence is visualized directly and compared with BrdU or PCNA staining as shown in Fig. 3.

Analysis of Replication of Pericentric Heterochromatin in ACF1-Depleted NIH 3T3 Cells

The analysis of ACF1-depleted cells indicates a replication defect late in S-phase (Collins et al.,[15] Fig. 2) when heterochromatin is replicated. This type of chromatin remains condensed throughout the cell cycle. To study the replication of heterochromatin more directly we focused on pericentromeric heterochromatin in NIH 3T3 mouse cells. The visualization of replication in this type of heterochromatin does not require elaborate methods such as fluorescence *in situ* hybridization (FISH) since it is readily visible in 4',6-diamidino-2-phenylindole (DAPI)-stained mouse cells as large, intensely stained foci in the nucleus[23,24] (see Fig. 4A). We also studied the influence of ACF1 on heterochromatin replication using 5-azadeoxycytidine, a nucleotide-analogue that inhibits DNA-methyltransferases. The reduction in DNA methylation leads to a specific decondensation of the heterochromatin.[25] The protocol below describes the experimental set-up and methodology used in our lab to study replication of pericentromeric heterochromatin in combination with *in vivo* ACF1 depletion by RNAi and 5-azadeoxycytidine treatment. A typical result is illustrated in Fig. 4B.

1. Seed NIH 3T3 cells at 20% confluence on coverslips in a 12-well dish as described for transfection of siRNA.
2. Transfect ACF1-siRNA or control siRNA as described above.
3. After 16–20 h, remove transfection medium, wash twice with medium and add fresh medium.
4. At 48 h posttransfection, 5-aza-2'-deoxycytidine (Sigma, 5-azadC) was added in fresh medium at a concentration of 2 μM (+5-azadC) or fresh medium without 5-azadC was added as a control (−5-azadC). A stock solution of 20 mM is made in 50% acetic acid and kept at −20°.
5. After a further incubation for 18 h, pulse label the cells with 10 μM BrdU for 10 min and stain with an anti-BrdU antibody, as described above.
6. After the BrdU staining the experiment is quantified by counting cells with different replication patterns. Since we are interested in the replication of pericentromeric heterochromatin (as compared to replication in the euchromatin), we scored for two types of cells for the experiment depicted in Fig. 4B:

   Cells that have a homogeneous and bright nuclear BrdU staining, as illustrated in Fig. 4A, "early S phase." These are cells that are in early to mid-late S (0 to 4 h in S-phase). This number of cells did not change upon ACF1-depletion and therefore served as a control for BrdU staining efficiency and potential effects of 5-azadC treatment.

   Cells that have large foci of BrdU staining that overlap with the areas of intense DAPI staining of pericentromeric heterochromatin (present at 4.5–5.5 h into S-phase), illustrated in Fig. 4A, "mid-late S phase." Only cells where the BrdU incorporation sites

[23] M. H. Fox, D. J. Arndt-Jovin, T. M. Jovin, P. H. Baumann, and M. Robert-Nicoud, *J. Cell Sci.* **99**, 247 (1991).
[24] R. T. O'Keefe, S. C. Henderson, and D. L. Spector, *J. Cell Biol.* **116**, 1095 (1992).
[25] T. Haaf, *Pharmacol. Ther.* **65**, 19 (1995).

completely overlaps with DAPI should be counted as replicating pericentromeric heterochromatin.

For reasons of clarity (euchromatin replicating cells are control for experimental variation and potential effects of depletion on general replication) this method ignores the cells that are in transition from one stage to another and also ignores cells that are in the very late S-phase (6–7 h

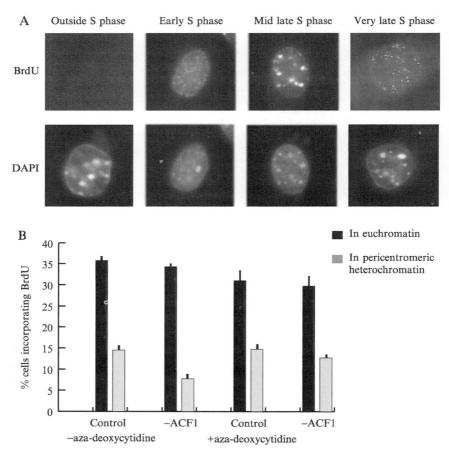

FIG. 4. Analysis of the effect of ACF1 depletion on DNA replication within pericentromeric heterochromatin in NIH 3T3 cells. (A) Characteristic BrdU incorporation patterns during S phase in 3T3 cells. (B) Quantification of ACF1 or mock depleted cells with early DNA replication pattern or replication within pericentromeric heterochromatin with or without aza-deoxycytidine treatment. Shown is the average result and standard deviation of four independent experiments. More than 250 cells were scored for each experiment.

in S-phase) and are characterized by small areas of BrdU incorporation (see Fig. 4A, "very late S phase").

Acknowledgments

We thank Debbie van den Berg for Westerns of Fig. 1A. Work in our laboratory is funded by the Marie Curie Cancer Care. R.P. was also supported by a grant from the Association for International Cancer Research (AICR), St. Andrews, Scotland.

[25] Preparation and Assays for Mammalian ISWI Complexes

By Orr G. Barak and Ramin Shiekhattar

The ATP dependent chromatin remodeling enzymes in mammals are defined by their catalytic subunits, which include orthologs of SWI/SNF, imitation switch (ISWI), the chromodomain helicases (CHDs), and others. Of these, the two mammalian ISWI gene products, SNF2H and SNF2L, incorporate into the widest range of unique complexes. SNF2H and SNF2L share approximately 80% identity at the amino acid level with the greatest variation existing within the first 70 amino acids. Both proteins share the SNF2 ATPase and SANT (*S*wi3, *A*da2, *N*CoR, *T*FIIIB) domains characteristic of the ISWI family. The level of redundancy between the two ISWI proteins *in vivo* remains unclear, however, work from Picketts and coworkers demonstrates a nonoverlapping expression pattern between the two mice ISWI proteins, both temporally and anatomically.[1] *In vitro*, both proteins contain intrinsic ATPase and chromatin remodeling activity stimulated by incorporation into the physiologic complexes or by nucleosomal substrates.[2] While they share some binding partners, it is becoming apparent that SNF2H or SNF2L are exclusive to certain remodeling machines exemplified by the SNF2H containing WSTF-ISWI chromatin remodeling complex (WICH).[3]

Our group has succeeded in the purification and analysis of a subset of the mammalian ISWI complexes. Here, we detail various methods for the biochemical isolation of electrophoretically pure Williams syndrome

---

[1] M. A. Lazzaro and D. J. Picketts, *J. Neurochem.* **77,** 1145 (2001).
[2] J. D. Aalfs, G. J. Narlikar, and R. E. Kingston, *J. Biol. Chem.* **276,** 34270 (2001).
[3] L. Bozhenok, P. A. Wade, and P. Varga-Weisz, *EMBO J.* **21,** 2231 (2002).

METHODS IN ENZYMOLOGY, VOL. 377

transcription factor-related chromatin remodeling factor (WCRF),[4] SNF2H-Cohesin,[5] and mammalian nucleosome remodeling factor (NURF).[6] We include protocols for both *in vitro* and *in vivo* analyses of these complexes. Importantly, other labs have characterized four other mammalian ISWI complexes CHRomatin-Accessibility Complex (CHRAC),[7] remodeling and spacing factor (RSF),[8] WICH,[3] and nucleolar remodeling complex (NORC).[9] We refer you to the primary literature for protocols detailing their isolation.

Materials, Reagents, and Instrumentation

Materials

HeLa (S3 strain) and HEK293 cells from American Type Culture Collection (Rockville, MD); fetal bovine serum and DMEM media from Gibco/BRL (Gaithersburg, MD); Spectra/Por dialysis tubing (10,000 MW cutoff) from spectrum (Houston, TX); pCDNA3-flag-BPTF; Poly-prep chromatography column (BioRad), Whatman chromatography paper.

Reagents

Phenylmethyl sulfonyl fluoride (PMSF), aprotinin, leupeptin pepstatin, potassium chloride, potassium phosphate, Tris–Chloride, β-mercaptoethanol, HEPES, glycerol, EDTA, Dithiothreitol (DTT), calcium chloride, NP-40, colloidal blue, ammonium sulfate, P11 phosphocellulose (Whatman), DEAE sephacel (Amersham Biosciences), flag beads (Sigma), flag peptide (5 mg/ml in ddH2O) (Sigma), 19:1 acrylamide to bisacrylamide (BioRad), ammonium persulfate, temed, urea, salI-high concentration (NEB), phenol/chloroform, DE81 chromatography paper (Whatman), DnaseI (BMB), PEI-cellulose TLC plates (Sigma), formic acid (Sigma), and lithium chloride.

[4] D. A. Bochar, J. Savard, W. Wang, D. W. Lafleur, P. Moore, J. Cote, and R. Shiekhattar, *Proc. Natl. Acad. Sci. USA* **97,** 1038 (2000).

[5] M. A. Hakimi, D. A. Bochar, J. A. Schmiesing, Y. Dong, O. G. Barak, D. W. Speicher, K. Yokomori, and R. Shiekhattar, *Nature* **418,** 994 (2002).

[6] O. Barak, M. A. Lazzaro, D. W. Speicher, D. J. Picketts, and R. Shiekhattar, *EMBO J.* **22,** 22 (2003).

[7] R. A. Poot, G. Dellaire, B. B. Hulsmann, M. A. Grimaldi, D. F. Corona, P. B. Becker, W. A. Bickmore, and P. D. Varga-Weisz, *EMBO J.* **19,** 3377 (2000).

[8] G. LeRoy, A. Loyola, W. S. Lane, and D. Reinberg, *J. Biol. Chem.* **275,** 14787 (2000).

[9] R. Strohner, A. Nemeth, P. Jansa, U. Hofmann-Rohrer, R. Santoro, G. Langst, and I. Grummt, *EMBO J.* **20,** 4892 (2001).

Buffers

BC Buffer: 20 mM Tris–Hcl, 0.2 mM EDTA, 10% glycerol, 10 mM β-mercaptoethanol pH 7.8 containing concentrations of KCl as indicated in protocol. All BC buffers are supplemented with 0.2 mM PMSF and 1 μg/ml of aprotinin, leupeptin, and pepstatin.

HT Buffer: 5 mM HEPES, 10 μM CaCl$_2$, 1 mM DTT, 0.5 mM PMSF, 10% glycerol, 40 mM KCl pH 7.6 supplemented with K$_x$PO$_4$ as indicated in protocol.

PS Buffer: 20 mM HEPES, 4 mM DTT, 0.5 mM EDTA, 10% glycerol, 0.2 mM containing concentrations of NH$_4$SO$_4$ as indicated in protocol. Buffer is supplemented with 0.2 mM PMSF and 1 μg/ml of aprotinin, leupeptin, and pepstatin.

IP buffer: 20 mM Tris–Hcl, 0.2 mM EDTA, 10% glycerol, 0.1 mM DTT pH-7.8 containing concentrations of KCl as indicated in protocol. All IP buffers are supplemented with 0.2 mM PMSF and 1 μg/ml of aprotinin, leupeptin, and pepstatin.

ATPase buffer: 20 mM Tris–Cl, 8 mM MgCl$_2$, 0.1 mM DTT, 50 mM KCl, 2% glycerol, 50 μg/ml of BSA, 50 μm cold ATP, 0.5 mCi of [γ-32P]ATP.

TBE: 100 mM Tris–borate, 2 mM EDTA

MDC buffer: 20 mM HEPES, 3 mM MgCl$_2$, 50 mM NaCl, 1 μM ZnCl$_2$, 2 mM DTT, 0.2 mM PMSF, 5% glycerol, 200 μg/ml BSA pH 7.5.

MDC stop buffer: 20 mM Tris–HCl, 50 mM EDTA, 2% SDS, 0.2 mg/ml proteinase K, 0.25 mg/ml yeast tRNA, pH 7.5.

Loading buffer: 90% formamide, 0.1% bromophenol blue and xylene cyanol.

RECCR buffer (4×): 10 mM MgCl$_2$, 40 mM Tris–HCl, pH 7.9.

TLC-Buffer: 1 M formic acid, 0.5 M lithium chloride.

Instrumentation

FPLC System from Amersham Biosciences (Uppsala, Sweden). Pre-poured columns—Amersham Biosciences (Uppsala, Sweden)—Superose6 HR 10/30, MonoQ HR 5/5, Phenyl Sepharose HR 10/10, HiPrep 16/10 Bu-tylFF. BioRad-Bio-Gel HT, BioScale CHT5-I, Bio-Sil Sec400. Toso Haas (Japan)—Heparin 5PW.

Purification of WCRF/hACF Complex

The first mammalian ISWI complex, WCRF/hACF, was purified using conventional chromatography from HeLa cell nuclear extract. Originally, we followed peak activities of ATP-dependent DNase I cleavage of mononucleosomes.[4] Due to simplicity of assay, we suggest a similar chromatographic separation scheme following immunoreactivity with anti-WCRF180 and anti-SNF2H antibodies rather than enzymatic activity. Our results suggest that 1.2 grams of nuclear extract should be sufficient for isolation of sufficient amounts of electrophoretically pure WCRF/hACF for subsequent enzymatic assays.

Protocol

1. Prepare HeLa nuclear extract ($>$1.2 g) using protocols described elsewhere[10] and dialyzed to a conductivity of $<$9 against BC-50 mM KCl (see Fig. 1A).

2. Dialyzed extract is then loaded on a 200 ml column of phosphocellulose and fractionated stepwise by increasing KCl concentrations of 0.1, 0.3, 0.5, and 1 M in BC buffer.

3. The 1.0 M P11 fraction (approx. 100 mg) containing the peak of SNF2H is dialyzed to a conductivity of $<$9 against BC-50 mM KCl. Load fraction onto a 20 ml DEAE-Sephacel column and elute with BC-0.35 M KCl.

4. The 0.35 M DEAE elution (approx. 50 mg) is dialyzed to 10 mM K_xPO_4 against HT Buffer and loaded onto a 20 ml BioGel HT column. Resolve the column using a linear 10 column volume gradient of 10–600 mM K_xPO_4 in HT buffer.

5. Pool SNF2H and WCRF180 immunoreactivity peak fractions ($\sim$0.275 M K_xPO_4, 5.0 mg) and dialyze against BC-0.1 M KCl. Load dialyzed fractions onto a Mono Q HR 5/5. Resolve column using a linear 10 column volume gradient of 0.1 to 0.5 M KCl in BC buffer.

6. Pool SNF2H and WCRF180 immunoreactivity peak fractions ($\sim$0.3 M KCl 0.7 mg) and dialyze against BC-0.1 M KCl. Load dialyzed fractions onto a Heparin 5PW. Resolve column using a linear 10 column volume gradient of 0.1 to 1 M KCl in BC buffer.

7. Pool SNF2H and WCRF180 immunoreactivity peak fractions ($\sim$0.6 M KCl) and fractionate on a Superose 6 HR 10/30 equilibrated in BC-0.7 M KCl containing 0.1% Nonidet P-40.

8. Electrophoretically pure WCRF180 with SNF2H can be visualized by SDS-PAGE followed by Colloidal blue staining. We identify peak fractions at 11–13 ml coeluting with the 670 kDa thyroglobulin sizing standard (see Fig. 1B).

---

[10] J. D. Dignam, *Methods Enzymol.* **182,** 194 (1990).

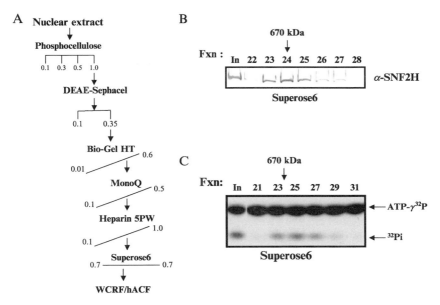

FIG. 1. WCRF/hACF-purification and ATPase assays. (A) Schematic outlining purification of the WCRF/hACF complex. Sloped lines indicate a gradient elution, horizontal lines indicate a stepwise elution. (B) Immunoblot for SNF2H in size exclusion fractions from the final purification step in the WCRF/hACF isolation. (C) Autoradiograph of ATPase enzymatic assay on fractions from (B). Fractions were incubated with labeled ATP and 100 ng of DNA followed by separation of free phosphate from ATP by PAGE.

Evaluation of WCRF by ATPase Assay (Gel Electrophoresis Assay)

The ISWI family of chromatin remodelers contains an intrinsic ATPase activity. This activity is unique from other cellular ATPases in that it is stimulated by and is virtually dependent on the presence of DNA or nucleosomes. The WCRF complex is responsive to both. Nuclear extract serves as an excellent positive control for this assay but will not exhibit DNA or nucleosome dependence.

1. Add purified WCRF to ATPase buffer. Amount of WCRF needed to assess activity varies by specific activity of each preparation. We find that 100–200 ng of the complex should be sufficient to assay activity, however, titration of WCRF may be necessary.

2. Incubate reaction at 30° for 30 min.

3. Analysis of ATPase activity can be visualized by either gel electrophoresis or thin layer chromatography. For WCRF, we will describe the

former. Pour a 12% polyacrylamide gel (19:1 acrylamide to bisacrylamide) containing 7 M urea in a $1\times$ TBE buffer.

4. Add 25 μl of loading buffer (90% formamide, 0.1% bromophenol blue and xylene cyanol) to ATPase reaction and load 10 μl onto gel.

5. Electrophorese samples at 150 V until bromophenol blue marker is one quarter to one third down the gel.

6. Place gel onto a square of chromatography paper and autoradiograph for 10–20 min.

7. An autoradiograph of the WCRF ATPase activity is demonstrated in Fig. 1C.

Evaluation of WCRF by Mononucleosome DNaseI Cleavage Assay

1. Protocol for reconstitution of labeled nucleosomes can be found in detail elsewhere.[11]

2. Perform the binding reaction for 1 h at 30° in a total volume of 20 μl. Reaction should include MDC buffer supplemented with 25 ng of nucleosomes, with or without 1 mM ATP, and 50 nM of WCRF.

3. Add 0.5 units of DNaseI to reaction and incubate for 1 min at room temperature.

4. Add 1 volume of MDC-stop buffer and incubate samples for 1 h at 50°.

5. Precipitate DNA with ammonium acetate, centrifuge sample for 10 min at maximum speed. Discard supernatant and wash pellet with 0.5 ml of 80% ethanol. Centrifuge sample and discard supernatant. Allow DNA pellet to dry.

6. Resuspend DNA pellet in 2 μl of water and 3 μl of loading buffer. Incubate at 90° for 5 min.

7. Electrophorese samples on an 8% polyacrylamide gel (19:1 acrylamide to bisacrylamide) with $1\times$ TBE and 8 M urea.

8. Autoradiograph gels as in ATPase assay.

Purification of the SNF2H-Cohesin Complex

The SNF2H-cohesin complex was isolated by conventional chromatographic separation following immunoreactivity to SNF2H. We chose the 0.5 M KCl elution from the phosphocellulose column, which contained SNF2H and minimal WCRF180 suggesting a unique ISWI complex. Indeed, the isolation and characterization of the SNF2H-cohesin was not

[11] R. T. Utley, T. A. Owen-Hughes, L. J. Juan, J. Cote, C. C. Adams, and J. L. Workman, *Methods Enzymol.* **274,** 276 (1996).

only unique, but the presence of the cohesins implied a function for ISWI. We were able to characterize this complex, not by *in vitro* enzymologic studies, but by careful *in vivo* dissection of occupancy by this complex at repetitive DNA sequences in a methylation sensitive manner. Here we describe the protocol for isolation of the SNF2H-cohesin complex. Important in this purification is the exclusion of WCRF180 immunoreactive fractions. While there appeared to be very little of the WCRF180 in the initial stages of fractionation, the presence of WCRF will persist throughout the purification and should be minimized whenever possible.

Protocol

1. Prepare HeLa nuclear extract (>2 g) using protocols described elsewhere[10] and dialyze to a conductivity of <9 against buffer BC-50 mM KCl (see Fig. 2A).

2. Dialyzed extract is then loaded on a 250 ml column of phosphocellulose and fractionated stepwise by increasing KCl concentrations of 0.1, 0.3, 0.5, and 1 M in BC buffer.

3. The 0.5 M P11 fraction (approx. 250 mg) containing the peak of SNF2H is dialyzed to a conductivity of <9 against BC-50 mM KCl. Load fraction onto a 45 ml DEAE-Sephacel column and elute with BC-0.35 M KCl.

4. The 0.35 M DEAE elution (approx. 140 mg) is dialyzed against 100 mM NH$_4$SO$_4$ in PS buffer. Load fraction onto a Phenyl Sepharose HR 10/10. Resolve column using a linear 10 column volume gradient of 0.7 to 0 M NH$_4$SO$_4$ in PS buffer.

5. Pool SNF2H immunoreactivity peak fractions and dialyze against 10 mM K$_x$PO$_4$ in HT buffer and load onto a BioScale CHT5-I column. Resolve the column using a linear 15 column volume of 10–600 mM K$_x$PO$_4$ in HT buffer.

6. Importantly, pool fractions with peak SNF2H immunoreactivity but lacking WCRF180 immunoreactivity and dialyze against BC-0.1 mM KCl. Load dialyzed fractions onto a Mono Q HR 5/5. Resolve column using a linear 10 column volume gradient of 0.1 to 0.5 M KCl in BC buffer.

7. Pool SNF2H immunoreactivity peak fractions ($\sim$0.6 M KCl) and fractionate on a Superose 6 HR 10/30 equilibrated in BC-0.7 M KCl containing 0.1% NP-40.

8. SNF2H-cohesin complex can be visualized by SDS-PAGE followed by silver stain (see Fig. 2B). We identify peak fractions at 9–10 ml off of the column.

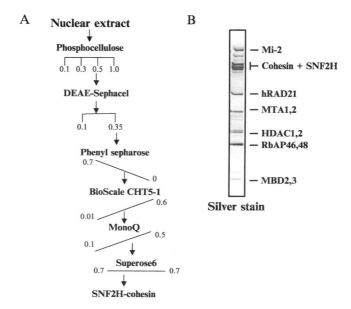

FIG. 2. SNF2H-cohesin purification. (A) Schematic outlining purification of the SNF2H-cohesin complex. Sloped lines indicate a gradient elution, horizontal lines indicate a stepwise elution. (B) Silver stain of SNF2H-cohesin complex from final purification step. Identities of polypeptides are indicated on the right.

Human NURF

The purification of human NURF requires compensating for two intrinsic features of this complex contradictory to traditional biochemical methods. The first is its relatively low abundance compared to the other human ISWI complexes described. The second is that the large subunit of human NURF, BPTF, like its *Drosophila* counterpart appears especially susceptible to proteolytic degradation. This protocol for isolation of the human NURF complex avoids these problems while bypassing the rigors of conventional chromatography. The strategy for this purification involves generating an HEK293 cell line expressing a Flag epitope tagged human BPTF, a subunit unique to the human NURF complex. The subsequent affinity purification results in the NURF complex along with two contaminating polypeptides also found in mock immunoprecipitations of naïve HEK293 extract. These polypeptides are unrelated to chromatin biology and pose no threat in performing subsequent enzymatic assays.

Protocol for Generating Stable Cell Line

A more complete protocol for generating stable cell lines can be found elsewhere.[12] In brief, we utilized a cotransfection protocol, whereby, one plasmid encodes the Flag-tagged cDNA (Flag-BPTF) of choice while the other contains a drug selection marker. We recommend puromycin selection over the commonly used neomycin selection due to its more potent toxicity to naïve cells as well as the lower cost of the reagent. Stable cell lines should be generated from a monoclonal population necessitating cell colony isolation. In our experience, polyclonal populations of cell lines stably expressing transgenes tend to lose expression of the transgenic cDNA. Finally, 10–20 colonies should be selected and screened by immunoprecipitation followed by Western blot. For the hNURF isolation protocol we generated a stable HEK293 cell line. We chose HEK293s because they express endogenous SNF2L, have a rapid doubling time, and are easily transfected.

Protocol for Preparing SNF2L Associated Complexes

Immunoprecipitation

Due to the extensive washes in our hNURF protocol, immunoprecipitation results in an electrophoretically pure (other than the aforementioned inert contaminating polypeptides), enzymatically active human NURF (see Fig. 3A). Subsequent chromatographic steps may be performed, however significant loss of BPTF by proteolysis often occurs. As with any affinity chromatography, an important control is immunoprecipitation of naïve cells for comparison.

1. We suggest purifying hNURF from no less than one hundred 15 cm culture plates of the stable Flag-BPTF HEK293 cell line grown to confluency. Collect cells by trypsinization, collect the nuclear extract[10] and dialyze against buffer IP-50 mM KCl until the conductivity is <11. The nuclear extract will have 50–100 mg of total protein. The subsequent steps should all be performed at 4°. Collect all fractions for Western blot analysis.

2. Immunoprecipitate hNURF by adding M2 anti-Flag agarose beads (Sigma) to nuclear extract. Use 0.5 ml of beads per 100 mg of nuclear extract. Beads should be washed three times with PBS prior to immunoprecipitation. Incubate beads with Flag-BPTF nuclear extract overnight with rotation.

3. Pellet beads by centrifugation at 1000 rpm for 5 min in a refrigerated centrifuge. Decant supernatant and save unbound fraction.

---

[12] J. Sambrook, E. F. Fritsch, and T. Maniatis, "Molecular Cloning: A Laboratory Manual." Cold Spring Harbor Laboratory Press, Cold Spring Harbor, New York, 1990.

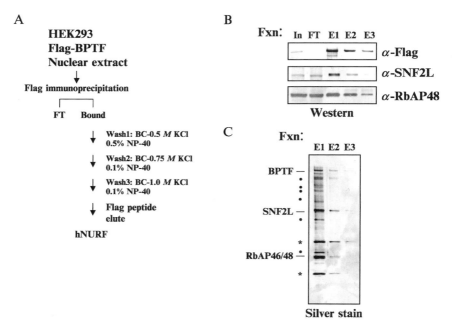

FIG. 3. hNURF purification. (A) Schematic outlining affinity purification of hNURF complex. (B) Immunoblot analysis of components of hNURF complex. In, input, FT, flow through, E1, E2, E3, elutions off of the M2 affinity beads. (C) Silver stain analysis of eluted fractions demonstrate an electrophoretically pure hNURF complex. BPTF, SNF2L, and RbAP46/48, components of the hNURF complex are indicated on the left. (•) Indicates proteolytic cleavage products of BPTF from inherent instability of the polypeptide. (*) Indicates polypeptides also present in mock immunoprecipitation from naïve HEK293 cells.

4. Wash 1—incubate beads with 10 column volumes of buffer IP-0.5 M KCl supplemented with 0.5% NP-40. Rotate for 10 min.

5. Wash 2—pellets beads, decant supernatant, and incubate beads with 10 column volumes of buffer IP-0.75 M KCl supplemented with 0.1% of NP-40. Rotate for 10 min.

6. Wash 3—pellets beads, decant supernatant, and incubate beads with 10 column volumes of buffer IP-1 M KCl supplemented with 0.1% of NP-40. Rotate for 10 min.

7. Load beads onto a 10 ml Poly-prep chromatography column (BioRad) and allow buffer to elute at the column's gravimetric flow rate.

8. Buffer exchange the beads into low salt by passing 5 column volumes of buffer IP-0.1 mM KCl over beads.

9. Flag-BPTF associated complexes can now be eluted using Flag-peptide. Add 1 column volume of buffer IP-0.1 M KCl supplemented with

400 μg/ml of flag peptide. Allow half of the bed volume of buffer to elute and cap column. Incubate beads and elution buffer for 15 min prior to elution.

10. Collect five more elutions with 15 min incubation intervals between each elution.

11. Fractions can be analyzed by Western blot assaying for immunoreactivity to SNF2L, RbAP48, and Flag or BPTF (Fig. 3B). Importantly, if unbound fraction contains a large quantity of Flag-BPTF, a second immunoprecipitation on the flow-through may be performed. Fractions should be analyzed for purity by silver stain (see Fig. 3C). While this protocol avoids much of the BPTF proteolysis, there may still be a plethora of polypeptides representing BPTF breakdown products.

Restriction Endonuclease Coupled Chromatin Remodeling for Assaying hNURF Activity

The restriction endonuclease coupled chromatin remodeling assay has proven effective in studying ISWI complex remodeling activity.[2] It utilizes two enzymatic reactions to assay for activity: Restriction endonuclease cleavage and chromatin remodeling. The advantages of the assay are reproducibility and speed. The disadvantages are a narrow window of appropriate conditions and the need for a high specific activity. The latter can also be an advantage given the virtual absence of background chromatin remodeling activity in mock immunoprecipitations. When performing this assay, it is important to include a nonATP control to demonstrate the ATP-dependence of this activity.

1. Protocol for preparation of DNA and reconstitution of nucleosomal array has been described in detail elsewhere.[13]

2. Affinity purified hNURF can be tested for enzymatic activity directly. Since the assay is sensitive to salt concentration, include the KCl (0.1 M) contribution of the elution buffer in calculating the components necessary for appropriate reaction conditions.

3. Set up a 20 μl reaction starting with the 4× RECCR buffer (5 μl) and supplementing with ATP (1 mM final), DTT (1 mM final), BSA (0.1 mg/ml final), SalI-high concentration (0.5 μl), labeled array ($\sim$10,000–50,000 cpm), and purified hNURF. Titrate in the hNURF accordingly.

4. The reaction is incubated at 37° for 90 min.

5. Raise the volume of each sample to 40 μl with water and add 30 μl of phenol/chloroform to deproteinate the DNA from the histones. Vortex briefly and centrifuge for 5 min at 13,000 rpm.

[13] L. M. Carruthers, C. Tse, K. P. Walker, III, and J. C. Hansen, *Methods Enzymol.* **304,** 19 (1999).

6. Carefully transfer 20 μl of the aqueous layer to a new tube containing 8 μl of 10× DNA loading buffer. Take extra care with avoiding the organic layer since residual phenol/chloroform will result in marked disturbance of the subsequent electrophoresis step.

7. Electrophorese samples using a 1% agarose gel in 0.5× TBE.

8. Transfer electrophoresed bands to DE81 chromatography paper using a gel dryer (Fisher). We recommend drying for 90 min at no more than 65° to avoid melting the agarose before vacuum mediated transfer occurs. This will result in a diffuse signal.

9. Expose dried gel to either phosphorimager or film and quantitate chromatin remodeling activity by comparing fraction of uncut DNA (see Fig. 4A).

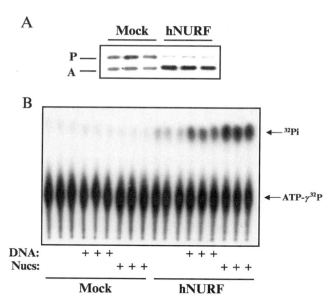

FIG. 4. Enzymatic analysis of hNURF complex. (A) Autoradiograph of restriction enzyme coupled chromatin remodeling assay demonstrating chromatin remodeling activity in purified hNURF complex. Addition of hNURF results in a decrease in the protected (P) DNA band and an increase in the accessible (A) band. Experiment, as shown, was performed in triplicate for assays of both mock and hNURF. (B) Autoradiograph of ATPase assay of hNURF complex indicating a potent DNA and nucleosome dependent ATPase activity. Reactions were incubated with labeled ATP as well as buffer, DNA (50 ng), or nucleosomes (50 ng). Free phosphate was separated from intact ATP using thin layer chromatography. Assays, as shown, were performed in triplicate.

Evaluation of hNURF ATPase Activity (Thin Layer Chromatography)

As mentioned previously, the ISWI family of chromatin remodelers contains an intrinsic ATPase activity. Here we describe a second means of analysis of ATPase activity. The assay conditions are identical to previous protocol, however, analysis of reaction products is performed using Thin Layer Chromatography (TLC) rather than PAGE.

1. Perform reaction as per previous protocol (WCRF/hACF analysis by ATPase) in a final volume of 10 μl.

2. Spot 1 μl of each reaction onto a PEI cellulose TLC plate on a line approximately 1 cm from one edge of the plate.

3. Allow spot to dry and immerse TLC plate in a TLC chamber. The edge closest to the reaction spots should be closest to TLC-buffer making sure to not immerse dried spots in TLC-buffer. Allow buffer to migrate up the plate by capillary action.

4. Once buffer front migrates 75–80% the total length of the plate, remove plate and allow to air dry.

5. Expose plate to either film or phosphoimager and perform densitometric analysis of released free phosphate (see Fig. 4B).

[26] Preparation and Analysis of the INO80 Complex

By XUETONG SHEN

The complete sequencing of genomes of model organisms has changed biological research dramatically. Genomic and proteomic approaches are now commonly used to analyze functions of unknown genes. For example, in the baker's yeast, *S. cerevisiae*, there are many uncharacterized open reading frames (ORFs). In order to study the function of these ORFs, one effective strategy is to combine both genetic and biochemical analyses. Typically, the ORF of interest is first analyzed using bioinformatic tools and is compared and grouped with similar genes of known function. The function of the ORF might also be inferred from genetic analysis, such as by studying the phenotypes of mutants. At the same time, the protein product of the ORF and associated proteins can be purified and studied biochemically to reveal functions. This strategy combines the strengths of genetic and biochemical approaches to provide a comprehensive view of biological function. On the technical side, the availability of a complete collection of yeast deletion mutants and a large collection of other types of existing mutants, such as temperature-sensitive mutants have greatly

METHODS IN ENZYMOLOGY, VOL. 377

facilitated genetic analysis. Moreover, two-hybrid assay, transcription profiling, phenotype test and synthetic lethal screens can also be done on a whole-genome scale. Biochemically, the use of epitope-tagging combined with immunoaffinity chromatography has drastically reduced the time and effort required for purifying proteins and protein complexes, compared to the use of multistep conventional chromatography. In addition, efficient and sensitive protein sequencing technologies, such as mass spectrometry, have greatly shortened the time needed for protein identification. These technologies have allowed the recent whole-genome protein-protein interaction studies in yeast to be conducted.[1,2]

Here, we describe the systematic analysis of INO80, an ATP-dependent chromatin remodeling complex in *S. cerevisiae*. To counteract the constraints imposed by chromatin organization, cells possess a number of multiprotein, ATP-dependent chromatin remodeling complexes, each containing an ATPase subunit from the Snf2/Swi2 superfamily. Chromatin remodeling complexes act by increasing nucleosome mobility and are most clearly implicated in transcription.[3,4] To identify remodeling complexes that potentially assist other DNA transactions, we undertook an analysis of the Snf2/Swi2-like proteins. Using the general strategy described earlier, we first analyzed the phylogeny of the Snf2/Swi2 family of proteins, and found that Ino80 is related to Snf2, yet belongs to a distinct subgroup. Biochemically, we epitope-tagged Ino80 and purified a large complex, called the INO80 complex, using single-step immunoaffinity chromatography. The INO80 complex contains ~12 polypeptides, including two proteins related to bacterial RuvB DNA helicase, which catalyzes branch migration of Holliday junctions during DNA recombination. The INO80 complex remodels chromatin and facilitates transcription from chromatin *in vitro*. Genetically, *ino80* mutants show hypersensitivity to agents that cause DNA damage, in addition to defects in transcription. These results suggest that chromatin remodeling driven by the INO80 complex is

[1] P. Uetz, L. Giot, G. Cagney, T. A. Mansfield, R. S. Judson, J. R. Knight, D. Lockshon, V. Narayan, M. Srinivasan, P. Pochart, A. Qureshi-Emili, Y. Li, B. Godwin, D. Conover, T. Kalbfleisch, G. Vijayadamodar, M. Yang, M. Johnston, S. Fields, and J. M. Rothberg, *Nature* **403,** 623 (2000).

[2] A. C. Gavin, M. Bosche, R. Krause, P. Grandi, M. Marzioch, A. Bauer, J. Schultz, J. M. Rick, A. M. Michon, C. M. Cruciat, M. Remor, C. Hofert, M. Schelder, M. Brajenovic, H. Ruffner, A. Merino, K. Klein, M. Hudak, D. Dickson, T. Rudi, V. Gnau, A. Bauch, S. Bastuck, B. Huhse, C. Leutwein, M. A. Heurtier, R. R. Copley, A. Edelmann, E. Querfurth, V. Rybin, G. Drewes, M. Raida, T. Bouwmeester, P. Bork, B. Seraphin, B. Kuster, G. Neubauer, and G. Superti-Furga, *Nature* **415,** 141 (2002).

[3] D. V. Fyodorov and J. T. Kadonaga, *Cell* **106,** 523 (2001).

[4] P. B. Becker and W. Horz, *Annu. Rev. Biochem.* **71,** 247 (2002).

connected to transcription and interestingly, DNA recombination or repair.[5] The INO80 complex also contains conventional actin, and three actin related proteins. Using similar strategies of combining biochemistry and genetics, we discovered that actin related proteins are functionally important to INO80 chromatin remodeling.[6] In addition, we extended our studies to other ATP-dependent chromatin remodeling complexes using the same strategy, and revealed a novel link between ATP-dependent chromatin remodeling and inositol polyphosphate production.[7] In summary, the combined use of biochemical and genetic analyses has allowed us to gain insights into the functions of the INO80 chromatin remodeling complex.

Principle

The methodologies we employed for analysis of the INO80 complex will be described in detail with emphasis on the epitope-tagging of INO80 and the immunoaffinity purification of the INO80 complex. The methods for making yeast whole cell extract and direct immunoaffinity purification were derived mainly from previous work by Tsukiyama et al.[8] The key strategy is to use the FLAG-epitope for immunoaffinity purification. Briefly, Ino80 was tagged chromosomally at its C-terminus with a double-FLAG tag. The INO80 complex was then coimmunoprecipitated from yeast whole cell extract using FLAG-immunoaffinity beads. Finally, the complex was eluted with a competing FLAG peptide. This method has resulted in a single-step purification of the INO80 complex to near homogeneity from yeast whole cell extracts. The choice of the FLAG epitope is critical. Currently, there are several available epitope tags, such as HA, myc, and TAP-tag. We have found that FLAG tag offers two major advantages. First, its small size (8 amino acids) makes it less likely to interfere with the function of the tagged protein, compared to more bulky tags. Second, FLAG-tagged proteins or complexes can be readily eluted under physiological salt and pH, by using a competing FLAG peptide, therefore preserving the biological activity of the proteins.

Another principle guiding our analysis is the use of mutants. For a multisubunit complex, such as the INO80 complex, each subunit contributes to the function of the entire complex in specific ways. Through analyzing mutants genetically and the corresponding mutant complexes biochemically, the function of each individual subunit can be dissected. We have

[5] X. Shen, G. Mizuguchi, A. Hamiche, and C. Wu, Nature 406, 541 (2000).
[6] X. Shen, R. Ranallo, E. Choi, and C. Wu, Mol. Cell 12, 147 (2003).
[7] X. Shen, H. Xiao, R. Ranallo, W. H. Wu, and C. Wu, Science 299, 112 (2003).
[8] T. Tsukiyama, J. Palmer, C. C. Landel, J. Shiloach, and C. Wu, Genes Dev. 13, 686 (1999).

used this strategy in the analyses of actin related proteins in the INO80 complex. In principle, other subunits can also be investigated similarly.

Materials and Reagents

Yeast Strains

We use the S288C yeast strain background for our studies. A consistent genetic background is highly important in interpreting data generated from genetic studies. The S288C strain is chosen because the complete and published baker's yeast genome is derived from it (see *Saccharomyces* Genome Database [SGD]). In addition, a complete collection of deletion mutants in this genetic background is commercially available from Research Genetics. Specifically, we use the so called "Designer" strains, the BY-series,[9] which feature genetic markers that have been completely removed, eliminating the chance of intragenic suppression in the marker genes (false positives in genetic selections or screens).

Buffers and Chemicals

The basic buffer is buffer H (25 mM HEPES-KOH pH 7.6, 1 mM EDTA, 10% glycerol, 0.02% NP-40). The choice of salt is KCl, buffer H-0.5 means 0.5 M KCl is in the buffer; buffer H-0.1 means 0.1 M KCl is in the buffer etc. To all buffers, add DTT (2.5 mM), $MgCl_2$ (2 mM), and 1× protease inhibitor cocktail fresh. Keep buffers at 4°C.

We have used a protease inhibitor cocktail to minimize protein degradation. An alternative method can also be used in concert with protease inhibitor, in which the key yeast protease, Pep4, can be deleted genetically. With protease inhibitor alone, we find that the INO80 complex remains stable and active in the whole cell extract even after prolonged incubation (6 h) at room temperature. To make 25 ml 100× protease inhibitor: add the following to 25 ml ethanol and keep at −20°C. 0.44 g PMSF, 0.79 g Benzamidine hydrochloride (Sigma), 3.43 mg Pepstatin (Sigma 10 mg/ml in DMSO), 0.64 mg Leupeptin (Sigma 10 mg/ml in water), 5 mg Chymostatin (Sigma 10 mg/ml in DMSO).

Hardware

Standard yeast culturing equipment is needed. In addition, a large-scale thermoshaker is needed to grow liters of yeast cultures at once. We use the INFOR thermoshaker with 12-L capacity. A household blender (Oster,

[9] C. B. Brachmann, A. Davies, G. J. Cost, E. Caputo, J. Li, P. Hieter, and J. D. Boeke, *Yeast* **14,** 115 (1998).

450 W) with a 250 ml Mini-Blend jar is needed for breaking yeast. A high-capacity (4 or 6 L) centrifuge (Beckman J6) is needed for the initial collection of yeast cells. Finally, an ultracentrifuge (Beckman) is needed to make the whole cell extract.

Protocol

Bioinformatics of INO80

The yeast gene search was performed in SGD (http://genome-www.stanford.edu/Saccharomyces/). Eight proteins closely related to Snf2 were identified by the BLAST program. The phylogeny of these proteins was generated using the PAUP program.[10] This analysis indicates that Ino80 is positioned in a distinct subgroup within the Snf2 family of proteins. BLAST searches for similar genes in other organisms revealed a human (hIno80) and a *Drosophila* (dIno80) ortholog.

Epitope-Tagging of INO80

In general, there are three ways to epitope-tag a gene in yeast. The simplest way is to put the gene of interest in an existing tagging plasmid and transform yeast with the resulting plasmid. The disadvantage of this method is that the copy number of the plasmid may be nonphysiological and therefore may cause artefact due to over-expression. In addition, the native promoter and terminator sequences are usually lost in the plasmid. A second way circumvents the pitfall of the plasmid-based method by tagging a gene chromosomally, using the so-called "knock-in" method. This method is usually limited to tagging a protein at the C-terminus, using PCR fragments generated from an existing template containing an epitope tag and a selectable marker. The disadvantage is that the selectable marker gene might interfere with the terminator sequence at the $3'$ end of the gene (in certain cases, the marker gene can be eliminated by recombination). Finally, a gene can be tagged chromosomally without introducing any foreign sequences except for the tag, by using a two-step gene replacement method.[11] This method generates little disruption to the gene structure and is most likely to maintain the gene function. One disadvantage is that a gene-specific plasmid has to be constructed for each individual gene; therefore requiring slightly more effort than using an existing plasmid.

[10] D. L. Swofford, Illinois Natural History Survey, Champaign, Illinois, 1993.
[11] C. Guthrie and G. R. Fink, eds., "Guide to Yeast Genetics and Molecular Biology." Academic Press, 1991.

To ensure the best results, we tagged the chromosomal *INO80* gene using the two-step gene replacement method. First, an integration plasmid, pRS406-INO80/2FLAG, was constructed by cloning the XhoI-HindIII fragment of *INO80* into the standard pRS406 plasmid.[9] In pRS406-INO80/2FLAG, two tandem copies of the FLAG sequence (DYKDDDDK) were linked by a KpnI site (5'-GACTACAAGGACGACGATGACAAGGG-TACCGATTACAAGGATGATGACGACAAG-3'), and were introduced right before the stop codon of *INO80*. In the first step, pRS406-INO80/2FLAG was linearized by limited Sca I digestion and integrated into the genome of BY4733 cells. Southern blotting was used to confirm the integration. At this point, the wild-type *INO80* gene was followed by a partial C-terminally FLAG-tagged *INO80* gene, with the *URA3* marker and rest of the integration plasmid in between. In the second step, the *URA3* marker was "popped out" via homologous recombination by selection on a 5-fluoroorotic acid (FOA) plate.[11] Depending on the location of recombination, a portion of the "pop-outs" contain the correct chromosomal FLAG-tag at the *INO80* locus, which was confirmed by PCR and sequencing of the region.

It is important at this stage to check whether the addition of the epitope-tag has affected the function of the gene. In some cases, addition of the HA-tag has caused the cells to become temperature sensitive, suggesting that tagging has altered the function of the protein. Phenotypic analyses were used to ensure that the INO80-FLAG strain behaves the same as the wild-type strain. A partial list of test conditions includes high temperature, DNA-damaging conditions, lack of inositol or phosphates, and different carbon sources. If C-terminal FLAG-tagging resulted in a nonfunctional protein, one can repeat tagging at a different position (the N-terminus of or internal to the protein). We find that about one-third of the genes require alternative tagging positions other than at the C-termini (unpublished observations). One should always be sensitive to the effects of tagging on protein function in various biological assays, and care should be taken to minimize these possible effects.

Making Yeast Whole Cell Extract

Yeast whole cell extract is made using the blender method. Cell breakage is achieved by mechanical blending in a household blender, in the presence of dry ice. Typically, 1–4 L of yeast culture can be handled at one time. However, some mutant strains grow slower; therefore requiring a larger starting culture. Here, we use a 4-L culture to describe the making of yeast whole cell extract.

1. Grow a patch of the tagged strain (INO80-FLAG) on a plate containing rich media (YEPD)[11] at 30°C.

2. Grow yeast in 100 ml YEPD preculture overnight. Inoculate 25 ml overnight culture to each 1 L YEPD in large flasks (total of 4 L), allowing at least 2/3 of air in the flask and shake vigorously for 1 to 2 days at $30°C$. The culture should be near saturation at this point.

3. Prechill SW28 rotor at $4°C$. All of the following spins should be carried out in a Beckman JS4.2 rotor at $4°C$. Spin down cells in 1-L bottles for 5 min at 4000 rpm, about 40 g of cells pellet is expected. Wash cell pellet once in two 50 ml tubes with a total of 50 ml H_2O (with 2.5 mM DTT, 2 mM $MgCl_2$, and $1\times$ protease inhibitor cocktail added fresh). Spin down in 50 ml tubes for 5 min at 3500 rpm. At this point, if the culture is slightly contaminated with bacteria, a dark top layer could be seen in the tube, and it should be removed as much as possible using a spatula. Wash cell pellet again with 50 ml buffer H-0.3 once and spin for 5 min, 30–40 g of wet cell pellet is expected. At this point, the pellet can be processed or frozen in liquid nitrogen and kept at $-80°C$ for later use.

4. Put wet or thawed cell pellet in a 60 ml syringe tube with a spatula, push cells through syringe opening (no needle involved) into a plastic container filled with liquid nitrogen to form "noodles." Break noodles into small pieces with spatula. Prechill Oster Mini-Blend jar on dry ice, do not prechill the blade unit (it might stick if frozen early). Pour residual liquid nitrogen out and transfer yeast noodle pieces into the chilled jar (one jar can handle up to 40 g of noodles, for better cell breakage, use up to 20 g per jar). Then add about a quarter jar of dry ice, put the blade unit in and screw on the black cap assembly (make sure to remove the inner rubber seal in the cap assembly to avoid build up of CO_2). Do not tighten the cap all the way to avoid explosion. Secure the cup onto the blender base. Break cells by blending at the highest setting for a total of 20 min. Wear protective eye wear or a face shield, and ear muff, and shake the whole blender vigorously and continuously back and forth (and side to side) by holding the base with one hand and pushing down the cup with the other hand. Shake the whole blender upside down occasionally. The seemingly violent manipulation ensures the continued flow of powderized cells in the jar. Every 5 to 10 min, add additional dry ice to ensure that cells remain as frozen powder and are not thawing or sticking to the cup. The addition of dry ice can be reduced or avoided if blending is carried out in a cold room. Allow a slight thawing during the last few minutes of blending so that it will take less time to thaw the cells after blending.

5. Thaw cells in the jar at room temperature. Add 40 ml (equal to cell volume) buffer H-0.3 when the cells become damp yet frozen. Use a

Pasteur pipette to collect residual cells from the blade unit and jar wall, as well as to help the thawing by mixing. Transfer completely thawed mixture into two SW28 centrifuge tubes (mineral oil can be used to fill up the remaining space in centrifuge tubes to avoid dilution of the extact). Balance and centrifuge in a SW28 rotor for 2 h at 25,000 rpm at $4°C$ in a Beckmen Ultra-centrifuge. Use paper wipes to lift off the white lipid layer (membrane) on the top of the tube (or use a Pasteur pipette to remove the lipid layer). Carefully pour out and collect the clear extract and avoid the cloudy residues near the bottom of the tube. Pool clear extracts and make 5 ml aliquots in 14 ml round-bottom polypropylene tubes. Typically 30 to 40 ml yeast whole cell extract can be obtained. Extracts appear yellow or light brown in color and should be clear (extract from *ade* mutant cells is red). Flash freeze the tubes in liquid nitrogen and store at $-80°C$.

Purification of INO80 Complex by FLAG-Immunoaffinity Chromatography

Prior to purification of INO80 complex, gel-filtration chromatography was performed to fractionate the yeast whole cell extract, using a Superose 6 column in buffer H-0.3 on an AKTAexplorer system (Pharmacia). Fractions were analyzed with a monoclonal FLAG antibody (M2, Sigma) to track the presence of the Ino80 protein. These analyses give estimates of the size of the INO80 complex, and also clarify the potential presence of multiple complexes of significantly different sizes containing the Ino80 protein. Although direct (single-step) purification is preferred, a preclearing step will reduce the background level and increase the purity slightly. The INO80 complex can be bound to an ion-exchange resin (DE52) in buffer H-0.1, and then eluted in two steps with buffer H-0.3 and buffer H-0.5. The combined eluants can be used as starting material for FLAG-immunoaffinity chromatography. Here, we describe the single-step purification using whole cell extract.

1. Thaw 5 ml whole cell extract on ice, add to 5 ml buffer H-0.5 in a 15 ml capped conical tube with gentle mixing to adjust the extract. Equilibrate 0.3 ml packed Sigma Anti-FLAG M2 agarose affinity gel (beads) by briefly washing the beads in 1 ml buffer H-0.5 for 5 times. Add beads to the adjusted yeast whole cell extract. Allow binding of the INO80 complex to the FLAG beads for 3 h at $4°C$. This can be done in a $4°C$ chamber or in a cold room, using a rotating device to ensure consistent mixing.

2. Spin down the beads at 2000 rpm in a clinical centrifuge for 30 s. Remove supernatant with a long tip Pasteur pipette. Transfer the beads to a 1.5 ml tube for subsequent washes. Wash the beads for 10 min each time with 1 ml buffer H-0.5 for a total of 8 times at 4°C. These repeated washes are important to lower the background contaminations. After the high salt washes, equilibrate the beads for elution by washing 3 times with 1 ml buffer H-0.1. Each wash is 5 min long at 4°C. Care should be taken to avoid taking an excess amount of beads in each wash to ensure yield.

3. Elute the INO80 complex by adding 0.2 ml FLAG peptide (Sigma) at 1 mg/ml in buffer H-0.1 to the beads and mix gently for 30 min at 4°C. Elution can be repeated up to 3 times to increase yield. Filter eluted complex through a low protein binding 0.45 μm microfilter unit (Millipore) to remove residue beads. To concentrate the complex, a Microcon filter (Millipore) with 30 kD MWCO can be used. Finally, small aliquots (5–10 μl) of eluted or concentrated complex were flash frozen in liquid nitrogen and stored at −80°C.

In Vitro *Assays for INO80 Functions*

To analyze the function of the purified INO80 complex, a series of *in vitro* assays can be used, most of which are described in detail elsewhere in this book. Here, we briefly outline these assays.

Subunit identification: We used silver-staining technique to visualize the subunits analytically (see Fig. 1). Freshly made SDS-PAGE is preferred to prepacked gels. For identification of proteins, $\sim$40 μg of INO80 complex was purified from 4 L of culture, and separated by 10% SDS-PAGE. The excised protein bands were subjected to in-gel tryptic digestion, and the peptide mixture was analyzed by μLC/MS/MS on a Finnigan LCQ quadrupole ion trap mass spectrometer. The MS/MS spectra generated were subjected to correlation analysis using the algorithm Sequest and results were confirmed manually for fidelity.[12]

ATPase assay: ATPase assay was carried out as described.[13] A standard ATPase reaction (10 μl) contained 20 mM HEPES-KOH (pH 7.6), 80 mM KCl, 0.8 mM EDTA, 6.6 mM MgCl$_2$, 2 mM DTT, 8% glycerol, 0.016% NP-40, 30 μM ATP, and 5 μCi α-P^{32}-labeled ATP. DNA and nucleosome were used for stimulation of ATPase activity. Reaction was 30 min at 30°C. Products were separated by thin-layer chromatography in 0.75 M KH$_2$PO$_4$. Signals were quantified on a Fuji Image Reader LAS-1000.

[12] J. K. Eng, A. L. McCormick, and J. R. Yates, III, *J. Am. Soc. Mass Spectrom.* **5,** 976 (1994).
[13] D. A. Gdula, R. Sandaltzopoulos, T. Tsukiyama, V. Ossipow, and C. Wu, *Genes Dev.* **12,** 3206 (1998).

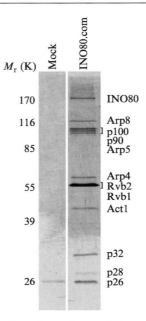

FIG. 1. The INO80 ATP-dependent chromatin remodeling complex. SDS-PAGE and silver stain showing the identity of subunits of the INO80 complex (INO80.com), purified by single-step FLAG-immunoaffinity chromatography from yeast whole cell extract. Adapted from X. Shen, G. Mizuguchi, A. Hamiche, and C. Wu, *Nature* **406,** 541 (2000) with permission.

Chromatin transcription: Transcription assay using chromatin templates was performed essentially as described.[14] GAL4-VP16 was used as a transactivator and purified *Drosophila* RNA polymerase II and purified or recombinant general transcription factors (TFIIA, TFIIB, TFIID, TFIIE, TFIIF, TFIIH, and TFIIS) were used. Chromatin template is assembled in *Drosophila* embryo extract and purified by gel filtration. Alternatively, chromatin template can be purified using a sucrose gradient. Restriction enzyme accessibility and primer extension assays were used to monitor chromatin remodeling and transcription, respectively.

Nucleosome mobilization assay: Mononucleosome are prepared as described.[15] Native *Drosophila* or recombinant yeast histones were used for octamer preparation. 359 bp *Drosophila hsp70* or yeast *INO1* promoter DNA fragments were used to assemble mononucleosomes, following a stepwise salt dialysis method. For some assays, reconstituted mononucleosomes

[14] G. Mizuguchi, T. Tsukiyama, J. Wisniewski, and C. Wu, *Mol. Cell* **1,** 141 (1997).
[15] A. Hamiche, R. Sandaltzopoulos, D. A. Gdula, and C. Wu, *Cell* **97,** 833 (1999).

were purified by glycerol gradient. A standard nucleosome mobilization assay (10 μl) contained 10 mM Tris-HCl (pH 7.5), 50 mM NaCl, 3 mM MgCl$_2$, 1 mM β-mercaptoethanol, 100 μg/ml BSA, 1 mM ATP, 5 to 6 nM mononuclesome, and 1.3 to 32 nM chromatin remodeling complexes. The reactions were incubated at 37°C for 30 min, and then separated in a 4% (5%) native PAGE gel in TE with extensive buffer circulation. Gels were stained with SYBR Green I and documented using a Fuji Image Reader LAS-1000.

Concluding Remarks

We have noted several issues related to the making of yeast whole extract and the single-step FLAG affinity purification. Using the simple blender method, about 30 to 50% of cells were broken. There are alternative methods for making yeast extract using large dedicated equipment such as a microfluidizer, and for immunoaffinity purifications (see Vary, Fazzio, and Tsukiyama, *Assembly of Yeast Chromatin Using ISWI Complexes*, in Chapter 6, volume 375 for details). During the blending process, chromatin is broken into small soluble pieces. We find that the DNA in the whole cell extracts range from 0.2 to 2 kb. Since the INO80 complex can bind to DNA, it is important to test for the presence of DNA in the final preparation. One way to remove DNA is by raising the salt concentration in the initial extract to buffer H-0.5; alternatively, DNase I digestion can be used before the final elution of the complex. For FLAG-tagging, single, double, or triple FLAG tags have all been successfully used. We also noted a common contaminant in FLAG-immunoaffinity purification, Pfk26, since it was common in several unrelated FLAG immunoaffinity purifications (possibly due to a cross-reactivity of the FLAG antibody). *PFK26* can be readily deleted if interference in assays is suspected. Moreover, we have used mutants and corresponding mutant INO80 complexes to study the function of the actin related proteins in the INO80 complex. We found a slight decrease in yields for the mutant complexes, which may be due to the slow growth of mutant cells. As a result, common contaminants were more pronounced in mutant complexes. Similarly, when purifying proteins or protein complexes, which are present at much lower levels than that of Ino80, the methods we described might not be sufficient and additional purification steps may be necessary.

Acknowledgments

We would like to thank Gaku Mizuguchi, Ali Hamiche, Ryan Ranallo, Hua Xiao, Wei-Hua Wu, and Eugene Choi for contributions to the analyses of the INO80 complex. X.S. was supported by fellowships from National Cancer Institute and American Cancer Society, and currently by Grants from the University of Texas M.D. Anderson Cancer Center. The methods reported here were developed while the author was a post-doctoral fellow in the laboratory of Carl Wu, and supported by the Intramural Research Program of the National Cancer Institute.

[27] Assay of Z-DNA Induction by Chromatin Remodeling Factors

By Hong Liu and Keji Zhao

Transcriptional activation requires modification of the repressive chromatin structure. Currently, two extensively studied mechanisms are implicated in modifying chromatin structure—one is the chemical modification of core histones including acetylation, methylation, phosphorylation, and ubiquitination and the other is the action of various ATP-utilizing enzymes prototyped by the yeast SWI/SNF complex.[1–3] We reported recently that the mammalian cell could also use formation of Z-DNA structure to modulate chromatin structure.[4]

The left-handed Z-DNA structure was identified two decades ago.[5] This energetically unfavorable conformation can be stabilized by negative supercoiling in plasmid DNA with the readiness $d(CG)n > d(TG)n > d(GGGC)n > d(TA)n$, reviewed by Herbert and Rich.[6] The promoter region of the murine c-Myc gene contains three Z-DNA-forming sequences. These sequences exist in Z-DNA conformation when the c-Myc gene is actively transcribed. However, the Z-DNA structure rapidly diminishes when the c-Myc gene is down-regulated.[7] The dynamic nature of the Z-DNA structure indicates that its formation is tightly regulated. It is estimated that the mammalian genome contains approximately 100,000 copies of potential Z-DNA forming units that do not have any known functions.[8] Recently, we found[4] that induction of Z-DNA structure by chromatin remodeling SWI/SNF-like BAF complexes may facilitate chromatin opening

[1] J. D. Aalfs and R. E. Kingston, *Trends Biochem. Sci.* **25,** 548 (2000).
[2] T. Jenuwein and C. D. Allis, *Science* **293,** 1074 (2001).
[3] C. L. Peterson and J. L. Workman, *Curr. Opin. Genet. Dev.* **10,** 187 (2000).
[4] R. Liu *et al.*, *Cell* **106,** 309 (2001).
[5] A. H. Wang *et al.*, *Nature* **282,** 680 (1979).
[6] A. Herbert and A. Rich, *Genetica* **106,** 37 (1999).
[7] B. Wittig, S. Wolfl, T. Dorbic, W. Vahron, and A. Rich, *EMBO J.* **11,** 4653 (1992).
[8] H. Hamada and T. Kakunaga, *Nature* **298,** 396 (1982).

    0076-6879/04 $35.00

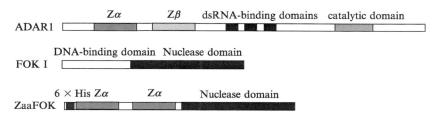

FIG. 1. Construction of ZaaFOK. The Z-DNA-binding domains of ADAR1 include $Z\alpha$, $Z\beta$, and a spacer region. To construct ZaaFOK, The nuclease domain of FOK I restriction enzyme was fused to the C-terminus of the Z-DNA-binding domains within which the $Z\beta$ domain was replaced with another copy of the $Z\alpha$ domain to increase the Z-DNA-binding affinity and reduce sequence specificity. The whole unit was cloned into pET15b for bacterial expression of a His-tagged fusion protein.

in the promoter of the colony-stimulating factor 1 gene (CSF1) that is implicated in the regulation of the proliferation, differentiation, and survival of macrophages.[9]

We used SW-13 cells, which do not have an active BAF complex because of the absence of the essential ATPase subunit BRG1 or its close homologue hBRM subunit,[10,11] to analyze the chromatin remodeling and induction of Z-DNA structure in the CSF1 promoter.[4] Ectopic expression of BRG1 in the cells reconstitutes the active BAF complex and activates about 80 genes including CSF1. The activation of the CSF1 promoter by BRG1 requires formation of proper chromatin structure and the presence of the Z-DNA forming TG repeat sequence. Since the frequently used pGL3 vectors do not form regular chromatin structure in SW-13 cells, we cloned the CSF1 promoter into the pREP4-luc episomal reporter vector that contains Epstein–Barr virus replication origin and encodes nuclear antigen EBNA-1. The vector replicates and forms regularly spaced nucleosomal structure in human cells.

Detection of Z-DNA structure in mammalian cells has been a technical challenge. c-Myc promoter was reported to contain Z-DNA structure using Z-DNA-specific antibodies and transcriptionally active nuclei embedded in agarose gel.[7] Recently, it was reported that the RNA-editing enzyme, ADAR1, contains a domain (called $Z\alpha$) that binds specifically to Z-DNA structure with high affinity. A fusion protein (named here as ZaaFOK shown in Fig. 1) generated by fusing 2 copies of the $Z\alpha$ domain with the

[9] E. R. Stanley, L. J. Guilbert, R. J. Tushinski, and S. H. Bartelmez, *J. Cell Biochem.* **21,** 151 (1983).
[10] C. Muchardt and M. Yaniv, *EMBO J.* **12,** 4279 (1993).
[11] J. L. Dunaief *et al., Cell* **79,** 119 (1994).

nuclease domain of FOK I endonuclease binds specifically to Z-DNA structure and produces double-stranded cleavages surrounding Z-DNA sequences *in vitro*.[12] Here, we present methods to analyze formation of Z-DNA structure in mammalian cells using formaldehyde cross-linking and ZaaFOK protein.

Methods

Overexpression and Purification of ZaaFOK

Materials and Solution

LB medium containing 60 μg/ml ampicillin

1 M isopropyl-β-D-thiogalactopyranoside (IPTG)

0.1 M phenylmethylsulfonyl fluoride (PMSF) in absolute ethanol

1 M imidazole, pH 8.0

Lysis buffer: 1× PBS, 0.1 mM EDTA, 0.3 M NaCl, 0.5% NP-40

Wash buffer: 1× PBS, 0.1 mM EDTA, 0.3 M NaCl, 0.5% NP-40, 15 mM imidazole

Elution buffer: 20 mM Tris, pH 8.0, 10% Glycerol, 0.4 M imidazole, 1 mM EDTA, 0.1% NP-40, 0.1 M NaCl

1000× stock solution of protease inhibitor cocktail in DMSO: 10 mg/ml leupeptin, 10 mg/ml chymostatin, 10 mg/ml pepstatin A, 5 mg/ml antipain

1 mM PMSF and 1× protease inhibitor cocktail are added to lysis and elution buffers just before use.

Procedure

1. Transform the BL21(DE3) plys S strain of *E. coli* bacteria with 0.05 μg pET-His-ZaaFOK plasmid. Plate on LB agar plate containing 100 μg/ml Ampicillin and incubate at 37° for 16 h.
2. Inoculate 25 ml of LB medium containing 60 μg/ml Ampicillin with a single colony and incubate at 37° overnight with shaking at 225 rpm.
3. Transfer 10 ml of the overnight culture into a sterile 2-liter flask containing 500 ml of LB medium containing 60 μg/ml Ampicillin. Shake at 225 rpm at 37° until the OD_{595} reaches about 0.45.
4. Add 0.5 ml of 1 M IPTG to final concentration of 1 mM. Continue incubation at 37° with shaking for 3 h.
5. Harvest the bacteria by centrifugation at 3000g for 15 min at 4°. Wash the pellet once with 100 ml of ice-cold lysis buffer.

[12] Y. G. Kim, K. Lowenhaupt, T. Schwartz, and A. Rich, *J. Biol. Chem.* **274,** 19081 (1999).

6. Resuspend the pellet in 5 ml of lysis buffer. Break the cells by freezing on dry ice for 10 min and thawing in cold water for 10 min. Then the bacterial lysate is sonicated with an Ultrasonic Processor XL-2020 (Heat Systems Inc.) for 6×20 s at 4.5 power setting with cooling for 30 s in ice water between each sonication.

7. Centrifuge the lysate in a Sorvall SS34 rotor at 9500 rpm for 15 min at $4°$.

8. Wash 0.5 ml of Ni-NTA agarose beads (bed volume) with 20 ml of wash buffer. Resuspend the beads in 20 ml of wash buffer. Transfer the lysate supernatant to the beads. Incubate for 30 min at $4°$ with rotation.

9. Spin down the beads at $1000g$ for 5 min at $4°$. Discard the supernatant.

10. Wash the beads with 25 ml of wash buffer for 5 min at $4°$ with rotation. Spin down the beads as in Step 9 and discard the supernatant.

11. Repeat the wash two more times as in Step 10.

12. Transfer the beads into one 1.5-ml Eppendorf tube. Elute ZaaFOK by adding 0.4 ml of elution buffer. Incubate for 10 min at $4°$ with rotation. Spin down the beads at 3000 rpm in a microfuge for 2 min at $4°$. Transfer the supernatant into a new Eppendorf tube.

13. Repeat the elution two more times. Combine all the three eluates.

14. Analyze the purified ZaaFOK protein by SDS-PAGE. The protein has a molecular weight of about 56 kDa. The protein can be stored by adding glycerol to 50%. It is stable at $-20°$ for at least 6 months. However, for extended period of storage, it should be kept at $-80°$.

Analysis of ZaaFOK Activity In Vitro

The activity of ZaaFOK can be assayed *in vitro* using the 18 GC and NoTG constructs which are derived from pREP4-CSF1pr-luc by replacing the TG repeats with either 18 GC repeats or a random DNA sequence.[4] Using diethylpyrocarbonate treatment, we showed previously that the 18 GC repeat sequence exists in Z-DNA conformation in supercoiled plasmid DNA, while the NoTG construct does not have Z-DNA structure in the same region. Relaxation of negative supercoiling by linearization of the plasmid DNA converts Z-DNA to the more stable B-DNA structure. Digestion of supercoiled 18 GC construct with ZaaFOK followed by BamH I digestion generates a fragment of about 2.1 kb, which is consistent with the cleavage at the 18 GC repeat sequence and the BamH I site. No cleavage at the 18 GC repeat sequence is observed if the plasmid is linearized with BamH I before ZaaFOK treatment.

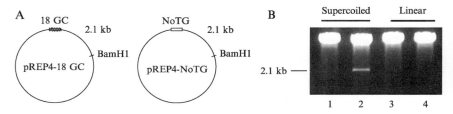

FIG. 2. ZaaFOK specifically binds and cleaves Z-DNA structure. (A) pREP4-18 GC contains a 18 GC repeat sequence located 2.1 kb away from the BamH I site. The 18 GC repeat sequence is replaced by a random DNA sequence in pREP4-NoTG. (B) Supercoiled pREP4-NoTG (lane 1) or pREP4-18 GC (lane 2) was digested with the ZaaFOK fusion protein, followed by digestion with BamH I. The DNA in lanes 3 (pREP4-NoTG) and 4 (pREP4-18 GC) was linearized by BamH I digestion before treatment with ZaaFOK. The reaction mixture was analyzed on 1% agarose gel.

Materials and Solutions

10× digestion buffer: 0.2 M Tris–HCl, pH 7.9, 0.1 M magnesium acetate, 0.5 M potassium acetate, 10 mM DTT, and 1 mg/ml BSA.
10× TAE buffer: 0.4 M Tris–acetate, pH 8.0, 10 mM EDTA.
1% Agarose/1× TAE

Procedure

The supercoiled plasmid DNA is purified with a Qiagen plasmid preparation kit. The digestion reaction contains 1 μg of 18 GC or NoTG DNA, 3 μl of 10× digestion buffer, 0.01–0.1 μg of ZaaFOK. Adjust to final volume of 30 μl and incubate for 1 h at 37°. The incubation is continued for another 30 min after addition of 1 μl of BamH I (20 units). The digestion products are analyzed on 1% agarose/1× TAE gel. Results of a typical digestion are shown in Fig. 2.

Detection of Z-DNA Structure in the Episomal CSF1 Promoter Construct in SW-13 Cells

The activity of the CSF1 promoter in pREP4-CSF1pr-luc construct is stimulated 5- to 10-fold by cotransfection of BRG1 expression vector into SW-13 cells.[4] The expression of BRG1 induces formation of Z-DNA structure in the TG repeat region in the CSF1 promoter. The Z-DNA structure can be detected by treatment of the transfected cells with ZaaFOK, following formaldehyde cross-linking and permeabilization with Triton X-100. The cleavage sites are detected by linker ligation-mediated PCR as described.[13]

[13] P. R. Mueller and B. Wold, *Science* **246,** 780 (1989).

Materials and Solutions

Complete medium: DMEM supplemented with 10% fetal calf serum and 1% Penicillin/streptomycin mix.

Digestion buffer: 20 mM Tris–HCl, pH 7.9, 10 mM magnesium acetate, 50 mM potassium acetate, 1 mM DTT, and 0.1 mg/ml BSA.

Stop buffer: 20 mM Tris–HCl, pH 7.5, 10 mM EDTA, 0.4% SDS.

1× *Klenow enzyme buffer:* 10 mM Tris–HCl, pH 7.5, 5 mM MgCl$_2$, 7.5 mM DTT.

10× T4 DNA *ligase buffer:* 0.5 M Tris–HCl, pH 7.5, 0.1 M MgCl$_2$, 0.1 M DTT, 10 mM ATP, 250 μg/ml BSA.

10× *annealing buffer:* 0.1 M Tris–HCl, pH 7.5, 50 mM MgCl$_2$, 250 mM NaCl, 10 mM DTT.

10× *PCR buffer:* 750 mM Tris–HCl (pH 8.8 at 25°), 200 mM (NH$_4$)$_2$SO$_4$, 0.1% Tween-20.

10× *PNK buffer:* 0.7 M Tris–HCl, pH 7.6, 0.1 M MgCl$_2$, 50 mM DTT.

10× *TBE running buffer:* 0.9 M Tris–borate, pH 8.0, 20 mM EDTA.

Formamide loading buffer: 90% formamide, 1× TBE, 0.05% bromophenol blue, 0.05% xylene cyanol.

Superfect transfection reagent (Qiagen)

50 mg/ml Hygromycin B

3 M sodium acetate (NaOAc), pH 5.2

25 mM MgCl$_2$

1 mM dNTPs

Phenol–chloroform

20 mg/ml glycogen

Ethanol

8 M Urea

40% bis:acrylamide (1:19)

Primers: Universal linker primers (*Wold Linkers*[13]):

WL1: 5′GCGGTGACCCGGGAGATCTGAATTC3′
WL2: 5′GAATTCAGATC3′
pGL3/169R: 5′TATGCAGTTGCTCTCCAGCGG3′
pGL3/150R: 5′GGTTCCATCTTCCAGCGGATAGAA3′

Procedure

1. Sw-13 cells are grown to 60% confluent in complete medium in a 10-cm petri dish.

2. Dilute 2 μg of pREP4-CSF1pr-luc and 8 μg of pBJ5-BRG1 in 300 μl of serum-free DMEM. Add 60 μl of Superfect to the DNA mixture. Mix well and incubate for 5 min at room temperature.

3. Aspirate medium from the SW-13 cells. Add 3 ml of complete medium to the DNA/Superfect mixture and transfer the mixture to the SW-13 cells. Incubate at $37°$ for 2 h.

4. Remove the transfection mixture by aspiration. Add 10 ml of fresh complete medium to the cells. Incubate at $37°$ for 24 h. Add 10 μl of 50 mg/ml hygromycin B to select for transfected cells. Incubate the cells at $37°$ for 72 h.

5. Add 0.27 ml of 37% formaldehyde to the petri dish (final concentration: 1%). Incubate for 10 min at $37°$.

6. Aspirate the medium. Wash the cells twice with $1\times$ PBS.

7. Collect the cells by scraping with a cell-lift.

8. Spin down the cells in an Eppendorf tube at 3000 rpm for 3 min in a microfuge at $4°$.

9. Permeabilize the cells by incubating the cells in 500 μl of digestion buffer $+0.5\%$ Triton X-100 for 5 min on ice.

10. Spin down the cells as in Step 8 and discard the supernatant.

11. Resuspend the cells in 200 μl of digestion buffer. Dispense into $5\times$ 50 μl aliquots.

12. Add 0, 10, 30, 100, and 300 ng of ZaaFOK, respectively. Mix well and incubate at $37°$ for 30 min with occasional mixing.

13. Stop the reaction by adding 150 μl of stop buffer. Add 5 μl of 20 mg/ml proteinase K. Incubate at $65°$ for 6 h or overnight.

14. Add 200 μl of phenol–chloroform solution to the DNA. Vortex vigorously for 30 s. Spin for 2 min at $4°$ in a microfuge. Transfer the supernatant into a new Eppendorf tube and repeat the extraction once.

15. Add 1 μl of 10 mg/ml RNase A to the DNA. Incubate at $37°$ for 20 min.

16. Extract with phenol–chloroform as in Step 14.

17. Add 20 μl of 3 M NaOAc, pH 5.2, 500 μl of ethanol. Mix well. Incubate in dry ice for 20 min. Spin for 15 min at 13,000 rpm at $4°$ in a microfuge.

18. Discard the supernatant. Wash the pellet once with 70% ethanol. Air-dry briefly and resuspend the pellet in 40 μl of $1\times$ TE. Determine the DNA concentration.

19. Blunt DNA ends with Klenow enzyme: the 20 μl of reaction mixture containing 3 μg of DNA, 0.1 mM dNTPs, 2 units of Klenow enzyme in $1\times$ Klenow enzyme buffer is incubated at room temperature for 20 min.

20. Add 180 μl of $1\times$ TE to reaction mixture. Extract with phenol–chloroform and precipitate the DNA as described from Steps 16–18. The DNA is resuspended in 30 μl of $1\times$ TE.

21. Anneal the universal linker WL1 and WL2 by mixing 2.5 micromoles of WL1 and 2.5 micromoles of WL2, 10 μl of 10× annealing buffer. Adjust to final volume of 100 μl. Heat at 94° for 5 min. Cool to 4° over a period of 2 h.

22. Linker ligation: 10 μl of DNA (1 μg), 5 μl of 10× T4 DNA ligase buffer, 30.5 μl of H_2O, 2.5 μl of 25 mM annealed universal linker WL(1 + 2), 2 μl of T4 DNA ligase (New England Biolabs, 400 U/μl). Incubate at 14°, overnight.

23. Add 50 μl of 1× TE to the reaction mixture. Extract with 100 μl of phenol–chloroform. Precipitate the DNA with 1 μl of 20 mg/ml glycogen, 10 μl of 3 M NaOAc, pH 5.2, and 250 μl of ethanol. Wash the pellet once with 70% ethanol. Resuspend the DNA in 20 μl of H_2O.

24. PCR amplification: PCR reaction mix: 10 μl of DNA, 5 μl of 10× PCR buffer, 5 μl of 25 mM $MgCl_2$, 5 μl of 1 mM dNTPs, 2 μl of 2.5 μM WL1 primer, 2 μl of 2.5 μM pGL3/169R primer, 20 μl of H_2O, 1 μl of Taq DNA polymerase (5 U/μl, Fermentas). Cycle: 94°, 30″; 58°, 30″; 72°, 30″; 25×.

25. While running the PCR, label pGL3/150R primer with T4 polynucleotide kinase (PNK) and γ-^{32}P-ATP by mixing 40 pmol of the primer, 2 μl of 10× PNK buffer, 9 μl of H_2O, 4 μl of γ-^{32}P-ATP, 1 μl of T4 PNK (10 U/μl, New England Biolabs). Incubate at 37° for 40 min. Heat at 94° for 10 min to inactivate PNK.

26. Add 1 μl of the labeled primer to the PCR reaction mixture. Continue cycling: 94°, 1 min; 60°, 5 min, 72°, 5 min; 2×.

27. Add 150 μl of 1× TE to the reaction mixture. Extract with 200 μl of phenol–chloroform. Precipitate the DNA with 1 μl of 20 mg/ml glycogen, 20 μl of 3 M NaOAc, pH 5.2, 500 μl of ethanol. Incubate for 10 min on dry ice. Spin for 15 min at 13,000 rpm at 4°. Wash the pellet once with 70% ethanol and air-dry briefly. Resuspend the pellet in 15 μl of formamide loading buffer. Heat at 94° for 4 min. Chill on ice.

28. Load 5 μl onto 8% acrylamide/6 M urea/1× TBE gel. Run at 260 V till xylene cyanol migrates to about 1 cm to the bottom.

29. Results of a typical experiment are shown in Fig. 3A.

Detection of Z-DNA Structure in the Endogenous CSF1 Promoter in SW-13 Cells

Even though the chromatinized CSF1 promoter reporter vector could be used to evaluate the effect of chromatin structure on the regulation of the promoter by the BAF complex, it is desirable to examine the induction

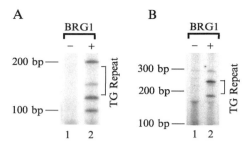

FIG. 3. Examples of experiments using the methods described to detect BRG1-induced Z-DNA formation in the episomal CSF1 promoter vector and endogenous CSF1 promoter. (A) Detection of Z-DNA structure in the episomal CSF1 promoter. SW-13 cells transfected with BRG1 (+) or a control vector (−) were cross-linked with formaldehyde, followed by ZaaFOK digestion. The cleavage sites were detected by LM-PCR using specific primers recognizing the vector sequence. The TG repeat region in the CSF1 promoter is indicated on the right side of the panel. (B) Detection of Z-DNA structure in the endogenous CSF1 promoter. The LM-PCR of the same DNA as in panel A was performed using PCR primers recognizing the endogenous CSF1 promoter sequence. The TG repeat region is indicated on the right side of the panel.

of Z-DNA structure by the BAF complex in the endogenous CSF1 promoter. The DNA obtained by ZaaFOK digestion as described above can be used for this purpose with a different set of PCR and labeling primers specific for the endogenous CSF1 promoter. This method could also be used to analyze Z-DNA structure at other gene loci using the corresponding specific primers.

Materials and Solutions

CSF1 promoter primers: CSF1/361F: 5′ACGATCATAGAGCGCT AGCACTGA3′
CSF1/318F: 5′AAGGAAAGGGTCGGT CCGCAGA3′

Procedure

1. PCR amplification: amplify the DNA ligated to the universal linker WL (1 + 2) with primers WL1 and CSF1/361F as described in Step 24 above.
2. Label primer CSF1/318F with ^{32}P as Step 25 above. Add 1 μl of the labeled primer to the PCR reaction mixture and continue cycling as in Step 26 above.
3. Purify the DNA and analyze as described in Steps 27 and 28 above.
4. A typical result is shown in Fig. 3B.

[28] Immuno-Depletion and Purification Strategies to Study Chromatin-Remodeling Factors *In Vitro*

By GILLIAN E. CHALKLEY and C. PETER VERRIJZER

The packaging of the eukaryotic genomes into repressive chromatin has important consequences for the mechanisms of gene expression control. Together with gene-specific regulators and the basal transcription factors, proteins that modulate chromatin structure form an integral part of the gene control machinery. There are two broad classes of enzymes that regulate chromatin structure: (1) factors that utilize the energy derived from ATP-hydrolysis to modulate the contacts between histones and DNA, resulting in an increased DNA accessibility within a nucleosomal array$^{1-6}$ (2) enzymes that mediate covalent modifications of the histones.$^{7-10}$ In particular the unstructured N-terminal histone tail domains are subjected to extensive modifications, including acetylation, phosphorylation and methylation. Furthermore, the carboxyl-terminal domains of histones H2A and H2B are targets for ubiquitination. The patterns of histone tail modifications are believed to form a so-called "histone code" that creates recognition sites for specific gene-regulatory factors and can influence the higher-order folding of chromatin.^{8,9} Both classes of chromatin modulating enzymes have been isolated from a variety of organisms ranging from yeast, to flies and man. Typically, they comprise large and highly conserved multiprotein complexes.

ATP-dependent chromatin-remodeling complexes contain a SWI2/SNF2-like ATPase and use the energy of ATP-hydrolysis to alter the structures of nucleosomes.^{11} As a result, they control gene expression by allowing transcription factors to gain access to the DNA. These complexes can be divided into several main families, each characterized by different core ATPase subunits and associated factors. Named after their constituent

^{1} P. B. Becker and W. Horz, *Annu. Rev. Biochem.* **71,** 247 (2002).
^{2} S. Kadam and B. M. Emerson, *Curr. Opin. Cell Biol.* **14,** 262 (2002).
^{3} K. Katsani, T. Mahmoudi, and C. P. Verrijzer, *Curr. Top. Microbiol. Immunol.* **274,** 113 (2003).
^{4} G. J. Narlikar, H. Y. Fan, and R. E. Kingston, *Cell* **108,** 475 (2002).
^{5} I. A. Olave, S. L. Reck-Peterson, and G. R. Crabtree, *Annu. Rev. Biochem.* **71,** 755 (2002).
^{6} T. Tsukiyama, *Nat. Rev. Mol. Cell Biol.* **3,** 422 (2002).
^{7} S. L. Berger, *Oncogene* **20,** 3007 (2001).
^{8} T. Jenuwein and C. D. Allis, *Science* **293,** 1074 (2001).
^{9} B. M. Turner, *Bioessays* **22,** 836 (2000).
^{10} Y. Zhang and D. Reinberg, *Genes Dev.* **15,** 2343 (2001).
^{11} A. Flaus and T. Owen-Hughes, *Curr. Opin. Genet. Dev.* **11,** 148 (2001).

ATPase they are referred to as the SWI2/SNF2, ISWI or Mi-2 family of chromatin-remodelers.[1,2,4] Recent studies have emphasized an extensive structural and functional diversification among chromatin-remodeling complexes. One of the current challenges in the study of chromatin-modulating enzymes is to define the functional differences between related factors that perform comparable enzymatic reactions *in vitro*.

Chromatin-remodeling factors have been identified by genetic screens and biochemical reconstitution and purification strategies. Here, we discuss two approaches involving the use of selective antibodies that complement and extend the classic genetic and biochemical studies of chromatin modulating enzymes: (1) Immunopurifications allow the relatively rapid identification of associated partner proteins of a given factor. (2) Immunodepletions allow the functional requirement of a given factor to be rigorously tested in the presence of other, known or unknown, chromatin regulators. This latter approach complements *in vitro* reconstitution reactions using purified components. Reconstitution of biochemical processes with highly purified components will establish which factors are sufficient for, for example, *in vitro* transcription activation. However, the selective removal of a given factor from a nuclear extract by immunodepletion will determine whether a factor is also necessary, or whether related proteins can compensate for its loss.

The generation of highly selective reagents is the most critical determinant of success for both immunopurification and immunodepletion experiments. Therefore, we provide a series of considerations and protocols for the generation of the necessary tools. These protocols cover the whole process from generation of antigens, antibody columns and their use in functional experiments. The power of immunodepletion and immunopurification strategies is then illustrated by our functional studies of the BRM complex. For a comprehensive description of the use and storage of antibodies we refer to Harlow and Lane.[12] Most methods described here are generic protocols, which can easily be adapted and optimized for a specific purpose.

Methods

General Buffers

HEMG: 25 mM N-2-Hydroxyethylpiperazine-N'-2-ethanesulfonic acid (HEPES)-KOH pH 7.6, 0.1 mM EDTA, 12.5 mM MgCl$_2$, 10% glycerol. Unless otherwise stated buffers based on HEMG also contain freshly

[12] E. Harlow and D. Lane, "Antibodies: A Laboratory Manual." Cold Spring Harbour Laboratory, Cold Spring Harbour, New York, 1988.

added: 1 mM dithiothreitol (DTT), and protease inhibitors: 0.2 mM Phenylmethylsulfonyl fluoride (PMSF), 1 mM $Na_2S_2O_5$ (NaMBS). For antigen production other protease inhibitors are not usually required.

PBS-Phosphate buffered saline: 137 mM NaCl, 2.7 mM KCl, 10 mM Na_2HPO_4, 2 mM KH_2PO_4, pH 7.4

TBS: 25 mM Tris buffer pH 7.5, 144 mM NaCl.

General Notes

- We use the abbreviation CV to denote column volumes. This is the volume occupied by the settled chromatography resin. We use it here to represent volume either in a column or the bed volume when using batch purifications. Volumes of buffers used in purifications are expressed as multiples of CV.
- As an alternative to column chromatography, most of the purification protocols we describe here can be carried out in batch by stepwise incubations and washes in a screw-capped tube. Following each incubation, the beads are recovered through low speed centrifugation. Typically, both techniques give comparable results and which approach to use is largely a matter of personal taste. See notes after GST fusion protein purification for further details.
- Typically, protein purification procedures are carried out at 4° or on ice, unless otherwise stated in the protocol.

Generation of Antigens for Immunization and Affinity Matrices

The purification of sufficient amounts of endogenous material for use as an antigen is often a costly and labor intensive process. Therefore, it is more convenient to use either synthetic peptides or purified recombinant proteins as sources of antigen. The synthesis of peptides is straightforward, though somewhat costly. We typically use peptides of about 15 residues in length as antigens. There are no hard and fast rules as to which protein sequences will make good antigens. The selection of suitable peptide sequences is essentially a process of trial and error. However there are a few points that should be taken in consideration. Firstly, conserved protein domains should be avoided as they are typically not very immunogenic and the resulting antiserum may show cross-reactivity with related proteins. Secondly, it makes sense to choose regions with hydrophilic amino acids as, in contrast to hydrophobic domains, they are more likely to be exposed in the native protein. Finally, to avoid problems with secondary protein structure in the native protein, it might be preferable to choose domains predicted to be unstructured.

When using expression in *E. coli* to produce antigens, the same considerations apply. The advantages of the use of recombinant proteins over peptides are their cost effectiveness and that it allows the expression of much
larger polypeptides. Wherever possible we express non-conserved protein
domains ranging from 20 to 40 kDa in size. Larger polypeptides often give
insufficient levels of expression. We typically express epitope-tagged
proteins, particularly Glutathione S-transferase (GST) or $6\times$ histidine
(HIS)-tagged proteins, because the tag permits a simple one step affinity purification. In our experience neither the GST nor the HIS-tag compromises
the selectivity of the resulting antisera. An additional advantage of HIS-
tagged proteins is that they can be purified from solubilized inclusion bodies
under denaturing conditions. Inclusion bodies, when comprised largely of
recombinant protein, can also be used directly as good sources of antigen.

Later we provide short protocols for recombinant protein expression
and purification from *E. coli*. It should be noted that these protocols are optimized for simplicity and protein yields without too much consideration
for biological activity of the recombinant proteins. Once the desired antigen has been prepared, immunization can be performed using standard
procedures.[12]

Protein Expression in E. Coli

1. Use a single colony of freshly transformed BL21-CodonPlus cells
   to inoculate 5–10 ml of 2YT (16 g/L Bacto-tryptone, 10 g/L Bacto-
   yeast extract, 5 g/L NaCl pH 7), containing 100 μg/ml ampicillin
   (or another appropriate antibiotic) and 0.4% glucose. Grow culture
   overnight at $37°$.
2. Inoculate 500 ml of 2YT containing 100 μg/ml ampicillin and 0.4%
   glucose, with 5 ml of overnight culture. Grow at $37°$ and monitor
   the OD_{600}.
3. When the OD_{600} reaches 0.5, induce the culture with 0.4 mM
   isopropyl-β-D-1-thiogalactopyranoside (IPTG final concentration).
4. Grow for a further 2 h at $37°$. (All subsequent steps are now carried
   out at $4°$ or on ice.)
5. Collect the cells from 500 ml culture by centrifugation at 3000g for
   15 min.
6. Wash pellet with phosphate buffered saline (PBS).
7. Resuspend cells in 20 ml of HEMG containing 500 mM NaCl,
   0.1% (v/v) NP-40, and 150 μg/ml lysozyme. Note to omit DTT and
   EDTA from the buffer if HIS-tagged proteins are to be
   subsequently purified by Ni-NTA chromatography.
8. Incubate on a rotator, for 20 min at $4°$.

9. Add DNase I and RNase A each to a final concentration of 1 μg/ml. Incubate for an additional 20 min at 4°.
10. Sonicate with a probe sonicator in an ice bath three times 30 s each with probe setting at about 40 W.
11. Freeze in liquid nitrogen and thaw in an ice bath to rupture the cells.
12. Centrifuge at 20,000g for 30 min at 4°. Collect the supernatant and pellet. Take aliquots and snap freeze these in liquid nitrogen. Store at −80°.
13. Analyze samples by SDS-PAGE followed by Coomassie Blue staining.

Experimental Notes

- Soluble protein can now be purified through either the GST- or HIS-tag as described below. The insoluble protein can either be used directly as antigen or HIS-tagged proteins can be purified from the pellet solubilized in 6 M urea. If required, solubility can often be improved by expression at 24° or 30° and induction with 0.1 mM IPTG.
- We prefer to use *E. coli* BL21-CodonPlus (DE3)RIL (Stratagene). This is a protease deficient strain that expresses a number of tRNAs corresponding to codons *argU, ileY,* and *leuW* rare in *E. coli,* thus avoiding possible problems with codon bias. The RIL strain is particularly suitable for AT-rich genomes. Stratagene also provides a −RP strain which contain the *argU* and *proL* tRNA genes which are therefore more suitable for GC-rich genomes.
- If the protein expressed is detrimental to the host cells it is advisable to inoculate the starter culture (Step 1) early in the morning and induce on the same day. Moreover, induction at a lower OD_{600} (e.g., 0.2) often leads to higher yields in these cases.

Preparation of Antigen from Pellet Fraction

When the protein of interest is highly expressed but insoluble, it can still be very useful as an antigen. The isolated inclusion bodies (largely comprising recombinant protein) are then simply injected as a homogenized particulate suspension.

Protocol

All steps are carried out at 4° or on ice.

1. The pellet fraction from 1 L of induced bacterial culture, is resuspended by dounce homogenization in 20 ml of HEMG

containing 500 mM NaCl, 0.1% (v/v) NP-40. Sonication can also be used as an aid to resuspension if the pellets prove to be particularly difficult to resuspend.

2. Centrifuge at 10,000g for 25 min.
3. Repeat wash step with PBS.
4. Resuspend the pellet in 2–5 ml of PBS.
5. Assess the concentration and purity of the final preparation by a protein concentration assay and SDS-PAGE gel electrophoresis followed by Coomassie staining with known amounts of BSA run in parallel.

Purification of GST Fusion Proteins

GST fusion proteins are purified by affinity purification using immobilized glutathione on a matrix, such as Sepharose. Although we have not found it necessary, the tag can be removed by cleavage with an appropriate site-specific protease if required. All steps are carried out at 4° or on ice. The concentration and purity of the final preparation should be assessed by a protein concentration assay and SDS-PAGE, with known amounts of BSA run in parallel, followed by Coomassie staining.

Protocol

1. Thaw crude extract in an ice bath and centrifuge at 20,000g for 25 min.
2. Equilibrate the glutathione beads (Glutathione Sepharose, Amersham Biosciences) in HEMG containing 500 mM NaCl, 0.01% (v/v) NP-40, then mix with the clarified extract. For 20 ml of crude extract, produced from 1 L of culture, we typically add 0.5–2 ml of beads. The capacity of the glutathione beads is about 5 mg/ml.
3. Incubate on a rotating wheel or nutator for about 1 h.
4. Pour the extract with beads in an empty column.
5. Wash the column with 20 CV of HEMG containing 500 mM NaCl, 0.1% (v/v) NP-40.
6. Wash the column with 5 CV HEMG containing 200 mM NaCl, 0.01% (v/v) NP-40. The column can either be developed by using a reservoir and gravity flow or, alternatively, a peristaltic pump or chromatographic system can be used.
7. Elute the column with 5 CV of HEMG containing 200 mM NaCl, 0.01% (v/v) NP-40, and 10 mM reduced glutathione. Please note that the elution step is omitted if the antigens are be used to affinity purify antibodies (see later).

Alternative Method—Purification in Batch Batch purification might be a more convenient method if equipment is limiting or if multiple protein preparations are to be made simultaneously. This method yields equally useful antigens. In general, all purification procedures described here can either be performed in batch or by column chromatography with equivalent results. When performing batch purifications the following points should be taken into consideration

1. A minimum of 5 wash steps are required. These involve resuspension of the bound beads in 10 CV of buffer, rotate the tube for 5 min to keep the beads in suspension during the wash, then collect beads using low speed centrifugation ($500g$ for 2 min). Remove supernatant and repeat wash.
2. To elute add 1.5 CV of elution buffer.
3. Rotate for 30 min at room temperature.
4. Separate beads and eluate by low speed centrifugation at $500g$, 2 min.

Experimental Notes

- If the antigen preparation is to be used for affinity purification of antibodies, it is not necessary to elute the protein. Following extensive washes, the antigen can be directly coupled to the glutathione peptide on the beads and used as an affinity matrix. (See later for method of direct coupling).

Purification of Soluble His-Tagged Fusion Proteins

Important: Make sure that the bacterial lysis buffer does not contain chelating agents such as EDTA or reducing agents such as DTT. All steps are carried out at 4° or on ice.

Protocol

1. Thaw 20 ml of crude extract (from 1 L of culture) in an ice bath and centrifuge at $20,000g$ for 25 min.
2. Add 100 μl of a 1 M imidazole pH 7.5 stock to the extract to a 5 mM final concentration.
3. Equilibrate Ni-NTA beads (Qiagen) in HG buffer (25 mM HEPES-KOH pH 7.6, 10% glycerol, 0.2 mM PMSF, 1 mM NaMBS) containing 500 mM NaCl, 5 mM imidazole pH 7.5. The capacity of Ni-NTA matrix is about 5–10 mg/ml. Depending on the level of

protein expression, we typically add 1 to 6 ml of beads to a 20 ml extract in a 50 ml screw cap tube.

4. Incubate for about 2 h at 4° on a rotating wheel.
5. Pour the slurry into an empty column and allow the beads to settle.
6. Wash the packed column with 20 CV of HG buffer containing 500 mM NaCl, 5 mM imidazole and 0.1% (v/v) NP-40.
7. Wash column with 2 CV of HG buffer containing 500 mM NaCl, 50 mM imidazole and 0.01% (v/v) NP-40.
8. Elute the HIS-tagged protein with 4 CV of HG buffer containing 200 mM NaCl, 500 mM imidazole, and 0.01% (v/v) NP-40. Collect fractions of about a fifth of the CV. Add DTT to the fractions from a 100 mM stock to a final concentration of 1 mM.

Preparation of Peptides Conjugated to Keyhole Limpet Hemocyanin

Peptides first have to be coupled to a carrier molecule before they can be used as antigens as, by themselves, peptides are too small to be immunogenic and raise an effective immune response. We describe here a method for coupling to Keyhole Limpet Hemocyanin (KLH) although other carriers can also be used.

Protocol

1. Weigh out an equal amount of peptide and KLH (Calbiochem) carrier protein. We typically use about 2 mg each per rabbit.
2. Dissolve peptide and KLH in 1 ml 0.1 M Na$_2$CO$_3$ pH 9.6, per mg of KLH in a glass centrifuge tube. Mix with a magnetic stirrer.
3. Add glutaraldehyde, from a 25% stock, to a final concentration of 0.05%.
4. Stir for several hours or overnight in the dark at room temperature. The solution frequently turns yellow or orange and precipitates may form.
5. Add 100 μl of 1 M glycine ethyl ester to a 0.1 M final concentration and incubate for an additional 30 min. Remove the stir-bar.
6. Precipitate the proteins by the addition of 5 ml of ice-cold acetone, incubate at $-70°$ for 30 min.
7. Centrifuge at 10,000g for 15 min at room temperature.
8. Remove acetone and air dry the pellet.
9. Resuspend the pellet in PBS or 0.9% (w/v) NaCl. Store at $-20°$ until use. The coupled carrier frequently becomes insoluble but this does not affect its quality as an antigen. If it is insoluble, dounce homogenize or sonicate the pellet before immunization.

Generation of Affinity Purified Antibodies

Once the desired antigen has been prepared immunization can be performed using standard procedures.[12] Obviously, the quality of the immune serum will determine the usefulness of all reagents generated from it. Therefore, when the antiserum is of a low quality, it is often advisable to repeat the immunization, perhaps using another protein domain (or domains) as an antigen, rather than embarking upon lengthy optimization of assays. Affinity purification retains only those antibodies directed against the protein of interest, thus ensuring that full advantage of the high specificity of the antibodies produced can be taken. The first step in this procedure is to generate an affinity matrix. Depending on the original antigen, a number of different methods (detailed later) are appropriate. The crude serum is then partially purified, by an ammonium sulphate precipitation step, to separate antibodies from other serum proteins. Next, the desired antibodies are purified by affinity column chromatography.

Preparation of Affinity Matrices

General Notes on Protein Coupling to Affi-Gel (BioRad)

Although there are several equally useful supports, we normally use either Affi-gel 10 or Affi-gel 15 for coupling of antigens. We aim to use an antigen concentration of about 20–25 mg/ml of gel, up to a maximum of 30 mg/ml. Above 30 mg/ml the coupling efficiency is reduced. If the amount of antigen is limiting good results can still be obtained with much lower amounts of protein. Although a variety of coupling buffers can be used, buffers containing Tris, glycine or any other molecules with primary amine groups should be avoided as they will also couple to the gel. Affi-gel 10 couples ligands best at a pH near or below their isoelectric point. Therefore, when the coupling pH is 6.5–7.5, Affi-gel 10 is best for ligands with isoelectric points of 6.5–11 (neutral or basic). Affi-gel 15 couples ligands best at a pH near or above their isoelectric points. Therefore at neutral pH use Affi-gel 15 for ligands with isoelectric points below 6.5 (acidic).

Generation of Affinity Matrices Using Soluble Antigen

The method described below assumes coupling of 40 mg of antigen to produce a 2 ml column. However, lower protein amounts or diluted samples will also yield useful affinity columns.

1. Dilute the prepared antigen to 8 ml in coupling buffer (for example PBS), transfer to a screw-capped tube and add 2 ml of equilibrated

Affi-gel beads. Incubate for 4 h on a rotating wheel at $4°$. (If speed is required and stability of the antigen allows then incubation at room temperature for 1 h is also possible.)

2. Block any remaining active ester groups by the addition of 200 μl (100 μl/ml beads) of 1 M ethanolamine pH 8.0 and incubate for a further hour.

3. Recover beads by centrifugation at 250g for 2 min. Take a sample of the supernatant to determine the binding efficiency.

4. Wash the beads with 5 CV of PBS. Centrifuge for 2 min at 250g to recover beads.

5. Pack the Affi-gel into a column, for example, C10/10 column (Amersham Biosciences).

6. Wash column with 10 CV PBS containing 2 M NaCl at 0.3 ml/min.

7. To remove any uncoupled protein, wash the column with the buffer that will eventually be used to elute bound antibodies from the column, that is, 100 mM Na-citrate buffer pH 2.5. Re-equilibrate with 20 CV TBS.

8. Store the column in TBS containing 0.02% sodium azide at $4°$.

9. The efficiency of binding can be determined by comparing samples of the starting material, and the unbound fraction using SDS-PAGE electrophoresis.

Generation of Affinity Matrices Using Insoluble Protein

The method described below assumes coupling of 40 mg of antigen to produce a 2 ml column. However, lower protein amounts will also yield useful affinity columns.

1. Solubilize the antigen in 5 M GuHCl buffer (5 M Guanidine hydrochloride in PBS) by dounce homogenization. (Sonication may be used as an aid to resuspension if necessary.) Incubate on a rotating wheel for about an hour at room temperature.

2. Estimate the protein concentration by SDS-PAGE using known amounts of BSA as a comparison. Dilute 40 mg of the resuspended antigen to a final volume of 8 ml in GuHCl buffer. More diluted preparations will also couple efficiently.

3. Centrifuge at 20,000g, $4°$ for 25 min, to remove any remaining aggregates. Transfer the supernatant to a screw-cap tube.

4. Add 2 ml of Affi-gel (BioRad) beads equilibrated in the appropriate buffer to the supernatant and incubate for 1 h at room temperature, or 4 h—overnight at $4°$.

5. To block any remaining active ester groups add 200 μl (100 μl/ml gel) of 1 M ethanolamine pH 8.0 and incubate for a further hour.

6. Recover beads by centrifugation at $250g$ for 2 min. Take a sample of the supernatant to determine binding efficiency and depletion of the extract.
7. Wash the beads three times with 10 ml of GuHCl buffer by incubating on a rotator for 5 min. After each wash recover the beads by centrifugation at $250g$ for 2 min.
8. Pack the Affi-gel into a column (e.g., C10/10 column [Amersham Biosciences]). Develop the column with a gradient from 5 M GuHCl buffer to PBS without GuHCl, at a flow rate of 0.2 ml/min over 3 h.
9. Wash the column with 5 CV of PBS.
10. Wash the column with 5 CV PBS containing 2 M NaCl.
11. To remove any uncoupled material, wash the column with the buffer that will eventually be used to elute bound antibodies from the column, that is, 100 mM Na-citrate buffer pH 2.5. Re-equilibrate with 20 CV TBS.
12. Store the column in TBS containing 0.02% sodium azide at 4°.
13. The efficiency of binding can be assessed comparing samples of the starting material, and the unbound fraction.

Experimental Notes

Alternatively, the whole procedure can be performed in batch by step-wise incubations and washes in a screw-capped tube.

Generation of Affinity Matrix by Direct Coupling of Purified GST-Fusion Proteins

1. Purify GST-fusion protein as described above but do not elute from the beads.
2. Wash beads batch-wise with 10 CV of 0.2 M sodium borate pH 9.0, repeat twice. Recover the beads from each wash step by centrifugation at $500g$.
3. Resuspend beads in 10 CV of 0.2 M sodium borate pH 9.0 and take a sample of slurry corresponding to 10 μl of packed beads.
4. Add dimethylpimelimidate (solid) to a final concentration of 20 mM.
5. Mix for 30 min at room temperature on a rocker or spinning wheel.
6. Stop the reaction by washing the beads once in 0.2 M ethanolamine pH 8.0, and then incubate for 2 h at room temperature in 0.2 M ethanolamine on a rotating wheel or rocker.
7. Take a sample of slurry corresponding to 10 μl of packed beads.

8. To remove any uncoupled material, wash the column with the buffer that will eventually be used to elute bound antibodies from the column, for example, 100 mM Na-citrate buffer pH 2.5.
9. Wash the beads twice with 10 CV of PBS, then store the beads in PBS containing 0.02% sodium azide. The beads are now ready for affinity purification of the antibody.
10. The efficiency of coupling can be determined by boiling samples of beads taken before and after coupling in Laemmli SDS-sample buffer. Analyze these by SDS-PAGE followed by Coomassie Blue staining. If the coupling has been efficient no protein should be present in the "after-coupling" sample.

Generation of Affinity Matrices by Coupling Peptides

We routinely use CNBr-activated Sepharose 4B (Amersham Biosciences) for peptide coupling. It should be noted that CNBr-activated Sepharose 4B is also a convenient general matrix for protein coupling. The recommended ligand concentrations for CNBr-activated Sepharose 4B are 1–10 μM ligand per ml of gel when using peptides or 5–10 mg protein per ml of gel. Buffers containing primary amine groups such as Tris or glycine should be avoided as they will also couple to the gel.

1. Weigh out the required amount of CNBr-activated Sepharose. One gram of sepharose will make about 3.5 ml final volume of gel, so first calculate how much serum you will need to purify. We routinely use 0.7 g of activated Sepharose to generate a 2 ml column, which we typically use to purify antibodies from 5 ml of crude serum.
2. Suspend 0.7 g of CNBr activated sepharose in 1 mM HCl. To remove additives, wash beads on a sintered glass filter with 1 mM HCl, use approximately 200 ml in total. When preferred, this step can also be performed in batch.
3. Wash with 20 ml of PBS
4. Collect matrix and transfer to a 15 ml centrifuge tube.
5. Add 230 μg of peptide in 5 ml PBS to matrix.
6. Mix gently at room temperature for 1–2 h for example, on a rotating wheel. To avoid fines do not use a stir bar.
7. Collect matrix by centrifugation at 250g.
8. Block the coupling reaction by the addition of ethanolamine pH 8.0 to 0.2 M final concentration followed by a 2 h incubation at room temperature on a rotating wheel or rocker. The following steps can either be performed by column chromatography or batch-wise using centrifugation at 250g to recover the beads after a 5 min incubation for each step.

9. Wash with 10 CV (20 ml) PBS.
10. Wash with 10 CV of 100 mM sodium acetate pH 4.
11. Wash with 10 CV of 2 M NaCl in PBS.
12. Wash with 10 CV of TBS.
13. Store matrix in TBS containing 0.02% sodium azide at 4°.

Affinity Purification of Antibodies from Serum

Prior to the actual affinity purification, antibodies are separated from other proteins in the crude serum by ammonium sulphate precipitation. Next, the desired antibodies are purified from the other antibodies by affinity column chromatography.

Preparation of Polyclonal Antiserum for Antibody Purification

1. To 5 ml of serum, add 25 μl 1 M MgCl$_2$ (5 mM final concentration).
2. Slowly add while mixing 4.1 ml 4 M ammonium sulphate pH 7. Mix gently for 15 min at room temperature.
3. Pellet the antibodies by centrifugation at 10,000g for 10 min at 4°.
4. Pour off supernatant and resuspend pellet in 5 ml TBS.
5. Add 50 μl 10% NP-40 (0.1% v/v final concentration).
6. Store at 4° until use.

Affinity Purification of Polyclonal Antibodies

Prior to getting started, it is useful to estimate how much Tris base will be required to neutralize the fractions in Step 11. Pack 2 ml of prepared affinity resin into a suitable column, e.g. a C10 column with adapter (Amersham Biosciences). Do not exceed the recommended flow rates for the matrix used. As for all the protocols in this chapter, the procedure described below can easily be adapted for a batch-wise purification.

1. Equilibrate with 10 CV of TBS + 0.1% (v/v) NP-40.
2. In the meantime, centrifuge the antibody solution at 10,000g for 15 min at 4° to remove particulates. Alternatively, pass over a 0.22 μm low protein binding syringe filter.
3. Pass 5 ml prepared antibody over the column using a slow flow rate of no more than 0.2 ml/min. Some researchers repeat this step and recycle the antibody preparation over the column a few times.
4. Wash with 20 ml TBS + 0.1% (v/v) NP-40.
5. Invert column and reverse direction of flow.
6. Wash 10 CV of TBS + 0.1% (v/v) NP-40.
7. Wash 5 CV of TBS.

8. Wash 5 CV of 0.9% (w/v) NaCl.
9. Elute with 5 CV of 100 mM Na-citrate buffer pH 2.5, collect 700 μl fractions. Collect fractions in tubes containing the pre-determined volume of 2 M Tris base (not pH-ed) to neutralize fractions.
10. Re-equilibrate column with TBS. Columns can be stored in TBS with 0.02% (w/v) sodium azide and used again multiple times.
11. Analyze fractions by SDS-PAGE; 10 μl fractions containing the purified antibodies should give Coomassie stainable bands of 55 KD (heavy chain) and 22 KD (light chain).

Preparation of Antibody Coupled Affinity Matrices

The species and subclass of the antibodies of interest determines which type of beads to use for coupling. Most commonly we use Protein A Sepharose (Amersham Biosciences) to couple rabbit antibodies and Protein G Sepharose (Amersham Biosciences) for mouse monoclonal antibodies. However, Protein A Sepharose also works well for some subclasses of monoclonal antibodies. For a comprehensive discussion of this subject see Harlow and Lane.[12] Monoclonal antibodies either present in the supernatant of cultured hybridoma cells or in mouse acytis can be directly coupled to either protein G or sometimes to protein A beads. Because they represent a homogeneous antibody population, an affinity purification on an antigen column is not required.

Protocol

1. Mix the purified antibody with pre-equilibrated Protein A (or G) Sepharose beads in a screw-capped tube. The recommended concentration for binding is approximately 2 mg of antibody per ml of wet beads.
2. Incubate on a rotating wheel or nutator with gentle mixing for 30 min at room temperature. Do not pass the antibody over a Protein A column as this will lead to a higher concentration of antibodies at the top of the column than at the bottom.
3. Wash twice with 10 CV of 0.2 M sodium borate pH 9.0 for 5 min on a rotating wheel and recover the beads by centrifugation at 500g for 2 min.
4. Resuspend beads in 10 CV of 0.2 M sodium borate pH 9.0. Remove a sample equivalent to 10 μl beads for later analysis.
5. To cross-link the antibodies to the Protein A, add dimethylpimelimidate (solid) to a final concentration of 20 mM and incubate on a rotating wheel for 30 min at room temperature.

6. To stop the coupling reaction, add ethanolamine pH 8.0 to a final concentration of 0.2 M. Incubate for 5 min. Recover the beads by centrifugation at 500g for 2 min. Gently resuspend the beads in 4 CV of 0.2 M ethanolamine and incubate for 1 h on a rotating wheel.
7. Wash the beads twice with 10 CV of PBS. Take a small sample of the beads and then store the beads in PBS containing 0.02% (w/v) sodium azide at 4°. The coupling efficiency can be verified by boiling the "before" and "after" coupling samples of beads in sample loading buffer and analyze by SDS-PAGE followed by Coomassie staining. The antibodies should only be visible in the "before" lane, but not in the "after" lane.

Immunodepletion and Purification Studies of Chromatin-Remodeling Factors

This section deals with the actual use of the immunoaffinity matrices. Below we first describe protocols for immunodepletion of transcription and chromatin-remodeling extracts, and for the immunopurification of proteins and their associated factors. Next, we illustrate the use of these techniques in chromatin research by describing two examples of our studies on the *Drosophila* ATP-dependent chromatin remodeling Brahma (BRM) complex.

Immunodepletion of Transcription and Chromatin Remodeling Extracts

Avoid reducing agents such as DTT in buffers as they may reduce the disulfide-bond between the heavy and light chain of the antibody and thus affect the quality of the affinity matrix.

1. Thaw the extract or column fractions of interest and remove potential aggregates by high speed centrifugation step or alternatively pass over a 0.22 μm filter. Take a small aliquot of the extract for later analysis.
2. Prepare microcentrifuge (Eppendorf type) tubes containing either coupled antibody beads or control beads (i.e., protein A beads that have been taken through the coupling procedure either without antibodies or using an unrelated antibody), equilibrated with a buffer matching that of the sample. The amounts of beads to be used depends on their quality and on the scale of the experiment.
3. Add at least an equal volume (compared with the bead volume) of extract or pooled column fractions to each tube. Incubate on a rotating wheel or nutator for 1 h at 4°.

4. Centrifuge at $500g$ for 2 min. Take a small aliquot for later analysis then transfer the supernatant to another tube containing fresh, equilibrated affinity beads (or control beads).
5. Repeat incubation on a rotating wheel or nutator for a further hour at $4°$.
6. Centrifuge at $500g$ for 2 min. Transfer the supernatant to a fresh tube, take a small aliquot for later analysis. Aliquot the depleted extract in small volumes, snap freeze in liquid nitrogen and store at $-80°$.
7. Determine the efficiency of the immunodepletion by analyzing the input sample and the samples after the first and second round of depletion by SDS-PAGE followed by Western immunoblotting. If the depletion is satisfactory, the stored fractions can now be used in functional assays.

Experimental Notes

- We find that a repetition of the immunodepletion step yields better results than either the addition of more affinity beads or longer incubation times.
- The selectivity of depletion is often best when very high concentrations of antibodies are coupled to the beads so that a small volume of beads can be added.

Immunopurification of Protein Complexes from Crude Extracts or Partially Purified Fractions

Avoid reducing agents such as DTT as they may reduce the disulfide-bond between the heavy and light chain of the antibody and thus affect the quality of the affinity matrix. Prior to starting, it is useful to estimate how much Tris base will be required to neutralize the fractions in Step 10.

Protocol

1. Thaw the extract or column fractions of interest on ice. Add NP-40 to a final concentration of 0.1% (v/v) and then centrifuge the extract at $100,000g$ for 30 min at $4°$ to remove any precipitates. Alternatively, the extract may be passed over a $0.22 \mu m$ filter. When volumes are very small a centrifugation step in a microcentrifuge at maximum speed will suffice. This step is critical for a successful purification!
2. Prepare tubes containing either coupled antibody beads or control beads, equilibrated with HEMG containing 200 mM NaCl (omit DTT). Obviously, the amounts of beads added is dependent on

their quality and on the scale of the experiment. 0.5 or 1.5 ml microfuge tubes can be used, for larger volumes use appropriately sized screw-capped centrifuge tubes. If immunopurified proteins are to be analyzed by mass spectrometry, for example, to identify subunit composition of complexes, we typically start with about 200 μl of beads.

3. Add at least an equal volume (to bead volume) of extract or pooled column fractions to each tube.

4. Incubate on a rotating wheel or nutator for 2 h at 4°. Longer incubations are sometimes required but typically, the longer the incubation time the more binding of nonspecific background proteins occurs. Thus, overnight incubations should be avoided.

5. Recover beads by centrifugation at 500g for 3 min. It is prudent to store the supernatant to be used again when it is not completely depleted or for another protein of interest. Take a small sample of the supernatant for later analysis of the "unbound fraction."

6. Wash 6× with 10 CV of HEMG containing 800 mM NaCl, 0.1% (v/v) NP-40.

7. Wash 2× with 10 CV HEMG containing 200 mM NaCl, 0.01% (v/v) NP-40, to reduce salt and NP-40 concentration. In cases where the detergent may interfere with later analysis, for example with mass spectrometry, more extensive wash steps are sometimes required to remove the NP-40.

8. Either collect the beads in a mini column or recover them by centrifugation.

9. Elute the bound protein by addition of 3–5 CV of NaCitrate buffer, pH 2.4 followed by short incubation on a rotating wheel for not more than 5 min. Alternatively, step elute from a column and collect fractions.

10. Recover beads by centrifugation and take supernatant. Add 2 M unbuffered Tris base until the pH has been neutralized to approximately pH 7.0. Check neutralization by spotting a small sample on pH paper.

11. Wash the beads two more times with the NaCitrate elution buffer and collect supernatant. Next, wash the beads several times with PBS, ensure that the pH is neutral and store the column in PBS with 0.02% (w/v) sodium azide. These columns can often be reused for well over 10 times.

12. Determine the binding selectivity by analyzing the input material, the unbound fraction and bound proteins by SDS-PAGE. The resulting immunopurified protein sample can be concentrated by TCA precipitation.

TCA Precipitation

1. Measure the volume of the sample and add one fourth volume of 100% TCA/DOC (100% [w/v] trichloroacetic acid, 0.1% [w/v] sodium deoxycholic acid).
2. Mix and incubate for 30 min on ice.
3. Centrifuge in an Eppendorf type centrifuge for 15 min at $4°$.
4. Remove supernatant and wash the pellet with 0.5 ml of $-20°$ acetone. Large samples or those containing NP-40 may produce "oily" pellets. It is advisable to repeat the acetone wash for these pellets 2–3 times.
5. Centrifuge again for 10 min at $4°$.
6. Remove the acetone and dry the pellet, either by air-drying or in a speed-vac without heat.
7. Resuspend sample in SDS-PAGE sample buffer. If the sample turns yellow on addition of the sample buffer, add 1–2 μl of 1 M Tris pH 7.5 or 8, until the sample turns blue.

Experimental Notes

- Instead of a batch purification as described above, some researchers prefer to prepare an affinity column and recirculate the extract up to 5 times. However, we obtain better results by performing the binding step in batch. Subsequent steps can then either be performed by column chromatography or in batch using a centrifuge to recover the beads. Which method to use is largely a matter of personal preference.
- Sometimes it helps to include a "precleaning" step prior to the actual immunopurification. This can be achieved by performing a mock purification using, for example, coupled protA or protG beads, either not coupled to an antibody, or coated with a control antibody.
- A variety of buffers can be used for the immunopurification. Often, high salt up to 1 M of NaCl or KCl, and 1% of nonionic detergents such as NP-40 or Triton X-100 can be used. The strength of the antibody-antigen binding as well as the stability of the complex are the key determinants of the washing conditions. Thus, the determination of optimal washing conditions is essentially a matter of trial and error. We recommend binding using relatively mild conditions followed by the combination of a series of long washes under relatively mild conditions followed by a few short, but more stringent washes.

Illustrative Results Using Immunodepletion and Immunopurification Techniques

Immunodepletion is a very powerful tool, which can be used to assess the functional relevance of a given factor in a complete assay system such as, for example, a crude nuclear extract. We have used this approach to determine which ATP-dependent chromatin-remodeling factor is required for transcriptional activation by the *Drosophila* transcription factor Zeste.[13] Four distinct ATP-dependent chromatin remodeling factors have been isolated from fly embryo extracts: BAP, NURF, ACF and CHRAC.[1-3] BAP refers to the Brahma (BRM) chromatin remodeling complex and stands for BRM associated proteins.[14] The BAP complex is the fly homologue of the yeast and human SWI/SNF complexes. NURF, ACF and CHRAC are three distinct ATP-dependent chromatin remodeling complexes, which all share the ISWI protein as the central ATPase subunit. To determine which of these *Drosophila* remodeling factors are required for Zeste function, we utilized affinity-purified antibodies against either BRM or ISWI to combine immunodepletion of a nuclear extract with *in vitro* transcription assays on chromatin templates (see Fig. 1). Whereas, the anti-BRM antibodies will remove only the BAP complex, the anti-ISWI antibodies will remove NURF, CHRAC and ACF from the embryo nuclear extract. As shown in the left panel, the depletion of BRM or ISWI from *Drosophila* embryo nuclear extracts was efficient and specific and did not affect other factors involved in transcription such as the TFIID subunit TAF80. Reconstituted *in vitro* transcription reactions on chromatin templates revealed that depletion of BRM abolished activation of transcription by Zeste, while depletion of the ISWI-containing complexes had no effect (right hand panel).

Selective, high affinity antibodies provide efficient purification of proteins and their associated factors. This is illustrated by the single step isolation of proteins associated with BRM from a crude protein fraction. We first fractionated a *Drososphila* embryo nuclear extract through glycerol gradient sedimentation. BRM displayed a broad range of migration during glycerol gradient sedimentation whereas OSA, one of the known BRM-associated proteins,[13,15] was identified in a subset of the BRM containing fractions. This may indicate that more than one BRM complex exists (Mohrmann and Verrijzer, unpublished data). To purify BRM

[13] A. J. Kal, T. Mahmoudi, N. B. Zak, and C. P. Verrijzer, *Genes Dev.* **14,** 1058 (2000).
[14] O. Papoulas, S. J. Beek, S. L. Moseley, C. M. McCallum, M. Sarte, A. Shearn, and J. W. Tamkun, *Development* **125,** 3955 (1998).
[15] R. T. Collins, T. Furukawa, N. Tanese, and J. E. Treisman, *EMBO J.* **18,** 7029 (1999).

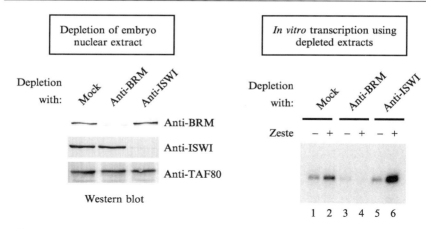

FIG. 1. Immunodepletion reveals that activation by Zeste requires the BRM complex, but not the ISWI factors. The efficiency and specificity of the immunodepletion of a *Drosophila* nuclear extract with, either mock, anti-BRM or anti-ISWI antibodies was verified by Western blot analysis using antibodies directed against BRM, ISWI and dTAF$_{II}$80. The ability of a nuclear extract to support activation by the transcription factor Zeste was tested after either mock depletion (lanes 1 and 2), depletion with affinity purified anti-BRM antibodies (lanes 3 and 4) or anti-ISWI antibodies (lanes 5 and 6). The chromatin templates were assembled using the S-190 system, sarcosyl treated, and purified over a sucrose gradient,[13,16] prior to use in transcription reactions, either in the absence (odd numbered lanes) or presence of Zeste (even numbered lanes). This figure was adapted from Kal *et al.*[13]

complexes from the OSA containing fractions, we used affinity purified anti-BRM antibodies coupled to protein A beads. The immunopurification was performed under stringent conditions involving washes with a buffer containing 800 m*M* NaCl and 0.1% NP-40. Analysis of the immunopurified BRM complex by SDS-PAGE followed by silver staining, consistently reveals over 8 protein bands, ranging from 250 to 45 kDa in size (see Fig. 2). The identity of proteins associated with BRM was determined by mass spectrometry microsequencing and are indicated in Fig. 2.

In conclusion, immunodepletions provide a rigorous way to establish whether a factor is strictly necessary for a functional process. These types of experiments are an important complement to *in vitro* reconstitution reactions using purified components. The purity of the system might create cofactor requirements, which sometimes do not reflect actual *in vivo* conditions. The very rapid developments in mass spectrometry and the genome

---

[16] M. Bulger and J. T. Kadonaga, *Meth. Mol. Genet.* **5,** 241 (1994).

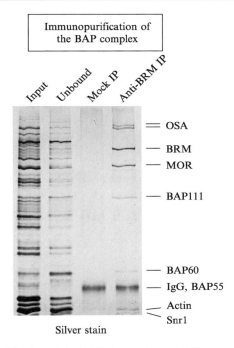

Silver stain

Fig. 2. Immunopurification of the BAP chromatin remodeling complex. Nuclear extracts from *Drosophila* embryos were fractionated by glycerol gradient sedimentation. Fractions were analyzed by SDS-PAGE followed by Western immunoblotting using antibodies directed against the BRM associated protein OSA.[16] Next, the OSA containing fractions were incubated with beads coated with affinity purified anti-BRM antibodies. Protein A beads were used as a negative control. Proteins retained on the beads after extensive washes with a buffer containing 800 mM KCl and 0.1 % NP-40, were resolved by SDS-PAGE on an 8% polyacrylamide gel and stained with silver. Proteins associated with BRM were identified by mass spectrometry microsequencing.

projects have now made it possible to rapidly determine the identity of proteins in complex protein samples. Therefore, in the proteomic era, immunopurifications have become an even more powerful and simple tool to identify the partner proteins of virtually any protein of biological interest. Finally, the use of immunodepletions on crude extracts, and immunopurifications provide a straightforward way to study the products and associated proteins of genes identified in genetic screens in biochemical assays.

Acknowledgments

We acknowledge the essential contributions made by our past and current colleagues to the methods described here. We thank A. Kal and L. Mohrmann for providing Figs. 1 and 2, respectively. We also thank D. Baker, E. Kalkhoven, L. Mohrmann and J. van der Knaap for comments on the manuscript. This work was supported in part by the Dutch Cancer Society (RUL2001–2519) and the EC (HPRN-CT-2000–00078).

Section IV

Transcription and Other Transactions on Chromatin Templates

[29]   Transcription Through the Nucleosome by mRNA-Producing RNA Polymerases

By W. WALTER, M. KASHLEV, and V. M. STUDITSKY

Eukaryotic genes that are actively transcribed by RNA polymerase II (Pol II) *in vivo* still retain their nucleosomal structure, indicating that even if the nucleosome is disrupted during transcription, this structure is recovered almost immediately after passage of the enzyme (Clark[1] and Orphanides and Reinberg[2] for reviews). Analysis of the outcome of this encounter resulted in the discovery that different RNA polymerases (RNAPs) use two very different mechanisms to transcribe through nucleosomes *in vitro*.[3,4] The first (Pol III-related) mechanism is utilized by bacteriophage SP6 and T7 RNAPs, as well as Pol III. This mechanism is characterized by a relatively low nucleosomal barrier to transcription[5-7] and by direct transfer of the complete histone octamer from in front of to behind the transcribing polymerase.[8-12] Pol II and *E. coli* RNAP use a very different mechanism for transcription through the nucleosome, which is characterized by a much stronger barrier,[3,13] and by the loss of an H2A/H2B dimer without changing the position of the histones on the DNA.[3,4]

The mechanism of transcription through chromatin and the regulation of this process are poorly studied. Until now, three types of Pol II-based experimental *in vitro* systems were available for analyzing the outcome of a Pol II-nucleosome encounter.[13-15] One type includes systems supporting

[1] D. J. Clark, *in* "The Nucleus." JAI Press, Greenwich, 1995.

[2] G. Orphanides and D. Reinberg, *Nature* **407,** 471 (2000).

[3] M. L. Kireeva, W. Walter, V. Tchernajenko, V. Bondarenko, M. Kashlev, and V. M. Studitsky, *Mol. Cell* **9,** 541 (2002).

[4] W. Walter, M. Kireeva, V. M. Studitsky, and M. Kashlev, *J. Biol. Chem.* **278,** 36148 (2003).

[5] V. M. Studitsky, D. J. Clark, and G. Felsenfeld, *Cell* **83,** 19 (1995).

[6] R. U. Protacio, K. J. Polach, and J. Widom, *J. Mol. Biol.* **274,** 708 (1997).

[7] J. Bednar, V. M. Studitsky, S. A. Grigoryev, G. Felsenfeld, and C. L. Woodcock, *Mol. Cell* **4,** 377 (1999).

[8] D. J. Clark and G. Felsenfeld, *Cell* **71,** 11 (1992).

[9] M. F. O'Donohue, I. Duband-Goulet, A. Hamiche, and A. Prunell, *Nucleic Acids Res.* **22,** 937 (1994).

[10] V. M. Studitsky, D. J. Clark, and G. Felsenfeld, *Cell* **76,** 371 (1994).

[11] V. M. Studitsky, *Methods Mol. Biol.* **119,** 17 (1999).

[12] V. M. Studitsky, G. A. Kassavetis, E. P. Geiduschek, and G. Felsenfeld, *Science* **278,** 1960 (1997).

[13] M. G. Izban and D. S. Luse, *Genes Dev.* **5,** 683 (1991).

promoter-dependent transcription initiation, either in crude extracts[13] or with highly purified proteins.[15] The main disadvantage of this approach is that only a small fraction of the templates is transcribed.[16] This low efficiency of template utilization makes analysis of the fate of nucleosomes after transcription nearly impossible. In contrast, DNA templates containing a single-stranded, 3'-extending oligo dC "tail" support efficient end-initiation by Pol II *in vitro*.[14] However, in this system, extended DNA:RNA hybrids can be formed, and determination of the fate of the nucleosome during transcription is complicated by the formation of extremely stable DNA-Pol II complexes at the end of DNA.[17] Moreover, end-initiated and promoter-initiated RNA polymerases differ in the way they progress through the nucleosome, most likely because the structures of the ECs are different.[17]

More recently, a novel approach for analysis of Pol II elongation complexes (ECs) was developed.[18,19] It employs assembly of "authentic" ECs using histidine-tagged yeast Pol II and synthetic RNA and DNA oligonucleotides. The structures and functional properties of the assembled and promoter-initiated ECs are very similar.[19,20] While this experimental system faithfully recapitulates many important properties of chromatin transcribed *in vivo* and allows analysis of the fate of nucleosomes during transcription,[3] only a single round of transcription can be conducted. Since a large fraction of assembled complexes is falling apart during transition into elongation phase of transcription,[3] the yield of ECs is relatively low, making their direct structural analysis very difficult.

Recently, it has been shown that *E. coli* RNAP and Pol II use very similar mechanisms for transcription through nucleosomes.[4] Later we describe an experimental system utilizing promoter-initiated *E. coli* RNAP that recapitulates general features of the Pol II-related mechanism of transcription through the nucleosome. This system allows analysis of multiple rounds of transcription through nucleosomes and direct structural analysis of the intermediates formed during this process. In this system, transcription through nucleosomes can be analyzed after ligation of the promoter-initiated ECs to positioned mononucleosomes that are assembled separately.[4]

---

[14] R. L. Dedrick and M. J. Chamberlin, *Biochemistry* **24,** 2245 (1985).

[15] G. Orphanides, G. LeRoy, C. H. Chang, D. S. Luse, and D. Reinberg, *Cell* **92,** 105 (1998).

[16] J. A. Knezetic, G. A. Jacob, and D. S. Luse, *Mol. Cell. Biol.* **8,** 3114 (1988).

[17] Y. V. Liu, D. J. Clark, V. Tchernajenko, M. E. Dahmus, and V. M. Studitsky, *Biopolymers* **68,** 528 (2003).

[18] I. Sidorenkov, N. Komissarova, and M. Kashlev, *Mol. Cell* **2,** 55 (1998).

[19] M. L. Kireeva, N. Komissarova, D. S. Waugh, and M. Kashlev, *J. Biol. Chem.* **275,** 6530 (2000).

[20] M. L. Kireeva, N. Komissarova, and M. Kashlev, *J. Mol. Biol.* **299,** 325 (2000).

Materials and Methods

In the approaches described below, the ECs and nucleosomes are assembled separately and then ligated together (see Fig. 1). The ECs can be assembled by two ways. ECs can be formed on the strong *E. coli* A1 promoter of bacteriophage T7 (T7A1) using σ^{70}-containing holoenzyme. Alternatively, authentic ECs are assembled on synthetic oligonucleotides using the core enzyme.[18] Both approaches produce ECs having very similar properties during transcription of DNA[18] or nucleosomal templates.[4] The ECs are immobilized on Ni-$^{2+}$NTA beads through a hexahistidine tag positioned on the C-terminus of β' subunit of RNAP. Below we only describe a protocol for promoter-dependent formation of ECs; assembly using the core enzyme has been described in detail elsewhere.[21]

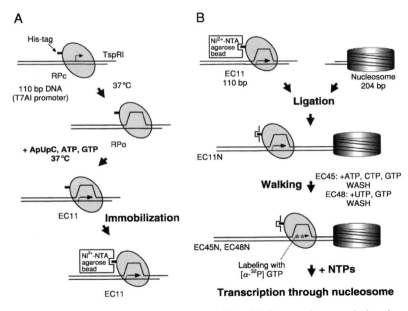

FIG. 1. Experimental system using promoter-initiated ECs to study transcription through the nucleosome by *E. coli* RNAP. (A) The formation of promoter-initiated ECs. The 110 bp TspRI-cut fragment of DNA containing the T7A1 promoter and hexahistidine-tagged RNAP are incubated at 37° to form open complexes. Transcription is initiated through the addition of ApUpC, ATP, and GTP and incubation at 37° to form EC11. EC11 is immobilized on Ni^{2+}-NTA agarose and washed. (B) The experimental approach. Immobilized EC11 is ligated to the 204 bp DNA or nucleosomal template, and the ECs are washed. The polymerase is walked to the +45 position and in some cases the +48 position using α[^{32}P]NTPs to label the RNA. The ECs are washed after each manipulation, and transcription is resumed by addition of all four NTPs.

Nucleosomes are reconstituted on well-characterized nucleosome positioning sequences from the *Xenopus* 5S RNA,[3,22] and are then ligated to the ECs in the absence of NTPs. Transcript elongation is resumed in the presence of NTPs in solid phase or in solution after elution with imidazole. The result of transcription through the nucleosome is then analyzed.

Design and Purification of the DNA Fragments Containing the T7A1 Promoter and the Nucleosome Positioning Sequence

A 110 bp DNA fragment containing the T7A1 promoter was used for transcription initiation. This fragment contains 70 bp upstream of the transcription start site (+1) and a long sticky end starting at the +40 position created by TspRI cleavage that allows for efficient ligation to the nucleosomal template containing a complimentary TspRI end. This DNA fragment was prepared by annealing 2 long overlapping oligos (83 nt each, with a 30 bp overlap) and filling in the ends with Vent (exo-) DNA polymerase (New England Biolabs, Beverly, MA). After digestion, the promoter fragment is ligated to the piece of DNA used for nucleosome reconstitution (pVT1 template[3,23]), and the entire template is PCR amplified. Finally, the template is separated into its separate components (promoter fragment and nucleosome positioning sequence) *via* TspRI digestion and gel purification. The DNA for nucleosome reconstitution is moderately endlabeled so that the quality of the reconstitutes can be monitored (see later). Then nucleosomes and the elongation complexes are assembled on the isolated DNA fragments separately to avoid nucleosome formation at the promoter. Note that nonadhesive tubes (USA Scientific, Ocala, F1) were used in all experiments.

1. The upper oligo (5′ AAAGGATCCAGATCCCGAAAATTTAT-CAAAAAGAGTATTGACTTAAAGTCTAACCTATAGGA-TACTTACAGCCATCGAGAGGG 3′) and lower oligo (5′ GTTTCCTGTGTGTGCCCAGTGCCGGTGTCGCTTGGGTT-GGCTTTTCGCCGTGTCCCTCTCGATGGCTGTAAG-TATCCTATAGG 3′, Invitrogen Corporation, Carlsbad, CA) were mixed in equimolar quantities (192 pmole each) in 1× annealing buffer (10 mM Tris-HCl, pH 7.5; 100 mM NaCl; 1 mM EDTA) at a final volume of 50 μl.

---

[21] N. Komissarova, M. L. Kireeva, I. Sidorenkov, and M. Kashlev, *Methods Enzymol.*, in press.

[22] J. J. Hayes, D. J. Clark, and A. P. Wolffe, *Proc. Natl. Acad. Sci. USA* **88,** 6829 (1991).

[23] W. Walter, M. L. Kireeva, V. Tchernajenko, M. Kashlev, and V. M. Studitsky, *Methods Enzymol.*, in press.

2. The oligos were annealed by heating them to $85°$ in a H_2O bath for 10 min and allowing them to cool to RT slowly.

3. The annealed oligos were filled in by incubating at $72°$ for 20 min in the presence of $1\times$ Vent DNA polymerase buffer [10 mM KCl, 20 mM Tris-HCl (pH 8.8), 10 mM $(NH_4)_2SO_4$, 2 mM $MgSO_4$, 0.1% Triton X-100], 200 μM dNTPs (Amersham Pharmacia Biotech, Piscataway, NJ), 0.1 mg/ml BSA, and 1.5 U of Vent (exo-) DNA polymerase (New England Biolabs, Beverly, MA). 10 μl of annealed oligo was used per 50 μl of fill in reaction.

4. The duplex was ethanol precipitated and digested with TspRI (New England Biolabs, Beverly, MA) overnight at $65°$.

5. The TspRI-cut 113 bp band was PAGE purified, extracted, and ethanol precipitated.

6. The 113 bp TspRI-cut T7A1 DNA fragment was ligated to the 204 bp TspRI-cut pVT1 DNA fragment.[3,23] The 317 bp fragment was gel purified, extracted, and ethanol precipitated.

7. The DNA was PCR amplified in preparative amounts using the primers (Invitrogen Corporation, Carlsbad, CA): upper-5' GGATC-CAGATCCCGAAAATTTATC 3' and lower-5' CCTTCC AAG-TACTAACCAGGCC 3'. Some of the lower primer is radiolabeled with $\gamma[^{32}P]ATP$ (7000 Ci/mmol, ICN Biomedicals, Irvine, CA) using T4 polynucleotide kinase (New England Biolabs, Beverly, MA, as per NEB recommendations) prior to the PCR.

8. The resulting 314 bp product is purified by ethanol precipitation and digested with TspRI as described above.

9. The sample is loaded onto an 8% (19:1) polyacrylamide gel containing $1\times$ TAE and 4 M urea (to prevent reassociation of the 9 nt, GC-rich sticky ends of the TspRI-digested fragments). The 110 and 204 bp TspRI-cut fragments are cut out of the gel, the gel slices are crushed, and the DNA is extracted overnight at $4°$ in 3–5 volumes of TE buffer, ethanol precipitated, and resuspended in dH_2O.

10. The template DNA is further "cleaned up" using QIAquick gel extraction kit columns (Qiagen, Chatsworth, CA, as per kit protocol).

Transcription Buffers

TB0 contains 20 mM Tris-HCl (pH 7.9), 5 mM $MgCl_2$, and 1–2 mM β-mercaptoethanol. TB40, TB150, TB300, and TB1000 contain 40 mM, 150 mM, 300 mM, and 1 M KCl, respectively. Acetylated BSA was purchased from Sigma (St. Louis, MO).

Formation of ECs by Promoter Initiation and Immobilization on Ni^{2+}-NTA Agarose

In this protocol, the ECs are formed separately in solution, immobilized on Ni^{2+}-NTA agarose (Qiagen, Chatsworth, CA) and washed, and then ligated to the nucleosomal template (see Fig. 1). If the templates were to be analyzed after transcription, the promoter fragment was 5' end labeled with $[\gamma$-$^{32}P]ATP$ (7000 Ci/mmol, ICN Biomedicals, Irvine, CA) and T4 polynucleotide kinase. It should be noted that each wash step involves adding the wash buffer to the resin, microcentrifuging for 10 s, and then collecting the supernatant. Large orifice pipet tips (USA Scientific, Ocala, FL) were used for pipetting the resin. The scheme for promoter initiation is illustrated in Fig. 1A.

1. To form open complexes, a 9 μl reaction containing 1 pmol (0.5 μg) of hexahistidine-tagged RNAP holoenzyme and 0.7 pmol of 110 bp T7A1 promoter fragment in TB40 was incubated at $37°$ for 5–10 min.
2. To form EC11 (numerical index indicates the length of the RNA in the stalled EC), 1 μl of $10\times$ start mix (200 μM ApUpC [*Oligos, Etc.,* Wilsonville, OR], 200 μM ATP, and 100 μM GTP [Amersham Pharmacia Biotech, Piscataway, NJ]) was added, and the reaction was incubated at $37°$ for 5 min.
3. Twenty microliters of Ni^{2+}-NTA agarose (50% suspension, Qiagen, Chatsworth, CA) was washed 3 times with 0.5 ml of TB40, incubated in the presence of 0.5 mg/ml of acetylated BSA (Sigma, St. Louis, MO) for 10 min, and washed 2 times with 0.5 ml TB40. The volume was adjusted to 25 μl.
4. EC11 was immobilized on the resin by constant shaking for 15 min at RT.
5. The ECs were washed 6 times with 0.25 ml of TB40.

Reconstitution and Analysis of Mononucleosomes for Transcription

It is highly advantageous to find an efficient method of making nucleosomal templates where the amount of free DNA is $<15\%$, and the nucleosomes do not have to be further purified after reconstitution. The most efficient protocol that we have found for reconstitution of mononucleosomes on the 204 bp pVT1 template involves the use of donor chromatin as a source of histone octamers for exchange onto the template DNA.[24,25] The detailed protocol for preparation of donor chromatin has been

[24] G. Meersseman, S. Pennings, and E. M. Bradbury, *EMBO J.* **11**, 2951 (1992).
[25] S. Pennings, G. Meersseman, and E. M. Bradbury, *Proc. Natl. Acad. Sci. USA* **91**, 10275 (1994).

published[23] and will not be discussed here. Nucleosomes prepared this way contain excess donor chromatin, but it can be removed by washing after the template is ligated to immobilized ECs. The method described below is for templates that are about 150–250 bp in size and, thus, allow for only one nucleosome per molecule of DNA. The quality of nucleosome preparations is determined by the amount of free DNA and subnucleosomal particles that contaminate the sample and how well the nucleosomes formed in the proper location on the template. The position of a nucleosome can be determined with about 10 bp resolution based on its mobility during native PAGE.[26] However, this method cannot discriminate between two symmetrically positioned nucleosomes or differently positioned nucleosomes formed on DNA ∼200 bp or less.[27] The exact position can be further narrowed down by restriction enzyme digestion or micrococcal nuclease mapping (not discussed here[11]). See Fig. 2 for an example of restriction enzyme mapping on the ligated, 314 bp template. The map of the nucleosomes on the template is illustrated in Fig. 2A. Although only one nucleosome can be formed on each individual template, a mixture of different nucleosome positions is obtained. Two regions of the template were preferred locations for nucleosome formation (N1 and N2), and each of these regions had two local heterogeneous positions (a and b), giving rise to four different nucleosome positions (N1a, N1b, N2a, and N2b). The N1 nucleosomes are resistant to cleavage with EcoRI, and N1a is sensitive to cleavage with EcoRV. N1b is only slightly sensitive to EcoRV because this restriction enzyme site is right on the nucleosomal border. The N2 nucleosomes are resistant to cleavage by EcoRV and sensitive to cleavage by EcoRI. The ligated nucleosomes are sensitive to MspI digestion, showing, in combination with the EcoRI and EcoRV digestion data, that ligation does not alter the nucleosome positioning (see Fig. 2B).[3,4]

1. 5′ end-labeled 204 bp TspRI-cut pVT1 DNA is prepared as described above. One to five micrograms of the DNA is mixed with long −H1 donor chromatin at a ratio of 1:60 (wt:wt), respectively, (sample volume is determined by donor chromatin concentration) in buffer containing 1 M NaCl and 0.1% Igepal CA-630 (Sigma, St. Louis, MO).
2. The sample is dialyzed overnight at 4° against a gradient (∼1 L) starting at 1 M NaCl and ending with no NaCl in buffer containing 10 mM Tris-HCl (pH 7.5), 0.2 mM EDTA, and 0.1% NP-40.

---

[26] S. Pennings, G. Meersseman, and E. M. Bradbury, *J. Mol. Biol.* **220,** 101 (1991).
[27] S. Pennings, G. Meersseman, and E. M. Bradbury, *Nucleic Acids Res.* **20,** 6667 (1992).

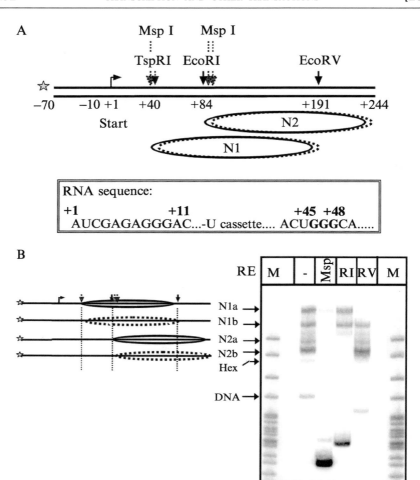

Fig. 2. Restriction enzyme mapping of the 314 bp nucleosomal template. (A) Nucleosome positioning on the ligated 314 bp template. The numerical labeling on the template represents the positions along the DNA relative to the transcription start site ($+1$). Locations of the TspRI, MspI, EcoRI, and EcoRV restriction sites are indicated. The template was labeled at the $5'$ end (indicated by a star). The nucleosome positions, N1a, N1b, N2a, and N2b are shown as ovals (a is a solid oval, b is a dashed oval) below the template. N2 occupies the 5S positioning sequence. The transcribed sequence is shown in the bottom panel, and important RNAP "walking" steps are indicated. Three hundred and fourteen base pair nucleosomes were obtained by ligating 204 bp nucleosomes to $5'$ end-labeled 110 bp T7A1 promoter fragment, and the templates were analyzed by native PAGE before and after restriction enzyme digestion. (B) Analysis of nucleosomes in a native gel. The mobilities of the nucleosomes (N1a, N1b, N2a, and N2b), hexasomes, and free DNA are indicated. Positions of

3. The concentration of the reconstituted nucleosomes is determined by the specific activity of the DNA.

4. Reconstitutes (10 ng aliquots) are supplemented with buffer providing 20 mM Na-HEPES (pH 7.8), 5 mM $MgCl_2$, 2 mM spermidine (Sigma, St. Louis, MO), and 0.5 mg/ml BSA.

5. One sample is not digested, while appropriate restriction enzymes (10 U) are added to the others, and digestion is performed at room temperature (RT) for 0.5–1 h.

6. Buffer is added providing a final concentration of 10 mM EDTA, 10% sucrose, and 50 μg/ml sheared herring testes DNA (Intergen, Purchase, NY).

7. The templates are resolved by native gel electrophoresis (4.5% acrylamide (39:1), 5% glycerol, 20 mM Na-HEPES (pH 8), 0.1 mM EDTA) at 100 V for 2.5–4 h (depending on the size of the DNA fragment and the degree of resolution desired) as described.[23]

8. Here, and in all other experiments described, quantitation is performed using a Cyclone Storage Phosphor System (Packard, Meriden, CT).

Ligation of the ECs to the 204 bp DNA or Nucleosomal Template

The scheme for ligation of the ECs to the nucleosomes is illustrated in Fig. 1B.

1. Washed EC11 (with the volume adjusted so that the final volume of the ligation reaction would be 25 μl) was incubated in the presence of 100 ng of TspRI-cut 204 bp template (nucleosome or DNA), 100 μM ATP, 1% PEG-8000, 0.5 mg/ml Acetylated-BSA, and 50 U T4 DNA ligase at 12° for 1–2 h.

2. The ligated EC was washed 3 times with 0.5 ml TB300, incubated in TB300 for 10 min, and washed 3 times with 0.5 ml TB40.

Transcription and Analysis of RNA

The scheme for transcription through the nucleosome is illustrated in Fig. 1B. The early transcribed sequence for this template is shown in Fig. 2A. To analyze RNAP pausing during transcription through the nucleosome, the RNA can be labeled in several different ways. For instance, the RNA can be labeled via incorporation of $\alpha[^{32}P]$ATP, CTP, or GTP

---

the restriction enzyme sites along the DNA are shown at the top. The location of DNA labeling is indicated by a star. Schematics illustrating the individual nucleosome positions and their corresponding restriction enzyme sensitivities are shown on the left.

during the formation of EC11. Essentially, the RNA can be labeled at any position by walking the RNAP along the template in the presence of different combinations of NTPs.[28] In the protocol below, the RNA is labeled from +46 to +48 by incorporation of $\alpha[^{32}P]GTP$ (see Fig. 2A). Walking RNAP past +40 moves the polymerase across the ligation junction (TspRI site). Thus, the RNAP runs off of the unligated template, and only the RNA being transcribed on ligated templates is labeled and analyzed. An example of transcription through the nucleosome at different salt concentrations is shown in Fig. 3.

1. EC45 was formed by incubating EC11 in the presence of 0.5 mg/ml Acetylated-BSA and 10 μM each of ATP, CTP, and GTP in TB40 for 10 min at RT. The final volume of the reaction was $\sim$50 μl.
2. EC45 was washed 6 times with 0.5 ml TB40.
3. EC48 was formed by incubating EC45 in the presence of 0.4 μM $\alpha[^{32}P]GTP$ (3000 Ci/mmol, PerkinElmer Life Sciences, Boston, MA) and 5 μM cold UTP for 5 min at RT. The final volume of the reaction was 30 μl.
4. EC48 was washed 6 times with 0.5 ml TB40.
5. The samples were supplemented with 0.5 mg/ml Acetylated-BSA, aliquoted, and the KCl concentration of each aliquot was adjusted as desired (40 mM, 150 mM, 300 mM, and 1 M KCl). The volume of each aliquot depends on the number of reactions (see # 6 below).
6. Transcription was performed at RT for the desired time point in the presence of 200 μM NTPs. The typical transcription reaction volume is about 10 μl, and this protocol provides enough material for 24–36 reactions.
7. The reaction was stopped with an equal volume of loading buffer containing 10 M urea and 25 mM EDTA.
8. The sample was boiled and 3 μl was loaded on a 10% (19:1) denaturing uria polyacrylamide gel.

Transcription in Solid Phase and Analysis of Released Templates

To analyze the fate of the nucleosomes during transcription, EC11 containing labeled DNA is obtained and ligated to nucleosomes as described previously. The templates are transcribed and analyzed in a native gel. The scheme for transcription through the nucleosome is illustrated in Fig. 1B. An example of the results one should obtain using this protocol is shown in Fig. 4 (lanes 1–3). It should be noted that during early elongation

[28] M. Kashlev, E. Nudler, K. Severinov, S. Borukhov, N. Komissarova, and A. Goldfarb, *Methods Enzymol.* **274**, 326 (1996).

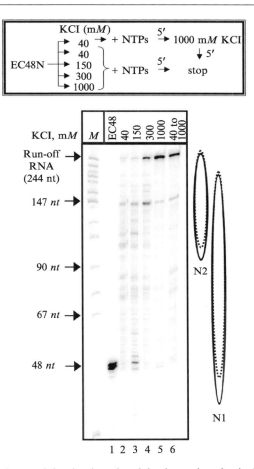

FIG. 3. The nucleosomal barrier is reduced by increasing the ionic strength of the transcription reaction. Markers are an MspI digest of pBR322. DNA marker lengths are indicated at the left in italics. A schematic for the experiment is at the top. RNA was labeled from positions +45 to +48 as described in the text. ECs were washed and the salt concentration was adjusted as indicated above the lane. Transcription was resumed for 5 min at RT with the addition of NTPs. The 48-mer starting material and 244 *nt* run-off RNA are indicated with arrows on the left. Nucleosome positions (N1 and N2) on the template are depicted on the right.

(before formation of EC45), some of the ECs are unstable and release non-transcribed templates into solution (see Fig. 4, lane 1). Thus, by washing EC45, any templates that were released by RNAP and not transcribed to completion were removed from the analysis.[3,4] A second incubation of

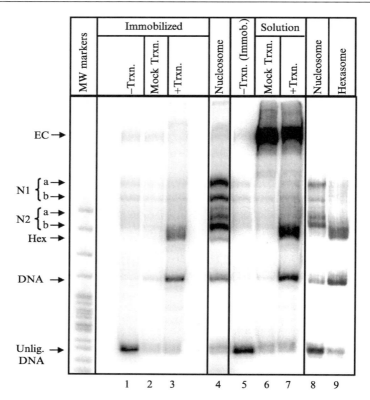

FIG. 4. Transcription through the nucleosome by promoter-initiated *E. coli* RNAP results in the loss of an H2A/H2B dimer. Markers are an MspI digest of pBR322. Labeled 314 bp templates were analyzed in a native gel before transcription ("−Trxn."; EC11 + ACG, templates released as a result of dissociation of early ECs during transcription of initial 44 nt of the template), after mock transcription (washed EC45 + ACG, templates released due to residual falling apart of the early ECs), and after transcription ("+Trxn."; washed EC45 + NTPs, templates released due to transcription). Lanes 1–3 are from transcription with immobilized RNAP. Lane 5 (and lane 1) is the supernatant from walking the immobilized RNAP to form EC45. EC45 is washed and eluted with imidazole to transcribe in solution (lanes 6 and 7). The nucleosome and hexasome mobility controls (lanes 4, 8, and 9) were created by ligating 204 bp reconstituted nucleosome and hexasome to 5′ end-labeled 110 bp T7A1 promoter fragment. The mobilities of the ECs, nucleosomes (N1a, N1b, N2a, and N2b), hexasomes (not resolved by position), free DNA, and unligated DNA are indicated.

EC45 with ATP, CTP, and GTP served as a control for nontranscribed templates released due to residual falling apart of the early ECs (lane 2). After the addition of all 4 NTPs, fully transcribed templates were released into the supernatant (lane 3). The release of full length transcript into solution after incubation of EC45 with all 4 NTPs can be confirmed in a control experiment with labeled transcripts rather than labeled templates.[4] Transcription through the nucleosome by *E. coli* RNAP resulted in the appearance of a faster migrating band (lane 3) as compared to the mobility of the original nucleosomes (lanes 1 and 4). This band had the same mobility in a native gel as the reconstituted hexasome control (lane 9). The identity of hexasome can also be confirmed using other approaches described previously.[23]

1. EC45 was formed by incubating EC11 in the presence of 0.5 mg/ml Acetylated-BSA, 20 μg/ml Rifampicin (Sigma, St. Louis, MO), and 100 μM ATP, CTP, and GTP in TB300 for 10 min at RT. The final volume of the reaction was 34 μl. Rifampicin was added to limit transcription to a single round. The reaction was mixed, centrifuged for 10 s, and 8 μl of supernatant was collected (-transcription control: EC11 + ACG, contains templates released as a result of EC instability during transcription of the initial 40 bp of the template rather than templates that were transcribed to completion).

2. EC45 was washed 3 times with 0.5 ml TB300 containing 20 μg/ml of Rifampicin (TB300 + Rif), incubated in 0.5 ml TB300 + Rif for 15 min at RT, and washed 3 times with 0.5 ml TB300 + Rif.

3. The sample was supplemented with 0.5 mg/ml Acetylated-BSA in a 70 μl volume. Thirty microliters was aliquoted into 2 tubes.

4. Transcription was performed in a final volume of 15 μl for 15 min at RT by the addition of NTPs.
   A. Mock Transcription control (EC45 + ACG, contains templates released as a result of dissociation of ECs during transcription of the initial 40 bp of the template rather than templates that were transcribed to completion): The 15 μl reaction contained 200 μM ATP, CTP, and GTP.
   B. Transcription (EC45 + NTPs, contains fully transcribed templates): The 15 μl reaction contained 200 μM of all 4 NTPs.

5. The sample was mixed, centrifuged for 10 s, and 8 μl of supernatant was collected from each sample.

6. The three supernatant species collected (EC11 + ACG, EC45 + ACG, and EC45 + NTPs) can be aliquoted for further analysis (such as restriction enzyme mapping). Each sample is enough for 1–6 aliquots.

7. Buffer is added providing a final concentration of 10 mM EDTA, 10% sucrose, 50 μg/ml sheared herring testes DNA (Intergen, Purchase, NY), and 350 ng of donor chromatin (to block nonspecific interactions with the wells of the gel). The donor chromatin was obtained as described.[29]

8. The templates are resolved by native gel electrophoresis (4.5% acrylamide [39:1], 5% glycerol, 20 mM Na-HEPES [pH 8.0], 0.1 mM EDTA) at 100 V for 2.5–4 h (depending on the size of the DNA fragment and the degree of resolution desired) as described.[23]

Transcription in Solution

Transcription in solution is an important control experiment to evaluate whether immobilization of RNAP changes the pathway of transcription through nucleosomes. Transcription in solution can be performed very similarly to what was described above. It is recommended to start with the reaction in the immobilized state so that excess promoter fragments and unligated nucleosomes can be removed from the reaction. Immobilized RNAP is then eluted from the resin using imidazole. In fact, if the imidazole elution is performed after the formation of EC45, the nontranscribed templates released as a result of EC dissociation during early transcription can also be removed from the reaction. See Fig. 4 (lanes 5–7) for an example of the results one should obtain using this procedure.

1. Washed EC45 was eluted with 100 mM imidazole (pH 7.5) in TB300 + Rif containing 0.5 mg/ml Acetylated-BSA for 10 min at RT. The final volume of the elution mix was 45 μl.
2. The sample was mixed, centrifuged for 10 s, and 22 μl of supernatant was collected. Ten microliters of the sample was aliquoted into 2 tubes. The sample was diluted 2-fold with TB300 + Rif containing 0.5 mg/ml Acetylated-BSA.
3. Transcription is performed in the presence of 200 μM NTPs, and a mock transcription control is performed in the presence of 200 μM ATP, CTP, and GTP for 15 min at RT.
4. The reactions are stopped by the addition of transcription stop/gel loading buffer providing 10 mM EDTA, 10% sucrose, 50 μg/ml sheared herring testes DNA, and $\sim$350 ng of donor chromatin.
5. Analysis of templates by native PAGE was performed as described earlier.

Reconstitution and Purification of Hexasomes from Purified Histones

Transcription through the nucleosome by Pol II and *E. coli* RNAP results in the loss of an H2A/H2B dimer and the formation of a subnucleosomal particle, the hexasome.[3,4] Thus, hexasomes are reconstituted on the 204 bp pVT1 template, gel purified, and ligated to the T7A1 promoter

---

[29] J. Ausio, F. Dong, and K. E. van Holde, *J. Mol. Biol.* **206,** 451 (1989).

fragment for use as a mobility control in native gel electrophoresis. Hexasomes are reconstituted from purified histones.[3,23] The reconstitution protocol described below is a slightly modified version of the method used by the Bradbury laboratory,[24,25] and is described in more detail in Walter and Kireeva.[23]

1. Five micrograms of DNA is mixed with 1.23 μg of H3/H4 and 0.67 μg of H2A/H2B (ratio of H3/H4:H2A/H2B = 1.82).
2. Dialysis is performed at 4° against the same buffer but with decreasing NaCl concentration (2 M, 1.5 M, 1 M, 0.75 M, 0.5 M, and 10 mM NaCl) for 1 h at each step.
3. Reconstitutes are collected in low adhesion microcentrifuge tubes (USA Scientific, Ocala, FL) and supplemented with buffer providing a final concentration of 10% sucrose.
4. The templates are resolved by native gel electrophoresis as described above.
5. The appropriate band is cut out of the gel, the gel is crushed, and the hexasomes are extracted overnight at 4° in 1–2 volumes of 10 mM Na-HEPES (pH 8.0), 0.1 mM EDTA, and 0.5 mg/ml BSA. Note that the hexasome preparation always contains nucleosomes, so nucleosomes can also be purified at this stage.
6. The supernatant is collected and the concentration of the sample is determined by the specific activity of the DNA.

Ligation of the 204 bp Nucleosomes or Hexasomes to the T7A1 Promoter Fragment for Use as Mobility Controls in Native PAGE

It is suggested that a nontranscribed nucleosome control for mobility during native PAGE be used. Then, the starting material can be directly compared to the transcribed template (see Fig. 4, lanes 4 and 8). This is also a useful tool for mapping of the nucleosomes after ligation to the promoter fragment (see Fig. 2). Also, since transcription leads to the formation of a hexasome, a reconstituted hexasome mobility control is usually provided in the analysis of the transcribed template (see Fig. 4, lane 9).

1. Gel-purified 204 bp hexasomes or nucleosomes from gel-purification or donor chromatin exchange are ligated to 5′ ^{32}P-labeled TspRI-cut T7A1 DNA as mobility controls for the 314 bp hexasomes or 314 bp nontranscribed nucleosomes. Equimolar amounts of the promoter fragment and the 204 bp hexasomes or nucleosomes are incubated in the presence of 100 μM ATP, 1% PEG-8000, and 50 units of T4 DNA ligase (New England Biolabs, Beverly, MA) at 12° for 1–2 h.

2. The samples are aliquoted and prepared for digestion or electrophoresis as described for the mononucleosome preparation above.

Concluding Remarks

It is a tremendous advantage that $E.$ $coli$ RNAP and Pol II use the same mechanism for transcription through the nucleosome.[4] The EC assembly protocol used for transcription with Pol II[19] and $E.$ $coli$ RNAP,[21] though highly useful, has its limitations as well. Assembled ECs tend to be more susceptible to dissociation during transcription of the early region of the template, whereas promoter-initiated ECs do not fall apart as readily.[3,4,23] Moreover, reactions with promoter initiated RNAP can be scaled up by transcribing in solution rather than being limited by the capacity of the Ni^{2+}-NTA agarose resin.

Acknowledgments

We would like to thank M. Kireeva, N. Komissarova, and J. Becker for providing purified $E.$ $coli$ core and holoenzyme RNAP and for helpful advice. The work was supported by the NIH grant GM58650 to V.M.S. The contents of this publication do not necessarily reflect the views or policies of the Department of Health and Human Services, nor does mention of trade names, commercial products, or organizations imply endorsement by the U. S. Government.

[30] Reconstitution and Transcriptional Analysis of Chromatin *In Vitro*

By WOOJIN AN and ROBERT G. ROEDER

The DNA in eukaryotic cells is packaged by histones and nonhistone proteins to form a highly repressive structure known as chromatin. The basic unit of chromatin is the nucleosome, which consists of 146 bp of negatively supercoiled DNA wrapped around an H3/H4 tetramer and a pair of H2A/H2B dimers.[1,2] It has become increasingly evident that covalent modifications of histones and ATP-dependent remodeling of chromatin structure play major roles in the regulation of a variety of nuclear processes involving DNA, but the underlying mechanisms are not well characterized.[3-7] This

---

[1] K. E. van Holde, "Chromatin." Springer, New York, 1988.
[2] K. Luger, A. W. Mader, R. K. Richmond, D. F. Sargent, and T. J. Richmond, *Nature* **389**, 251 (1997a).

chapter describes methodology for the *in vitro* assembly, modification and functional analysis of recombinant chromatin templates. This allows a detailed analysis of independent versus cooperative functions of histone-modifying coactivators, and corresponding histone modifications, in transcriptional activation. The protocol involves: (i) assembly of chromatin using the recombinant ACF-based system described by Ito *et al.*,[8] and recombinant histones (wild type or mutated) expressed according to the procedure described by Luger *et al.*,[9,10] (ii) incubation with specific DNA binding activators and various histone modifying cofactors to allow targeted modifications, and (iii) incubation with nuclear extract and nucleoside triphosphates to allow preinitiation complex assembly and function (transcription). The model template used in initial studies is shown in Fig. 1A and the overall protocol is summarized in Fig. 1B.

Preparation of Recombinant and Native Histones

The use of recombinant wild type and mutant histones expressed in bacteria allows functional analysis of specific histone domains as well as specific histone modifications.[11,12] Recombinant *Xenopus* histones are prepared exactly according to the protocol described by Luger *et al.*[9,10] Briefly, individual histones are expressed in bacteria and purified by gel filtration and ion exchange chromatography under denaturing conditions. Pure individual histones are then reconstituted into an octamer that is further purified and separated through a final gel filtration step at high salt. In our experience the histone octamer, rather than separate H2A/H2B dimers and H3/H4 tetramers or individual histones, gives the best assembly since the stoichiometry of the octameric histone complex is established prior to deposition on DNA. The ability to introduce specific mutations in histones has allowed us, thus far, to document the requirements for specific tails and corresponding acetylation events in transcriptional activation,[11,12] but

[3] T. Jenuwein and C. D. Allis, *Science* **293,** 1074 (2001).
[4] J. L. Workman and R. E. Kingston, *Annu. Rev. Biochem.* **67,** 545 (1998).
[5] S. L. Berger, *Curr. Opin. Genet. Dev.* **12,** 142 (2002).
[6] P. Cheung, C. D. Allis, and P. Sassone-Corsi, *Cell* **103,** 263 (2000).
[7] G. J. Narlikar, H. Y. Fan, and R. E. Kingston, *Cell* **108,** 475 (2002).
[8] T. Ito, M. E. Levenstein, D. V. Fyodorov, A. K. Kutach, R. Kobayashi, and J. T. Kadonaga, *Genes Dev.* **13,** 1529 (1999).
[9] K. Luger, T. J. Rechsteiner, A. J. Flaus, M. M. Waye, and T. J. Richmond, *J. Mol. Biol.* **272,** 301 (1997).
[10] K. Luger, T. J. Rechsteiner, and T. J. Richmond, *Methods Enzymol.* **304,** 3 (1999).
[11] W. An, V. B. Palhan, M. A. Karymov, S. H. Leuba, and R. G. Roeder, *Mol. Cell* **9,** 811 (2002).
[12] W. An and R. G. Roeder, *J. Biol. Chem.* **278,** 1504 (2003).

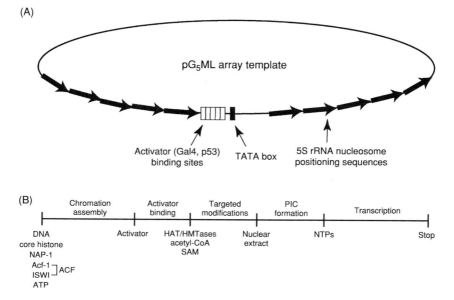

FIG. 1. (A) Schematic illustration of DNA template used for *in vitro* chromatin assembly. (B) Schematic summary of chromatin assembly, modification, and transcription protocol.

obviously has broad applications for many types of modifications. As an alternative to the use of recombinant histones, it is also possible to use native core histones. Common sources include chicken erythrocyte nuclei[13] and HeLa nuclei.[14] However, it is important to note that these may contain prior *(in vivo)* modifications that could influence the particular histone modifications or transcription parameters being assayed. Indeed, slightly lower transcription was observed with chromatin assembled with native HeLa histones versus recombinant histones in our recent study,[11] possibly reflecting the effect of preexisting modifications of native histones. In all cases, prepared histones are stored at $-80°$ in $2\ M$ NaCl, 30% glycerol, TE buffer (20 mM Tris, 0.25 mM EDTA, pH 8.0) at a minimal concentration of 1 mg/ml. Histone concentrations are determined by Coomassie Blue staining with BSA standards.[15]

[13] T. D. Yager, C. T. McMurray, and K. E. van Holde, *Biochemistry* **28,** 2271 (1989).
[14] J. Cote, R. T. Utley, and J. L. Workman, *Methods Mol. Genet.* **6,** 108 (1995).
[15] J. Sambrook, E. F. Fitsch, and T. Maniatis, "Molecular Cloning." Cold Spring Harbor Laboratory Press, Cold Spring Harbor, NY, 1989.

Preparation of DNA Template

Two types of DNA templates are most frequently used for *in vitro* transcription: those with G-less cassettes downstream of the core promoter, which allow direct analysis of synthesized RNA,[16] and those with natural sequences downstream of the core promoter, which require primer extension analyses of RNA synthesis.[15,17] Our assays have generally employed chromatin templates assembled from DNA templates containing the adenovirus major late (AdML) core promoter in front of a 380-bp G-free cassette, upstream activator (GAL4 or p53) binding sites, and 5′ and 3′ flanking regions that each contain 5 repeats of a nucleosome positioning sequence from a sea urchin 5S rRNA gene[18,19] (see Fig. 1A). Because plasmid DNA purity is critical in chromatin assembly and subsequent transcription assays, DNA templates are prepared with a Qiagen plasmid kit followed by two successive rounds of CsCl equilibrium gradient purification.[15]

Chromatin Reconstitution *In Vitro*

Systems commonly employed for *in vitro* assembly of chromatin templates include: (i) S190 *(Drosophila)*[20] or S150 *(Xenopus)*[21] extracts, (ii) salt dilution of long DNA fragments with histone or histone-donor chromatin[18,22] and (iii) the recombinant ACF-based system[8] described here. The extract-based systems allow assembly of high quality chromatin with regularly spaced nucleosomes, although the crude extracts contain many factors that may contribute to subsequent functional assays. The salt-dilution method also generates nucleosomal arrays, but these typically are not regularly spaced in the absence of nucleosome positioning sequences. The ACF system, by contrast, is comprised of a limited number of well characterized recombinant proteins (Acf1, ISWI, and NAP1), generates a chromatin template with regularly spaced nucleosomes, and allows incorporation of pure histones (intact or mutant) with no prior modifications. Importantly, chromatin assembled from plasmid DNAs and recombinant histones with the ACF system also exhibits the expected functional properties—complete repression in a transcription assay in the absence of a transcriptional

[16] M. Sawadogo and R. G. Roeder, *Proc. Natl. Acad. Sci. USA* **82,** 4394 (1985).
[17] S. J. Triezenberg, "Current Protocols in Molecular Biology," Vol. 1, 4.8.1–4.8.5, New York, 1995.
[18] R. T. Simpson, F. Thoma, and J. M. Brubaker, *Cell* **42,** 799 (1985).
[19] D. J. Steger, A. Eberharter, S. John, P. A. Grant, and J. L. Workman, *Proc. Natl. Acad. Sci. USA* **95,** 12924 (1998).
[20] M. Bulger and J. T. Kadonaga, *Methods Mol. Genet.* **5,** 241 (1994).
[21] G. C. Glikin, I. Ruberti, and A. Worcel, *Cell* **37,** 33 (1984).
[22] J. C. Hansen, J. Ausio, V. H. Stanik, and K. E. van Holde, *Biochemistry* **28,** 9129 (1989).

activator and a large induction (and net activity approaching that observed with DNA templates) dependent upon transcriptional activators and interacting histone modifying factors.

The recombinant ACF system developed by Kadonaga and co-workers[8] employs recombinant *Drosophila* Acf1, ISWI, and NAP1. Our adapted system also utilizes *Drosophila* Acf1 and ISWI, expressed via baculovirus vectors and purified exactly as described. However, we use mouse NAP-1[23] rather than *Drosophila* NAP-1, as it has proved to provide chromatin with a more complete transcriptional repression. A typical preparation of His-tagged mNAP-1 employs large-scale expression in bacteria, followed by purification through Ni-NTA and Q-sepharose chromatography (Pharmacia). Although NAP1 purified only through the Ni-NTA step has been used in chromatin assembly, our experience indicates that the additional purification of multimeric NAP-1 with Q-Sepharose substantially improves the structural features of assembled chromatin. The mNAP-1 protein is eluted at 300 mM NaCl from Q-Sepharose and kept in BC100 buffer at $-80°$. The most critical parameter in the chromatin assembly reaction is the ratio of histone to DNA. DNA concentrations are estimated by $A_{260 \text{ nm}}$ and protein concentrations by Coomassie staining (together with BSA standards). We do not recommend using direct UV spectroscopy to measure the concentration of mutant histones, especially tailless histones, since UV absorption is found to be dependent on the amino acid content in tails as well. Prior to inclusion in chromatin assembly reactions, the supercoiled DNA of interest is relaxed in BC100 buffer by recombinant topoisomerase I, and the extent of DNA relaxation is monitored by agarose gel electrophoresis.

Chromatin Assembly

1. Prepare a NAP1/histone mix by sequential additions as follows: 1 μl 0.5 M KCl, 2.5 μg (40 pmol) mNAP1, 0.35 μg (3.5 pmol) of the core histone octamer diluted to 0.1 μg/μl in HEG buffer (25 mM HEPES, pH 7.6, 0.1 mM EDTA, 10% glycerol), and finally HEG buffer to a final volume of 55 μl.
2. Incubate on ice for 30 min.
3. Add 1 μl of an ACF mix containing 30 ng (0.2 pmol) Acf1 and 40 ng (0.2 pmol) ISWI per microliter in HEG buffer.

[23] T. K. Kundu, V. B. Palhan, Z. Wang, W. An, P. A. Cole, and R. G. Roeder, *Mol. Cell* **6,** 551 (2000).

4. Add 10.5 μl AM (ATP/Mg^{++}) mix prepared by combining 2.4 μl 0.1 M ATP, 3.4 μl 0.1 M MgCl$_2$, and 4.7 μl HEG to establish final concentrations of 3.4 mM ATP and 4.8 mM MgCl$_2$.
5. Add 3.5 μl of DNA template in HEG buffer at a concentration of 100 ng/μl (0.35 μg or 0.17 pmol of a 5.5 Kb plasmid, also try 0.3 μg and 0.4 μg for titration experiment).
6. Add 1 μl (2 ng) of topoisomerase I (50 ng/μl stock), 3.5 μl (30 mM final) 0.6 M phosphocreatine (prepared in 20 mM HEPES, pH 7.6, and keep at $-20°$; can be freeze-thawed several times) and 1 μl (20 μg/ml final) 1 mg/ml creatine phosphokinase (prepared in 0.1 M potassium phosphate buffer, pH 7.0 containing 50% glycerol and kept in small aliquots in $-80°$ for up to one year).
7. Allow assembly at 27° for 3 h. Assembled chromatin can be purified by sucrose gradient centrifugation or gel filtration chromatography[24,25] but this is not essential for the modification/transcription assays.

Structural Analysis of Assembled Chromatin

Two methods of determining the quality of chromatin assembly are in routine use in our laboratory: DNA supercoiling analysis and digestion with MNase. We first confirm nucleosome formation by a DNA supercoiling analysis, assuming that the formation of each nucleosome introduces a change in linking number of -1 in a circular DNA template in the presence of topoisomerase I.[18,26] This analysis provides information on the extent of chromatin assembly in the circular plasmid. More importantly, the assembly of regularly spaced nucleosomes is then examined by partial digestion with MNase, which preferentially cleaves free linker DNA between adjacent core nucleosomes. The nucleosome repeat length of assembled chromatin varies depending on the salt or divalent cation concentration but, under the conditions described here, exhibits an average length of 200 bp when analyzed by gel electrophoresis. It is worth noting that the MNase digestion assay *per se* does not allow an assessment of the extent of chromatin assembly. Therefore, supercoiling and MNase analyses both are minimally required to confirm the quality of assembled chromatin prior to any functional application. Typical reactions for supercoiling and MNase digestion

[24] R. T. Kamakaka, M. Bulger, and J. T. Kadonaga, *Genes Dev.* **7**, 1779 (1993).
[25] G. Orphanides, G. LeRoy, C. H. Chang, D. S. Luse, and D. Reinberg, *Cell* **92**, 105 (1998).
[26] J. E. Germond, B. Hirt, P. Oudet, M. Gross-Bellark, and P. Chambon, *Proc. Natl. Acad. Sci. USA* **72**, 1843 (1975).

analysis of ACF-assembled chromatin are shown in Fig. 2 and indicate complete occupancy of plasmid DNA by regularly-spaced nucleosomes.

Supercoiling Assay

1. After chromatin assembly with 350 ng plasmid DNA in a 70 μl reaction, add 70 μl of MNase stop buffer (20 mM EDTA, 200 mM NaCl, 1% SDS, 0.25 mg/ml glycogen).

2. Add 50 μg of proteinase K (25 mg/ml stock, Sigma) and incubate for 1 h at 37°.

3. Extract with an equal volume of phenol : chloroform : isoamylalcohol (25:24:1) and precipitate the DNA with 0.25 M ammonium acetate and 2.5 volume of 100% ethanol.

4. Wash the plasmid DNA with 70% ethanol and run the DNA sample on a one-dimensional 1% agarose gel in 1× TBE at 5 V/cm.

5. Visualize the DNA bands by staining the gel with ethidium bromide (0.5 μg/ml), which requires a minimum of about 0.2 μg of plasmid DNA for each lane.

Micrococcal Nuclease Digestion Assay

1. Assemble 700 ng (two 70 μl assemblies) of DNA into chromatin (a minimum of 0.5 μg of chromatin DNA is required to visualize MNase digests by ethidium bromide staining) as described for chromatin reconstitution and add 110 μl of HEG to a final volume of 250 μl.

2. Add CaCl$_2$ to a final concentration of 2 mM and digest with MNase for 8 min at room temperature. It is important to note that the optimal concentration of MNase varies according to its activity and the experimental conditions must usually be titrated. In general 0.1–2 mU of MNase (Sigma, stored in small aliquots of 0.5 U/μl at −20°) works well in our hands.

3. Stop the digestion quickly by adding 250 μl of stop mix (20 mM EDTA, 200 mM NaCl, 1% SDS, 0.25 mg/ml glycogen) and vortexing briefly.

4. Add 100 μg of proteinase K and incubate at 37° for 1 h.

5. Extract with an equal volume of phenol : chloroform : isoamylalcohol (25:24:1) and precipitate the DNA with 0.25 M ammonium acetate and 2.5 volume of 100% ethanol.

6. Wash the DNA pellet with 70% ethanol and dissolve the pellet in TE (20 mM Tris, 0.25 mM EDTA, pH 8.0) containing 10% glycerol.

7. Electrophorese the DNA samples on a 1% agarose gel in 1× TBE at 5 V/cm and stain the gel with ethidium bromide. We typically see 6–10 bands with 200 base repeat length.

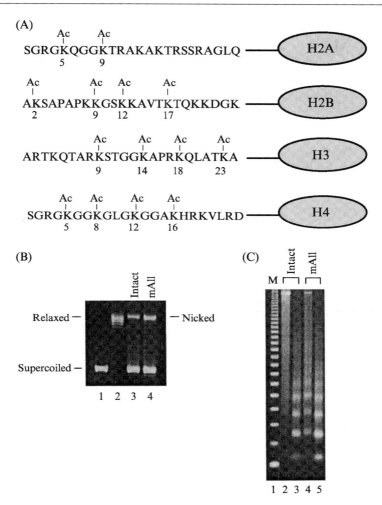

FIG. 2. Supercoiling and micrococcal nuclease analysis of assembled chromatin. (A) Schematic of histone tail sequences. "Ac" indicates the major acetylation sites for p300. (B) Chromatin was assembled with relaxed circular DNA and recombinant histone octamers using recombinant *Drosophila* ACF (Acf-1 and ISWI) and mouse NAP1 in the presence of topoisomerase I as described. The plasmid DNA (200 ng) was purified from chromatin assembled with intact (Fig. 3A, lane 1) or mutant (Fig. 3A, lane 8) core histones and analyzed by 1% agarose gel electrophoresis and ethidium bromide staining. Intact indicates an octamer with all histones intact and mAll indicates an octamer with lysine to arginine substitutions at the major acetylation sites (see Fig. 2A) in all four core histones. (C) Chromatin assembled either with intact histones (lanes 2 and 3) or with mutant histones (lanes 4 and 5) as in Fig. 2B was treated with 0.1 mU (lanes 2 and 4) or 0.5 mU MNase (lanes 3 and 5) and analyzed by 1.2% agarose gel electrophoresis and ethidium bromide staining. A 123 bp was used as a size marker (M, lane 1).

Histone Modification Assay with Assembled Chromatin

Using the circular nucleosomal array template, which is sufficiently long to mimic the natural repressive action of chromatin, we have been able to correlate activator targeted histone modifications with transcriptional activation. The modification assay is done with recombinant activators and histone modifying cofactors and measures the incorporation of radioisotopic acetyl- (HAT assay) or methyl- (HMT assay) groups into amino acid substrates of histone tails. The properties of long nucleosome arrays differ significantly from those of short arrays or mononucleosomes in that histone modifications by various cofactors are completely dependent upon activator binding to the nucleosomal template.[11,23,27] Consistent with the activator-dependency of the observed histone modifications, we recently have shown that Ga14-VP16 directly recruits p300 and that acetylation is targeted to the promoter region.[11,23] Moreover, a detailed study demonstrated that the substrate specificities displayed by a specific HAT (p300) vary significantly for free histones versus histones within a chromatin template.[12] Given the context-dependent specificities of histone modifications, such assays are best done with long nucleosomal arrays that better reflect the $in\ vivo$ substrates. Note also that pH and salt concentrations can affect the overall modification activities and histone substrate specificities and thus should be optimized for each case.[28] Figure 3 shows a representative chromatin modification assay for Ga14-VP16 dependent acetylation by p300 (see figure legend for more detail). More recently, similar activator-dependent modifications by histone methyltransferases (HMTs) PRMT1 and CARM1 have been demonstrated (unpublished data). Note that the chromatin assembly factors are maintained in this assay, as it has been shown that ATP-dependent remodeling is required for optimal p300-dependent histone acetylation[11,27] (see also Fig. 3B).

[27] T. Ito, T. Ikehara, T. Nakagawa, W. L. Kraus, and M. Muramatsu, *Genes Dev.* **14,** 1899 (2000).
[28] A. Mizzen, J. E. Brownell, R. G. Cook, and C. D. Allis, *Methods Enzymol.* **304,** 675 (1999).

FIG. 3. Activator-dependent acetylation of nucleosomal histones by p300. (A) Reconstituted histone octamers were analyzed by SDS-15% PAGE and Commassie blue staining. Histone octamers contained all intact histones (lane 1) or individual mutant ("m") histones (as indicated) and complementary intact histone(s) (lanes 2–8). (B) Chromatin (300 ng) assembled with intact histones was subjected to HAT assays with Gal4-VP16 (20 ng), p300 (30 ng) and ^{3}H-acetyl-CoA (5 μCi) addition as indicated. As described previously, and consistent with an activator-dependent recruitment of p300, all four histones were acetylated in an activator-dependent manner. Consistent with a role for ATP-dependent remodeling (via ACF) in histone acetylation,[27] addition of apyrase alone or in combination with the p300 inhibitor lysyl-CoA significantly inhibited acetylation. (C) Assays were identical to those in an activator-dependent manner. Consistent with a role for ATP-dependent remodeling (via

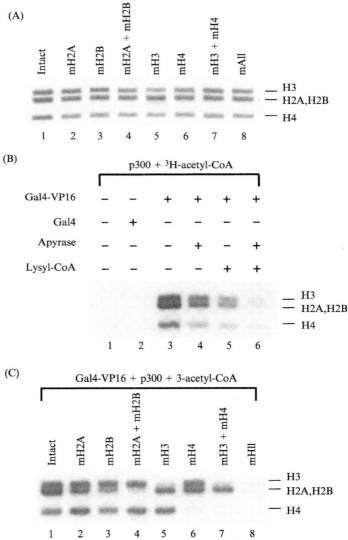

ACF) in histone acetylation,[27] addition of apyrase alone or in combination with the p300 inhibitor lysyl-CoA significantly inhibited acetylation. (C) Assays were identical to those in Fig. 3B (lane 3) except that chromatin substrates were reconstituted with mutant histones ("m" indicates the substitution of primary lysine substrates, summarized in Fig. 2A, to arginine) as indicated (lanes 2–8). Note that lysine to arginine mutations at the major *in vivo* acetylation sites drastically reduced acetylation by p300 (lane 1 versus lanes 2–8), indicating that the *in vitro* modifications are appropriately restricted to discrete amino acids. Figure is reproduced from W. An, V. B. Palhan, M. A. Karymov, S. H. Leuba, and R. G. Roeden, *Mol. Cell* **9**, 811 (2002).

Histone Modification Reactions

1. On completion of a 30 μl chromatin assembly reaction containing 10 ng DNA/μl, add 20 ng activator (Ga14-VP16 or p53), 30 ng p300, 5 μCi [^{3}H] acetyl-coenzyme A (^{3}H-acetyl-CoA) and/or S-adenosyl-L-[methyl ^{3}H] methionine (^{3}H-SAM). The optimal concentrations of (co) activators vary depending on their activities. Typically, 10–30 ng activator and 20–60 ng histone modifying protein (coactivator) are tested for activator-dependent modification of chromatin. The use of pure components in this assay excludes a requirement for HDAC inhibitors.

2. Incubate for 30 min at 30°. When analyzing the interdependency of multiple modifications by sequential cofactor additions, we allow 20 min for each reaction.

3. Stop the reaction by adding SDS-PAGE loading buffer and separate the proteins on SDS-15% PAGE (10 × 10 × 0.15 cm; acrylamide : bis = 37.5:1).

4. Fix the gel in 10% methanol/10% acetic acid (v/v) for 30 min and soak in amplifier (Amersham) with gentle shaking for 30 min.

5. Dry the gel and subject to autoradiographic analysis. A typical exposure takes 1–2 days.

In Vitro Transcription Assay with Assembled Chromatin

In contrast to what is observed with DNA templates, where the general transcription machinery shows a significant basal level transcription that can be markedly elevated by a DNA binding activator, the uniform packaging of DNA into chromatin templates (as described here) represses all transcription (basal and activated) and necessitates additional processes (chromatin modification and remodeling) that facilitate subsequent binding and function of the transcriptional machinery. This section describes transcription reactions with coactivator-modified recombinant chromatin templates that use a low concentration of HeLa nuclear extract as a source of the general transcription factors and allow assessment of the functional consequences of various histone modifications. The transcription reaction containing the chromatin template is incubated sequentially with activator and chromatin remodeling factors prior to the sequential addition of nuclear extract (to allow preinitiation complex formation) and the four ribonucleoside 5'-triphosphates (to allow transcription) as summarized in Fig. 1B. Although most reaction parameters have been standardized in our studies, the level of transcription can be affected by several parameters that include the quality of chromatin templates, (co) activators, and nuclear extracts. Therefore optimal conditions should be established empirically by careful titrations of all components. As controls, equimolar amounts of corresponding plasmid DNA templates are transcribed under conditions identical to those used for chromatin templates. Typically, final transcription reactions contain 20 mM HEPES (pH 7.8), 5 mM MgCl$_2$, 60 mM KCl, 5 mM DTT, 5% glycerol, and ~50 μg of HeLa nuclear extract protein in addition to chromatin template/assembly factors, activators, and coactivators as described for the modification assays.

Figure 4A shows the results of representative transcription analyses that compare basal and activator-dependent transcription on DNA versus chromatin templates in the presence and absence of p300 and acetyl-CoA. As emphasized previously,[11,23] the DNA template shows a low basal activation that is dramatically stimulated by Ga14-VP16 and completely independent of p300 or acetyl-CoA. In contrast, the corresponding chromatin template shows little or no basal activity, but a high level of Ga14-VP16 dependent activity that is also dependent upon p300 and acetyl-CoA. Moreover, and importantly, the Ga14-VP16- and p300/acetyl-CoA-dependent activity in the chromatin template reflects not just derepression (recovery of the basal activity observed with the DNA template) but also a large net increase in activity. This is evident from the fact that the activator/coactivator-dependent chromatin activity greatly exceeds the basal activity and, indeed, approaches the activity observed with the activator on the DNA template. This indicates that the activator functions not just to make the promoter accessible to the general transcription machinery, through recruitment of chromatin remodeling/recruitment factors, but through subsequent steps (via other cofactors such as Mediator) as well.

Figure 4B shows a representative analysis of the effects of lysine-to-arginine substitutions at the major acetylation sites (see Fig. 2A) in the core histone tails. As indicated, and as recently published,[11] mutations in H3 and H4 tails, but not in H2A and H2B tails, greatly impair p300-dependent transcriptional activation by Ga14-VP16. These results extend prior correlations of histone acetylation with p300-dependent transcription by demonstrating that histones are obligatory substrates for derepression and subsequent activation. As such they demonstrate the power of recombinant chromatin/*in vitro* transcription analyses, since one cannot otherwise (as in most *in vivo* analyses) eliminate the possibility that nonhistone substrates (well documented for many HATs and HMTs) are the essential substrates and that the histone modifications, while highly correlative, are nonessential. Our preliminary results also indicate dynamic roles for histone methylation in p53-dependent transcription and, more intriguingly, interdependencies between histone acetylation and methylation (e.g., by PRMT1 and CARM1) for optimal transcription activity.

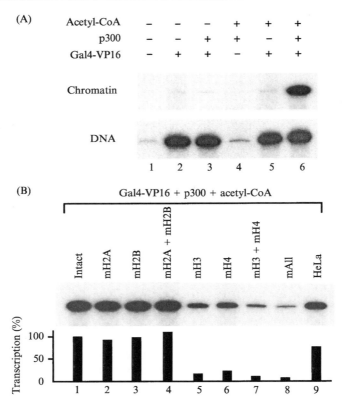

FIG. 4. p300-mediated histone acetylation effects Gal4-VP16 dependent transcription from chromatin, but not DNA, templates. (A) Transcription of a free DNA template and a corresponding chromatin template assembled with intact recombinant histones. Assays followed the protocol of Fig. 1B and contained (as indicated) DNA or chromatin (40 ng DNA), Gal4-VP16 (10 ng), p300 (10 ng), and acetyl-CoA (10 μM). (B) Transcription from chromatin templates reconstituted with intact (lane 1) or mutant (lanes 2–8) recombinant histones or with natural HeLa histones (lane 9) as indicated. The intact/mutant histone octamers used for chromatin reconstitution are shown in Fig. 3A.

Transcription Reaction

1. Prepare a 20 μl reaction containing chromatin template (40 ng, 11 fmol of 5.4 kb DNA) and activator (10 ng in general, but more for a weak activator) in 0.5× HAT buffer (10 mM HEPES, pH 7.8, 30 mM KCl, 2.5 mM DTT, 0.25 mM EDTA, 5 mM sodium butyrate). Incubate at 27° for 20 min.

2. Add 1 μl of 0.25 M DTT, 2.5 μl HM buffer (400 mM HEPES, pH 7.8, 100 mM $MgCl_2$) and BC200 to effect a concentration of 60 mM KCl in the final (50 μl) transcription reaction.

3. Add chromatin remodeling factors (approximately 20 ng of each factor in BC200) with 1 μl of 10 mM ATP (for ATP-dependent remodelers), 1 μl of 0.5 mM acetyl-CoA (for HATs), and/or 50 μM S-adenosyl-L-[methyl] methionine (SAM) (for HMTs). Add H_2O to bring the volume to 40.5 μl and mix by gentle pipeting several times.

4. Incubate at 30° for 30 min (if studying the order of multiple remodeling/modification processes, allow a minimum of 20 min for each reaction).

5. Add 5 μl HeLa nuclear extract (10 mg/ml protein in BC100), prepared as described,[29] and incubate at 30° for 15 min.

6. Add 1 μl RNasin (10 U/μl, Promega), 2.5 μl of 20× nucleotide mixture (12 mM each of ATP and CTP, 0.5 mM UTP, 2 mM 3′-O-methyl-GTP in 20 mM HEPES, pH 7.8) and 10 μCi (1 μl) of [α-^{32}P] UTP (3000 Ci/mmol). Incubate at 30° for 40 min.

7. Add 1 μl RNase T1 (20 U/μl) and incubate at 30° for 15 min (this digests radiolabeled transcripts other than those from the G-less cassette).

8. Add 150 μl of transcription stop buffer (1.5% SDS, 27 mM HEPES, pH 7.8, 27 mM EDTA, 270 mM NaCl, 30 ng/μl tRNA) and 2 ul of proteinase K (25 mg/ml) to the transcription reaction and incubate at 37° for 1 h.

9. Extract the radiolabeled RNA with 200 μl (equal volume) of phenol : chloroform : isoamylalcohol (25:24:1) and, after recovery of the upper phase (~180 μl), add 18 μl of 3 M sodium acetate (pH 5.2) and 400 μl of 100% (v/v) ethanol.

10. Place the sample in powdered dry ice for 30 min and recover the RNA precipitate by microcentrifugation for 20 min at 4°. Quickly dry the RNA pellet in a Speed-Vac and dissolve in 20 μl of RNA loading buffer (50 mM Tris–HCl, pH 7.5, 98% deionized foramide, 0.003% Bromophenol blue, 0.003% Xylene cyanol).

11. Run the radiolabeled RNA on a 5% polyacrylamide (19:1)—8 M urea gel (GIBCO BRL Model V16) in 1× TBE buffer for 2 h at 280 V.

12. Wash the gel for 30 min in distilled water to remove urea and free α-^{32}P-UTP, and vacuum dry for 1 h.

13. Autoradiograph the gel with an intensifying screen (the exposure time varies depending on the reaction but the general exposure time is 24 h).

[29] J. D. Dignam, R. M. Lebovitz, and R. G. Roeder, *Nucleic Acids Res.* **11**, 1475 (1983).

Note that the protocol described here is for the use of templates with G-less cassette with transcript quantitation by incorporation of radiolabeled UTP. When primer extension is used to quantitate transcription, the $20\times$ nucleotide mixture in step 6 is replaced with a mixture giving 0.5 mM of each nucleoside triphosphates in the assay, step 7 is omitted, and transcripts are amplified by PCR using a ^{32}P-labeled probe and reverse transcriptase after RNA purification (step 10). In our experience, identical results are obtained with either transcription methodology.

[31] Techniques Used to Study Transcription on Chromatin Templates

By ALEJANDRA LOYOLA, SHU HE, SANTAEK OH, DEWEY G. MCCAFFERTY, and DANNY REINBERG

Research conducted over the last 10 or so years has presented investigators with a novel view of chromatin structure and function. It has become increasingly clear that chromatin must be considered when studying reactions that have DNA as a substrate. For example, chromatin has been shown to play a key role in the regulation of transcription. The most basic RNA polymerase II *in vitro* transcription system, which carries out transcription on naked DNA, does not support transcription on chromatin templates. This implies that chromatin-specific factors must be added to the system. At least two factors are minimally necessary, the chromatin remodeling factor RSF and the chromatin elongation factor FACT. This review describes methods for the assembly and analysis of chromatin, as well as for the *in vitro* transcription of chromatin templates. We also describe a new technique for the synthesis of chromatin from synthetic histones prepared by optimized native chemical ligation. This technique will be particularly useful in the analysis of post-translational modification of histones, because a homogeneous population of histones carrying a particular post-translational modification can be obtained.

During the 1980s and early 1990s, biochemists determined the minimal requirements for DNA polymerase II transcription on naked DNA templates *in vitro*. The collection of proteins required for such a transcription reaction consists of the general transcription factors TFIIB, TFIIF, TFIIE, and TFIIH, along with the TATA binding protein (TBP) and RNA polymerase II.[1,2] Subsequently, transcription was carried out in a more physiological context with the use of a DNA template that was packaged as chromatin. Very soon, researchers discovered that the packaging of DNA into chromatin inhibits RNA polymerase II transcription, and these results indicated that other chromatin-specific factors must be required in the reaction.[3,4] Therefore, a number of investigations have now focused on the identification of such factors that allow transcription to occur on chromatin templates.

The basic repeating unit of chromatin, the nucleosome, includes two copies of each of the four core histones H2A, H2B, H3, and H4 wrapped by 147 base pairs (bp) of DNA.[5] Each core histone is composed of two domains, (i) a three-helix domain located in the C-terminal region, called the histone fold that is important for histone/histone and histone/DNA interactions, and (ii) a positively charged N-terminal tail, the site of several post-translational modifications.[6]

In general terms, there are two points in the transcription reaction where chromatin can block the transcription process: (i) the initiation of transcription and (ii) the elongation of transcripts (see Fig. 1). Two families of protein complexes have been identified that function in the initiation of transcription on chromatin templates: (a) ATP-dependent-chromatin remodeling factors[7,8] and (b) enzymes that covalently modify mainly the N-terminal tail of histones.[9–11] In addition to these families of proteins, a histone H2A/H2B chaperone, called FACT (*Fa*cilitates *c*hromatin *t*emplate), is required to facilitate the elongation of transcripts on chromatin templates.[4]

ATP-Dependent Chromatin Remodeling Complexes

There are three families of ATP-dependent chromatin remodeling complexes—SWI/SNF (*swi*tch/*s*ucrose *n*onfermenting), ISWI (*I*mitation of *Swi*tch), and CHD (*c*hromodomain *h*elicase/ATPase *D*NA binding protein)—all of which have been highly conserved throughout evolution. Each family contains a subunit with ATPase activity, which provides the

[1] R. G. Roeder, *Trends Biochem. Sci.* **21**, 327 (1996).
[2] G. Orphanides, T. Lagrange, and D. Reinberg, *Genes Dev.* **10**, 2657 (1996).
[3] G. Orphanides and D. Reinberg, *Nature* **407**, 471 (2000).
[4] G. Orphanides, G. LeRoy, C. H. Chang, D. S. Luse, and D. Reinberg, *Cell* **92**, 105 (1998).
[5] K. Luger, A. W. Mader, R. K. Richmond, D. F. Sargent, and T. J. Richmond, *Nature* **389**, 251 (1997).
[6] A. Wolffe, "Chromatin: Structure and Function." Academic Press, San Diego, 1998.
[7] M. Vignali, A. H. Hassan, K. E. Neely, J. L. Workman, *Mol. Cell. Biol.* **20**, 1899 (2000).
[8] A. P. Wolffe and D. Guschin, *J. Struct. Biol.* **129**, 102 (2000).
[9] P. Cheung, C. D. Allis, and P. Sassone-Corsi, *Cell* **103**, 263 (2000).
[10] S. Y. Roth, J. M. Denu, and C. D. Allis, *Annu. Rev. Biochem.* **70**, 81 (2001).
[11] Y. Zhang and D. Reinberg, *Genes Dev.* **15**, 2343 (2001).

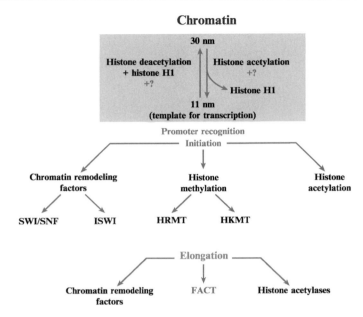

Fig. 1. Description of the various families of protein complexes involved in RNA polymerase II transcription initiation and elongation. (See color insert.)

energy for chromatin remodeling. These subunits are members of the SNF2 family of helicases, although none of them possess helicase activity.[12,13] The three types of complexes remodel chromatin by distinct mechanisms. SWI/SNF, for example, creates an altered nucleosomal state that is topologically different from nonremodeled nucleosomes. The nature of this change, which renders the nucleosome more accessible to the transcription machinery, is not well understood, but it does not involve changes in the histone composition. ISWI, on the other hand, "remodels" chromatin by moving nucleosomes, rather than by affecting the topological state of the nucleosome. It has been proposed that through this nucleosomal movement some promoters (that is, the transcriptional regulatory DNA sequences associated with genes) become accessible to the transcriptional machinery. For details about the various models proposed for chromatin remodeling, the reader is referred to recent reviews.[13,14]

[12] K. Havas, I. Whitehouse, and T. Owen-Hughes, *Cell Mol. Life Sci.* **58**, 673 (2001).
[13] P. Varga-Weisz, *Oncogene* **20**, 3076 (2001).
[14] G. Langst and P. B. Becker, *J. Cell Sci.* **114**, 2561 (2001).

Histone Modifying Enzymes

Regarding histone tail modifications (for instance, phosphorylation, acetylation, methylation, and ubiquitination), it is becoming clear that (i) these modifications do not work in isolation and (ii) interplay between these modifications can affect chromatin structure through the recruitment and/or removal of specific factors, as proposed by the histone code hypothesis.[15,16] There are several examples that illustrate these points:

1. The phosphorylation of serine-10 on histone H3 promotes the acetylation of lysine-14 on histone H3, which results in the activation of several specific genes in yeast. In addition, these two modifications inhibit the methylation of lysine-9 on histone H3, a modification that is associated with the formation of heterochromatin.[17]

2. Suv39H1-mediated methylation of H3-K9 is inhibited if H3-K4 is methylated. Moreover, the preferential association of the NuRD chromatin remodeling and deacetylase complex with the histone H3 tail is inhibited by H3-K4 methylation, but not by H3-K9 methylation.[18]

3. Another example is the stimulation of transcription mediated by the methylase PRMT1 (Protein Arginine Methyltransferase-1). PRMT1 methylates arginine-3 on histone H4, which results in the stimulation of H4 tail acetylation.[19]

4. The methylation of lysine-20 on histone H4 is negatively modulated by H4-K16 acetylation.[20]

In Vitro Transcription Systems

To better characterize the mechanisms of transcriptional regulation, it is essential to have defined minimal systems that support the assembly of chromatin and regulated transcription *in vitro*. In such a system, scientists can decipher the interplay between various chromatin-related factors as well as distinct histone tail modifications.

Recently, several defined systems for chromatin assembly were described. In these systems, certain members of the ISWI family were used

[15] B. M. Turner, *Bioessays* **22**, 836 (2000).

[16] B. D. Strahl and C. D. Allis, *Nature* **403**, 41 (2000).

[17] S. L. Berger, *Science* **292**, 64 (2001).

[18] P. Zegerman, B. Canas, D. Pappin, and T. Kouzarides, *J. Biol. Chem.* **277**, 11621 (2002).

[19] H. Wang, Z. Q. Huang, L. Xia, Q. Feng, H. Erdjument-Bromage, B. D. Strahl, S. D. Briggs, C. D. Allis, J. Wong, P. Tempst, and Y. Zhang, *Science* **293**, 853 (2001).

[20] K. Nishioka, J. C. Rice, K. Sarma, H. Erdjument-Bromage, J. Werner, Y. Wang, S. Chuikov, P. Valenzuela, P. Tempst, R. Steward, J. T. Lis, C. D. Allis, and D. Reinberg, *Mol. Cell* **9**, 1201 (2002).

to assemble chromatin *in vitro*. Human and *Drosophila* ACF (*A*TP-utilizing *C*hromatin Assembly and Remodeling *F*actor) and CHRAC (Chromatin Accessibility Factor) complexes, as well as the yeast ISWI1 and ISWI2 complexes all were shown to be capable of assembling chromatin in combination with the histone chaperone NAP-1 (*N*ucleosome *A*ssembly *P*rotein-1).[21] RSF (*R*emodeling and *S*pacing *F*actor), another member of the ISWI family, is able to assemble chromatin *in vitro*, although it does not require NAP-1.[22] Indeed, interaction studies suggest that the largest subunit of RSF has histone chaperone activity. Before the identification of these assembly factors, chromatin was obtained by several different methods, such as polyglutamic acid and salt dialysis.[6] The disadvantage of these methods is that they do not generate chromatin with regularly spaced nucleosomes. Crude extracts derived from *Drosophila*[21] or *Xenopus* eggs[23] are very effective in assembling physiologically spaced chromatin, but the chromatin obtained is contaminated with many proteins that remain associated with the chromatin template even after purification by standard procedures, including gel filtration columns or sucrose and glycerol gradients.

The minimal requirements for RNA polymerase II transcription on chromatin templates were also recently described. In addition to the general transcription factors, two chromatin-specific factors are necessary, RSF and FACT. RSF allows the transcription machinery to access the DNA and permits formation of competent transcription preinitiation complexes. FACT facilitates the elongation of transcripts by RNA polymerase II on chromatin templates.[24]

This review focuses on three parameters in studying transcription through chromatin *in vitro:* (1) the formation of recombinant chromatin by RSF, (2) the formation of homogeneous chromatin populations carrying a particular modification by chemically synthesizing histones that contain site specific modifications, and (3) the minimal transcription system needed to transcribe chromatin templates.

Purification of Factors

FACT and RSF

Native FACT and RSF

RSF is a heterodimer composed of hSNF2H, an ATPase that belongs to the SWI2/SNF2 family of helicases, and Rsf-1 (p325), a protein encoded by a novel human gene.[24] FACT is also composed of two subunits, the human homologue of the yeast STP16/Cdc68 protein and the high mobility group-1-like *s*tructure *s*pecific *r*ecognition *p*rotein-1 SSRP1.[25]

RSF was purified from the nuclear pellet fraction derived from HeLa cells using an assay that specifically scores for activator-dependent chromatin remodeling and the formation of competent transcription initiation complexes on chromatin templates.[24] FACT was initially purified from a nuclear extract fraction derived from HeLa cells using an assay that specifically scores for RNA polymerase II transcription elongation on chromatin templates.[4] However, it was found later that the nuclear pellet fraction is a rich source of FACT. Therefore, the purification scheme for FACT and RSF described here starts with the nuclear pellet fraction.

Nuclear pellet is obtained as previously described.[26] Briefly, HeLa cells are harvested by centrifugation at 3000 rpm (rotor: H6000A, Sorvall) for 10 min at $4°$, and washed with phosphate-buffered saline (PBS). The cells are osmotically swollen by incubation in 5 volumes (with respect to the cell pellet) with buffer A (10 mM Tris-HCl, pH 7.9; 1.5 mM MgCl$_2$; 10 mM KCl; 0.5 mM DTT; and 0.2 mM PMSF) for 10 min on ice and then centrifuged at 10,000 rpm (rotor: SLA-3000, Sorvall) for 10 min at $4°$. The pellet is resuspended in 2 volumes (with respect to the cell pellet) of buffer A and homogenized with 10 strokes using a glass pestle B homogenizer. The homogenized suspension is centrifuged at 10,000 rpm (rotor: SLA-3000, Sorvall) for 10 min at $4°$. The supernatant is the cytosolic fraction and the pellet contains the nuclei. Nuclei are resuspended in 3 ml of buffer C per 10^9 cells (20 mM Tris-HCl, pH 7.9; 0.42 M NaCl; 1.5 mM MgCl$_2$; 0.2 mM EDTA; 0.5 mM PMSF; 0.5 mM DTT; and 25% glycerol) and homogenized with 10 strokes using a glass pestle B homogenizer, and stirred for 30 min at $4°$. The suspension is centrifuged at 15,000 rpm (rotor: SLA-3000, Sorvall) for 30 min at $4°$. The supernatant contains the nuclear extract fraction, while the pellet contains the nuclear pellet fraction.

Approximately 5 g of the nuclear pellet are resuspended in 36 ml of buffer B (50 mM Tris-HCl, pH 7.9; 5 mM MgCl$_2$; 0.5 mM EDTA; 5 mM DTT; 0.2 mM PMSF; and 25% glycerol). A 3 M stock solution of ammonium sulfate is added to a final concentration of 0.3 M, and the solution is mixed quickly by inverting the tube several times. The solution will became dense; after this occurs, the solution is solubilized by sonication on ice (Sonic Dismembrator 550, Fisher Scientific), with 8–12 strokes in cycles of 33 s for each. The suspension is cleared of debris by centrifugation

[21] J. K. Tyler, *Eur. J. Biochem.* **269,** 2268 (2002).
[22] A. Loyola, G. LeRoy, Y. H. Wang, and D. Reinberg, *Genes Dev.* **15,** 2837 (2001).
[23] R. A. Laskey, A. D. Mills, and N. R. Morris, *Cell* **10,** 237 (1977).
[24] G. LeRoy, G. Orphanides, W. S. Lane, and D. Reinberg, *Science* **282,** 1900 (1998).
[25] G. Orphanides, W. H. Wu, W. S. Lane, M. Hampsey, and D. Reinberg, *Nature* **400,** 284 (1999).
[26] J. D. Dignam, R. M. Lebovitz, and R. G. Roeder, *Nucleic Acids Res.* **11,** 1475 (1983).

at 40,000 rpm (rotor: T-647.5, Sorvall) for 1 h at $4°$. A saturated ammonium sulfate solution, pH 7.9, is then added drop wise with continuous stirring to a final concentration of 68% saturation. The solution is then stirred for 30 min and centrifuged at 35,000 rpm (rotor: T-647.5, Sorvall) for 1 h at $4°$. The supernatant is discarded, and the pellet is resuspended in buffer D (50 mM Tris-HCl, pH 7.9; 0.1 mM EDTA; 2 mM DTT; 0.2 mM PMSF; and 25% glycerol) to adjust the ammonium sulfate concentration to 0.05 M. The solubilized nuclear pellet is then loaded onto a 600-ml DEAE-52 column (Whatman, Catalog number 4057200, capacity 10 mg/ml) equilibrated with buffer D containing 0.05 M ammonium sulfate. The column is washed with buffer D/0.05 M ammonium sulfate, and the proteins are eluted with a 4-column volume, linear gradient of ammonium sulfate (from 0.1 to 0.6 M, in buffer D). About 50–70% of the proteins flow through the column, while FACT and RSF elute together between 0.1 and 0.25 M ammonium sulfate. The fractions containing FACT/RSF, assayed as described below, are pooled and dialyzed against 0.2 M ammonium sulfate in buffer D (buffer D now contains 10% glycerol from this point on) and then loaded onto a 140-ml Heparin-agarose type I column (Sigma, Catalog number H6508, capacity, 5 mg/ml) equilibrated with buffer D/0.2 M ammonium sulfate. The column is washed with buffer D/0.2 M ammonium sulfate and proteins are eluted with a 5-column volume, linear ammonium sulfate gradient (from 0.2 to 0.7 M, in buffer D). At this point in the purification, about 50% of the proteins flow through the Heparin-agarose column. FACT and RSF are eluted together, between 0.23 and 0.3 M ammonium sulfate.

The fractions containing FACT and RSF are then pooled and dialyzed against buffer D/0.05 M ammonium sulfate and loaded onto a Tosoh Biosep FPLC TSK-DEAE-5PW (21.5 mm × 15 cm, capacity, 10 mg/ml) column equilibrated with buffer D/0.1 M ammonium sulfate. The column is washed with buffer D/0.1 M ammonium sulfate and proteins are eluted with a 7-column volume, linear ammonium sulfate gradient (from 0.1 to 0.7 M, in buffer D). This column separates FACT and RSF; RSF elutes first in a sharp peak at a conductivity equivalent to 0.18 M ammonium sulfate. FACT is also eluted in a sharp peak at a conductivity equivalent to that of 0.27 M ammonium sulfate.

The fractions containing RSF are pooled and concentrated to 250 μl with a Millipore micro-concentrator (10 kDa cut off, Catalog number UFV4BGC00) and then loaded onto a gel filtration column (Amersham Pharmacia Biotech FPLC Superose 6 HR 10/30). The column is equilibrated with buffer BC500 (20 mM Tris-HCl, pH 7.9; 0.2 mM EDTA; 10 mM β-Mercaptoethanol; 0.2 mM PMSF; 10% glycerol; and 0.5 M KCl). RSF elutes with a relative molecular mass of 500 kDa. The final yield is approximately 0.5 mg from 2×10^{11} cells, and the preparation is 90% homogeneous (see Fig. 2A).

The fractions containing FACT are pooled and dialyzed against buffer D/1.1 M ammonium sulfate and loaded onto an 8-ml FPLC Phenyl Superose HR 10/10 column (Amersham Pharmacia Biotech, 5 mg/ml), that had been equilibrated with buffer D/1.1 M ammonium sulfate. The column is washed with buffer D/1.1 M ammonium sulfate, and proteins are eluted with an 8-column volume, linear ammonium sulfate gradient (from 1.1 to 0 M in buffer D). FACT is eluted in the flow through. This fraction is dialyzed against buffer BC100 (20 mM Tris-HCl, pH 7.9; 0.2 mM EDTA; 10 mM β-Mercaptoethanol; 0.2 mM PMSF; 10% glycerol; and 0.1 M KCl) and applied to a 1-ml phosphocellulose-P11 column (Whatman) that had been equilibrated with buffer BC100. The column is washed with BC100, and proteins are eluted with a 15-ml linear KCl gradient (from 100 to 800 mM) in buffer BC. The majority of the FACT activity is eluted at a KCl concentration of $\sim$400 mM. The final yield is approximately 1.5 mg of FACT, and the preparation is about 90% homogeneous (see Fig. 2B).

At each step of the purification, the fractions are assayed by Western blot with antibodies specific to the RSF subunits hSNF2H and Rsf-1 and to the FACT subunits hSPT16/p140 and SSRP1. Fractions from the final column are visualized by silver staining, and also assayed for chromatin assembly in the case of RSF and transcription elongation on chromatin templates in the case of FACT.

Recombinant FACT

Recombinant FACT is purified from SF9 cells coinfected with Flag-tagged hSPT16/p140 and His-tagged SSRP1 baculoviruses. The purification of the FACT complex consists of two affinity chromatographic steps, M2-agarose and Ni-agarose.

Log phase SF9 cells are grown on 150-mm culture plates at a confluence of 2×10^7 cells/plate. Ten plates worth of these cells are infected with the hSPT16/p140 and SSRP1 baculoviruses at a multiplicity of infection (m.o.i.) of 10. Seventy-two hours after infection, the cells are harvested, pelleted, and washed twice with 50 ml of cold PBS. The cells are resuspended in 10 ml lysis buffer (20 mM Tris-HCl, pH 7.9; 0.6 M NaCl; 4 mM MgCl$_2$; 0.4 mM EDTA; 2 mM DTT; 20 mM β-glycerophosphate; 20% glycerol; 0.4 mM PMSF; 1 mM benzamidine-HCl; 4 μg/ml leupeptine; and 2 μg/ml aprotinin) and disrupted with three quick freeze/thaw cycles. The insoluble material is removed by centrifugation at 11,000 rpm (rotor: SLA-600TC, Sorvall) for 10 min at 4°. Anti-FLAG M2-agarose beads (1 ml, Sigma, Catalog number A-1205), equilibrated with lysis buffer, are added to the supernatant. After incubation for 4 h at 4°, the

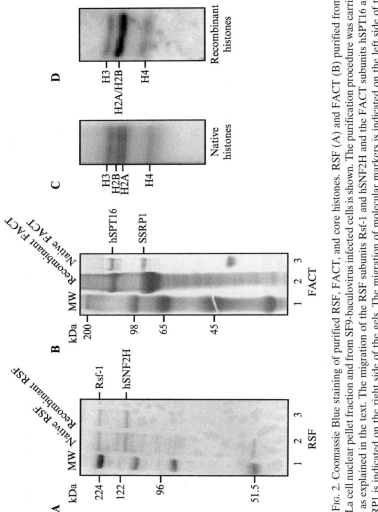

FIG. 2. Coomassie Blue staining of purified RSF, FACT, and core histones. RSF (A) and FACT (B) purified from a HeLa cell nuclear pellet fraction and from SF9-baculovirus infected cells is shown. The purification procedure was carried out as explained in the text. The migration of the RSF subunits Rsf-1 and hSNF2H and the FACT subunits hSPT16 and SSRP1 is indicated on the right side of the gels. The migration of molecular markers is indicated on the left side of the gels. Native core histones (C) purified from HeLa cells or purified recombinant Xenopus laevis core histones expressed in bacteria (D) are shown. The purification procedure was carried out as explained in the text. The migration of each of the core histones is indicated on the left side of the gels. Note that recombinant H2A and H2B migrate together due to the high similarity in the molecular weight and the lack of post-translational modifications.

beads are washed four times with 15 ml of the lysis buffer, and the bound proteins are eluted by incubation for 30 min with 1 ml of 0.2 mg/ml Flag peptide (Sigma, F3290) dissolved in lysis buffer. The elution process is repeated three times. The eluate is analyzed by SDS-polyacrylamide gel electrophoresis (SDS-PAGE) and Coomassie Blue staining as well as by Western blot with antibodies to the FACT subunits. The fractions containing STP16 and SSRP1 are pooled and adjusted to 30 mM imidazole and incubated with 1 ml of Ni-NTA agarose beads (Qiagen) with rotation for 4 h to overnight at 4°. The beads are washed four times with 15 ml of lysis buffer containing 30 mM imidazole, and the bound material is eluted by 30 min incubations with 1 ml of lysis buffer containing 480 mM imidazole. The elution process is repeated four times. The extent of purification is analyzed by Coomassie Blue staining as well as by Western blot with antibodies to both subunits of FACT. The FACT containing fractions are pooled and dialyzed against BC50 (20 mM Tris-HCl, pH 7.9; 0.2 mM EDTA; 10 mM β-Mercaptoethanol; 0.2 mM PMSF; 20% glycerol; and 50 mM KCl). The final yield is approximately 1 mg of FACT per 10 plates of 150 mm size (see Fig. 2B).

It is important to mention that it is not recommended to reverse the affinity purification procedure. The Ni-agarose purification is performed as the last column because it allows for the removal of the Flag peptide derived from the M2-column purification. The Flag peptide interferes in several reactions including transcription.

Recombinant RSF

Recombinant RSF is purified from SF9 cells co-infected with Flag-tagged Rsf-1 and hSNF2H baculoviruses. Log phase SF9 cells are grown in a 500 ml flask to a confluence of 10^6 cells/ml. These cells are infected with the Rsf-1 and hSNF2H baculoviruses at a m.o.i. of 10. Forty-eight hours post infection, the cells are harvested, pelleted and washed twice with cold PBS. The cells are resuspended in 10 ml lysis buffer (20 mM Tris-HCl, pH 7.9; 0.6 M NaCl; 4 mM MgCl$_2$; 0.4 mM EDTA; 2 mM DTT; 20 mM β-glycerophosphate; 20% glycerol; 0.4 mM PMSF; 1 mM benzamidine–HCl; 4 μg/ml leupeptine; and 2 μg/ml aprotinin) and disrupted with three quick freeze/thaw cycles. The insoluble material is removed by centrifugation at 11,000 rpm (rotor: SLA-600TC, Sorvall) for 10 min at 4°. Anti-FLAG M2-agarose beads (1 ml, Sigma, Catalog number A-1205), equilibrated with lysis buffer, are added to the supernatant. After incubation for 4 h at 4°, the beads are washed four times with 15 ml of the lysis buffer, and the bound proteins are eluted by incubation for 30 min with 1 ml of 0.2 mg/ml of Flag peptide (Sigma, F3290) dissolved in lysis buffer. The elution process is repeated three times. The eluate is analyzed by

SDS-PAGE and Coomassie Blue staining as well as by Western blot with antibodies to the RSF subunits. The fractions containing RSF are pooled, concentrated to 250 μl with a Millipore microconcentrator (10 kDa cut off, Catalog number UFV4BGC00) and then loaded onto a gel filtration column (Amersham Pharmacia Biotech FPLC Superose 6 HR 10/30). The column is equilibrated with buffer BC500 (20 mM Tris-HCl, pH 7.9; 0.2 mM EDTA; 10 mM β-Mercaptoethanol; 0.2 mM PMSF; 10% glycerol; and 0.5 M KCl). Recombinant RSF elutes with a relative molecular mass of 500 kDa. The final yield is approximately 0.3 mg of RSF per 0.5 $\times$ 10^8 SF9-infected cells (see Fig. 2A).

Core Histones

Native Histones

Native histone octamers are purified from HeLa cells by a method adapted from Bulger and Kadonaga.[27] Briefly, HeLa nuclei are isolated by incubation with 5$\times$ pellet volume of hypotonic buffer as described[27] and digested with 200 U/ml of micrococcal nuclease to obtain oligonucleosomes ranging between 1 and 2 kb. The oligonucleosomes are purified away from the nuclear proteins through a 5 to 39% sucrose gradient of 36 ml centrifuged at 20,000 rpm (rotor: SW28, Beckman) for 16 h at 4°. Fractions of 2 ml are collected and 20 μl aliquots of each fraction are analyzed by Coomassie blue staining. Finally, histones are removed from the DNA by chromatography on a hydroxyapatite column (Bio-Gel HT Gel, BioRad, Catalog number 130-0150, 5 mg/ml). This purification scheme yields highly purified, hypoacetylated core histones (see Fig. 2C). To obtain hyperacetylated histones, the cells are treated with histone deacetylase inhibitors (0.1 μM Trichostatin A [TSA] and/or 10 mM Butyric acid) for 24 h prior to harvesting. In addition, buffers must contain histone deacetylase inhibitors during all of the purification steps described above. Core histone octamers can be separated into H2A/H2B dimers and H3/H4 tetramers by MonoS column chromatography.[28]

Recombinant Histones

HISTONE POLYPEPTIDES. Recombinant *Xenopus laevis* histones were expressed and purified by a method adapted from Luger.[29] Although the original purification scheme has three chromatographic steps, we found that the isolation of inclusion bodies results in a highly enriched histone fraction

[27] M. Bulger and J. T. Kadonaga, *Methods Mol. Genet.* **5,** 241 (1994).
[28] C. H. Chang and D. S. Luse, *J. Biol. Chem.* **272,** 23427 (1997).
[29] K. Luger, T. J. Rechsteiner, and T. J. Richmond, *Methods Mol. Biol.* **119,** 1 (1999).

that can be used to reconstitute core histone octamers *in vitro*. After the final sizing column, the core histones are highly purified.

BL21 (DE) pLysS bacteria are transformed with the pET-histone expression plasmids and plated on Luria broth (LB) with 100 μg/ml of ampicillin and 25 μg/ml of chloramphenicol. The plates are incubated at 37° overnight. The next day, a 5 ml aliquot of LB medium containing ampicillin and chloramphenicol is inoculated with a single bacterial colony, and the culture is grown overnight at 37°. It is recommended that the induction with isopropyl-β-thiogalactopyranoside (IPTG) be tested on five different colonies. The next morning, a 5 ml aliquot of LB medium containing ampicillin and chloramphenicol is inoculated with 50 ul of the overnight culture, and the suspension is then grown until the OD_{600} is 0.5–0.7. At this point, histone expression is induced with 0.2 mM IPTG, and the cultures are incubated for an additional 2 h at 37°. The level of induction is checked on an 18% SDS-polyacrylamide gel, where the expression levels of histones in the IPTG-induced cultures are compared with uninduced levels.

Once a colony with high levels of induction is selected, 1 L of LB medium containing ampicillin and chloramphenicol is inoculated with 1 ml of the overnight culture. The 1 L culture is then induced for 2 h with 0.2 mM IPTG when it reaches an OD_{600} of 0.5–0.7. The bacteria are then harvested by centrifugation at 5000 rpm (rotor: SLA-3000, Sorvall) for 10 min, and the pellet is resuspended in 100 ml wash buffer (50 mM Tris-HCl, pH 7.5; 100 mM NaCl; 1 mM Na-EDTA; and 1 mM benzamidine) and frozen in liquid nitrogen. The solution is then thawed quickly in a warm water bath. The viscosity resulting from bacterial lyses is reduced by 3 s of sonication (Sonic Dismembrator 550, Fisher Scientific), and the resulting solution is centrifuged at 12,000 rpm (rotor: SLA-600TC, Sorvall) for 20 min at 4°. The pellet, which contains the inclusion bodies of the corresponding histone protein, is washed twice with 100 ml wash buffer plus 1% Triton X-100, washed twice with wash buffer without Triton, and centrifuged at 12,000 rpm (rotor: SLA-600TC, Sorvall) for 10 min at 4°. The pellet is resuspended in 30 ml of unfolding buffer (7 M guanidinium–HCl; 20 mM Tris-HCl, pH 7.5; and 10 mM DTT) and mixed gently for 1 h at room temperature. The undissolved material is removed by centrifugation at 12,000 rpm (rotor: SLA-600TC, Sorvall) for 15 min at 20°. The final yield is approximately 100 mg.

HISTONE OCTAMERS. Histone octamers are obtained by mixing the four unfolded *X. laevis* recombinant histones isolated as described earlier in equimolar amounts to approximately 4 mg of total protein (in about 1 ml volume). The mixture is dialyzed at 4° against 2 L of refolding buffer (2 M NaCl; 10 mM Tris-HCl, pH 7.5; 1 mM Na-EDTA; and 5 mM β-Mercaptoethanol) with at least 3 buffer changes. Either the second or third dialysis step should be performed overnight. The sample is centrifuged to remove any precipitate and concentrated to 250 μl with a Millipore microconcentrator (10 kDa cut off, Catalog number UFV4BGC00). The sample is loaded onto a HiLoad Superdex 200 HR 10/30 column (Amersham Pharmacia Biotech) equilibrated with refolding buffer. Recombinant core histone octamer elutes with a relative molecular mass of 100 kDa. Aliquots of the column fractions are analyzed by Coomassie blue staining. After pooling the histone octamer containing-fractions, the concentration is determined spectrophotometrically ($A_{276} = 0.45$ for a solution of 1 mg/ml of core octamer). The purified octamers are stored at 4°, at a concentration of 1 mg/ml. They are stable for at least six months (see Fig. 2D).

The H2A/H2B dimers and the H3/H4 tetramers can be obtained through the same procedure, by mixing equimolar amounts of H2A and H2B or equimolar amounts of H3 and H4 followed by dialysis in the refolding buffer and purification over the Superdex 200 column. Recombinant H2A/H2B dimers elute with a relative molecular mass of 25 kDa, while H3/H4 tetramers with a relative molecular mass of 50 kDa.

Synthetic Histones Containing Site-Specific Modifications

Native chemical ligation (NCL) is a methodology by which two unprotected peptide fragments can be chemoselectively condensed in aqueous solution, forming a stable peptide bond.[30–34] In classical cysteine NCL,[30] one peptide fragment contains a C-terminal thioester and the other peptide possesses an N-terminal cysteine residue. Upon mixing the two together in aqueous solution (generally under denaturing conditions), a chemoselective transthioesterification occurs rapidly, joining the two peptides via a thioester bond. In a second slow step, an intramolecular *S*-to-*N* acyl shift ensues, creating a peptide bond and leaving an internal cysteine residue at the junction site. Since most histone posttranslational acetyl and methyl modifications are found on lysine or arginine residues within the first 30 amino acids of the highly conserved N-terminus, we devised an efficient modular NCL strategy for histone synthesis. This strategy involves synthesis using Boc-SPPS of peptides corresponding to this highly conserved N-terminus region followed by their ligation to a recombinant protein fragment corresponding to the histone C-terminal globular domain (see Fig. 3). After ligation, the NCL junction site can be made effectively traceless by desulfurization with H_2/Raney nickel.[35] Although we illustrate this general

[30] P. E. Dawson, T. W. Muir, I. Clark-Lewis, and S. B. H. Kent, *Science* **266,** 776 (1994).
[31] P. E. Dawson and S. B. H. Kent, *Ann. Rev. Biochem.* **69,** 923 (2000).
[32] J. P. Tam, J. X. Xu, and K. D. Eom, *Biopolymers* **60,** 194 (2001).
[33] C.-F. Liu and J. P. Tam, *J. Am. Chem. Soc.* **116,** 4149 (1994).
[34] C. F. Liu and J. P. Tam, *Proc. Natl. Acad. Sci. USA* **91,** 6584 (1994).

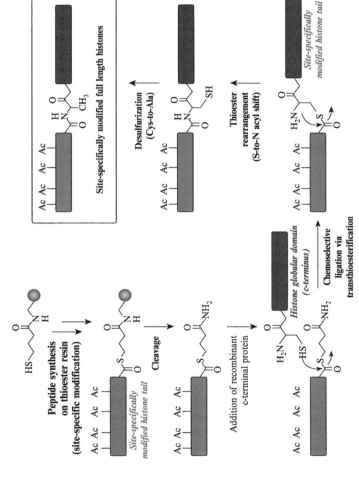

FIG. 3. Synthetic histones containing site-specific modifications. (A) Scheme for the synthesis of site-specifically modified histones using native chemical ligation combined with desulfurization. Peptide thioesters corresponding to the N-terminal 15–30 amino acids of histones are prepared on 3-mercaptopropionyl-MBHA resin. N-Terminal side chain modifications (e.g., Lys(Ac), Lys(Me)$_3$, etc.) are introduced site-specifically using suitably protected amino acid modules introduced via standard Boc-SPPS methods. Following assembly and HF-mediated cleavage from the solid support, the modified histone termini thioester peptides are ligated to a companion recombinant histone C-terminal protein fragment that terminates in a Cys residue. The ligation site may be subsequently rendered traceless by desulfurization. (See color insert.)

method of histone synthesis using *Xenopus laevis* histones H3 and H4, it is compatible with the preparation of histones 2A and 2B from this and other organisms.

SYNTHESIS OF HISTONE N-TERMINAL PEPTIDE THIOESTERS. Site specifically-acetylated histone H3- and H4-derived N-terminal thioester peptides (ca. 15–30 amino acids) are assembled by manual solid-phase peptide synthesis (SPPS). Trifluoroacetic acid (TFA) labile N^α-Boc protected amino acids are employed for the stepwise synthesis, and hydrofluoric acid (HF) is used to facilitate simultaneous peptide side chain deprotection and cleavage from the solid support. All amino acids and coupling reagents are commercially available (Applied Biosystems, Bachem, Novabiochem or Advanced Chemtech) with the exception of Boc-L-Lys(Me)$_3$-OH, which was prepared according to the method of Chen and Benoiton.[36] Peptide synthesis-grade solvents are from Applied Biosystems or Aldrich Chemicals. HF was purchased from Matheson gases. HF cleavage reactions were performed in a Kel-F apparatus (Peninsula Laboratories). Acetyl lysine residues are introduced into the sequence as N^α-Boc-L-Lys(Ac)-OH.

Peptide synthesis was conducted on a 0.25–1.0 mmol scale on a methylbenzhydrylamine resin (MBHA, 0.3–0.7 mmol/g amine substitution, Advanced Chemtech) that had been previously derivatized with a 3-mercaptopropionic acid linker.[32] Peptide loading and subsequent extension reactions were conducted with a fourfold excess of Boc-amino acid, O-benzotriazole-N,N,N',N'-tetramethyl-uronium-hexafluoro-phosphate (HBTU) and 1-hydroxybenzotriazole hydrate (HOBt) using Kent's rapid *in situ* neutralization synthetic protocol.[37] Coupling efficiencies were monitored by quantitative ninhydrin analysis[38] and double couplings were introduced as necessary for difficult sequences. Although we are describing a manual SPPS method, we note that the thioester synthesis can also be conducted easily using an automated peptide synthesizer equipped to perform Boc/TFA chemistry such as the Applied Biosytems 433A instrument. Once assembled, histone-derived peptides can be cleaved from the resin using anhydrous hydrogen fluoride (10 ml) at 0° (1 h) with 10% *p*-cresol (v/v), precipitated by the addition of ice cold Et$_2$O, filtered to collect the crude peptide, redissolved in 1% AcOH, and lyophilized. Crude peptide thioesters were purified by preparative reversed-phase HPLC on octadecyl silica (Vydac C$_{18}$) using linear gradients of Buffer A (99.9 : 0.1 H$_2$O/CF$_3$COOH, v/v)

[35] L. Z. Yan and P. E. Dawson, *J. Amer. Chem. Soc.* **123**, 526 (2001).

[36] C. M. Chen and N. L. Benoiton, *Can. J. Chem.* **54**, 3310 (1976).

[37] M. Schnolzer, P. Alewood, A. Jones, D. Alewood, and S. B. Kent, *Int. J. Peptide Prot. Res.* **40**, 180 (1992).

[38] V. K. Sarin, S. B. H. Kent, J. P. Tam, and R. B. Merrifield, *Anal. Biochem.* **117**, 147 (1981).

and Buffer B ($90:10:0.1$ $CH_3CN/H_2O/CF_3COOH$, v/v/v). Fractions containing the desired peptide thioesters, judged as pure by analytical HPLC, were combined, concentrated by rotary evaporation to remove excess CH_3CN, lyophilized, and confirmed for composition by MALDI-TOF mass spectrometry.

PRODUCTION OF XENOPUS LAEVIS HISTONES H3 AND H4 C-TERMINAL GLOBULAR DOMAINS CONTAINING N-TERMINAL CYS RESIDUES. The second requirement for NCL is a peptide or protein containing an N-terminal cysteine, which serves as the nucleophilic thiol. Our general approach is to engineer the thiol into the N-terminal region of histones (in this example H3 and H4) using site-directed mutagenesis, then create a truncated histone C-terminal fragment in which the engineered cysteine immediately follows the fMet initiator. The fMet residue is either cleaved during recombinant expression, as was observed for the H3 and H4 constructs described below, or, if necessary this residue can be removed by coexpression of the histone protein with hexahistidine-tagged E. coli methionine aminopeptidase (His_6-MetAP) or by treatment of the histone protein with purified His_6-MetAP.[39] In any case, the histone C-terminal domain is left with an unprotected N-terminal cysteine residue for NCL. In the following examples, we chose to engineer the cysteine residue at positions Ala15 of H4 and Ala25 of H3 since NCL followed by desulfurization would generate a native-like primary sequence. We note that this method is not alanine position limited in that the engineered cysteine may be incorporated into other non-alanine sites within the histone N-terminal of ca. 30 amino acids with little adverse effect on biological activity.[40,41]

Recombinant C-terminal domains of Xenopus laevis histones H3 and H4 containing N-terminal Cys residues were prepared by site-directed mutagenesis of full length histones H3 and H4, followed by PCR deletion of the N-termini and subcloning into a pET system protein expression vector (Novagen). Using Luger's plasmids encoding for Xenopus laevis histones H3 and H4 as the templates,[29] H3(A25C) and H4(A15C) point mutants were prepared using the Quickchange™ site-directed mutagenesis method (Stratagene) according to the manufacturer's directions. Next, the N-terminal sequences of histones H3 (ΔN24) and H4 (ΔN14) were deleted by PCR amplification with primers encoding an N-terminal Met initiation codon adjacent to the Ala-to-Cys mutation. Following PCR, the insert was

---

[39] W. T. Lowther, D. A. McMillen, A. M. Orville, and B. W. Matthews, Proc. Nat. Acad. Sci. USA **95,** 12153 (1998).

[40] M. A. Shogren-Knaak, C. J. Fry, and C. L. Peterson, J. Biol. Chem. **278,** 15744 (2003).

[41] S. He, D. Bauman, J. Gronlund, N. L. Kelleher, and D. G. McCafferty, "Proceedings of the 17th American Peptide Symposium." San Diego, California, 2001.

digested with $NdeI$ and $HindIII$ restriction endonucleases and ligated into a pET 30 b vector (Novagen) that had been previously treated with $NdeI$, $HindIII$ and calf alkaline intestinal phosphatase. The C-terminus of histone H3 maintained a native sequence due to the preservation of its TAG stop codon prior to its incorporation into the $HindIII$ ligation site. To illustrate the utility of this method for generating variant histones, the TAA stop codon was removed from histone H4 during PCR amplification in order to introduce a hexahistidine tag sequence (KLAAALEHHHHHH) at the C-terminus. $E.\ coli$ BL21-codon plus (DE3)-RIL cells harboring the expression plasmids for histone H3 (ΔN24) A25C and H4 (ΔN14) A15C C-terminal domains were grown in LB medium containing 50 $\mu g/\mu l$ kanamycin at 37° with shaking at 225 rpm. Cells were induced at $OD_{600} = 0.6$ by the addition of IPTG (1 mM) and harvested after 3 h by centrifugation. Cell pellets were quickly frozen in liquid nitrogen and stored at $-70°$ until needed. Expression levels and assessment of protein solubility was monitored by SDS-PAGE analysis.

For protein purification, the following steps were performed at 4°. Frozen cell pellets (2 L cell of culture) were thawed and resuspended in cold 50 mM Tris-HCl buffer (pH 8, 50 ml) containing 1 mM EDTA (Buffer C). The cell suspension was lysed by passage through a French press and subsequently clarified by ultracentrifugation. The pellet containing the inclusion bodies was suspended in 50 ml of buffer C, centrifuged for 15 min, suspended in 50 ml of buffer C containing 2 M urea, centrifuged, suspended in 50 ml of buffer C containing 1% Triton X-100, and ultracentrifuged. The pellet was suspended in buffer D (6 M guanidinium–HCl, 100 mM sodium phosphate buffer pH 7.5, 10 ml) and stirred overnight at room temperature. Following ultracentrifugation to clarify the urea solution, the pH of the supernatant was lowered to 5.5 by addition of 1 M NaH_2PO_4 and the resultant solution was preparatively purified by reversed-phase HPLC over octadecyl silica and a linear gradient of Buffer A and Buffer B over 60 min. Fractions containing the desired proteins were combined, lyophilized and confirmed for composition by high-resolution Fourier transform electrospray mass spectroscopy.

NATIVE CHEMICAL LIGATION REACTION AND DESULFURIZATION. The native chemical ligation reactions were performed under denaturing conditions by a modification of the thiophenol-catalyzed method of Dawson.[42] Recombinant H3 or H4 protein thiols were added to an appropriate histone H3 or H4 N-terminal peptide thioester, premixed with thiophenol

[42] P. E. Dawson, M. Churchill, M. R. Ghadiri, and S. B. H. Kent, $J.\ Amer.\ Chem.\ Soc.$ **119,** 4325 (1997).

(2%, v/v) in phosphate buffer solution (pH 7.5) containing 6 M guanidinium-HCl to give a final concentration of 2 mM of each peptide. Ligation reactions were typically carried out with 1.5-fold molar excess of thioester and the reaction progress was monitored by analytical HPLC. The ligation reactions involving H4 (15-102)-His$_6$ A15C were performed at room temperature. However, ligations involving histone H3 (25-133) A25C were performed at 60° in order to overcome the limited solubility of one or more components of the ligation reaction mixture at 25°. Ligations with both H3 and H4 thiols typically resulted in 50–60% conversion to products based on yields of product isolated by reversed-phase HPLC over butyl-silica.

Native-like full-length synthetic H3 and H4 proteins were generated by desulfurization of the ligation products.[35] Typically, histone proteins (20 mg) were dissolved in argon-degassed phosphate buffer solution (pH 5.8) containing 6 M guanidinium–HCl to a final concentration of 0.5–2 mg/ml. Freshly-prepared Raney nickel was added (10-fold excess w/w) and the reaction was placed under an atmosphere of hydrogen. Yields after ligation, desulfurization and reversed-phase HPLC purification typically ranged from 10 to 35%. The molecular composition of all histone proteins and intermediates were verified using Fourier transform electrospray MS (N. Kelleher, University of Illinois) or MALDI-TOF MS analysis. Purity was assessed by analytical HPLC and SDS-PAGE with silver staining (see Fig. 4).

ASSEMBLY AND PURIFICATION OF A HYBRID SYNTHETIC H3/RECOMBINANT H4 TETRAMER. Recombinant *Xenopus laevis* histone H4 was expressed and purified as described in Histone polypeptides section Hybrid Synthetic H3/Recombinant H4 Tetramer were folded and purified as described in Histone octamers section (Fig. 4).

Chromatin Assembly and Analysis

The techniques described in this section can be applied equally well when chromatin has been assembled with either native, recombinant, or chemically synthesized core histones.

Chromatin Assembly

Chromatin assembly reactions are performed as previously described[22] using the 3-kb plasmid pG5MLP, which contains five binding sites for the transcriptional activator Gal4 upstream of the Adenovirus Major Late promoter and a 390-base pair (bp) G-less cassette. The conditions described here can be applied to different plasmids. In a 0.6 ml siliconized Eppendorff tube, 2 μg of the template are combined with 1.8 μg of native

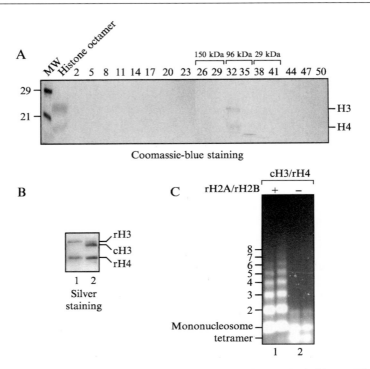

FIG. 4. Assembly of histone tetramers and chromatin using synthetic histone H3 (cH3, prepared by NCL) and recombinant histone H4 (rH4). (A) SDS-PAGE analysis of the size exclusion chromatographic purification of cH3/rH4 tetramer assembly. (B) SDS-PAGE gel analysis of the cH3/rH4 tetramers visualized by silver staining. Eighteen percent SDS-PAGE analysis of cH3/rH4 tetramer visualized by Coomassie Blue staining. (C) Agarose gel analysis of the products from micrococcal nuclease digestions of RSF-mediated chromatin assembly reactions using the cH3/rH4 tetramer, recombinant rH2A/rH2B tetramer and a plasmid DNA template. (See color insert.)

or recombinant core histones (or equimolar amounts of H2A/H2B dimers plus H3/H4 tetramers), 0.3 μg of RSF, 150 μg of bovine serum albumin, 3 mM ATP, 30 mM phosphocreatine (Sigma, Catalog number P-6502), 0.2 μg of phosphocreatine kinase (Sigma, Catalog number C-3755), 5 mM MgCl$_2$, 50 mM KCl, 10 mM HEPES, pH 7.6, 0.2 mM EDTA, and 5% glycerol in a final reaction volume of 150 μl. The reaction mixture is typically incubated at 30° for 5 h, but can also be incubated overnight without affecting the efficiency of chromatin assembly. Chromatin assembled by this method can be stored at 4° for at least two weeks without loss of nucleosomes.

The ratio between DNA and core histones in the chromatin assembly reaction is critical and needs to be titrated for every new DNA and core histone preparation. If the ratio is not correct, chromatin assembly will not be successful.

When chromatin is assembled for transcription experiments, the chromatin assembly reaction can be performed in the presence of a transcriptional activator, and the resulting chromatin then will be remodeled. For the sample reaction described here, the activator, Gal4VP16, is added to the chromatin assembly reaction in a 10–20-fold molar excess over the DNA. The activator can also be added to previously assembled chromatin and incubated for 1 h at 30° in the presence of 0.1 μg of RSF and 3 mM ATP.

When recombinant histones are used to assemble chromatin, the RSF-dependent chromatin assembly reaction requires acetylation of the histones H2A and H2B.[22] Therefore, the RSF-dependent-chromatin assembly reaction is not efficient when recombinant histones are used. A way to circumvent this problem is to acetylate H2A/H2B dimers with p300 in the presence of the cofactor acetyl-CoA. The acetylated-dimers can be purified on a size exclusion column (Superdex 200, or Superdex 75, Amersham Pharmacia Biotech). Chromatin assembly can then be performed as described earlier, except that, rather than using core histone octamers, acetylated recombinant H2A/H2B dimers are combined with equimolar amounts of recombinant H3/H4 tetramers. Another strategy is to mix equimolar amounts of native MonoS-purified H2A/H2B dimers (derived in the presence of deacetylase inhibitors) with recombinant H3/H4. Both methods result in efficient chromatin assembly.

Partial Micrococcal Nuclease Digestion and Southern Blotting

The efficiency of chromatin assembly as well as chromatin remodeling is evaluated by partial micrococcal nuclease digestion of the reactions products. Micrococcal nuclease digestion is performed with 1 μg of DNA in the following buffer: 10 mM HEPES, pH 7.6; 50 mM KCl; 5 mM MgCl$_2$; 0.2 mM EDTA; and 10% glycerol supplemented with 3 mM CaCl$_2$, in a final reaction volume of 80 μl. In order to get the most information out of this kind of experiment, typically, two different empirically determined concentrations of micrococcal nuclease are used to characterize a single chromatin preparation. For example, one can use 10 μl of a 20-unit/ml preparation of micrococcal nuclease (Sigma, Catalog number N-5386) in one reaction and an order of magnitude less enzyme in a second reaction.

The micrococcal nuclease digestion is carried out for 10 min at 30° and stopped by the addition of 80 μl of stop buffer (20 mM EDTA; 0.2 M NaCl; 1% SDS; and 0.25 mg/ml glycogen) containing 0.07 mg/ml of

Proteinase K. The reaction is then incubated for an additional 15 min at 37°. The DNA is then extracted with 160 ul phenol/chloroform, and 120 μl of the aqueous phase is collected. Ammonium acetate (5 μl of a 2.5 M stock) is added to the solution and the DNA is precipitated with 600 μl of cold ethanol (stored at $-80°$). The pellet is then washed with 70% ethanol, air dried, and resuspended in 10 μl of sample buffer (10 mM Tris, pH 7.9; 1 mM EDTA; 0.1% bromophenol blue; and 10% glycerol). The DNA products are resolved on a 1.3% agarose gel (Bio-Rad, Sub-Cell Model 192, 25 × 15 cm^2 trays) in 50 mM Tris/0.38 M glycine. The gel is subjected to electrophoresis at 170 V for 4 h or at 50 V overnight. After electrophoresis, the gel is rapidly stained for 10 min with ethidium bromide and then destained with several washes of running buffer. At this step, the chromatin assembly reaction can be evaluated (see Fig. 5). When chromatin is regularly spaced, the DNA fragments appear as a DNA ladder. If chromatin is not formed or it is not periodically spaced, DNA fragments appear as a smear, with strong bands at the bottom of the gel, representing mononucleosomes.

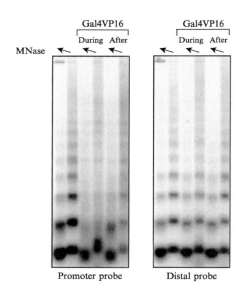

FIG. 5. Southern blot of micrococcal nuclease digestions of RSF-assembled and remodeled chromatin. Chromatin was assembled with RSF in the presence and absence of the activator Gal4VP16, added during the assembly or after the assembly, as indicated on top of the figure. The membrane was probed with a promoter probe (left side), followed by stripping off the probe and re-probing with a distal probe (right side). Each reaction denotes two different concentrations of micrococcal nuclease used in the assays.

The remodeling reaction is evaluated by Southern blotting, as described in Sambrook et al.[43] Briefly, the gel is soaked for 30 min in denaturing buffer (0.5 M NaOH; 1.5 M NaCl), then soaked for another 30 min in neutralizing buffer (0.5 M Tris-HCl, pH 7.0; 1.5 M NaCl), and finally, soaked for 30 min in 20× SSC buffer (3 M NaCl; 0.3 M NaCitrate; adjusted to pH 7.0, at room temperature). The gel is then transferred to a nitrocellulose membrane overnight. The next morning, the membrane is subjected to UV-crosslinking (UV Stratalinker™ 1800, Stratagene) and incubated in blocking solution (6× SSC; 0.5% nonfat milk; 0.005% NP-40) for 1 h at 65°. The radiolabeled-probe complementary to the region of interest is then added to a fresh blocking solution, and the membrane is incubated with the probe for 4 h at 65°. After 5 washes, for 20 min each at 65°, with washing solution (1.5× SSC, 0.05% SDS), the membrane is exposed onto X-ray film.

In order to test for RSF-mediated activator-dependent chromatin remodeling, the Southern blot is first performed with a promoter probe and then the membrane is stripped with 10% SDS and probed with a labeled DNA fragment distal to the promoter site. In the case of pG5MLP plasmid, the promoter probe consists of the following oligonucleotide, 5′-GGG-GCTATAAAAGGGGGTGGGGGCGCGTTC-3′, and the distal probe consists of the following oligonucleotide containing sequence 1000 bp upstream of the promoter, 5′-CTCCGATCGTTGTCAGAAGTAAGTTG-GGCG-3′, as described previously.[4] The probe is radiolabeled using polynucleotide T4 DNA kinase (Ambion), according to the manufacturer's instructions. The membrane should not be dried; otherwise the probe will not be able to be removed.

Chromatin remodeling is observed as a loss of regularity in the DNA ladder. Because chromatin remodeling occurs near the promoter, loss of regularity in the DNA ladder should be detected only with the promoter probe and it should be activator-dependent. The distal probe should display regularity in the DNA ladder, indicating the presence of regularly spaced chromatin outside of the activator-binding site (see Fig. 5).

Chromatin Purification

Some experiments require clean chromatin. While RSF-mediated chromatin assembly yields highly purified chromatin, RSF is still present in the reaction along with ATP and in some cases, free histones. By using gel filtration chromatography, chromatin can be purified of these unwanted

---

[43] J. Sambrook, E. F. Fritsch, and T. Maniatis, "Molecular Cloning: A Laboratory Manual." Cold Spring Harbor Laboratory, Cold Spring Harbor, New York, 1989.

components. Of the many options available, we prefer to use small gel filtration columns.

Chromatin is purified over a Sepharose™ CL-4B column (Amersham Pharmacia Biotech). The resin (4 ml) is packed in a BioRad 0.5 × 20 cm glass Econo-column and equilibrated with 10 mM HEPES, pH 7.5; 10 mM KCl; and 0.5 mM EGTA. The column can be run at room temperature. Chromatin (4 μg in a volume of 300 μl) is loaded onto the column bed and allowed to completely sink into the resin, collecting the buffer. Then 200 μl of buffer are loaded onto the column and collected together with the previous fraction. Buffer aliquots of 250 μl are then added sequentially and collected in different test tubes. This process is repeated 6 times. The chromatin usually elutes after 1.4 ml of buffer have been added to the column. To analyze the profile of chromatin elution, 5 μl aliquots from each column fraction are mixed with 5 ul H$_2$O, 0.7 ul 10% SDS, and 2 μl of 5×-sample buffer (50 mM Tris, pH 7.9; 5 mM EDTA; 0.5% bromophenol blue and 50% glycerol) and incubated for 10 min at 65°. The samples are then analyzed by electrophoresis on a 1% agarose gel.

Transcription from Chromatin Templates

Transcription by RNA polymerase II from a chromatin template can be performed with nuclear extracts as well as with a reconstituted transcription system. For transcription with nuclear extract, chromatin is remodeled during the transcription reaction by chromatin remodeling activities present in the extract. Remodeling is dependent upon the presence of the transcriptional activator protein. To carry out transcription in the reconstituted system, we recommended remodeling the chromatin during the chromatin assembly reaction, as described in the previous section. In this way, transcription will only be dependent upon FACT, as chromatin will already be remodeled. Chromatin remodeling can also be performed before the formation of the transcription preinitiation complex, but if previously remodeled chromatin is used, the transcription assay results are more reproducible with respect to the FACT effect.

Chromatin Transcription Derived from Nuclear Extract

HeLa nuclear extracts are obtained as previously described[26] and as summarized in the FACT/RSF purification section. Transcription is performed in a 40 μl reaction containing the following components: 20–40 ng chromatin templates (measured in terms of DNA content), 100 ng Gal4VP16 (see later), and 80 μg of HeLa nuclear extract in a reaction buffer containing 0.5× BC100 buffer supplemented with 2.6% polyethylene glycol (average MW 8000 Da), 3.75 mM ammonium sulfate, and

5 mM MgCl$_2$. The reaction mixture is incubated for 30 min at 30° to allow the formation of the transcription preinitiation complex. Nucleotide triphosphates (0.8 mM ATP, 0.8 mM CTP, 0.6 μM UTP [FPLC-purified, Amersham Pharmacia Biotech] and 5 μCi ^{32}P-UTP [NEN]) are then added, and the mixture is incubated for an additional 30 min at 30°. The RNA transcripts are digested with 10 U of RNase T1 for 5 min at 30° (in the case in which the template used contains a G-less cassette, as described here), and then transcription is stopped with 120 ul of stop buffer (250 mM sodium acetate, pH 5.5; 10 mM EDTA; 0.2% SDS; 1 mg/ml yeast tRNA). The transcription products are purified by phenol/chloroform extraction, ethanol precipitated, and washed with 75% ethanol. After the sample is air-dried, the pellets are resuspended in 8 ul of 9 M urea/1% NP-40 and 8 ul of 95% formamide containing 0.01% bromophenol-blue. The RNA products are resolved in a denaturing gel consisting of 6% acrylamide/6 M urea, and electrophoresis is performed at 220 V for 30–40 min. The gel is washed in hot water for 10 min to remove the urea, then dried and exposed to an X-ray film at −80° or with a phosphorimager screen (Storm 860, Molecular Dynamics).

Transcription is performed in the presence and absence of the activator protein. Because the transcription reaction is inhibited by the presence of nucleosomes, transcription will only occur in the presence of the activator (see Fig. 6).

The order of addition of the activator in the transcription process is not a trivial matter. As an example, it has been shown that the histone acetyltransferase p300 is recruited to the promoter by interaction with Gal4VP16.[44] Therefore, if p300 is used in the transcription reaction, it is recommended that the chromatin templates be preincubated with p300, acetylCoA, and Gal4VP16 prior to the incubation with nuclear extract.

Reconstituted Transcription

RNA polymerase II transcription reconstituted with recombinant or highly purified transcription factors (for purification procedures of general transcription factors and RNA polymerase II, the reader is referred to Maldonado *et al.*[45]) is performed as previously described.[46] The preinitiation complex is formed in a 40-μl reaction containing 20–40 ng of chromatin templates (measured in terms of DNA content), 50 ng TBP, 10 ng TFIIB, 50 ng TFIIE, 10 ng TFIIF, 150 ng TFIIH (Phenyl Superose fraction),

[44] T. K. Kundu, V. B. Palhan, Z. Wang, W. An, P. A. Cole, and R. G. Roeder, *Mol. Cell* **6,** 551 (2000).
[45] E. Maldonado, R. Drapkin, and D. Reinberg, *Methods Enzymol.* **274,** 72 (1996).
[46] O. Flores, H. Lu, and D. Reinberg, *J. Biol. Chem.* **267,** 2786 (1992).

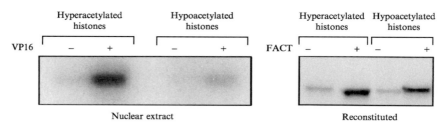

FIG. 6. RNA polymerase II transcription on RSF-assembled chromatin. (Left side) Transcription was carried out with hyperacetylated or hypoacetylated chromatin templates using HeLa nuclear extract in the presence and absence of the activator Gal4VP16. The same amount of DNA was used in all lanes. Products of the transcription reactions were separated by electrophoresis on a denaturing polyacrylamide gel. (Right side) Transcription was carried out with hyperacetylated or hypoacetylated chromatin templates using a reconstituted system, in the presence and absence of FACT. All reactions contained the activator Gal4-VP16. Products of the transcription reactions were separated by electrophoresis on a denaturing polyacrylamide gel.

100 ng of highly purified RNA polymerase II, and 250 ng FACT (native or recombinant, purified as described above). Transcription is then carried out under the same conditions as described above for nuclear extract.

As mentioned earlier, reconstituted RNA polymerase II transcription on chromatin templates requires two chromatin-specific factors, RSF and FACT. RSF will allow formation of the preinitiation complex, and FACT will allow transcript elongation.[24] The method described herein starts with remodeled-chromatin; therefore, formation of the preinitiation complex will take place in the absence of RSF or any other chromatin-remodeling factor. In the absence of FACT, only short mRNAs are obtained; however, in the presence of FACT, full-length, 390-nucleotide transcripts will be generated by the transcription reaction (see Fig. 6).

Concluding Remarks and Future Directions

The techniques described in this review provide a blueprint for generating a defined *in vitro* system for chromatin assembly and transcription assays. Because the system described herein is highly defined, it can be used to analyze the role of individual chromatin-related factors in transcription, as well as in other reactions involving DNA. As an example, we first used nuclear extract to study the role of p300-mediated acetylation of histones in chromatin transcription by RNA polymerase II, and found that p300 acetylation stimulates transcription. However, this stimulation was only observed when transcription was carried out with unfractionated nuclear extract. Transcription reactions carried out in the reconstituted system described in this article were refractory to stimulation by p300.[22] These experiments suggest that acetylation may result in the recruitment of some other factors that are present in the extract, but not in the defined system, such as bromodomain-containing proteins shown to interact with acetylated lysine residues on the histone tails.[47] The purified system thus provides a powerful biochemical assay for the identification of these other putative factors that can influence transcription.

In a similar manner, the activity of chromatin modifiers can be assayed in conjunction with chromatin remodeling factors and other components of the transcription machinery thereby providing grist to further understand the interplay between different chromatin-related factors in transcription by RNA polymerase II.

Moreover, the application of the synthetic histones containing site-specific modifications will be a very powerful tool to investigate the functional role of particular histone modifications in the future.

Acknowledgments

We thank Rimma Belotserkovskaya and Rushad Pavri for helpful comments and suggestions on the manuscript. We also thank Dr. Lynne Vales for critical reading of the manuscript. This work was supported by grants from NIH GM 65539 to DGM and GM37120 and the Howard Hughes Medical institute to DR.

[47] L. Zeng and M. M. Zhou, *FEBS Lett.* **513,** 124 (2002).

[32] Analysis of DNA Repair on Nucleosome Templates

By BRIAN C. BEARD and MICHAEL J. SMERDON

The plethora of damage inflicted on eukaryotic genomes by exogenous and endogenous sources has required the cell to develop a variety of biochemical processes designed to repair various DNA lesions.[1,2] The substrate for DNA repair enzymes in eukaryotes is the DNA molecule packaged into a chromatin hierarchy consisting of looping, twisting, and

[1] E. C. Friedberg, G. C. Walker, and W. Siede, "DNA Repair and Mutagenesis." ASM Press, Washington, DC, 1995.
[2] J. H. J. Hoeijmakers, *Nature* **411,** 366 (2001).

general packaging.[3,4] These structures play critical roles in processes such as gene expression, DNA replication, and DNA repair.[3,5,6] For all of these processes, the fundamental packaging unit, the nucleosome, must be manipulated for proteins to bind to their cognate DNA sequences and for enzymes to catalyze reactions with DNA.

The obstacle that chromatin presents to DNA repair enzymes is illustrated by model systems showing inhibited repair of nucleosomal substrates using cell extracts[7–10] and purified enzymes.[11–17] The distribution of DNA damage *in vivo* is also modulated by the chromatin environment[5] and the difficulty associated with measuring repair of randomly damaged DNA in chromatin is exacerbated by differences in damage location, relative to the nucleosome surface, and DNA sequence context of the damage site.

To investigate protein-nucleosome interactions of DNA excision repair pathways at a molecular level, "designed" nucleosomes can be assembled from homogeneously damaged DNA fragments, prepared by solid-phase synthesis or by other means. Using site-specific damaged DNA and nucleosome positioning elements, allows one to select the nucleosome "environment" for the lesion and to analyze the effect of this environment on DNA repair. An important consideration for these studies is that the location and type of DNA damage can play a critical role in the ability of DNA to form stable nucleosomes and may affect DNA positioning on the histone octamer.[15,18] Therefore, characterization of the damaged DNA positioning

[3] A. P. Wolffe, "Chromatin: Structure and Function." Academic Press, San Diego, 1998.

[4] K. E. van Holde, "Chromatin." Springer, New York, 1988.

[5] M. J. Smerdon and A. Conconi, *in* "Progress in Nucleic Acid Research and Molecular Biology" (K. Moldave, eds.), p. 227. Academic Press, New York, 1999.

[6] T. Melendy and R. Li, *Front. Bios.* **6,** 1048 (2001).

[7] K. Sugasawa, C. Masutani, and F. Hanaoka, *J. Biol. Chem.* **268,** 9098 (1996).

[8] Z. Wang, X. Wu, and E. C. Friedberg, *J. Biol. Chem.* **266,** 22472 (1991).

[9] J. V. Kosmoski, E. J. Ackerman, and M. J. Smerdon, *Proc. Natl. Acad. Sci. USA* **98,** 10113 (2001).

[10] K. Ura, M. Araki, H. Saeki, C. Masutani, T. Ito, S. Iwai, T. Mizukoshi, Y. Kaneda, and F. Hanaoka, *EMBO J.* **20,** 2004 (2001).

[11] D. R. Chafin, J. M. Vitolo, L. A. Henricksen, R. A. Bambara, and J. J. Hayes, *EMBO J.* **19,** 5492 (2000).

[12] J. V. Kosmoski and M. J. Smerdon, *Biochemistry* **38,** 9485 (1999).

[13] H. Nilsen, T. Lindahl, and A. Verreault, *EMBO J.* **21,** 5943 (2002).

[14] C. F. Huggins, D. R. Chafin, S. Aoyagi, L. A. Henricksen, R. A. Bambara, and J. J. Hayes, *Mol. Cell* **10,** 1201 (2002).

[15] X. Liu, D. B. Mann, C. Suquet, D. L. Springer, and M. J. Smerdon, *Biochemistry* **39,** 557 (2000).

[16] U. Schieferstein and F. Thoma, *EMBO J.* **17,** 306 (1998).

[17] B. Suter, M. Livingstone-Zatchej, and F. Thoma, *EMBO J.* **16,** 2150 (1997).

[18] D. B. Mann, D. L. Springer, and M. J. Smerdon, *Proc. Natl. Acad. Sci. USA* **94,** 2215 (1997).

and stability in nucleosomes should be measured concomitantly with repair experiments. Finally, model chromatin substrates also allow manipulation of the histones (e.g., modification of histone tails via acetylation of lysines or removal of histone tails by proteolysis) for detailed studies on the effects of these changes on DNA repair.[11,19]

Preparation of DNA Templates

Sequence Selection

The choice of oligonucleotide sequence used for designing nucleosome core particles (NuCP) *in vitro* is an important consideration (for a review see Widlund *et al.*[20]). An elegant method for sequence selection (SELEX) has been utilized to extract genomic sequences that strongly position nucleosomes.[21] Prior to these studies, a nucleosome positioning element, known as the TG motif,[22,23] has been utilized to facilitate nucleosome formation during competitive reconstitution.[12,23–25] The TG motif consists of $(G/C)_3(N)_2(A/T)_3(N)_2$ (where N refers to any nucleotide) and bends DNA via sequential compression of the minor groove and expansion of the major groove.[22] When polymers of the TG motif are utilized, in tandem, sequential bending of DNA occurs. This bending affords the TG motif sequence an advantage of forming nucleosomes more efficiently than a sequence yielding a relatively linear DNA stucture.[22]

Incorporation of transcription factor binding sites within a "designed" nucleosome allows protein binding to be monitored relative to the sequence location on nucleosomes, as well as the effect of DNA damage on protein affinity. For protein binding studies, a segment of the glucocorticoid hormone receptor binding site (from the long terminal repeat of mouse mammary tumor virus) called GRE (see Fig. 1, shown in gray), has been used as a sequence bracketed by tandem polymers of the TG motif (see Fig. 1, shown in black). This sequence construction was used to position the GRE at the nucleosome dyad, and the effect of both nucleosome structure and DNA damage on protein binding can be probed.[12,23,25]

[19] X. Liu and M. J. Smerdon, *J. Biol. Chem.* **275,** 23729 (2000).
[20] H. R. Widlund, H. Cao, S. Simonsson, E. Magnusson, T. Simonsson, P. E. Nielsen, J. D. Kahn, D. M. Crothers, and M. Kubista, *J. Mol. Biol.* **267,** 807 (1997).
[21] P. T. Lowary and J. Widom, *J. Mol. Biol.* **276,** 19 (1998).
[22] T. E. Shrader and D. M. Crothers, *Proc. Natl. Acad. Sci. USA* **86,** 7418 (1989).
[23] Q. Li and O. Wrange, *Mol. Cell. Biol.* **8,** 4375 (1995).
[24] T. E. Shrader and D. M. Crothers, *J. Mol. Biol.* **216,** 69 (1990).
[25] Q. Li and O. Wrange, *Genes Devel.* **7,** 2471 (1993).

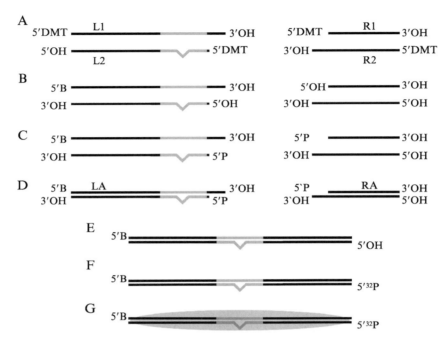

Fig. 1. Sequence assembly of model nucleosome templates for DNA repair experiments. (A) solid-phase synthesis of nucleosome positioning element (TG motif, black lines) flanking a hormone response element (GRE cassette, gray lines) with a site-specific damage (5'DMT [5'dimethoxytrityl] and 3'OH [3'hydroxyl]), (B) purified oligonucleotides following RP-HPLC and preparative gel electrophoresis (oligonucleotides have 5'B [5'blocking group] and 5'OH [5'hydroxyl]), (C) phosphorylation of interior 5' sites for ligation (5'P [phosphorylated 5']), (D) annealing of L1/L2 (LA) and R1/R2 (RA), (E) ligation of single stranded oligonucleotides to create the full-length double stranded product, (F) end-labeling of the damaged strand [5'³²P (phosphorylated with [γ-³²P] ATP)], and (G) nucleosome reconstitution of the full-length oligonucleotide (nucleosome represented by gray oval). Assembly of undamaged substrate is identical except the GRE cassette (gray bars) is the native sequence.

Synthetic DNA-Based Approach

The procedures outlined here utilize synthetic, site-specific damaged oligonucleotides for DNA repair experiments. These oligos can be made from commercially available phosphoramidites, such as 8-oxo-2'-deoxyGuanosine, O^6-Me-2'-deoxyGuanosine, and 2'-deoxyUracil (Glen Research, Sterling, VA) or chemically synthesized phosphoramidites, such as the *cis-syn*-cyclobutane thymine dimer (*cs*CTD, Ref. 12). Our strategy for synthesis and assembly of DNA substrates has been described in detail.[12]

Fragment Design

The basis for design of the oligonucleotide fragment is the ligation of two asymmetric oligonucleotides to yield a full-length product. The nomenclature used here refers to the upper and lower single strands as L1, L2, R1, and R2, and for the annealed double strand forms LA (Left Arm) and RA (Right Arm) (see Fig. 1). This procedure assumes that the full-length product is too long to synthesize in one continuous strand by solid-phase DNA synthesis. When assaying repair of DNA damage, it is important to specifically end label one strand of the final double stranded product, and the shorter of the two fragments is preferable. This will aid in gel purification of the full-length product because the majority of the contaminating (unligated) fragment will be the shorter radioactively labeled fragment.

Specific labeling of synthetic oligonucleotides can be achieved in one of three ways: (1) add a blocking group in the last step during DNA-synthesis (i.e., 5'-Amino-Modifier 5 [Glen Research]) that will prevent T4 polynucleotide kinase (TPK) from incorporating a radioactive nucleotide onto one of the strands, (2) radioactively label the desired fragment (see Fig. 1, step C R2) before full assembly of the oligonucleotide, and (3) leave recessed 3'-ends that can be filled in with the Klenow fragment of DNA polymerase I and $[\alpha\text{-}^{32}P]dNTP$. Option (1) is the preferred method due to ease of labeling and limited handling of radioactive material relative to option (2). Importantly, if monitoring repair intermediates that contain strand breaks, labeling of the continuous undamaged strand may be required for various assays using option (2) or (3).

The rest of this protocol will assume that the radioactive labeling is of the fully assembled oligonucleotide (see Fig. 1, step F) using option (1). We note that when using this option the blocking group added to the 5'-end of L1 is "irreversible." Therefore, this option is not compatible with procedures that necessitate 5'-end labeling of the undamaged strand (e.g., DNase I footprinting, see below).

In using option (1), the interior 5'-sites (L2 and R1) that will be used for ligation are recessed leaving the 3' end of the complimentary strand (L1 and R2) exposed. This is to inhibit radioactive labeling of the interior 5'-sites of unligated products in the final mixture. The overhang on L1 and R2 is a complimentary 4 nt nonpalindromic sequence. A nonpalindromic sequence is used so that no 'self-ligation' products can form (see Fig. 1, step E, LA : LA or RA : RA). A 1.5-fold molar excess of RA-oligonucleotide is used during the ligation reaction to deplete the mixture of unligated LA. This is to avoid radioactive labeling of unligated LA that would complicate gel purification. The full-length oligonucleotide fragment is labeled specifically on the 5'-end of RA with TPK and $[\gamma\text{-}^{32}P]ATP$. The full-length radioactively labeled oligonucleotide is purified on a native gel (10% polyacrylamide and 0.5% bisacrylamide in $1\times$ TBE). This product can be used for nucleosome reconstitutions, nucleosome characterization, and DNA repair assays.

Procedure

1. RP-HPLC of 5′-dimethoxytrityl oligonucleotides (5′DMT): linear gradient from 95% Buffer A (0.1 M triethylammonium acetate [pH 7.0]) and 5% Buffer B (acetonitrile) to 60% Buffer A and 40% Buffer B in 30 min at 5 ml/min on a Vydac® C4 semipreparative column. 5′DMT have a retention time of approximately 17 min and as the length of the oligonucleotide increases the resolution of the 5′DMT from incomplete synthesis products decreases. A second RP-HPLC purification step is carried out as above once the 5′DMT group has been removed with 80% acetate that changes retention time to about 13 min.[12]

2. The ligation of the oligonucletide fragment uses the interior 5′ hydroxyl groups (L2 and R1). These 5′hydroxyls are phosphorylated using the TPK forward reaction (Gibco-BRL).

3. Complimentary sequences are annealed via a slow cooling method.[12] During the annealing of L1 and L2 to form the double stranded LA (see Fig. 1, step D), L1 is at 1.5-fold molar excess and R1 and R2 are at equal molar concentrations. The excess L1 is to assure that no L2 single stranded product remains with an available 5′-end that can be labeled by TPK. The annealed products are separated on native gels (10% polyacrylamide and 0.5% bisacrylamide in $1\times$ TBE) and visualized with ethidium bromide.

4. LA and RA are ligated (T4 Ligase; Gibco-BRL) using the complimentary nonpalindromic 4 nt overhang on L1 and R2 oligonucleotides. This yields full-length products of undamaged and damaged oligonucleotides (see Fig. 1, step E; Ref. 12). The ligated products can be visualized on native gels as above.

5. Approximately 100 ng of fully assembled DNA is 5′-end labeled (R2 is labeled L1 is blocked) with $[\gamma-^{32}P]$ATP and TPK (Gibco BRL). Radiolabeled DNA is extracted with phenol and separated from unincorporated radionucleotides on a G-25 spin column (Amersham Biosciences). Full-length oligonucleotides are then gel purified on a native gel (10% polyacrylamide and 0.5% bisacrylamide in $1\times$ TBE). Specificity of end labeling, which is crucial to analyses of DNA damage and repair in nucleosomes, was analyzed by digestion at an asymmetrically positioned restriction enzyme cut site (nonspecific labeling appears as a doublet).

6. Site-specific DNA damage is analyzed with DNA damage-specific enzymatic cleavage (e.g., T4 endonuclease V cleavage at csCTD's or uracil DNA glycosylase [UDG]/AP endonuclease [APE] cleavage at guanine:uracil [G:U] mismatches). The digestion products are visualized on $7 M$ urea gels (10% polyacrylamide and 0.5% bisacrylamide in $1\times$ TBE) (see Fig. 2, right panel).

Nucleosome Stability and Characterization

DNA Damage: Effect on Nucleosome Formation

Various types of DNA damage differentially modulate packaging of DNA into chromatin. This is due to the distortion imposed on the DNA molecule which is dependent upon the type of damage (e.g., csCTD's, (+) -7R,8S-dihydroxy-9S,10R-epoxy-7,8,9,10-tetrahydro-benzo[a]pyren bound to the N^2 position of guanine, and G:U mismatch).[12,15,18,26] To test the nucleosome stability of damaged DNA oligonucleotides, undamaged and damaged DNA substrates can be reconstituted into mononucleosomes via histone octamer transfer,[18,27] using an excess of chicken erythrocyte core particles (CE-CP) at varying concentrations. Nucleosome formation is visualized by electrophoretic mobility gel shift assays (EMSA) on native gels (6% polyacrylamide and 0.3% bisacrylamide in $0.5\times$ TBE). At the various CE-CP concentrations ratios of the radioactively labeled DNA (NuCP and naked DNA) can be quantified, allowing for comparison of undamaged and damaged oligonucleotides (see Fig. 2, left panel). The

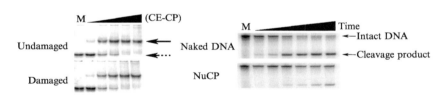

FIG. 2. Undamaged and damaged nucleosome reconstitutions and UDG/APE digestion. Left panel: Native gels of nucleosome reconstitutions at varying CE-CP concentrations (naked DNA, dashed arrow; NuCP, solid arrow). Lane M is naked DNA only and black triangle at top of gel indicates increasing CE-CP concentration (from $\sim$30 to 300 ng/ml). Right panel: Denaturing gels of site-specific G:U mismatch-containing naked DNA (upper gel) and NuCP DNA (lower gel) following incubation with UDG and APE, each at 1 nM. Incubation time is from 30 s to 1 h and M is mock treated.

[26] B. C. Beard, S. H. Wilson, and M. J. Smerdon, $Proc. Natl. Acad. Sci. USA$ 100, 7465 (2003).
[27] R. T. Simpson and P. Kunzler, $Nucleic Acids Res.$ 6, 1387 (1979).

difference in the free energy (ΔG) of formation between undamaged and damaged DNA samples can be calculated from the equation:

$$\Delta\Delta G = -\mathrm{RT}\ln(K_D/K_U) \tag{1}$$

where K_D is the ratio of NuCP to naked damaged DNA and K_U is the ratio of NuCP to naked undamaged DNA.[18] Depending on the nucleosome positioning element and DNA damage, reconstitution may not be complete. To help circumvent this problem, NuCP can be purified via nucleoprotein gels[28] or sucrose gradients.[18]

Restriction Enzyme Site Accessibility Versus DNA Damage Accessibility

Transient exposure of DNA away from the histone octamer is dependent upon location of the site relative to the nucleosome dyad center.[29] Restriction enzymes can be used as probes for nucleosome stability because cleavage of a specific sequence requires the restriction enzyme to bind nearly the entire circumference of the DNA molecule.[30] This is in stark contrast to certain repair enzymes (e.g., UDG and APE) that bind fairly discrete loci around the damage site to carry out catalysis.[31,32] Thus, restriction enzymes act as an effective probe to test if the digestion products visualized during incubation with repair enzymes (see Fig. 2, right panel) are due to (a) transient exposure of the DNA away from the histone surface or (b) catalysis of the DNA while tightly associated with the histone octamer.[26] Finally, restriction enzyme incubation of NuCP both before and during repair assays reveals the stability of the nucleosome with and without repair enzymes present.[14]

Footprinting

DNA damage can severely distort the DNA molecule.[1] This distortion can affect nucleosome reconstitution[12,18] and possibly affect the rotational setting of the DNA molecule.[33] The rotational settings of the damaged and undamaged oligonucleotide can be assayed by footprinting techniques (i.e., hydroxyl radical or DNase I).[34] If footprinting of the undamaged strand is desired using the labeling procedure outlined above, option (1) cannot be

[28] S.-Y. Huang and W. T. Garrard, Methods Enzymol. 170, 116 (1989).
[29] J. D. Anderson, A. Thastrom, and J. Widom, Mol. Cell. Biol. 22, 7147 (2002).
[30] K. J. Polach and J. Widom, Methods Enzymol. 304, 278 (1999).
[31] G. Slupphaug, C. D. Mol, B. Kavli, A. S. Arvai, H. E. Krokan, and J. A. Tainer, Nature 384, 87 (1996).
[32] C. D. Mol, T. Izumi, S. Mitra, and J. A. Tainer, Nature 403, 451 (2000).
[33] J. V. Kosmoski, Ph.D. thesis, Washington State University, 1999.
[34] A. Revzin, "Footprinting of Nucleic Acid Complexes." Academic Press, New York, 1993.

utilized because the blocking group is irreversible. To label the undamaged strand, recessed 3'-ends can be used and filled in with $[\alpha\text{-}^{32}P]$dNTP's and DNA polymerase I Klenow fragment [option (3)].

Conclusion

With specifically damaged chromatin substrates, investigators have begun to address explicit questions regarding the effect of chromatin on DNA repair. Moreover, the ability of various DNA repair enzymes to carry out their activities on chromatin, while others cannot, provides insight into the catalytic activities that can proceed on intact nucleosomes.[11,13,14,26,33] The assembly process described here lends itself to incorporation of a variety of lesions into the same nucleosome background using one approach. Expansions of the research described here include nucleosome reconstitution with specifically modified histones, and site-specific damage introduction into dinucleosomes or nucleosome arrays. These and other applications of this system should advance our understanding of the roles that chromatin and chromatin modifications play in DNA repair.

Acknowledgments

We would like to thank current and past members of the Smerdon lab, especially Drs. Joe Kosmoski and Antonio Conconi for helpful discussions and guidance during the initial stages of this work. This work was made possible by NIH grants ES04106 and ES02614 from the National Institute of Environmental Health Sciences (NIEHS). Its contents are solely the responsibility of the authors and do not necessarily represent the official views of the NIEHS, NIH.

Author Index

Numbers in parentheses are footnote reference numbers and indicate that an author's work is referred to although the name is not cited in the text.

A

Aalfs, J. D., 389, 412
Aasland, R., 12, 18, 19(27), 34, 249
Abbondandolo, A., 50
Abraham, J., 38
Ackerman, E. J., 500
Adams, C. C., 394
Adashi, E. Y., 86, 90(12)
Adler, P. N., 281
Agapite, J., 55
Agarwal, S., 53
Aggarwal, A. K., 153
Ahmed, A. A., 85
Ahn, N. G., 197, 202(11)
Ajiro, K., 198, 199, 201
Al-Abed, Y., 183(8), 184, 187(8)
Alami, R., 97, 98
Albig, W., 100
Alewood, D., 488
Alewood, P., 488
Alexander, M. K., 38
Alkema, M. J., 285, 287(5; 6)
Allard, S., 13, 24, 30(62), 159
Allegra, P., 200
Allfrey, V. G., 200
Allis, C. D., 5, 11(25; 39), 12, 13, 14, 16, 18,
    19, 19(28), 22, 24, 25, 29, 30, 31, 31(48),
    32(48), 35, 71, 81, 81(43), 82, 111, 131,
    132, 133(6; 7), 135, 137(6; 7; 9), 139(6; 7),
    145(7; 11; 12), 154, 154(6), 155, 158(6),
    161, 162, 162(20), 163, 180, 197, 198, 199,
    199(12), 200, 201, 207, 207(16; 28), 212,
    213, 215, 218, 218(14; 16), 219, 219(13),
    222(16), 223, 223(20), 224, 227, 228,
    228(5), 233(22), 236, 236(25), 237, 266,
    282, 412, 421, 460(3; 6), 461, 468, 475, 477
Allshire, R. C., 55, 153
Almer, A., 57

Alon, T., 75, 255
Alt, E., 88
Altheim, B. A., 25, 243
Ambros, P., 47
An, W., 460, 461, 462(11), 464, 468,
    468(11; 12; 23), 471(11; 23)
Anderson, C. W., 25
Anderson, H. J., 199(21), 200
Anderson, J. D., 506
Anderson, K., 51, 52
Anderson, S., 13, 22
Andrada, R., 54
Andre, B., 51, 52
Andrejka, L., 49
Andrew, D. J., 79
Andrulis, E. D., 25
Anest, V., 198, 201
Annan, R. S., 29
Ansari, A., 29
Ansorge, W., 78
Aoyagi, S., 500, 506(14)
Aparicio, O. M., 38, 235
Aprile, A., 50
Araki, M., 500
Archambault, J., 28
Arents, G., 25, 31, 31(71), 227
Argiropoulos, B., 268, 277(15)
Armstrong, C. M., 13, 38, 176, 236
Armstrong, J. A., 70, 71, 72(16), 76,
    81(14; 16), 82(14; 16), 83(16), 84(14),
    377, 378(21)
Arndt-Jovin, D. J., 387
Arneric, M., 48
Arvai, A. S., 506
Astromoff, A., 51, 52
Attree, O., 47
Aubin, R., 197
August, A., 48
Aurias, A., 47

509

Ausio, J., 213, 457(29), 458, 463
Auston, D., 13, 18
Ausubel, F. M., 28, 328, 342(18)
Avalos, J., 168
Avery, S. V., 28, 38
Avrutskaya, A. V., 29
Avvakumov, N., 48
Axel, R., 269
Axelrod, A., 235
Ayala, J., 89(23), 90, 92(23), 94(23), 95(23), 96(23), 100(23)
Ayer, D. E., 169
Ayyanathan, K., 223(23), 224

B

Badenhorst, P., 71
Bahr, A., 224
Bailey, A., 78
Bailey, L. C., 47
Bailly, V., 232
Bakayev, V. V., 318
Balakrishnan, R., 54
Balasubramanian, R., 158, 159(13)
Baldwin, A. S., 198, 201
Bambara, R. A., 500, 501(11), 506(14)
Band, D., 25
Banerjee, M., 38, 236
Bangham, R., 51, 52
Bannister, A. J., 11(32), 12, 18, 21(32), 153, 197
Barak, O. G., 170, 188, 377, 389, 390
Baran, N., 327, 329(14)
Barbaric, S., 29, 56, 57(13), 59(13)
Barnes, D. M., 50
Barnett, P., 290
Barra, J. L., 85(6), 86
Barratt, M. J., 197, 198, 200(4; 30), 201, 202
Bartelmez, S. H., 413
Barthmaier, P., 49
Baskerville, C., 236
Basrai, M. A., 28
Bassett, D. E., Jr., 42
Bastiaens, P. I., 54
Bastuck, S., 402
Bauch, A., 402
Baudat, F., 202
Bauer, A., 402

Bauer, U. M., 132, 137(8), 145(8), 150(8), 151(8)
Bauman, D., 489
Baumann, P. H., 387
Bavykin, S., 33
Bazett-Jones, D. P., 198, 199(12)
Beard, B. C., 499, 505
Becker, P. B., 71, 72, 76(18), 105, 344, 345, 347, 348, 353, 354, 364, 368, 371(7), 376, 390, 402, 421, 422(1), 439(1), 476
Beck-Sickinger, A. G., 194
Beddington, R., 88, 90(22), 92(22)
Bednar, J., 445
Beek, S. J., 66, 71, 72, 255, 256(6), 439
Beeser, A., 198, 201
Beisel, C., 256
Bell, A. C., 286
Belmont, A. S., 29
Belotserkovskya, R., 11(24), 16, 18, 132, 135(10), 145(10), 151(10), 154(5), 155
Bender, M. T., 269
Bender, W., 69, 256, 267, 268, 285
Benes, V., 78
Benito, R., 51, 52
Benoiton, N. L., 488
Berger, L., 132, 133(6), 137(6), 139(6)
Berger, S. L., 11(24), 13, 16, 17, 18, 33, 50(169), 51, 55, 71, 111, 130, 131, 132, 135, 135(10), 137(8), 145(8; 10), 150(8), 151(8; 10), 154, 154(5; 6), 155, 158(6), 159, 207, 421, 460(5), 461, 477
Bergson, C., 62
Berloco, M., 71, 72(16), 81(16), 82(16), 83(16), 377, 378(21)
Berns, A., 91
Bernstein, B. E., 4(5), 5, 11(32; 37), 12, 18, 21, 21(32)
Beroud, C., 50
Berrios, M., 29
Besse, A., 98
Beuchle, D., 69
Bhandoola, A., 202
Bhaumik, M., 88
Bhaumik, S. R., 18
Bi, X., 28, 38
Bickmore, W. A., 42, 376, 390
Biggar, S. R., 299
Biggins, S., 32
Billington, B. L., 38, 235
Binkley, G., 54

Y

Yager, T. D., 462
Yalcin, A., 377
Yamamoto, Y., 198, 201
Yan, L. Z., 488, 491(35)
Yan, Y. X., 49
Yan, Z., 299
Yanagida, M., 29
Yang, C.-C., 268, 282(16)
Yang, D., 306, 312
Yang, K., 88
Yang, L., 223(22), 224
Yang, M., 402
Yang, W.-M., 167, 169, 170
Yang, X., 316, 325
Yang, X. J., 128, 168
Yang, Y. F., 48
Yaniv, M., 47, 299, 413
Yao, Y.-L., 167
Yasui, Y., 198
Yates, J. L., 290, 291(22)
Yates, J. R., 238
Yates, J. R. III, 13, 18, 25, 29, 135, 158, 409
Yates, J. R. R., 22
Yau, P. M., 111, 117, 128
Yeh, E., 49
Yen, K., 42, 45
Ying, C. Y., 50(169), 51
Yokomori, K., 377, 390
York, J. D., 59
Yoshizak, i. F., 23, 34(56)
Young, I., 200
Young, M. K., 164, 299, 306, 311, 312
Young, R. A., 12, 19, 21(36), 316
Youngman, E., 8, 23
Yu, D. Y., 13, 24
Yu, J., 173
Yu, K., 50

Yu, L., 85, 199
Yu, Q., 28

Z

Zackai, E., 200
Zak, N. B., 71, 75, 255, 439, 440(13)
Zakian, V. A., 29, 38
Zavari, M., 33
Zegerman, P., 153, 477
Zeng, L., 153, 499
Zhang, K., 117, 128
Zhang, M., 316, 318(5), 320(5), 321(6), 322(5; 6), 338, 339(22; 24), 340, 354
Zhang, N., 42, 45
Zhang, W., 12, 16
Zhang, X., 129
Zhang, Y., 12(45), 13, 22, 24, 81, 81(42), 82, 120, 132, 137(9), 170, 212, 213(1), 215, 216, 218, 218(15), 219, 219(13), 222, 222(1; 10; 15), 223(22), 224, 225(18), 256, 285, 421, 475, 477
Zhang, Z., 23, 28, 48
Zhao, K., 299, 313(6), 315, 412
Zheng, B., 100
Zhong, S., 198, 201
Zhou, H., 198, 199, 207(16)
Zhou, J., 5, 12, 14, 18, 24, 154
Zhou, M. M., 153, 499
Zhou, S., 173, 299, 303(3), 311(3)
Zhou, Y., 377
Zhu, H., 54
Zhu, J., 202
Zijderveld, A., 91
Zink, B., 278, 280
Zink, D., 267
Zlatanova, J., 86
Zurita, M., 69

Subject Index

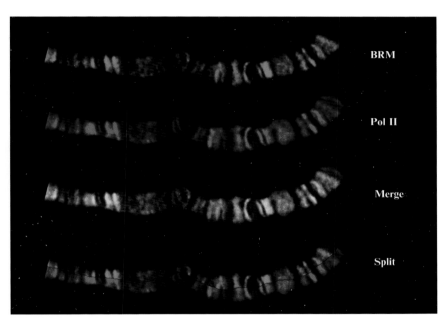

Labels: BRM, Pol II, Merge, Split

Corona ET AL., CHAPTER 4, FIG. 4. "Split" chromosomes reveal that BRM and RNA Pol II colocalize on polytene chromosomes. Salivary glands from wild type larvae were stained with rabbit anti-BRM and goat anti-RNA Pol II (subunit IIc). To generate a "split" image in Photoshop, individual images (BRM and IIc) were put in separate layers. The "lasso tool" was used to select half the chromosome arm in the upper layer and delete the selection to reveal the underlying layer. Displaying multiple staining patterns in a "split" format helps avoiding visual artifacts when high levels of one protein can mask low levels of another.

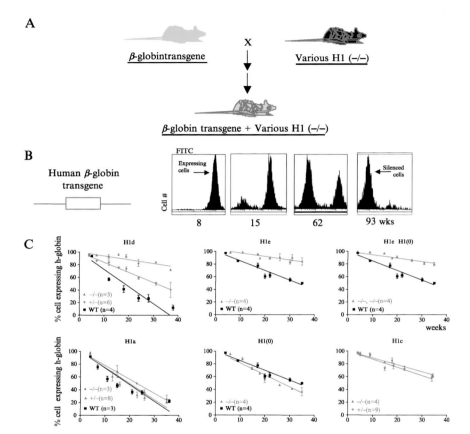

A

β-globintransgene X Various H1 (–/–)

β-globin transgene + Various H1 (–/–)

B

Human β-globin transgene

FITC

Expressing cells → | | | | ← Silenced cells

Cell #

8   15   62   93 wks

C

% cell expressing h-globin

H1d
▲ –/–(n=3)
▼ +/–(n=6)
■ WT (n=4)

H1e
▲ –/–(n=4)
■ WT (n=4)

H1e H1(0)
▲ –/–, –/–(n=4)
■ WT (n=4)
weeks

% cell expressing h-globin

H1a
▲ –/–(n=3)
▲ +/–(n=8)
■ WT (n=3)

H1(0)
▲ –/–(n=4)
■ WT (n=4)

H1c
▲ –/–(n=4)
▼ +/–(n=9)

FAN AND SKOULTCHI, CHAPTER 5, FIG. 4. Age-dependent silencing is modulated by specific linker histone deletions. (A) Scheme of experimental strategy for generating mice containing the human β-globin transgene and deletion of a single H1 subtype. (B) The 4.4 kb human β-globin gene transgene is subject to an age-dependent variegating position effect. Permeabilized RBCs were stained with FITC-labelled anti human-β-globin antibodies and periodically analysed by flow cytometry. The proportion of RBCs expressing the human transgene decreases with age. (C) Linker histone subtypes differentially affect transgene expression. Silencing of the human-β-globin transgene in mice homozygous or heterozygous for deletions of linker histones was monitored over time by flow cytometry and compared to controls. Deletion of H1d, or H1e or (H1e and H1^0) dramatically attenuates the rate of silencing. Deletions of H1a, H1c, and H1^0 has no effect on the rate of silencing. n is the number of mice analyzed. (Fig. 4B, C are reproduced from Alami *et al.* Copyright National Academy of Sciences, USA.)

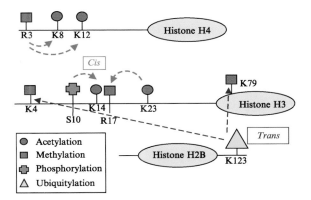

Lo ET AL., CHAPTER 7, FIG. 1. Schematic display of histone modification patterns. *Cis* relationships are modifications on a single histone tail, that include histone H3 and H4. *Trans* relationships are modification patterns between two tails, that include H2B and H3. The modifications include acetylation, methylation, phosphorylation and ubiquitylation.

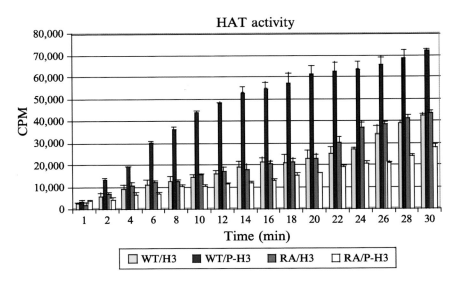

Lo ET AL., CHAPTER 7, FIG. 3. Gcn5 HAT activity over a time course on unmodified or P-Ser-10 modified histone H3 peptides. WT, HAT domain of Gcn5; RA, Gcn5 HAT domain bearing a substitution mutation at Arg-164.

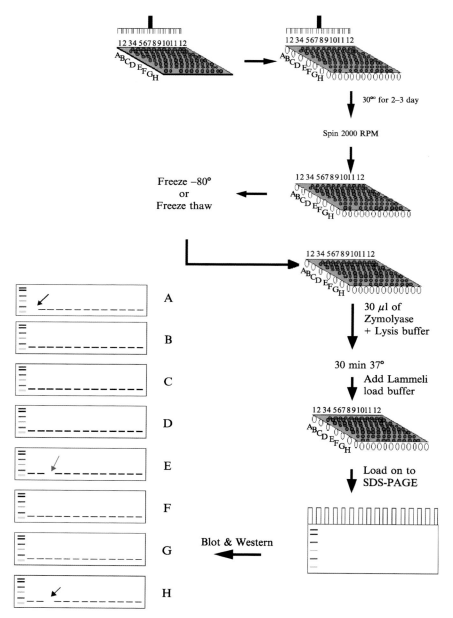

Schneider *et al.*, Chapter 13, Fig. 1. Global Proteomic analysis of *S. cerevisiae* (GPS). Schematic representation of GPS employing polyclonal antibody specific to lysine 4 methylated histone H3. As described under the detailed procedure section, employing a 96-well pinning device, the entire collection of nonessential gene deletion mutants in yeast were inoculated into liquid culture and allowed to grow. Cells were collected by centrifugation and resuspended in lysis buffer containing Zymolyase. Following lysis, each extract was applied to a 16% SDS-PAGE. The gels were then blotted to nitrocellulose paper and the presence of the methylated histone H3 were detected by the incubation of each blot in antimethyl K4 histone H3 antibodies.

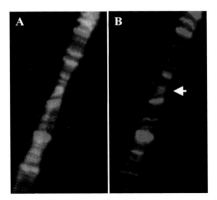

Huang and Chang, Chapter 16, Fig. 3. Colocalization of Psc and Psq at the Bithorax complex on polytene chromosomes. Polytene chromosomes from wild-type third instar larvae were simultaneously stained with antibodies against Psq and Psc. The chromosome arm shown here contains the Bithorax complex (indicated by arrow in B) where three homeotic genes including *Ubx* reside. (A) Triple color image is shown with DNA in green, Psq in red and Psc in blue. The region where three signals merge is shown as turquoise color. (B) Double color image of the same chromosome arm. The merged signal is shown as purple color.

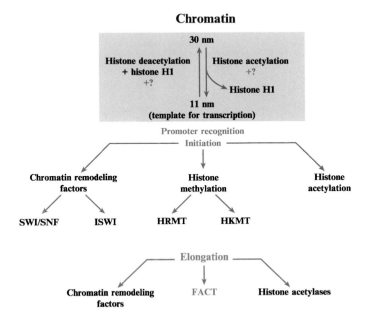

Loyola *et al.*, Chapter 31, Fig. 1. Description of the various families of protein complexes involved in RNA polymerase II transcription initiation and elongation.

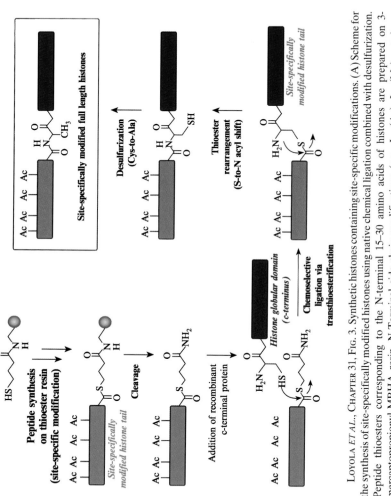

LOYOLA *ET AL.*, CHAPTER 31, FIG. 3. Synthetic histones containing site-specific modifications. (A) Scheme for the synthesis of site-specifically modified histones using native chemical ligation combined with desulfurization. Peptide thioesters corresponding to the N-terminal 15–30 amino acids of histones are prepared on 3-mercaptopropionyl-MBHA resin. N-Terminal side chain modifications (e.g., Lys(Ac), Lys(Me)$_3$, etc.) are introduced site-specifically using suitably protected amino acid modules introduced via standard Boc-SPPS methods. Following assembly and HF-mediated cleavage from the solid support, the modified histone termini thioester peptides are ligated to a companion recombinant histone C-terminal protein fragment that terminates in a Cys residue. The ligation site may be subsequently rendered traceless by desulfurization.

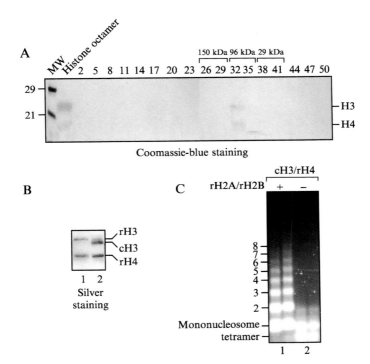

LOYOLA ET AL., CHAPTER 31, FIG. 4. Assembly of histone tetramers and chromatin using synthetic histone H3 (cH3, prepared by NCL) and recombinant histone H4 (rH4). (A) SDS-PAGE analysis of the size exclusion chromatographic purification of cH3/rH4 tetramer assembly. (B) SDS-PAGE gel analysis of the cH3/rH4 tetramers visualized by silver staining. Eighteen percent SDS-PAGE analysis of cH3/rH4 tetramer visualized by Coomassie Blue staining. (C) Agarose gel analysis of the products from micrococcal nuclease digestions of RSF-mediated chromatin assembly reactions using the cH3/rH4 tetramer, recombinant rH2A/rH2B tetramer and a plasmid DNA template.